Forest Operations, Engineering and Management

I0005658

Special Issue Editor

Raffaele Spinelli

MDPI • Basel • Beijing • Wuhan • Barcelona • Belgrade

MDPI

Raffaele Spinelli
University of the Sunshine Coast
Australia

Editorial Office
MDPI
St. Alban-Anlage 66
Basel, Switzerland

This is a reprint of articles from the Special Issue published online in the open access journal *Forests* (ISSN 1999-4907) from 2017 to 2018 (available at: http://www.mdpi.com/journal/forests/special_issues/IUFRO_2017_forests_engineering)

For citation purposes, cite each article independently as indicated on the article page online and as indicated below:

LastName, A.A.; LastName, B.B.; LastName, C.C. Article Title. *Journal Name* **Year**, *Article Number*, Page Range.

ISBN 978-3-03897-184-9 (Pbk)
ISBN 978-3-03897-185-6 (PDF)

Cover image courtesy of Raffaele Spinelli.

Contents

About the Special Issue Editor

Raffaele Spinelli obtained a Forest Science degree from the Tuscia University in Viterbo, and a PhD degree in Forest Engineering from the National University of Ireland, Belfied College. Since 1988, he has been a tenured researcher for the National Research Council (CNR) and leads the 'Wood Harvesting Research Group' at the Timber and Tree Institute (IVALSA) in Florence. His fields of expertise include forest technology, system efficiency, wood biomass supply chains and the environmental impacts of wood harvesting operations. He has worked as the Italian representative in 11 EU R&D projects, 5 EU Concerted Actions and 4 EU COST Actions, all on forest operations and biomass. He has co-authored 160 peer-review papers in international journals with IF, and over 200 papers on technical magazines. He also works on the editorial board of nine international scientific journals covering forestry and forest engineering subjects.

Preface to "Forest Operations, Engineering and Management"

Forest operations represent the active element of productive forest management, where costs are incurred and revenues accrued. Of course, there is much scope for minimizing the former and maximizing the latter, and a very strong interest in doing so. As one of the many instances of human activity, forest operations shape the environment and affect the lives of people, for better or for worse. Efficiently managed forest operations offer the highest benefit and the lowest cost in all fields: financial, social and environmental.

Increasing the efficiency of forest operations is the main goal of forest engineering, which represents a special sector of interest within the field of forestry, while maintaining a cross-disciplinary character, which is necessary for covering the many facets of forest work. Indeed, forest engineering draws from a number of different disciplines, which obviously include forestry and engineering, but also economics, medicine, geography and biology, to just mention a few.

Forest engineering generally deals with practical issues that have a strong economic impact, concentrated within a relatively short time span. For this reason, the industry has always had a strong interest in forest engineering, which explains the close connection between scientists and economic operators. Supported by the industry, forest engineering research has always been very active, advancing technological progress in forestry.

This book collects a representative sample of the most recent papers on the subject, which come from many different countries and cover a large variety of subjects, confirming the wide scope covered by forest engineering. The book contains some of the classic productivity studies that are at the foundation of forest engineering, as well as a number of site impact studies—by now a confirmed pillar of forest engineering research. Other important papers included in this collection cover ergonomics, worker health and safety and business issues. Additional works address the harvesting, processing and storage of wood biomass, indicating the strategic place occupied by wood energy research in modern forest engineering.

Raffaele Spinelli
Special Issue Editor

Article

Deadwood Decay in a Burnt Mediterranean Pine Reforestation

Carlos R. Molinas-González [1,*], **Jorge Castro** [1,*] and **Alexandro B. Leverkus** [1,2]

1   Departamento de Ecología, Facultad de Ciencias, Universidad de Granada, E-18071 Granada, Spain; alexandro.leverkus@uah.es
2   Departamento de Ciencias de la Vida, Edificio de Ciencias, Campus Universitario, Universidad de Alcalá, Alcalá de Henares E-28805, Spain
*   Correspondence: molinas.ca@gmail.com (C.R.M.-G.); jorge@ugr.es (J.C.); Tel.: +34-633-718-095 (C.R.M.-G.); +34-958-241-000 (ext. 20098) (J.C.)

Academic Editor: Raffaele Spinelli
Received: 6 March 2017; Accepted: 26 April 2017; Published: 8 May 2017

Abstract: Dead wood remaining after wildfires represents a biological legacy for forest regeneration, and its decay is both cause and consequence of a large set of ecological processes. However, the rate of wood decomposition after fires is still poorly understood, particularly for Mediterranean-type ecosystems. In this study, we analyzed deadwood decomposition following a wildfire in a Mediterranean pine plantation in the Sierra Nevada Natural and National Park (southeast Spain). Three plots were established over an elevational/species gradient spanning from 1477 to 2053 m above sea level, in which burnt logs of three species of pines were experimentally laid out and wood densities were estimated five times over ten years. The logs lost an overall 23% of their density, although this value ranged from an average 11% at the highest-elevation plot (dominated by *Pinus sylvestris*) to 32% at an intermediate elevation (with *P. nigra*). Contrary to studies in other climates, large-diameter logs decomposed faster than small-diameter logs. Our results provide one of the longest time series for wood decomposition in Mediterranean ecosystems and suggest that this process provides spatial variability in the post-fire ecosystem at the scale of stands due to variable speeds of decay. Common management practices such as salvage logging diminish burnt wood and influence the rich ecological processes related to its decay.

Keywords: deadwood management; decay rate; decomposition; density loss; Mediterranean

1. Introduction

Deadwood decomposition is a key process for ecosystem functioning and structure. Throughout the time of decomposition, decaying wood provides shelter and habitat for a large number of organisms [1–4], guarantees nutrient availability and turnover [5–7], defines carbon residence time and sequestration [8,9], enhances soil moisture [10], and determines the vertical and horizontal physical structure of the habitat as snags or fallen logs [11–14]. All these processes, both singly and in synergic combination, deeply influence other ecosystem processes, ranging from the performance of individual plants to landscape-scale biodiversity and even biogeochemical cycles [15–20]. Knowledge of the factors that determine the rate of wood decomposition is therefore relevant for understanding the residence time of logs, with broad implications for numerous ecosystem functions and services [20–22].

The rate of wood decomposition is also of paramount importance for forest management and planning, particularly after severe disturbances that create large amounts of dead wood, such as fires, pest outbreaks, or windstorms [1,23]. Particularly in the case of burnt forests, the rapid loss of economic value of the wood due to decomposition and the difficulties that it imposes for transit and management are often-claimed arguments for the quick implementation of post-fire management [24–27]. In this sense, extensive post-fire salvage logging—i.e., the removal of the logs, usually accompanied with the in situ elimination of the rest of coarse woody debris—is a widely implemented post-fire management action that seeks to recover part of the capital of the forest as well as to prepare the terrain for post-fire restoration [25,26]. However, post-fire salvage logging may impact ecosystem functioning and the capacity for natural regeneration through a variety of processes, such as reducing nutrient and moisture availability, decreasing the necessary substrate for saproxylic organisms, diminishing advance regeneration, or increasing soil erosion, among others [25,28–32]. Understanding the rate of wood decomposition after a fire is thus of great relevance to properly balance the economic benefit of quick salvage operations against the potential benefits for conservation and natural regeneration of nonintervention approaches. However, studies on wood decomposition are scarce and mostly concentrated in certain types of ecosystems such as boreal forests [33,34]. In particular, studies in Mediterranean-type ecosystems are very scarce [34], except for some that have focused on the decomposition of standing snags [35,36].

Wood decomposition is affected by abiotic and biotic factors, as well as the interactions and feedbacks between them. The speed of decomposition depends on moisture and temperature [1,22,37,38], and hence it can be expected to vary across environmental gradients where these factors gradually change, such as elevational or latitudinal gradients [1,21,39]. Decomposition rates may also be affected by species identity and log diameter, as these factors determine the proportion between heartwood and sapwood [9], and heartwood resists decay for longer than sapwood [40].Trunk diameter can also determine the identity of detritivorous species that colonize the log, and these may, in turn, affect the species assemblages of decomposers [5,41,42]. In short, decomposition is a complex process whose understanding requires proper control of the starting conditions and stand characteristics.

In this study, we seek to determine the rate of wood decomposition in a burnt pine reforestation under Mediterranean conditions. Three experimental plots were distributed across an elevational gradient spanning some 800 m, and logs with a standardized length but variable diameter were marked, spread on the ground, and sampled over 10 years. Given the marked differences in climatic conditions and the change in species across the elevational gradient, we hypothesized that decomposition rate would vary across elevations (hypothesis 1). Furthermore, the proportion of hartwood to sapwood tends to increase with log diameter [43], so we hypothesized that decomposition would be faster in logs with smaller diameters (hypothesis 2). Given the large amount of conditions that may affect decomposition rates and their variability across time, we expected potential interactions between elevation and diameter to affect decomposition rates (hypothesis 3). Overall, we expect this study to contribute to the understanding of the speed of wood decomposition in Mediterranean-type ecosystems, which should ultimately provide input to make informed post-fire decision-making.

2. Methods

2.1. Study Site

The study was conducted in the Sierra Nevada Natural and National Park (southeast Spain), in an area that burned in September 2005 (the Lanjarón fire). The fire burned around 1300 ha of 35 to 45 year-old reforested pine stands on a southwest-oriented mountainside. It was a high-intensity crown fire that consumed all the leaves, twigs, and litter and charred the bark of the trunks [30]. After the fire, the Forest Service established three plots across an elevational gradient within the context of a long-term research program devoted to study the effect of salvage logging with respect to other post-fire burnt wood management alternatives on ecosystem restoration and regeneration ([17,26,44]; Table 1). The three plots were similar in terms of pre-fire tree density (1000–1500 trees ha^{-1}), fire intensity (high), bedrock (micaschist), aspect (southwest), soil type (haplic phaeozems), and other soil characteristics ([30,45]; Table 1). However, the plots differed in climatic conditions, as expected from the increasing elevational gradient: mean rainfall increased and temperature decreased with elevation.

This influenced the species of pine that had been planted at each site (Table 1). For this study, we made use of areas in which 90% of the burnt trees were felled, the trunks were separated from their main branches and cut in pieces of ca. 2 m, and all the wood was left on the ground [46]. The climate is Mediterranean, with rainfall concentrated in spring and autumn, alternating with hot, dry summers. Snow is common during the winter, persisting up to 2 months at the highest elevation.

Table 1. Location and characteristics of the study plots.

	Plot		
	1	2	3
Coordinates [1]	36°57′12.1″ N 03°29′36.3″ W	36°58′11.9″ N 03°30′1.7″ W	36°58′6.5″ N 03°28′49.1″ W
Elevation (m above sea level [1]	1477	1698	2053
Mean daily minimum temp. (°C) [2]	6.8 ± 0.2	5.6 ± 0.2	3.4 ± 0.2
Mean daily maximum temp. (°C) [2]	17.1 ± 0.2	16.2 ± 0.2	13.4 ± 0.2
Mean ann. precip. (mm) [2]	536 ± 41	550 ± 40	630 ± 42
Dominant species	*Pinus pinaster*	*P. nigra*	*P. sylvestris*
Mean log diameter (cm) [3]	12.6 ± 0.4	12.8 ± 0.3	10.0 ± 0.2

[1] Measured at the centroid of each plot. [2] Data obtained from interpolated maps of Sierra Nevada (1981–2010) generated at the Centro Andaluz de Medio Ambiente (CEAMA). [3] Estimated from the logs that were used in this study, mean ± 1 SE.

2.2. Sampling Design

Six months after the fire (March to April 2006), 50 sampling points were randomly established within an area of 2 ha at each elevation to monitor wood decomposition. The sampling points were sufficiently away from standing trees so as to avoid their collapse over the point. At each of the sampling points, five logs were cut with a chainsaw to a standardized length of 75 cm (experimental logs, hereafter) and spread over an area of ca. 1 × 1 m, resulting in 250 experimental logs per elevation (Figure 1). All the logs had the bark charred to a similar extent as a result of the even-aged and even-spaced nature of the stands, and they were only superficially affected by the fire (Figure 1). The experimental logs belonged to *Pinus pinaster* in plot 1 (lower elevation), *P. nigra* in plot 2 (intermediate elevation), and *P. sylvestris* in plot 3 (higher elevation), which were the main pine species at each elevation according to their climatic requirements. This variation in species is a normal situation in reforested (and natural) pine stands across marked elevational gradients. Although such sampling design does not allow the effect of pine species and elevation on decay rates to be separated, it does provide the opportunity to measure wood decomposition under three realistic forest scenarios. Each experimental log came from a different tree and was cut from a random height along the tree trunk. Therefore, the logs constitute a representative sample of trunk characteristics in the study site in terms of diameter and sectional origin along the trunks. At the same time, we cut one wood disc of 6–8 cm height from each of 50 logs that were randomly selected at each elevation (initial discs, hereafter). These discs were brought to the laboratory, and their volume was estimated after measuring two perpendicular diameters from both sides and four heights. The discs were then oven-dried at 40 °C to constant weight. The initial density of the wood was estimated from the known volume and weight of these discs, which did not present any sign of decay.

Figure 1. Wood samples used in the study. **Left**: standardized 75 cm length experimental logs that were spread through the three elevations (plots) since the beginning of the experiment (March–April 2006); a metal tag can be observed in one of the logs. **Center**: Wood disc from an experimental log after some years (two in this case) of decomposition. **Right**: a highly decomposed wood disc after 10 years; when discs showed a high degree of fragmentation, as in this case, we used the two longest available arcs of the circle to find the perpendicular bisectors and the center of the disk at their intersection, from which the diameter was measured.

2.3. Wood Decomposition

Wood decomposition was estimated by cutting one random subsample of each of 20 randomly-chosen experimental logs per elevation in June 2008, 2010, 2014, and 2016, thus at 2, 4, 8, and 10 years after the experimental setup. The experimental logs were brought to the laboratory, and afterwards a disc of 6–8 cm length was obtained from the central part of each log by using a manual saw. Log dimensions and dry mass were measured as indicated above for the initial discs. As decay progresses, dead wood pieces become more elliptical. This is why we used the conic-parabolic formula proposed by Fraver and colleagues [47] to estimate the volume of the logs. Wood density (g cm^{-3}) was then calculated by dividing wood dry weight by its volume for each disc sample.

2.4. Statistical Analyses

We analyzed wood density with mixed models in R version 3.3.1 [48], with the "nlme" package [49]. We initially fitted a model with Plot (a categorical factor with three levels), Year, and Log Diameter, as well as all the possible interactions between these factors, as fixed effects, and we included plot as a random effect to account for pseudoreplication [50]. The response variable was wood density. Model simplification was carried out by sequentially eliminating interaction terms from the model and performing likelihood ratio tests (LRTs) to assess their significance [50]. Heteroscedasticity was assessed with the varIdent function, with the use of LRTs [51]. Assumption checking through plotting of residuals and random effects was carried out as suggested by Pinheiro and Bates [51].

As our response variable was wood density measured at a specific point in time but we were interested in assessing the factors that changed the speed at which density was reduced, in the Results we mainly focus on the factors that changed the effect of Year on wood density (i.e., we interpreted an interaction between Year and another factor as an effect of that factor on the speed of decomposition).

3. Results

Wood density showed an initial decline from an average of 0.482 ± 0.004 g cm^{-3} (mean ± 1 standard error (SE)) in 2006 to 0.420 ± 0.007 g cm^{-3} in 2008. Density then increased slightly until 0.445 ± 0.007 g cm^{-3} in 2010, and then dropped again to 0.415 ± 0.008 g cm^{-3} in 2014 and 0.370 ± 0.013 g cm^{-3} in 2016 (all plots pooled). Our results show that wood density was affected by three interactions between the studied factors. First, there was an effect of plot on wood decomposition rates (i.e., Plot × Year interaction affecting wood density; Table 2): the wood decomposed slower at the highest-elevation plot than at the other plots (Figure 2). Second, the temporal change in wood density was also modified by the size of the log (i.e., significant Year × Diameter interaction; Table 2): larger logs decomposed faster than smaller logs (Figure 3). Third, there was a Diameter × Plot interaction, indicating different effects of log diameter on wood density across plots when years are pooled (Table 2).

4

The Year × Diameter × Plot interaction was not significant, indicating that diameter and plot affected the speed of decomposition independently. Note that the Plot factor includes differences in elevation and species.

Table 2. Results from linear mixed effects models on the effect of year, log diameter, and plot (defined by the elevation/species gradient) on wood density.

Term Removed from Model	Likelihood Ratio	p-Value
Year × Diameter × Plot [1]	1.63	0.44
Year × Diameter [2]	8.26	<0.01
Year × Plot [2]	6.48	<0.05
Diameter × Plot [2]	10.78	<0.01

[1] Tested by removal from the model containing all possible interactions among factors. [2] Tested by removal from the model containing all two-way interactions between the factors.

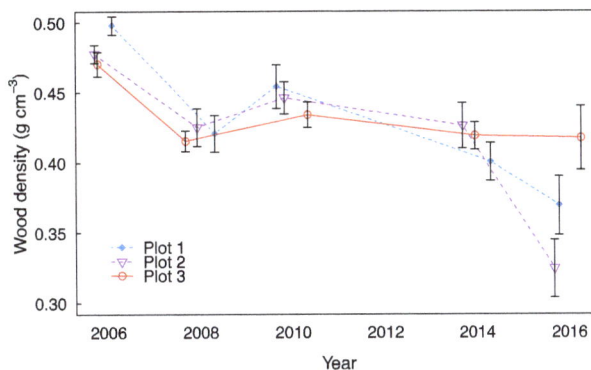

Figure 2. Temporal changes in wood density across the three study elevations (plots). Plot 1 was located at an elevation of 1477 m, Plot 2 at 1698 m, and Plot 3 at 2053 m (note that tree species varied across elevations too; Table 1). Plot did not significantly affect wood density in 2016 according to ANCOVA ($F_{2,66} = 0.74$, $p = 0.48$), but log diameter did (see Figure 3). Error bars indicate ± 1 SE of the mean.

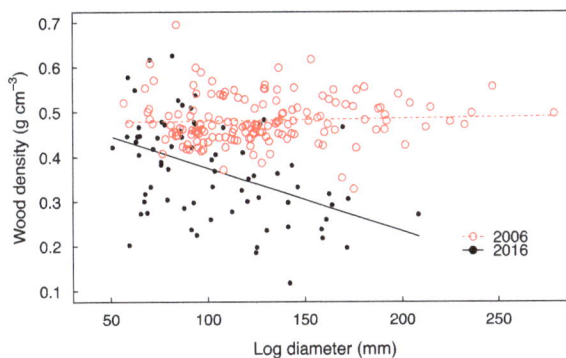

Figure 3. Effect of log diameter on wood decomposition. The density of wood was independent of diameter in 2006 (ANCOVA; $F_{1,70} = 0.28$, $p = 0.6$). In 2016, the average density was lower than in 2006 for all the ranges of diameters considered in the study and negatively affected by log diameter (ANCOVA; $F_{1,70} = 18.4$, $p < 0.001$), indicating that larger-diameter logs decomposed faster than smaller logs. The figure shows the measured values of each log and simple linear regression lines for each year.

4. Discussion

We found that the burnt wood lost nearly one-fourth of its mass after 10 years of decomposition. The rate of decomposition was lowest at the greatest elevation and for small-diameter logs. The differences across elevations might be related to several interacting factors that we cannot rule out, such as differences in climatic conditions and species, or even to an indirect effect of log diameters, which were lowest at the highest elevation (Table 1). This is, in any case, a normal situation under natural conditions, where variability in forest conditions across elevational gradients exists even within even-aged stands. Overall, our study provides novel results concerning burnt wood decomposition in Mediterranean mountains, and it represents one of the longest wood decomposition time-series available for a Mediterranean-type ecosystem.

Despite the overall decline in wood density, the rate of this process changed over time—and even reversed from the second to the fourth year. This has also been observed in previous studies of wood decomposition (e.g., [52]), and it might be related to the often-reported peak in colonization and nutrient immobilization by detritivorous organisms and decomposers in substrates with high carbon to nutrient ratios at the initial stages of decomposition [4,53]. On the other hand, our final values of wood density loss are clearly lower than those reported for other ecosystem types with higher rainfall. For example, Olajuyigbe and colleagues [23] found 50% of decomposition for *Picea sitchensis* after 12 years in Ireland, Mackensen and Bauhus [54] found a 71% decomposition for *Pinus radiata* after 10 years in southeastern Australia, Brown and colleagues [55] reported 25% of decomposition for *Pinus pinaster* after 5 years in places of Western Australia with average rainfall around 1000 L m^{-2} year^{-1}, and Yang and colleagues [56] found density losses greater than 50% after 9 years for three species in an old-growth tropical forest. The lower decomposition rate in our study might be explained by the uncoupling between moisture and temperature during the characteristic summer drought in Mediterranean-type ecosystems [57]. Still, the wood lost up to 26% and 32% of its mass in plots 1 and 2 (lowest and intermediate elevations, respectively) after 10 years, which supports that decomposition, despite being slower than in other temperate ecosystems, remains fast enough to ensure nutrient turnover, increase soil fertility, and reduce the fuel potential of the burnt logs [24,30,31]. Although the logs laid out for this study likely decomposed faster than the remaining wood due to their direct contact with the soil, the decomposition of the standing snags was likely not much slower, as all of them had collapsed and were mostly touching the ground 5.5 years after the fire [46].

The results also show an effect of log diameter on the speed of decomposition. The effect reported for the diameter in the literature is variable. For instance, several studies found no relationship between diameter and decay rate [52,58,59], and other studies found that decomposition rate declined with increasing log diameter because of reduced surface-to-volume ratios [59–61]. Interestingly, our results show the opposite trend. Although we did not study the mechanism behind this diameter effect, we consider that it was likely mediated by an effect of diameter on the deadwood-inhabiting organisms involved in the decay process. In particular, we observed larger holes produced by the larvae of xylophagous insects in the thicker logs (Figure 1). In fact, it is well known that the larvae of larger species tend to inhabit logs of greater diameter [42]. The galleries they create increase bole fragmentation and respiration and can be used by other detritivorous organisms and decomposers that further accelerate decay [4,62]. Logs with a larger diameter also retain more humidity, which is especially beneficial for the colonization of microbial fungi during drought periods [4,63] and thus accelerates decomposition [64]. Another explanation lies in the phenomenon known as "case-hardening", which refers to solar radiation heating and hardening the outer wood layers [65,66], so that a larger surface-to-volume ratio would induce a greater loss of moisture rather than enhanced decomposer colonization [64]. Under Mediterranean climate, characterized by a long, hot, and dry summer, the retention of moisture inside large logs may represent an important factor speeding up wood decomposition. Our results thus support that log size may accelerate decay under Mediterranean climate, and they highlight the need to carefully control log diameter to correctly interpret the speed of decay across environmental gradients.

Salvage logging is a common silvicultural practice after fires in Mediterranean pine reforestations, as well as other parts of the world [25,26,29]. The most obvious consequence of this approach is the subsequent large-scale absence of decomposing wood. While ecosystems and the species that constitute them are resilient to historical disturbance regimes, this resilience hinges on the existence of material legacies of the previous ecosystem that set the scene for regeneration; changes in the post-disturbance environmental conditions compared to those under which the ecosystem historically regenerated can seriously undermine resilience [67]. The present study documents relatively fast and heterogeneous decay rates in burned pine plantations under Mediterranean-type conditions, a process that involves myriads of species ranging from fungi to mammals and that virtually disappears after post-fire logging. The final decision concerning burnt wood should ultimately depend on the balance between the economic value of the wood, the cost of wood removal, the risks posed by the presence of decaying logs, and the ecological processes that dominate post-fire ecosystems under different management scenarios.

5. Conclusions

Our study shows that burnt wood decay in Mediterranean mountains is slower than in other temperate ecosystems, but still fast enough to be considered a process that may support nutrient cycling and ecosystem regeneration. Wood decay changed across an altitudinal gradient, a fact that is likely linked to changes in both abiotic (climatic conditions) and biotic (wood characteristics, decomposer and detritivorous communities) factors. Overall, we conclude that burnt wood is a biological legacy that should be partially or totally kept in situ after fires.

Acknowledgments: We thank the Consejería de Medio Ambiente, Junta de Andalucía, and the Direction of the Natural and National Park of Sierra Nevada for fieldwork permission, constant support, and facilities. This study was supported by Project 10/2005 from the Organismo Autónomo de Parques Nacionales (Spanish Government), CGL2008-01671 from the Spanish Ministerio de Ciencia e Innovación, and P12-RNM-2705 from Junta de Andalucía. A.B.L. acknowledges funding from Juan de la Cierva grant by Ministerio de Economía y Competitividad (FJCI-2015-23687). C.R.M-G. had a Ph.D. grant from the National University of Asunción (Paraguay) and Carolina Foundation (Spain). We are grateful to S. Marañón-Jiménez, who performed valuable fieldwork. Also, we thank Sergio Cortés-Merino and Fernando Bravo for their help in field work.

Author Contributions: J.C. conceived and designed the experiment; C.R.M-G. and J.C. performed the fieldwork; C.R.M-G., A.B.L., and J.C. analyzed the data; C.R.M-G., A.B.L., and J.C. wrote and edited the manuscript.

Conflicts of Interest: The authors declare no conflict of interest.

References

1. Harmon, M.E.; Franklin, J.F.; Swanson, F.J.; Sollins, P.; Gregory, S.V.; Lattin, J.D.; Anderson, N.H.; Cline, S.P.; Aumen, N.G.; Sedell, J.R.; et al. Ecology of coarse woody debris in temperate ecosystems. *Adv. Ecol. Res.* **1986**, *15*, 133–302.
2. Franklin, J.F.; Shugart, H.H.; Harmon, M.E. Tree death as an ecological process. *BioScience* **1987**, *37*, 550–556. [CrossRef]
3. Chamber, C.L.; Mast, J.N. Ponderosa pine snag dynamics and cavity excavation following wildfire in northern Arizona. *For. Ecol. Manag.* **2005**, *216*, 227–240. [CrossRef]
4. Stokland, J.N.; Siitonen, J.; Jonsson, B.G. *Biodiversity in Dead Wood*; Cambridge University Press: Cambridge, UK, 2012.
5. Swift, M. The ecology of wood decompositiopn. *Sci. Prog.* **1977**, *64*, 175–199.
6. Ganjegunte, G.K.; Condron, L.M.; Clinton, P.W.; Davis, M.R.; Mahieu, N. Decomposition and nutrient release from radiata pine (*Pinus radiata*) coarse woody debris. *For. Ecol. Manag.* **2004**, *187*, 197–211. [CrossRef]
7. Palviainen, M.; Finér, L.; Kurka, A.M.; Mannerkoski, H.; Piirainen, S.; Starr, M. Decomposition and nutrient release from logging residues after clear-cutting of mixed boreal forest. *Plant Soil* **2004**, *263*, 53–67. [CrossRef]
8. Russell, M.B.; Woodall, C.W.; D'Amato, A.W.; Fraver, S.; Bradford, J.B. Technical Note: Linking climate change and downed woody debris decomposition across forests of the eastern United States. *Biogeosciences* **2014**, *11*, 6417–6425. [CrossRef]

9. Cornwell, W.K.; Cornelissen, J.H.C.; Allison, S.D.; Bauhus, J.; Eggleton, P.; Preston, C.M.; Scarff, F.; Weedon, J.T.; Wirth, C.; Zanne, A.E. Plant traits and wood fates across the globe: Rotted, burned, or consumed? *Glob. Chang. Biol.* **2009**, *15*, 2431–2449. [CrossRef]

10. Means, J.E.; MacMillan, P.C.; Cromack, K.J. Biomas and nutrient content of douglas-fir logs and other detrital pools in a old-growth forest, Oregon, U.S.A. *Can. J. For. Res.* **1992**, *22*, 1536–1546. [CrossRef]

11. Schiegg, K. Are there saproxylic beetle species characteristic of high dead wood connectivity? *Ecography* **2000**, *23*, 579–587. [CrossRef]

12. Vodka, S.; Konvicka, M.; Cizek, L. Habitat preferences of oak-feeding xylophagous beetles in a temperate woodland: Implications for forest history and management. *J. Insect Conserv.* **2009**, *13*, 553–562. [CrossRef]

13. Angelstam, P.K.; Bütler, R.; Lazdinis, M.; Mikusinski, G.; Roberge, J.-M. Habitat thresholds for focal species at multiple scales and forest biodiversity conservation—dead wood as an example. *Ann. Zool. Fenn.* **2003**, *40*, 473–482.

14. Lassauce, A.; Paillet, Y.; Jactel, H.; Bouget, C. Deadwood as a surrogate for forest biodiversity: Meta-analysis of correlations between deadwood volume and species richness of saproxylic organisms. *Ecol. Indic.* **2011**, *11*, 1027–1039. [CrossRef]

15. Rajandu, E.; Kikas, K.; Paal, J. Bryophytes and decaying wood in hepatica site-type boreo-nemoral *Pinus sylvestris* forests in Southern Estonia. *For. Ecol. Manag.* **2009**, *257*, 994–1003. [CrossRef]

16. Marzano, R.; Garbarino, M.; Marcolin, E.; Pividori, M.; Lingua, E. Deadwood anisotropic facilitation on seedling establishment after a stand-replacing wildfire in Aosta Valley (NW Italy). *Ecol. Eng.* **2013**, *51*, 117–122. [CrossRef]

17. Leverkus, A.B.; Lorite, J.; Navarro, F.B.; Sánchez-Cañete, E.P.; Castro, J. Post-fire salvage logging alters species composition and reduces cover, richness, and diversity in Mediterranean plant communities. *J. Environ. Manag.* **2014**, *133*, 323–331. [CrossRef] [PubMed]

18. Chmura, D.; Żarnowiec, J.; Staniaszek-Kik, M. Interactions between plant traits and environmental factors within and among montane forest belts: A study of vascular species colonising decaying logs. *For. Ecol. Manag.* **2016**, *379*, 216–225. [CrossRef]

19. Cadieux, P.; Drapeau, P. Are old boreal forests a safe bet for the conservation of the avifauna associated with decayed wood in eastern Canada? *For. Ecol. Manag.* **2017**, *385*, 127–139. [CrossRef]

20. Serrano-Ortiz, P.; Marañón-Jiménez, S.; Reverter, B.R.; Sánchez-Cañete, E.P.; Castro, J.; Zamora, R.; Kowalski, A.S. Post-fire salvage logging reduces carbon sequestration in Mediterranean coniferous forest. *For. Ecol. Manag.* **2011**, *262*, 2287–2296. [CrossRef]

21. Shorohova, E.; Kapitsa, E. Influence of the substrate and ecosystem attributes on the decomposition rates of coarse woody debris in European boreal forests. *For. Ecol. Manag.* **2014**, *315*, 173–184. [CrossRef]

22. Russell, M.B.; Fraver, S.; Aakala, T.; Gove, J.H.; Woodall, C.W.; D'Amato, A.W.; Ducey, M.J. Quantifying carbon stores and decomposition in dead wood: A review. *For. Ecol. Manag.* **2015**, *350*, 107–128. [CrossRef]

23. Olajuyigbe, S.O.; Tobin, B.; Gardiner, P.; Nieuwenhuis, M. Stocks and decay dynamics of above- and belowground coarse woody debris in managed Sitka spruce forests in Ireland. *For. Ecol. Manag.* **2011**, *262*, 1109–1118. [CrossRef]

24. Passovoy, M.D.; Fulé, P.Z. Snag and woody debris dynamics following severe wildfires in northern Arizona ponderosa pine forests. *For. Ecol. Manag.* **2006**, *223*, 237–246. [CrossRef]

25. Lindenmayer, D.B.; Burton, P.J.; Franklin, J.F. *Salvage Logging and Its Ecological Consequences*; Island Press: Washington, DC, USA, 2008.

26. Castro, J.; Moreno-Rueda, G.; Hodar, J.A. Experimental test of postfire management in pine forests: impact of salvage logging versus partial cutting and nonintervention on bird-species assemblages. *Conserv. Biol.* **2010**, *24*, 810–819. [CrossRef] [PubMed]

27. Ritchie, M.W.; Knapp, E.E.; Skinner, C.N. Snag longevity and surface fuel accumulation following post-fire logging in a ponderosa pine dominated forest. *For. Ecol. Manag.* **2013**, *287*, 113–122. [CrossRef]

28. Donato, D.C.; Fontaine, J.B.; Campbell, J.L.; Robinson, W.D.; Kauffman, J.B.; Law, B.E. Post-Wildfire logging hinders regeneration and increases fire risk. *Science* **2006**, *311*, 352. [CrossRef] [PubMed]

29. Castro, J.; Allen, C.D.; Molina-Morales, M.; Marañón-Jiménez, S.; Sánchez-Miranda, Á.; Zamora, R. Salvage Logging versus the use of burnt wood as a nurse object to promote post-fire tree seedling establishment. *Restor. Ecol.* **2011**, *19*, 537–544. [CrossRef]

30. Marañón-Jiménez, S.; Castro, J. Effect of decomposing post-fire coarse woody debris on soil fertility and nutrient availability in a Mediterranean ecosystem. *Biogeochemistry* **2013**, *112*, 519–535. [CrossRef]

31. Marañón-Jiménez, S.; Castro, J.; Fernández-Ondoño, E.; Zamora, R. Charred wood remaining after a wildfire as a reservoir of macro- and micronutrients in a Mediterranean pine forest. *Int. J. Wildland Fire* **2013**, *22*, 681–695. [CrossRef]

32. Thorn, S.; Bässler, C.; Brandl, R.; Burton, P.J.; Cahall, R.; Campbell, J.L.; Castro, J.; Choi, C.-Y.; Cobb, T.; Donato, D.C.; et al. Impacts of salvage logging on biodiversity—a meta-analysis. *J. Appl. Ecol.* **2017**, in press.

33. Sippola, A.; Siitonen, J.; Kallio, R. Amount and quality of coarse woody debris in natural and managed coniferous forests near the timberline in Finnish Lapland. *Scand. J. For. Res.* **1998**, *13*, 204–214. [CrossRef]

34. Rock, J.; Badeck, F.-W.; Harmon, M.E. Estimating decomposition rate constants for European tree species from literature sources. *Eur. J. For. Res.* **2008**, *127*, 301–313. [CrossRef]

35. Lombardi, F.; Lasserre, B.; Tognetti, R.; Marchetti, M. Deadwood in relation to stand management and forest type in central apennines (Molise, Italy). *Ecosystems* **2008**, *11*, 882–894. [CrossRef]

36. Lombardi, F.; Cherubini, P.; Tognetti, R.; Cocozza, C.; Lasserre, B.; Marchetti, M. Investigating biochemical processes to assess deadwood decay of beech and silver fir in Mediterranean mountain forests. *Ann. For. Sci.* **2013**, *70*, 101–111. [CrossRef]

37. Liu, W.; Schaefer, D.; Qiao, L.; Liu, X. What controls the variability of wood-decay rates? *For. Ecol. Manag.* **2013**, *310*, 623–631. [CrossRef]

38. Herrmann, S.; Bauhus, J. Effects of moisture, temperature and decomposition stage on respirational carbon loss from coarse woody debris (CWD) of important European tree species. *Scand. J. For. Res.* **2013**, *28*, 346–357. [CrossRef]

39. Fravolini, G.; Egli, M.; Derungs, C.; Cherubini, P.; Ascher-Jenull, J.; Gómez-Brandón, M.; Bardelli, T.; Tognetti, R.; Lombardi, F.; Marchetti, M. Soil attributes and microclimate are important drivers of initial deadwood decay in sub-alpine Norway spruce forests. *Sci. Total Environ.* **2016**, *569*, 1064–1076. [CrossRef] [PubMed]

40. De Aza, C.H.; Turrión, M.B.; Pando, V.; Bravo, F. Carbon in heartwood, sapwood and bark along the stem profile in three Mediterranean *Pinus* species. *Ann. For. Sci.* **2011**, *68*, 1067–1076. [CrossRef]

41. Boddy, L. Fungal community ecology and wood decomposition processes in angiosperms: From standing tree to complete decay of coarse woody debris. *Ecol. Bull.* **2001**, *49*, 43–56.

42. Ulyshen, M.D. Wood decomposition as influenced by invertebrates. *Biol. Rev.* **2016**, *91*, 70–85. [CrossRef] [PubMed]

43. Yang, K.; Hazenberg, G. Sapwood and heartwood width relationship to tree age in *Pinus banksiana*. *Can. J. For. Res.* **1991**, *21*, 251–525. [CrossRef]

44. Leverkus, A.B.; Puerta-Piñero, C.; Guzmán-Álvarez, J.; Navarro, J.; Castro, J. Post-fire salvage logging increases restoration costs in a Mediterranean mountain ecosystem. *New For.* **2012**, *43*, 601–613. [CrossRef]

45. Leverkus, A.B.; Castro, J.; Delgado-Capel, M.J.; Molinas-González, C.; Pulgar, M.; Marañón-Jiménez, S.; Delgado-Huertas, A.; Querejeta, J.I. Restoring for the present or restoring for the future: Enhanced performance of two sympatric oaks (*Quercus ilex* and *Quercus pyrenaica*) above the current forest limit. *Restor. Ecol.* **2015**, *23*, 936–946. [CrossRef]

46. Molinas-González, C.R.; Leverkus, A.B.; Marañón-Jiménez, S.; Castro, J. Fall rate of burnt pines across an elevational gradient in a Mediterranean mountain. *Eur. J. For. Res.* **2017**. [CrossRef]

47. Fraver, S.; Ringvall, A.; Jonsson, B.G. Refining volume estimates of down woody debris. *Can. J. For. Res.* **2007**, *37*, 627–633. [CrossRef]

48. R Core Team. *R: A language and environment for statistical computing*; R Foundation for Statistical Computing: Vienna, Austria, 2013.

49. Pinheiro, J.; Bates, D.; DebRoy, S.; Sarkar, D. Linear and Nonlinear Mixed Effects Models. Available online: https://CRAN.R-project.org/package=nlme (accessed on 10 December 2016).

50. Crawley, M.J. *The R Book*, 2nd ed.; John Wiley & Sons: West Sussex, UK, 2013.

51. Pinheiro, J.C.; Bates, D.M. *Mixed effects models in S and S-Plus*; Springer: New York, NY, USA, 2000.

52. Foster, J.R.; Lang, G.E. Decomposition of red spruce and balsam fir boles in the White Mountains of New Hampshire. *Can. J. For. Res.* **1982**, *12*, 617–626. [CrossRef]

53. Coleman, D.C.; Crossley, D.A. *Fundamentals of Soil Ecology*, 2nd ed.; Academic Press: Waltham, MA, USA, 2003.

54. Mackensen, J.; Bauhus, J. Density loss and respiration rates in coarse woody debris of *Pinus radiata*, *Eucalyptus regnans* and *Eucalyptus maculata*. *Soil Biol. Biochem.* **2003**, *35*, 177–186. [CrossRef]

55. Brown, S.; Mo, J.; McPherson, J.K.; Bell, D. Decomposition of woody debris in western Australian forest. *Can. J. For. Res.* **1996**, *26*, 954–966. [CrossRef]

56. Yang, F.-F.; Li, Y.-L.; Zhou, G.-Y.; Wenigmann, K.O.; Zhang, D.-Q.; Wenigmann, M.; Liu, S.-Z.; Zhang, Q.-M. Dynamics of coarse woody debris and decomposition rates in an old-growth forest in lower tropical China. *For. Ecol. Manag.* **2010**, *259*, 1666–1672. [CrossRef]

57. Aschmann, H. Distribution and Peculiarity of Mediterranean Ecosystems. In *Mediterranean Type Ecosystem Origin and Structure*; Di Castri, F., Mooney, H.A., Eds.; Springer: Berlin/Heidelberg, Germany; New York, NY, USA, 1973; pp. 11–19.

58. Laiho, R.; Prescott, C.E. The contribution of coarse woody debris to carbon, nitrogen, and phosphorus cycles in three rocky mountain coniferous forests. *Can. J. For. Res.* **1999**, *29*, 1592–1603. [CrossRef]

59. Mackensen, J.; Bauhus, J.; Webber, E. Decomposition rates of coarse woody debris—A review with particular emphasis on Australian tree species. *Aust. J. Bot.* **2003**, *51*, 27–37. [CrossRef]

60. Jonsell, M.; Hansson, J.; Wedmo, L. Diversity of saproxylic beetle species in logging residues in Sweden—Comparisons between tree species and diameters. *Biol. Conserv.* **2007**, *138*, 89–99. [CrossRef]

61. Weedon, J.T.; Cornwell, W.K.; Cornelissen, J.H.C.; Zanne, A.E.; Wirth, C.; Coomes, D.A. Global meta-analysis of wood decomposition rates: A role for trait variation among tree species? *Ecol. Lett.* **2009**, *12*, 45–56. [CrossRef] [PubMed]

62. Kitchell, J.F.; O'Neill, R.V.; Webb, D.; Gallepp, G.W.; Bartell, S.M.; Koonce, J.F.; Aumus, B.S. Regulation of nutrient cycling. *Bioscience* **1979**, *29*, 28–34. [CrossRef]

63. Harvey, A.E.; Jurgensen, M.F.; Larsen, M.J. Seasonal distribution of ectomycorrhizae in a mature douglas-fir/larch forest soil in Western Montana. *For. Sci.* **1978**, *24*, 203–208.

64. Erickson, H.E.; Edmonds, R.L.; Peterson, C.E. Decomposition of logging residues in douglas-fir, western hemlock, pacific silver fir, and ponderosa pine ecosystems. *Can. J. For. Res.* **1985**, *15*, 914–921. [CrossRef]

65. Kimmey, J.W.; Furnis, R.L. *Deterioration of Fire-Killed Douglas-Fir*; Tecnical Bulletin USDA, U.S. Department of Agriculture: Washington, DC, USA, 1943.

66. Yatskov, M.; Harmon, M.E.; Krankina, O.N. A chronosequence of wood decomposition in the boreal forests of Russia. *Can. J. For. Res.* **2003**, *33*, 1211–1226. [CrossRef]

67. Johnstone, J.F.; Allen, C.D.; Franklin, J.F.; Frelich, L.E.; Harvey, B.J.; Higuera, P.E.; Mack, M.C.; Meentemeyer, R.K.; Metz, M.R.; Perry, G.L.W.; et al. Changing disturbance regimes, ecological memory, and forest resilience. *Front. Ecol. Environ.* **2016**, *14*, 369–378. [CrossRef]

Article

Soil Erosion and Forests Biomass as Energy Resource in the Basin of the Oka River in Biscay, Northern Spain

Esperanza Mateos [1,*], José Miguel Edeso [2] and Leyre Ormaetxea [3]

[1] Department of Chemical and Environmental Engineering, University of the Basque Country UPV/EHU, Rafael Moreno 'Pitxitxi', n 3, 48013 Bilbao, Spain

[2] Department of Mining, Escuela Técnica de Ingenieros de Minas, University of the Basque Country UPV/EHU, E-01006 Vitoria, Spain; josemiguel.edeso@ehu.eus

[3] Department of Mathematics, Faculty of Science and Technology, University of the Basque Country UPV/EHU, Barrio Sarriena s/n, 48940 Leioa, Spain; leyre.ormaetxea@ehu.eus

* Correspondence: esperanza.mateos@ehu.eus; Tel.: +34-946-10-43-43; Fax: +34-946-10-43-00

Academic Editor: Raffaele Spinelli
Received: 10 April 2017; Accepted: 12 July 2017; Published: 19 July 2017

Abstract: The aim of this work has been the development of a methodology for the evaluation of residual forest biomass in Biscay, a province in northern Spain. The study area is located in the Oka river basin, an area of great ecological value qualified by UNESCO (United Nations Educational Scientific and Cultural Organization) in 1984 as a Biosphere Reserve. The project tries to determine the potential, available and usable as energy resource, residual forests biomass, after the treatments of forest species in the area. Soil erosion was modeled using the USLE (Universal Soil Loss Equation) and MUSLE (Modified USLE) methods by estimating rainfall erosivity factor (R), the soil erodibility factor (K), the topographic factors (L and S), cropping factor (C), and the conservation practice factor (P). By means of these models, it will be possible to determine the current soil erosion rate and its potential evolution due to different forest treatments. Soil erodibility, slope of the terrain and the loss of SOC (Soil Organic Carbon) were the restrictive indicators for the bioenergy use of forest biomass, taking into account principles of sustainability. The amount of residual forestry biomass useable for energy purposes has been estimated at 4858.23 Mg year^{-1}.

Keywords: soil erosion; GIS (Geographic Information System); forest residues; energy valuation; harvesting management

---

1. Introduction

Forest biomass plays an important role in the overall carbon cycle. This is due, among other reasons, to the contribution it makes to fixing atmospheric carbon and, as a consequence, bringing about a reduction in the greenhouse effect. The Kyoto protocol officially recognised the role of forests as carbon sinks in the mitigation of global climate change factors, basically by reducing the atmospheric concentrations of CO_2 [1,2]. Forestry activities generate significant quantities of wastes and by-products that are suitable for diverse energy purposes. The most viable alternatives for energy use from forest biomass residues in the area of study are the production of solid bio-fuel and electricity generation, especially in cogeneration plants. The use of forest biomass as an energy source helps to reach the compromises acquired by the European Union in the Kyoto protocol by 2020, 20% of all energy consumption must come from renewable sources. In Spain, the 2011–2020 Renewable Energy Plan set the target of 20% of total primary energy needs to be met by renewable sources and about 10% of these by bioenergy [3]. The use of such residual biomass has become crucial for several reasons.

Firstly, the combustion of biomass plays a virtually neutral role in the carbon cycle and produces minimum emissions of SO_2, particles and NO_x. Moreover, due to the absence of chlorine compounds, the formation of dioxins is avoided. Secondly, its removal is the most effective way to prevent forest fires and plagues of insects, and, finally, it favours employment and profitability of forestry activities. These advantages make biomass a potential source of employment for the future, converting it into an element of great importance for regional stability, especially in rural areas [4]. Nevertheless, in order for biomass to be considered a true renewable resource, the speed with which it is used must not be greater than the speed with which it is regenerated as a resource [5]. Sustainable forestry management seeks to strike a balance between present and future needs, involving environmental, social and socio-economic factors [6]. The study of the influence of forestry biomass on soil quality is fundamental in order to predict the evolution of forests and to develop adequate management strategies.

The bibliography contains a number of different investigations into the effect of intensive forestry practices on the productivity of forestry species. Managed forests generally contain much less biomass compared to unmanaged natural forests [7,8]. Other researchers analyze the influence of forestry management practices on soil quality and carbon retention in pine plantations. Management practices differ considerably with respect to impacts on carbon sequestration in forests [9,10].

In recent decades and with the aim of satisfying industrial needs, the forestry sector of the study area (Figure 1b) has increased the surface area of plantations of rapid-growth species, basically *P. radiata* D. Don and *Eucalyptus globulus* Labill. The latter plays a fundamental role in supplying the cellulose and paper industry.

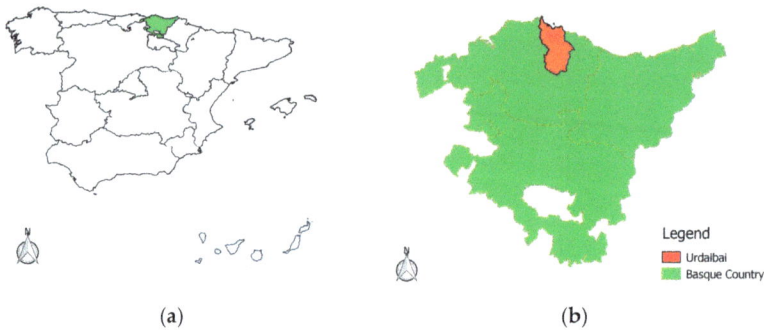

(a)   (b)

Figure 1. The landscape of Urdaibai. (a) Spain; (b) Urdaibai: Basque Country.

It is therefore necessary to develop a methodology to evaluate the potential and useable residual forestry biomass for energy production in order to be able to set up installations to use this fuel for energy purposes in the most suitable locations. Until now, the lack of a specific methodology to evaluate the precise amount of residual forestry waste produced by forestry activities has avoided an adequate use of these products for energy purposes. However, a number studies in different parts of Europe and the United States have determined the potential of the existing residual biomass for energy valuation [11,12]. However, few surveys assess the amount of useable forestry biomass taking soil loss into account. Soil quality is considered a key element os sustainable land management. Soil erosion is a major environmental problem in worldwide. Rainfall intensity, vegetation, soil, topography and geology and human beings are usually the main causes of soil erosion [13].

One of the main hindrances to biomass energy management is the difficulty involved in ensuring a steady supply for heat/electricity generating plants. As a result, the amount of biomass and the potential influence on soil loss from its use are essential factors required to quantify if sustainable biomass energy and other products are to be developed. The objectives of this study were as follows:

1. Validate the soil map of the Oka River Basin obtained in the Third National Forestry Inventory [14].
2. Predict the spatial distribution of soil erosion in order to prepare maps of soil erodibility factor (K) and soil losses of the Oka River Basin in order to draw up a soil loss map.
3. Determine the potential forestry biomass in the Oka River Basin using a GIS (Geographic Information System).
4. Determine the available biomass while taking environmental factors (erodibility and organic matter in soil) and economic factors (slope of land) into consideration. The terrain slope has been used as an influencing factor for the operations cost due to the difficulties of machinery to take all the material lying on the forest floor [12].

By estimating the amounts of available biomass that could be generated by forestry activity, it would be possible to plan the use of biomass, taking into consideration both economic and environmental aspects in order to determine the optimum location for the installation of an energy production centre using a GIS.

2. Methodology

2.1. Description of the Study Area

This study was carried out in the province of Biscay, located in the north of Spain (Figure 1a). The study covers the Oka River Basin and includes the Urdaibai estuary, a site of great ecological value. In 1984, it was declared a "Biosphere Reserve" by UNESCO (United Nations Educational Scientific and Cultural Organization). The river basin is located in the northern/central sector of the Province of Bizkaia (43′12″ N, 2′33″ W) (Figure 1b). With a temperate, very wet climate, its mean annual temperature is 12–14 °C, with precipitations that range between 1200–1800 mm year^{-1}. This river basin covers an area of approximately 178.1 km^2, the main basin being 27 km long and with a perimeter of 68 km.

According to data from the Third National Forest Inventory (NFI3) [14], forestry land covers a surface area of 14,709 ha, representing 66.86% of the Urdaibai Biosphere Reserve (UBR). 80% of the forestry mass consists of plantations of *P. radiata* and *E. globulus*, thus giving rise to the dominant vegetation type [15]. Morphologically, the river basin is oval in shape and has a general N-S orientation. The area is Mesozoic in origin and basically cretaceous, in which two, clearly contrasting sectors can be differentiated:

1. A southern section, head of the river Oka, with deep-v-shaped valleys, the sides of which are dominated by steep slopes.
2. A northern section organised round the Urdaibai estuary, with an average length of 10 km and a width of between 1 and 2 km, generating a broad, flat valley lying just over sea level, forming a typical landscape of marshes, sandbanks and cultivated areas.

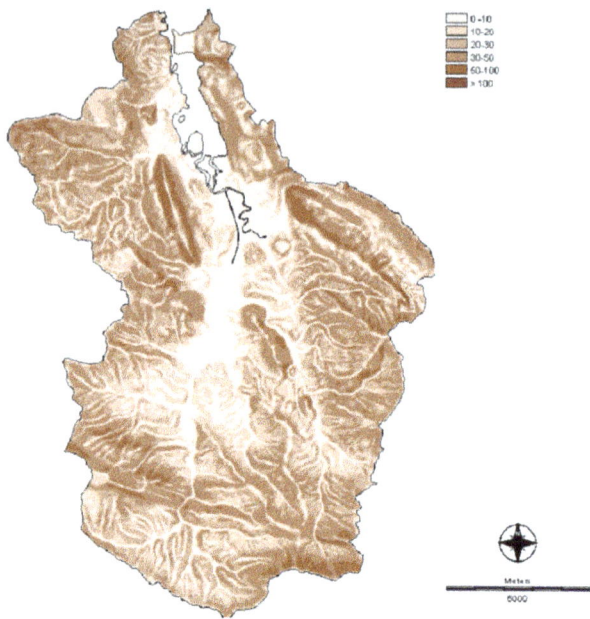

(a)

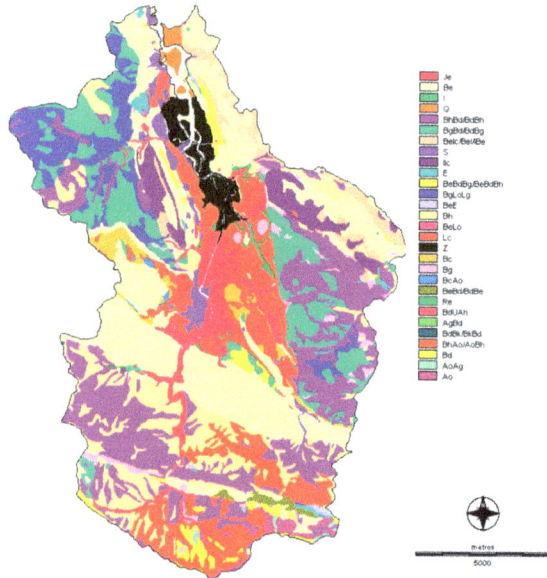

(b)

Figure 2. Maps of the Oka River Basin. (**a**) slopes map; (**b**) soils map.

The vegetation map of Urdaibai (Figure 3) shows that about 70% of the surface area is covered with trees, the dominant species being *Pinus radiata* D. Don (8764 ha), coinciding with the ecological niche of the oak forest [16], followed by *Quercus ilex* (1555 ha) and *Eucalyptus globulus* Labill (1321 ha). The rapid growth and the high productivity of *P. radiata* have led it to be, in practice, the dominant forestry species in the Urdaibai Reserve. Forestry activities are performed mainly on private land, as publicly-owned land accounts for less than 500 ha of the Oka River Basin as a whole. Despite their widespread distribution, based on the high market value of by-products from pruning operations, the areas of *Quercus* have not been considered as a potential source of residual biomass based on the high value added reached by by-products coming from forest operations. Scrub surfaces have not been considered in this study; since the surface they occupy in the territory is rather average (8.6%), it presents low productivity, and it is a very scattered resource. The study concentrates on the pine and eucalyptus forests made up of *P. radiata* and *E. globulus*.

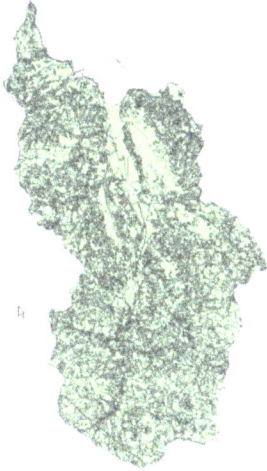

Figure 3. Urdaibai forestry map [16].

2.2. Validating the Soil Map

The aim of the first stage of the project was to validate the soil map of the Oka River Basin obtained in the Third National Forest Inventory (NFI3). For this reason, work had focused on gathering and reviewing bibliographical sources and base material (geology, vegetation, orthophotos, etc.). The 1:10,000 forestry map of the Basque Autonomous Community (2004–2005), based on the cartographical standards of the NFI3 in the UTM (Universal Transverse Mercator) reference system (time zone 30) European Datum (1950), was used.

A detailed soil map was drawn up based on information supplied by the 1:25,000 cartography of the Basque Government. Once imported, these were rasterised using a 2 m pixel size. The coverage was validated based on a laboratory analysis of 345 samples obtained in 115 plots (3 samples per plot) of the study area. The vegetation coverage was also validated with the help of orthophotos and aerial photography. A study was made of the composition of the soil in the area by means of sampling and subsequent laboratory analysis.

The Universal Soil Loss Equation (USLE) was used to estimate the rate of soil erosion in the study area. The USLE was adjusted based on the results obtained in exhaustive field work carried out between 1992 and 1996. During this period, a control was made of the erosion rate, and, for this purpose, 168 experimental plots were installed [17]. The slope map (Figure 2a) and the soils map of the study area (Figure 2b) were drawn up prior to initiating the study. Geologically, the basin comprises eighteen types of soils, cambisols being those that have the greatest extension (60%), followed by luvisols (8.9%), acrisols (7.1%) and fluvisols (5.8%).

In order to evaluate the erosion rate of the Oka River Basin, several simulations with GIS tools have been used to analyze the possible soil loss that might be caused by different types of forestry management used in *P. radiata* plantations. The results obtained by Edeso et al. [18] suggest that, in order to maintain the sustainability of the ecosystem of the area, the intensive soil preparation must be reduced in accordance with the sustainable forestry management proposal. The simplified form of the USLE equation is summarised in the following expression [19]:

$$A = R \cdot K \cdot S \cdot L \cdot C \cdot P \tag{1}$$

where A is the predicted annual rate of soil loss (Mg ha^{-1}); R is the rainfall erosivity factor (MJ mm ha^{-1} h^{-1}), K is the soil erodibility factor (Mg ha h ha^{-1} MJ^{-1} mm^{-1}), S is the slope steepness/slope gradient factor, L is the slope length factor, C is the crop or cover management factor and conservation or support practice factor (P) is the ratio of the soil loss from land having specified conservation practices to that from land ploughed in a direction parallel to the slope, if all other conditions remain unchanged [17,18,20].

2.3. Assessment of Annual Rate of Soil Erosion

Soil loss prediction is necessary for policy makers in land management and ecosystem protection [21]. We set out to simulate the soil loss produced in the Oka River Basin under a number of different conditions. For this purpose, we used the information obtained by direct experimentation and that deriving from the cartographical analysis, as well as from photo-interpretation and the analysis of images obtained through the use of remote sensors.

The erosion rate under the current edaphic and vegetation conditions in the Oka River Basin was simulated, determining the soil loss produced under these circumstances, as well as locating those areas of greatest risk of the production of sediments. To do this, the following documents supplied by the Basque Government were used: Digital Elevation Model (DEM) (1:10,000), soil map (1:25,000); vegetation map (1:25,000); slopes map obtained from the DEM; and map showing the location of meteorological stations and several additional maps/thematic layers. The rain factor (R) has been determined from the data supplied by 5 observatories situated in the study area, shown in Table 1.

Table 1. Annual rainfall erosivity factor (R) for the five study rain gauge stations.

Station	Altitude	X *	Y *	R
Muxika	20	525,325	4,793,010	210
Ereño	502	532,950	4,799,975	223
Mundaka	93	524,145	4,806,115	181
Bermeo	15	522,434	4,808,060	191
Amorebieta	65	521,787	4,785,950	209

* X and Y are UTM (Universal Transverse Mercator) coordinates.

Accurate values of R were obtained in the five stations in which meteorological data are available. These data were implemented through the digitisation of the mean values obtained in other soils with identical characteristics. These values have been extended in a continuous manner over the study area applying geostatistical techniques [17,18].

The K factor was evaluated based on a random sample made based on the different soils existing in the Oka River Basin (see Section 2.4). This factor is calculated based on a regression equation in accordance with the representative variables of the physical properties of the soil (percentages of organic matter and clay, and the structure and permeability indices).

In some cases, we used the mean values obtained in prior studies of other soils with identical characteristics [17,18]. Once the land losses had been obtained in each one of the cells, we re-classified the results by establishing different land categories, shown in Table 2.

Table 2. Priority scales for soil erosion.

Soil Erosion Class	Soil Loss (Mg ha^{-1} year^{-1})
Slight	(0–5]
Moderate high	(5–25]
Very high	(25–50]
Very severe	(50–100]

We did not include more groups as the maximum soil loss estimated in previous studies, even when these were subjected to intensive treatments, never exceeded a value of 77 Mg ha^{-1} year^{-1} [18].

With the data obtained, we set up a database, which, in turn, has been implemented in a GIS. All of the available information was coded in raster format, geo-referencing with a UTM coordinates system and with a space resolution of 2 × 2 m in order to determine the erosion rate of the Oka River Basin.

2.4. Soil Sampling and Analysis

We selected 115 plots with a minimum surface area of 0.5 ha and length of 60–80 m. These plots were selected so that they covered the main types of soils and the range of slopes of the study area. Soil sampling was carried out from December 2010 to March 2012. At each plot, 9 points were selected at random, in which a surface mineral horizon was taken of the soil (0–10 cm) through the use of an unaltered soil sampler. With the 9 samples, three samples of equal mass were gathered to build up three composite samples. These were air-dried and sieved at 2 mm in order to characterise these. At the same time, using a 40 mm long × 55 mm-internal diameter steel cylinder, 3 samples were taken to determine the bulk density. These were dried at 105 °C to a constant weight. In the air-dried soil samples, the following physiochemical parameters were determined: soil texture (percentage of sand, silt and clay), level of stone content (fractions: <2 mm, 2–50 mm and >50 mm), bulk density, pH, soil organic carbon (SOC), N and S, assimilable phosphorous, iron and aluminum, basic cations as well as Ca, Mg, Na and K and the capacity of the effective cationic exchange (ECEC). The soil texture was determined by means of the Robinson pipette method, according to the international method. The bulk density of each horizon was determined in each test pit following the cylinder methodology [22]. The pH was determined in KCl 0.1 N with a glass electrode, using a soil: dissolution ratio of 1:2.5 [22]. The concentrations of total C, N and S were determined by means of combustion in a LECO (Corporation St. Joseph, Michigan, MI, USA) automated analyzer.

2.5. Determining the Potential Residual Forestry Biomass

The forestry biomass capable of being used for the generation of electrical and/or heat energy is the residual biomass or forestry residue that remains on site following forestry treatments. Residual forestry biomass is considered as that part of the tree not used by the sawmill (branches and leaves or needles). Forest residues are those materials removed in timber-yielding exploitation which are not usually extracted due to the fact that they cannot be converted into by-products. These residues come from the remains left in the forest after forestry, pruning, clearing, and final cutting remains in forest production (cleaning, pruning, tree felling). They can be utilized for energy uses due to their excellent features as fuels.

In general, stand improvements (thinning and clearing) applied to pine forests involve the extraction of a third of the existing trees. With regard to the frequency of the clearing operations, the minimum intensity to be applied to *P. radiata* involves only three clearings, applied at approximately 10-year intervals until logging, which is done at 30 years [23]. Regarding eucalyptus forests, the conventional treatments applied to these plantations are reduced to logging, which is done at 10 years [23,24]. Table 3 shows the forestry treatments of the main tree species in the Oka River Basin.

Table 3. Forest biomass generating silvicultural operations.

P. radiata	**Stand improvements (thinning and clearing)** Two thinnings: 1st at 10 years, 2nd at 20 years Action is taken on 1/3 of current trees **Logging** Logging in 30 years
E. globulus	**Logging** Logging at 10 years

A good approach to estimating the stock of biomass is to use the concept of stratum, as defined in the Third National Forestry Inventory (NFI3). A stratum is formed by grouping the forest surfaces of similar features according to the species present, their states in terms of mass and the fraction of tree cover per area [12,25,26]. Specific information concerning such strata is available at the Autonomous Community of the Basque Country (ACBC) Forest Map [27]. Once strata in which *P. radiata* and *E. globulus* are the dominant species are selected, the most adequate forest treatments that can be carried out in a ten-year-old horizontal stratum are identified, thus the amounts of residual forest biomass that might be obtained in each stratum considering such treatments are estimated [2].

The methodological process followed for estimating the potential biomass is shown schematically in Figure 4.

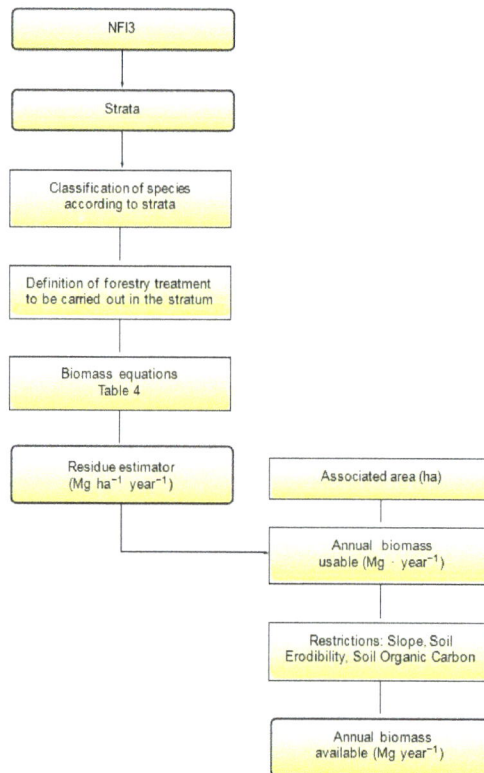

Figure 4. Schematic diagram of the method followed to determine the available residual biomass.

The methodology used to determine the annual quantity of forest biomass in the area of study consists primarily of determining two factors:

1. The forest residue per unit of surface and time derived from a forest mass (R_e, Mg ha^{-1} year^{-1}) according to the species and forest treatment to which each mass has been subjected to.
2. The surface (A, ha) occupied by the forest mass that is going to generate such residue.

The residue estimator R_e was calculated using an indirect methodology based on mathematical models that relate the aerial arboreal residual biomass (branches and leaves) with biophysical variables that are easy to measure in forestry inventories such as the normal diameter and sometimes the height [26,28–30].

The most common procedure para la estimación is establishing biomass functions via regression analysis between the biomass of a tree and easily measurable variables such as stem diameter and height. These relations can be expressed mathematically as allometric power models [31,32]. In this project, the methodology followed is based on the biomass fractions introduced by Montero et al. [28]. Such a method develops a logarithmic model, which relates the normal diameter of the tree to the net dry biomass and has the following analytical form:

$$\ln W = a + b \ln D \tag{2}$$

where W represents the biomass for each tree (expressed in kg dry mass), D is the diameter at breast height (DBH) or trunk diameter at 1.30 m, a and b are two specific regression parameters.

This logarithmic equation introduces a bias in the calculations and therefore it is necessary to introduce a correction factor (CF) calculated on the basis of the standard error of the estimation (SEE) [2,28]:

$$CF = e^{\frac{SEE^2}{2}} \tag{3}$$

$$W = CF \cdot e^a \cdot D^b. \tag{4}$$

The species-specific allometric equations of [33] for *P. radiata* and [34] for *E. globulus* were used to calculate the residual biomass Table 4. These equations are applied to trees of the diameter at breast height (DBH) of over 7.5 cm.

Table 4. Allometric equations to estimate the potential residual forestry biomass.

Species	Equations	R^2	References
P. radiata	$\ln W* = -2.47 + 1.95 \ln D$	0.81 ($p < 0.001$)	[33]
E. globulus	$W* = \frac{0.1785 \times D^{17564}}{2.110}$	0.81 ($p < 0.01$)	[34]

*W = Kg forestry residual dry matter; D = Normal diameter (cm).

Once the estimator of forest residue has been estimated (Re, Mg ha^{-1} year^{-1}) generated by the main forest species in Bizkaia in the strata in which those species are predominant, the following step consists of determining to what surface from the NIF3 can be applied each estimator of residue obtained. As the data used are sampling data, the value of the estimator in a point different from the sampling plot is unknown. To overcome this, Thiesen's polygons from the sampling points are used in order to assign the value of the residue estimator to an influential area of the tree stratum. Using Thiessen polygon analaysis from a layer of points, each Thiessen polygon defines an area of influence around its sample point, so that any location inside the polygon is closer to that point than any of the other sample points. The final result is a division of space in polygons whose name is Voronoi telesation, and in which the limits between polygons are equidistant with respect to their neighbor points. This way of assigning values to the residue estimator to the rest of the non-sampled surface is the most adequate according to several authors [24,35].

The quantity of annual RFB (residual forestry biomass) susceptible to being useful RFB ($Mg\ year^{-1}$), using the surface of the resultant polygons and the tree residue estimators, is estimated as:

$$RFB = \sum_n A_n Re_{ni} \tag{5}$$

Figure 5a,b show the methodological diagram for obtaining the residue estimator Re of the two forestry species analyzed.

The corresponding density of each sampled tree (d, $kg\ ha^{-1}$) is obtained in the calculation of the tree residue estimator. The RFB of tree species are established as the result of the forestry treatments of the forest species, taking into consideration that the rotation period of P. radiata is 30 years and every 10 years it is subjected to a thinning processs of $1/3$ of its mass, while the rotation period of E. globulus was considered to be 10 years without prior thinning (Table 5). In this study, the Ciemat methodology was followed [36].

Table 5. Total RFB ($Mg\ year^{-1}$).

Species	Forestry Management	RFB
P. radiata	Final felling	$RFB = RFB_T/30$
	Thinning and clearing	$RFB = \frac{RFB_T}{3 \times 10}$
E. globulus	Final felling	$RFB = RFB_T/10$

* RFB_T = Total residual forestry biomass obtained throughout the tree rotation period.

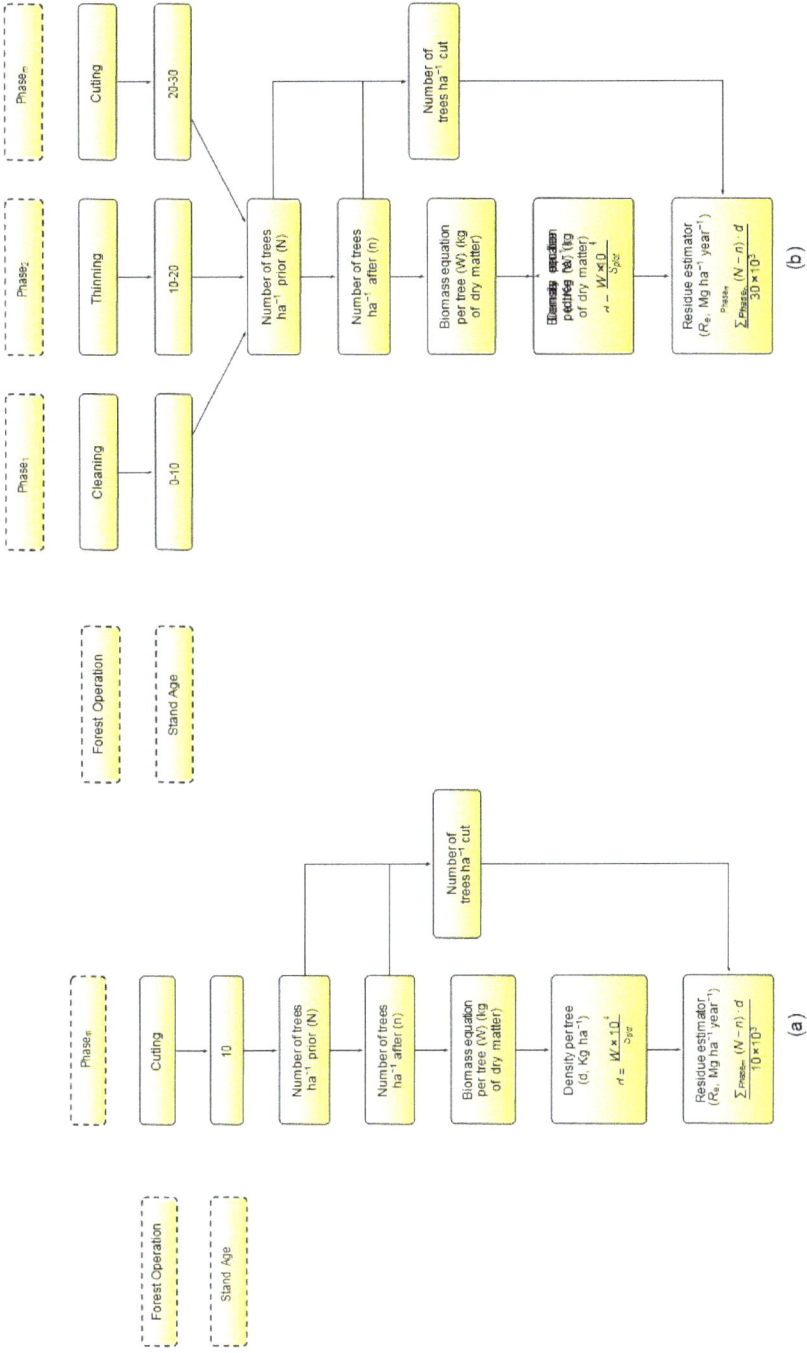

Figure 5. Methodological diagram for determining forestry residue. (**a**) *E. globulus*; (**b**) *P. radiate*.

The available residual biomass is found after applying a number of specific restrictions that impede the extraction of biomass (Figure 2a). The restrictions considered in this paper have been determined by the slope of the terrain, the content of organic matter in the upper layer of soil and the erosion, the last of these being measured by the K factor (susceptibility of the soil to be eroded) of the USLE/MUSLE model. Soil loss and soil erosion patterns may differ with slope gradient, slope aspect, and elevation. In general, increases in slope gradient can produce higher overland flow velocities, and, correspondingly, higher erosion. The slope gradient factor (S) is the ratio of the soil loss from given gradient of the slope to that from land having a 9% slope [37].

Nevertheless, in previous studies made in other river basins near the study area, soils with erosion in steep slopes were found [17]. For this reason, this study considers both parameters (slope and soil erodibility factor (K)) separately.

Slope has a direct effect, restricting the type of machinery used to collect and extract biomass. With an increase in slope, the biomass extraction process becomes increasingly expensive and also reduces biomass harvesting productivity [18,38].

With regard to soil loss, the erosion map Figure 6b shows that the soils of the Oka River Basin are, in general, free of erosion, of acceptable quality and with mean annual averages of soil loss of under 5 $Mg\ ha^{-1}\ year^{-1}$.

As shown in the slopes map, 32.69% of the surface area of the study area has steep slopes of over 50% and only 16.55% of the land is flat and corresponds largely to marshes and sand banks. In this study, three slope ranges have been marked out:

1. <35%, corresponds to the most adequate topographical area for supplying biomass with a lower risk of erosion.
2. [35, 60)%, presents moderate suitability with some risk of erosion.
3. ≥60%, would correspond to less adequate areas for the extraction of biomass due to their high erosion risk.

Carbon is one of the principal components of soil organic matter (SOM), a key component of soil that plays an important role in many biological, chemical, and physical properties [39,40] Regarding the organic matter of the soil, fertile soils are those that have organic carbon content (dry base) of over 2%, an average of 1–2% of SOC, and soils very poor in organic matter are those that have a organic carbon content of less than 1%. In previous research, an inversely proportional ratio has been obtained in the SOC and K factors.

Taking into consideration previous studies carried out in the sampling plots [17,41,42], we have considered soil erodibility factor intervals: for values of K ≤ 0.015, 0.015 < K ≤ 0.030, 0.030 < K < 0.045 and K ≥ 0.045 considering respectively low, medium and high erosionability soils. In this way, restricting values have been established for each parameter in order to define the availability of the forestry biomass (Table 6).

(a)

(b)

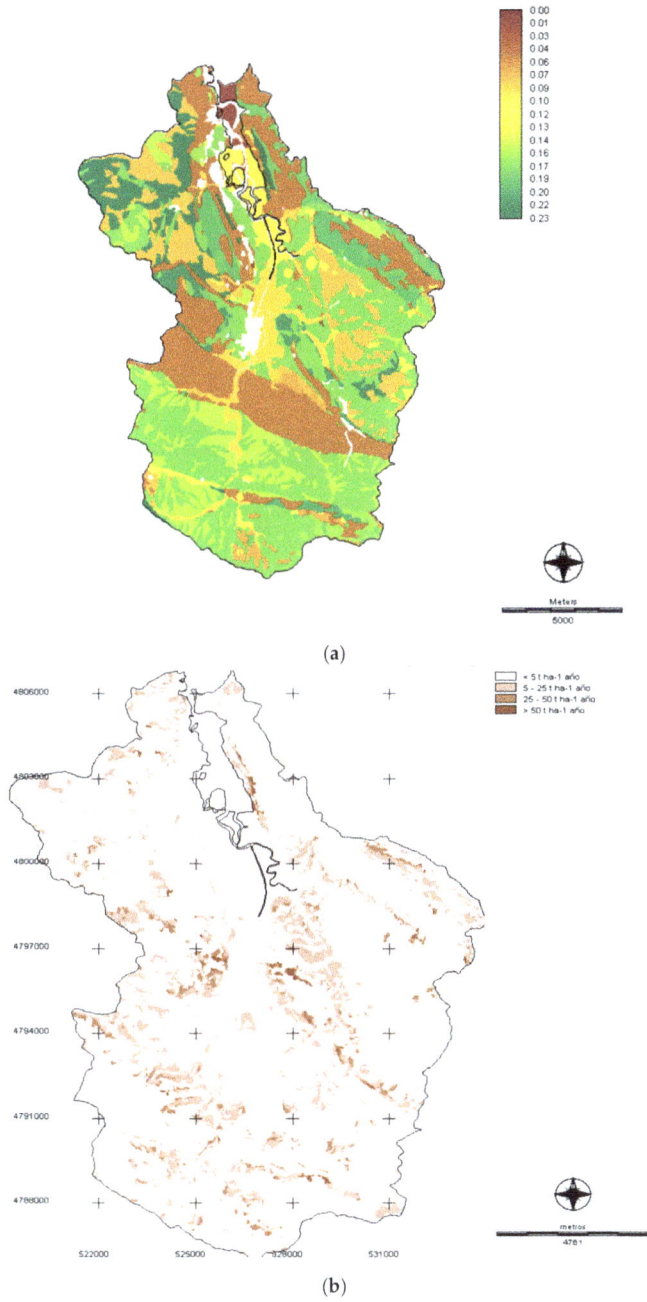

Figure 6. Maps of the Oka River Basin. (**a**) soil erodibility map; (**b**) soil loss map.

Table 6. Biomass available (%) in different conditions of slope (%), soil erodibility factor (K, Mg ha h ha^{-1} MJ^{-1} mm^{-1}) and soil organic carbon SOC (%).

	Slope	≤35	35–60	≥60
	0–0.15	100	80	60
K	0.15–0.30	90	70	50
	0.30–0.45	0	0	0
	>2	100	80	60
SOC	1–2	80	50	0
	0–1	0	0	0

2.7. Adjusting Predictive Models for Estimating Available Forestry Waste Biomass

A model is fitted in which the residual forestry biomass (RFB) is estimated as a variable dependent upon predictive variables: slope and erosionability of each plot. In order to study these relationships, linear regression models were used in Equation (6):

$$y_j = \beta_0 + \sum \beta_i x_{ij} + \varepsilon_j \tag{6}$$

where y_j is the dependent random variable (in our case, the RFB); β_0 is the linear regression independent coefficient; β_i are the predictive variable coefficients, which represent known values of the predictive variables, where ε_j is the random disturbance (error term). The models were fitted by means of regression, using the ordinary minimum squared method [43]. The fits were evaluated statistically by means of the determination coefficient corrected by the number of parameters (R^2aj). In order to verify the significance of each one of the fit parameters, an individual contrast was applied with the Student's t-test. For the choice of equations, the cases of variance normality and homogeneity were checked [11].

2.8. Implementing the Method in a GIS

Geographic Information Systems are useful tools for understanding the geographic context of a wide range of issues pertinent to bioenergy, especially energy demand and biomass supplies. Idrisi Selva and Quantum GIS using the different algorithms of the programme (Interpol and TIN for the interpolations) were used in the present study.

3. Results

3.1. Modelling the (RFB) in Accordance with the Slope Erodibility and Soil Organic Carbon

Parameter estimates, standard errors of the fitted models according to Equation (7):

$$RFB = \beta_0 + \beta_1 \times K + \beta_2 \times S + \beta_3 \times SOC. \tag{7}$$

The parameter estimates and goodness of fit statistics of the best linear model obtained for the main stand and canopy variables using the stepwise variables selection method and the NFI3 data as regressors are summarized in Table 7. The behaviour of the model we obtained is good, explaining more than 86% of the variability observed. The results obtained in the statistical analysis showed that there was a negative correlation between the variables K and SOC ($r = 0, 71, p < 0.01$).

Table 7. Estimated parameters, standard errors of the fitted models.

Tested Model	Dependent Variable	Independent Variable	R^2aj	Estimated Parameter	Standard Error	t-Value	$p > \lvert t \rvert$
Equation (7)	RFB	Intercept	0.86	0.033531	0.133103	108.998	0.046
		K		3.397875	0.012501	402.276	<0.001
		S		0.014747	2.004521	364.702	<0.01

The fitted equation Equation (7) shows good fit statistics. All the parameters were significant at a confidence level of 95%. In the evaluation of the cases of variance normality and homogeneity of the fitted model, no evidence was found that suggested any non-compliance of the cases. The statistical analysis indicates that there is a relationship between K and S. The greater the erosion, the lower the amount of SOC.

3.2. Physical-Chemical Characterisation of Soils

Tables 8 and 9 show the mean values obtained in the physical-chemical characterisation in 2 of the 115 plots analyzed: A (forest floor soil samples) and B (soil samples subjected to site following felling performed 6 months earlier.) The physical-chemical analysis made of the 115 plots revealed very different erodibility values (K) of the soil before and after felling, e.g., Table 8.

Table 8. Mean values of the physical characteristics of soils.

Forestry Management	Depth (cm)	Erodibility K	Soil Class FAO	Material	D.a. ‡ (g cm^{-3})	Depth (%)	Sand % Gr	Sand % Fin	Silt% Gr	Silt% Fin	Clay (%)	Texture
A §	0–10	<0.02	Alisols Haplico ALh	Argilitas	1.16	18	4.3	12.2	14.4	35.8	33.2	fpl †
B $	0–10	(0.15,0.30)	Anthrosols Arico	Sandstones and Clays	1.4	30	12.6	30.9	13.2	18.6	24.8	f *

* loamy texture; † silty clay loam; ‡ bulk density; § *P. radiata* forest sawtimber stage; $ forestry management after felling (without subsoiling).

The data obtained exhibited higher apparent density values in soils in which the tree biomass has been taken away after felling and the soil has been prepared for the following rotation compared to soils under plantations. Bulk density increased following site preparation practices. Such is the case of B soils where bulk density was 21% higher compared to A soils under mature plantation (Table 8). The results are in agreement with prior studies made in northern Spain, in which increase in apparent density of up to 1.5 g·cm^3 were found on land from which tree biomass had been removed and in which heavy machinery had been introduced [18,33].

Table 9. Mean values of the chemical parameters of the soils.

Forestry Management	Depth (cm)	pH	SOC * %	N %	Ca	Mg	Na	K	Al	ECEC †	Al. Sat. ‡%
					cmol$_c$ kg^{-1} %						
A	0–10	4.56	7.72	0.26	4.99	1.44	0.44	0.31	4.45	11.63	38.26
B	0–10	3.96	3.10	0.12	0.27	0.21	0.15	0.12	6.28	7.03	89.33

* soil organic carbon; † effective exchange capacity; ‡ aluminum saturation.

The results obtained from the analysis of the physical-chemical parameters show that the soils of this basin of the Oka River are in general highly acidic (average pH between 4.1–4.8). Irrespective of the type of soil and the initial geological material, the content of SOC of the surface soil horizons with mature forests is always high, with average carbon contents of between 6.2–7.9%. The results of SOC obtained in all the plots analyzed show high SOC contents, much higher than the 2% shown within the restrictions (Table 6). In this study, these SOC values were taken as environmental restriction following the recommendations of previous research carried out in the Mediterranean region [12]. In spite of the fact that the Oka River Basin is within the Atlantic zone, in view of the steep slopes, we decided to carry out the study with the recommended SOC values, in view of the orography and the steep slopes of the river basin. However, the results show that the SOC of the soils of the basin exhibit very high values and therefore the SOC would not mean in any case a restriction in the extraction of residual forestry biomass for energy use.

Irrespective of the type of soil and the initial geological material, the content of SOC of the surface soil horizons with mature forests is always high, with average carbon contents of between 6.2–7.9%. This data indicates that these are erosion-free soils, for which very low erosionability factor values have been obtained (K < 0.02) and that the slope of the land does not have a significant effect on the SOC. However, those soils that have been subjected to work done in order to prepare the terrain after the logging suffer a significant drop in the content of SOC (average contents of 3.05–5.45%) and significant increases in bulk density. This effect increases considerably when highly-mechanised forestry management systems are used and terrain is prepared using deep ploughing.

For this reason, felling and land preparation work make have a serious effect on the soil. It has been found that, for K < 0.15, the soil has an adequate texture, with SOC in excess of 5–5.4%, good permeability and high hydraulic conductivity, indicating that its conditions are adequate for withdrawing RFB for use. As the value of K increases, the content in SOC decreases. K values above 0.3 point to unstructured soils (lithosols, anthrosols, etc.) or decapitated soils (lacking surface horizon), and therefore may exhibit low percentages of SOC (between 1–2%) and different types of textures. In practical terms, there are no soils in the Oka River Basin with such a high level of erosion. The K values obtained in the river basin are in all cases lower than 0.3 except in a few unrepresentative cases.

Our study indicates that, in general, the soils of the Oka River Basin show low erosion levels. In addition, 63.45% of the soil analyzed exhibits erosion rates of under 10 Mg ha^{-1} year^{-1}, 25.67% show erosion rates of 10–20 Mg ha^{-1} year^{-1}, 8.61% show erosion rates of 20–30 Mg ha^{-1} year^{-1} and only 2.27% of the soil is heavily eroded with soil losses of over 50 Mg ha^{-1} year^{-1}. The last case corresponds to soils subjected to intensive land preparations after felling.

These results confirm the observations made by several members of this research team in soil erosion studies with regard to the drop in SOC associated with the intensity of the soil management following logging. Wet Biomass refers to the humidity of the RFB when this is collected, which is generally estimated at 30%, as it remains in piles for a time in the countryside after felling. As in the case of prior research, in our case, soil samples have been taken some 6 months following felling.

3.3. Estimating the Potential and Useable Residual Forestry Biomass in the Oka River Basin

The estimators of the residue obtained for P. radiata fluctuate between 0.286 and 0.885 Mg ha^{-1} year^{-1} of dry matter. With respect to E. globulus, the residue estimators obtained values from 0.346 to 2.156 Mg ha^{-1} year^{-1} dry matter. Similar studies carried out in other areas of Spain, e.g., [5] estimated 1.3 Mg ha^{-1} year^{-1} of wet matter residual biomass in P. radiata in Navarra (Spain), and [24] found values of estimated residual biomass of between 0.68 and 1.41 Mg ha^{-1} year^{-1} (dry mater) or 0.97 and 2.020 Mg ha^{-1} year^{-1} (wet matter) or in E. globulus plantations in Huelva (Spain). A humidity of 30% has been considered, as this is the humidity accepted by a number of different researchers when the residual biomass is gathered for energy valuation [24].

The amount of residual forestry biomass estimated in the Oka River Basin was 7775.32 Mg year^{-1} (dry mass), of which 6334.12 Mg year^{-1} (dry mass) correspond to the remains of P. radiata and 1441.20 Mg year^{-1} (dry mass) to those of E. globulus (Figure 7a).

The amount of residual forestry biomass useable for energy purposes taking into consideration technical and environmental constraints—erodibility (K), slope of the terrain and soil organic carbon (SOC), has been estimated at 4858.23 Mg year^{-1} (dry mass), which corresponds to 6940.33 Mg year^{-1} (Figure 7b). The results of the study show that the useable residual biomass represents 62.5% of the available potential residual biomass. The remaining unused forestry waste could offset the soil losses in the most highly eroded horizons in view of the fact that a large part of the nutrients assimilated by the tree are accumulated in the latter's waste or non-timber fraction [33].

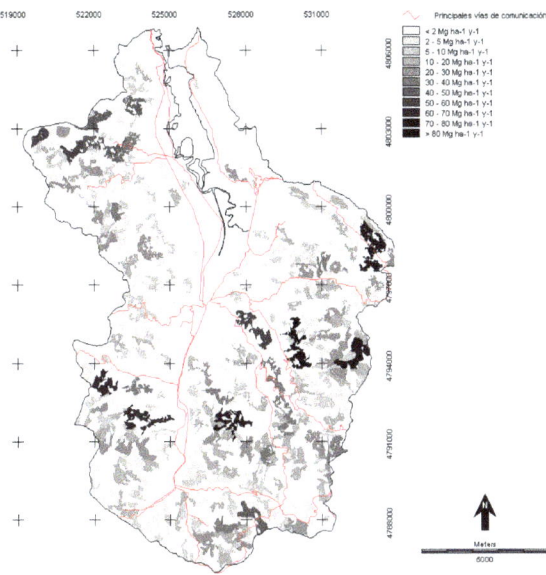

(a)

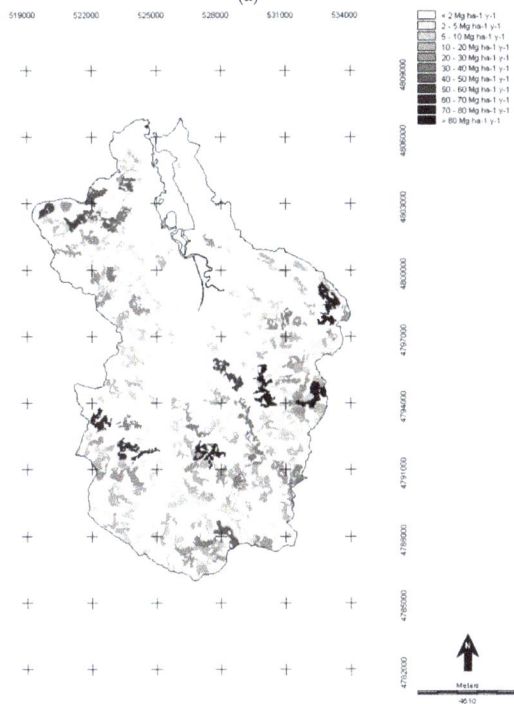

(b)

Figure 7. Residual biomass. (**a**) usable; (**b**) available.

4. Conclusions

The results show that the plots with less biomass correspond to those that have been subjected recently to intensive treatments of the land for the next rotation after felling. These areas are the ones that show higher soil loss values. Similar conclusions were obtained in soil studies made in the Arroyo de la Vega River Basin (Madrid) [44].

The forest residues considered in the present study could replace, at least in part, some of the fuels that are currently in use. We believe that, in the Basque Country, the most promising applications are cogeneration in medium-sized power stations and combustion in small and medium-sized plants for heat and steam generation. The specific features of the economy of scale restrict the design of energy plants to those with a minimum production capacity of at least 1 MW. This corresponds to a need for 4700 Mg year^{-1} of raw material (dry mass). This value is similar to the annual amount of useable residual biomass estimated in this paper (4858.23 Mg year^{-1}). The use of forestry waste for energy purposes could provide employment and business opportunities in rural communities. An increase in the use of residual forest biomass as a source of energy instead of conventional fossil fuels will represent important environmental advantages and will help member states to meet the targets of EU policies in terms of energy rationalisation.

Information on the amount of available forestry biomass waste, taking sustainability principles into consideration, represents a valuable tool for forestry management, due to its association with soil conservation, spread of pests and fires. Through the harvesting of forestry biomass, a large reduction in forest fires is expected. At this time, forest fires are a source of concern in the study area (of great ecological value and a UNESCO Biosphere Reserve), in view of the enormous potential environmental impact.

Acknowledgments: This work has been supported by the Basque Government and by the Office of Research of the University of the Basque Country grant by Project SAI10/147-SPE10UN90 and by Project NUPV14/11, respectively.

Author Contributions: Esperanza Mateos developed conceptual ideas, designed the study, conducted data analysis in field experiments and wrote the paper. José Miguel Edeso and Leyre Ormaetxea developed conceptual ideas and designed the study.

Conflicts of Interest: The authors declare no conflict of interest.

References

1. Amichev, B.Y.; Burger, J.A.; Rodriguez, J.A. Carbon sequestration by forests and soils on mined land in the midwestern and Appalachian coalfields of the US. *For. Ecol. Manag.* **2008**, *256*, 1949–1959. [CrossRef]
2. Mateos, E.; Garrido, F.; Ormaetxea, L. Assessment of Biomass Energy Potential and Forest Carbon Stocks in Biscay (Spain). *Forests* **2016**, *7*, 75. [CrossRef]
3. IDAE. Resumen del Plan de Energías Renovables 2011–2020. Available online: http://www.minetur.gob.es/energia/es-ES/Novedades/Documents/Resumen_PER_2011-2020.pdf (accessed on 1 November 2016).
4. Cabrera, M.; Vera, A.; Cornejo, J.M.; Ordás, I.; Tolosana, E.; Ambrosio, Y.; Martínez, I.; Vignote, S.; Hotait, N.; Lafarga, A.; et al. *Evaluación del Potencial de Energía de la Biomasa. Estudio Técnico PER 2011–2020*; Instituto para la Diversificación y Ahorro de la Energía (IDAE): Madrid, Spain, 2011.
5. Domínguez, J. *Los Sistemas de Información Geográfica en la Planificación e Integración de Energías Renovables*; CIEMAT: Madrid, Spain, 2003.
6. Diaz-Balteiro, L.; Alfranca, O.; Bertomeu, M.; Ezquerro, M.; Giménez, J.C.; González-Pachón, J.; Romero, C. Using quantitative techniques to evaluate and explain the sustainability of forest plantations. *Can. J. For. Res.* **2016**, *46*, 1157–1166. [CrossRef]
7. Fleming, R.L.; Leblanc, J.D.; Hazlett, P.W.; Weldon, T.; Irwin, R.; Mossa, D.S. Effects of biomass harvest intensity and soil disturbance on jack pine stand productivity: 15-year results. *Can. J. For. Res.* **2014**, *44*, 1566–1574. [CrossRef]

8. Nadeau Fortin, M.A.; Sirois, L.; St-Laurent, M.H. Extensive forest management contributes to maintain suitable habitat characteristics for the endangered Atlantic-Gaspésie caribou. *Can. J. For. Res.* **2016**, *46*, 933–942. [CrossRef]

9. Butnor, J.R.; Johnsen, K.H.; Sanchez, F.G.; Nelson, C.D. Impacts of pine species, stump removal, cultivation, and fertilization on soil properties half a century after planting. *Can. J. For. Res.* **2012**, *42*, 675–685. [CrossRef]

10. Cannell, M.G. *Forests and the Global Carbon Cycle in the Past, Present and Future*; European Forest Institute: Joensuu, Finland, 1995.

11. Fonseca, G.; Alice, G.; Rey, B. Models for biomass estimation in native forest tree plantations and secondary forests in the Costa Rican Caribbean Region. *Bosque* **2009**, *30*, 36–47.

12. Esteban, L.; García, R.; Ciria, P.; Carrasco, J. *Costs in Spain and Southern EU Countries. Clean Hydrogen-Rich Synthesis Gas Report, Chrisgras Fuels from Biomass*; Colección Documentos CIEMAT; CIEMAT: Madrid, Spain, 2009.

13. Carter, M.R. Soil quality for sustainable land management. *Agron. J.* **2002**, *94*, 38–47. [CrossRef]

14. Third National Forestry Inventory. Available online: http://www.nasdap.ejgv.euskadi.eus/informacion/inventarios-forestales/r50--7212/es/ (accessed on 3 June 2016).

15. Rodríguez-Loinaz, G.; Amezaga, I.; San Sebastián, M.; Peña, L.; Onaindia, M. Análisis del paisaje de la reserva de biosfera de Urdaibai. *Forum de Sostenibilidad* **2007**, *1*, 59–69.

16. Ibarrondo, M.; González-Amuchastegui, M. Gestión de montes en la reserva de la biosfera de Urdaibai: Una oportunidad perdida. *Boletín de la A.G.E.* **2008**, *46*, 329–344.

17. Edeso, J.M.; González, M.J.; Marauri, P.; Merino, A. Determinación de la tasa de erosión hídrica en función del manejo forestal: La cuenca del río santa lucía (Gipuzkoa). *Lurralde: Investigación y Espacio* **1997**, *20*, 67–104.

18. Edeso, J.M.; Merino, A.; González, M.J.; Marauri, P. Manejo de explotaciones forestales y pérdida de suelo en zonas de elevada pendiente del País Vasco. *Cuaternario y Geomorfología* **1998**, *12*, 105–116.

19. Kirkby, M.; Morgan, R. *Erosión de Suelos*; Editorial Limusa: Ciudad de México, Mexico, 1894.

20. Marcos, M.; Núñez, M. Biomasa forestal: Fuente energética. *Energética XXI* **2006**, *4*, 52.

21. Karamage, F.; Shao, H.; Chen, X.; Ndayisaba, F.; Nahayo, L.; Kayiranga, A.; Omifolaji, J.K.; Liu, T.; Zhang, C. Deforestation Effects on Soil Erosion in the Lake Kivu Basin, D.R. Congo-Rwanda. *Forests* **2016**, *7*, 281. [CrossRef]

22. Guitián Ojea, F.; Carballas Fernández, T. *Técnicas de Análisis de Suelos. 70 (Monografías de Ciencia Moderna)*; Editorial Pico Sacro: Madrid, Spain, 1976.

23. Dans del Valle, F.J.; Romero García, A.; Fernández, F.J. *Manual de Selvicultura del Pino Radiata en Galicia*; Asociación Forestal de Galicia: Galicia, Spain, 1997.

24. Zabalo, A. Modelo de Estimación del Potencial Energético de la Biomasa de Origen Forestal en la Provincia de Huelva. Ph.D. Thesis, Universidad de Huelva, Huelva, Spain, 2006.

25. Mateos, E.; Edeso, F.; Bastarrika, A.; Torre, L. Estimación de la biomasa residual procedente de la gestión forestal en Bizkaia. *Lurralde: Investigación y Espacio* **2012**, *35*, 13–30.

26. López-Rodríguez, F.; Pérez Atanet, C.; Blázquez, F.; Ruiz Celma, A. Spatial assessment of the bioenergy potential of forest residues in the western province of Spain, Cáceres. *Biomass Bioenergy* **2009**, *33*, 358–366. [CrossRef]

27. GeoEuskadi. Mapa Forestal del País Vasco. Available online: http://www.geo.euskadi.eus/mapa-forestal-del-pais-vasco-ano-2010/s69-geodir/es/ (accessed on 7 February 2017).

28. Montero, G.; Ruiz-Peinado, R.; Muñoz, M. *Producción de Biomasa y Fijación de CO_2 en los Montes Españoles*; Instituto Nacional de Investigación y Tecnología Agraria y Alimentaria (INIA): Madrid, Spain, 2005.

29. Alvarez, J.; Rodriguez Soalleiro, R.; Rojo, A. A management tool for estimating bioenergy production and carbon sequestration in *Eucalyptus globulus* and eucalyptus nitens grown as short rotation woody crops in north-west Spain. *Biomass Bioenergy* **2011**, *35*, 2839–2851.

30. Ruiz-Peinado, R.; Montero, G.; Del Rio, M. Biomass models to estimate carbon stocks for hardwood tree species. *For. Syst.* **2012**, *21*, 42–52. [CrossRef]

31. Zianis, D.; Mencuccini, M. On simplifying allometric analyses of forest biomass. *For. Ecol. Manag.* **2004**, *187*, 311–332. [CrossRef]

32. Stark, H.; Nothdurft, A.; Bauhus, J. Allometries for Widely Spaced *Populus* ssp. and *Betula* ssp. in Nurse Crop Systems. *Forests* **2013**, *4*, 1003–1031. [CrossRef]

29

33. Merino, A.; Rey, C.; Brañas, J.; Rodriguez, R. Biomasa arbórea y acumulación de nutrientes en plantaciones de Pinus radiata. *Investigaciones Agrarias Sistemas y Recursos Forestales* **2003**, *12*, 85–98.

34. Silva, R.; Tavares, M.; Pascoa, F. Residual Biomass of Forest Stands, *Pinus pinaster* Ait. and *Eucalyptus globulus* Labill. In Proceedings of the Congreso Forestal Mundial 1991, Paris, France, 17–26 September 1991.

35. Gutiérrez Puebla, J.; Gould, M. *SIG: Sistemas de Información Geográfica*; Sintesis: Madrid, Spain, 1994.

36. Ciemat. *Desarrollo de una Aplicación en Base de Sistema de Información Geográfica (SIG) para la Evaluación de Recursos de Biomasa a través de Internet*; Especificaciones Técnicas; Centro de Desarrollo de Energías Renovables: Soria, Spain, 2009.

37. Wischmeier, W.; Smith, D.D. *Predicting Rainfall Erosion Losses—A Guide to Conservation Planning*; Agriculture Handbook; Department of Agriculture: Washington, DC, USA, 1978.

38. Merino, A.; Fernández, A.; Solla, F.; Edeso, J. Soil changes and tree growth in intensively manager *Pinus radiata* in northern Spain. *For. Ecol. Manag.* **2004**, *196*, 393–404. [CrossRef]

39. James, J.; Harrison, R. The Effect of Harvest on Forest Soil Carbon: A Meta-Analysis. *Forests* **2016**, *7*, 308. [CrossRef]

40. Brady, N.; Weil, R. *The Nature and Properties of Soils*; Macmillan Publ. Co.: New York, NY, USA, 2002; Volume 13, p. 992.

41. Merino, A.; González, M.; Edeso, J.; Marauri, P. Modificación en los Caracteres de los Suelos de la Vertiente Cantábrica del país vasco Producidos por Prácticas Forestales. Available online: http://www.ingeba.org/lurralde/lurranet/lur18/merino18/merino18.htm (accessed on 10 March 2016).

42. Merino, A.; Edeso, J.; Gonzalez, M.; Marauri, P. Soil properties in a hilly area following different harvesting management practices. *For. Ecol. Manag.* **1997**, *103*, 235–246. [CrossRef]

43. SAS-Institute. *The SAS System for Windows (Release 8.02)*; SAS-Institute: Cary, NC, USA, 2016.

44. Casermeiro, M.A.; de la Cruz Caravaca, M.T.; Costa, J.H.; Hernando-Massanet, M.I.; Abril, J.M.; Sánchez, P. El papel de los tomillares (*Thymus vulgaris* L.) en la protección de la erosión del suelo. *Anales de Biología* **2002**, *24*, 81–87.

Article

Air Curtain Burners: A Tool for Disposal of Forest Residues

Eunjai Lee [1,*] and Han-Sup Han [2]

1   Department of Forest Sciences, Seoul National University, Seoul 08826, Korea
2   Department of Forestry and Wildland Resources, Humboldt State University, Arcata, CA 95521, USA;
     han-sup.han@humboldt.edu
*   Correspondence: ejay0512@gmail.com; Tel.: +82-2-880-4768

Received: 17 July 2017; Accepted: 10 August 2017; Published: 14 August 2017

Abstract: Open pile burning (OPB) forest residues have been limited due to several concerns, including atmospheric pollution, risk of fire spread, and weather conditions restrictions. Air Curtain Burner (ACB) systems could be an alternative to OPB and can avoid some of the negative effects that may result from OPB. The main objective was to compare the burning consumption rates and costs of two types of ACB machines, the S-220 and BurnBoss. In addition, we tested a hand-pile burning (HPB) consumption rate for a comparison with BurnBoss unit. The S-220's burning consumption rates ranged between 5.7 and 6.8 green metric ton (GmT)/scheduled machine hour (SMH) at a cost between US $12.8 and US $10.8/GmT, respectively. Costs were 70% higher when using the BurnBoss unit. Burning residue consumption rates and cost of disposal were considerably different: they were highly dependent on machine size, species, and fuel age of forest residues. Particularly, BurnBoss test burned over 40% more than HPB method and produced clean burn by airflow. The results from this study suggest that ACBs can be a useful tool to dispose of forest residues piled in many forests areas with less concerns of air quality and fire escape risks.

Keywords: open pile burning; burning consumption rates; costs; hand-pile burning; clean burn

---

1. Introduction

Forest residues include tree tops, limbs, and other tree parts generated from forest operations and can provide opportunities for production of bioenergy and bioproducts such as briquettes or biochar [1–3]. However, a low level of market demand for wood-based energy in the Northwestern U.S. have caused forest residues to be piled and left in the forests and sawmills [4]. It is also often financially unviable to use forest residues due to high costs of collection and transportation, and low market price [5,6]. In addition, leaving large piles of forest residues near houses or within public parks have been a concern due to high risk of fire hazard and other forest management issues (i.e., growing and rehabilitation). For this reason, open pile burning (OPB) has been widely used in the Western U.S. to dispose of forest residues, to reduce wildfire hazard, and improve forest and productivity [7,8]. This forest residues disposal method has been extensively used as it provides a cost-effective option for disposal of forest residues [9,10].

However, OPB could be potentially damaging to forests by increasing the wildfire hazard and obstructing regeneration [11,12]. This method has not only been shown to generate greenhouse gas (i.e., carbon monoxide (CO) and carbon dioxide (CO_2)) emissions and release particulate matter (PM) to the health hazard levels, but it also emits nuisance of smoke and objectionable odor to an ambient air [10,13]. For this reason, intentional burning of slash piles was often strictly controlled in public and residential areas [13]. OPB also requires the prevention of embers from escaping as well as monitoring weather conditions hourly: burning is allowed only if extremely narrow conditions are met [14,15]. For example, when planning a pile burn, ambient temperatures have to be less than 32 °C, the maximum wind speed should not exceed 8 km/h and relative humidity ought to be below 35%. These requirements would prohibit OPB from 1 June to 14 November in most areas in the Western U.S. Another drawback of OPB is the severe, undesirable effects on forest soil properties compared to wildfires or broadcast burning [12,16]. The extreme flaming temperatures (400 °C) with 60 h can be intense and penetrate at 0–20 cm soil depth which can destroy the chemical and biological soil properties [17]. Overall, the OPB method can result in poor air quality, smoke production, fire escape, and soil damage [7,18].

An alternative technology to dispose of forest residues is to use an Air Curtain Burner (ACB), which is designed by Air Burners Inc., Palm City, FL, USA (also called as Air Curtain Destructor or Incinerator; Figure 1). ACBs are divided into two main types, stationary (positioned at the centralized landing area) and mobile applications (half-ton pick-up truck mounted system). These machines were developed in compliance with US Environmental Protection Agency's 40 Code of Federal Regulation Part 60 regulation that determines allowable emissions from biomass burning [19,20].

(a) (b) (c)

Figure 1. The Air Curtain Burners used to test burning of forest residues in (**a**) S-220 in Jacksonville, Florida; (**b**) BurnBoss in Groveland, California; and (**c**) BurnBoss installed with ember screen in Volcano, California. The loader is used to load the forest residues in S-220 operation while BurnBoss operation is loaded by hand.

ACBs have been primarily used to dispose of woody residues, such as stumps and root wads [21], waste wood and landscape wastes [22], debris generated by hurricanes [23] and floating (water-borne) woody debris from natural disasters (e.g., tsunami and heavy rainfall) [19]. Details on how the ACBs FireBox efficiently burn woody materials is shown in Figure 2. These machines operate by blocking various air pollutant emissions including greenhouse gases and PM by using a high velocity (1600–2000 revolution per minute; RPM) of airflow from the air blower part which is referred to as "air curtain". In addition, air pollutant emissions is returned by circulation of air flow. Past studies showed that ACBs can reduce CO and PM emissions by 80% compared to OPB and reduce smoke opacity [23,24]. In addition, it also minimizes escaping embers, soil damage, and burn scars by creating an air curtain across the box [23,25,26]. Air Burners Inc. stated that adding air into the FireBox effectively improved burning consumption rate since by adding more oxygen to the forest residue pile in the FireBox during the burning [19]. As a result, ACB's forest residues treatment method is considered as a clean and air pollution control burning method to dispose of forest residues, but burning consumption rates and cost of disposal of forest residues for different ACB systems are largely unknown.

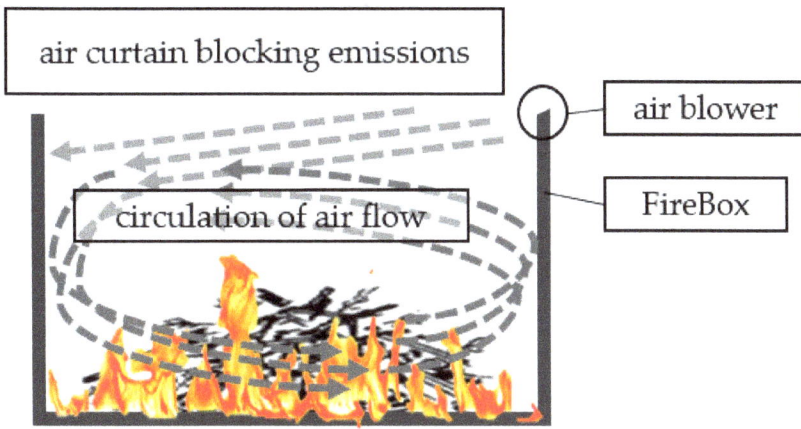

Figure 2. Principle of an Air Curtain Burner: the air curtain blocking emission and circulating the air inside FireBox is created when turning on the air blower.

Therefore, in this study, we examined an alternative method of disposing of forest residues using an ACB (Figure 3). The overall objective was to determine the performance of ACBs and evaluate the economic feasibility of burning slash piles using the S-220 and BurnBoss, which are ACB models that are commonly used for forest residue disposal. Specifically, this study sought to: (1) determine the forest residues burning consumption rate (green metric ton (GmT)/scheduled machine hour (SMH)) and cost of disposing forest residues ($/GmT) through field-based experiments, and (2) establish the logistics to an ACB burning operation. This study focused on disposal of forest residues resulting from fuels reduction treatments in green waste yards near residential areas and public parks. In order to assess the potential of BurnBoss units, we observed hand-built-pile burn's consumption rate next to the BurnBoss test at the same time.

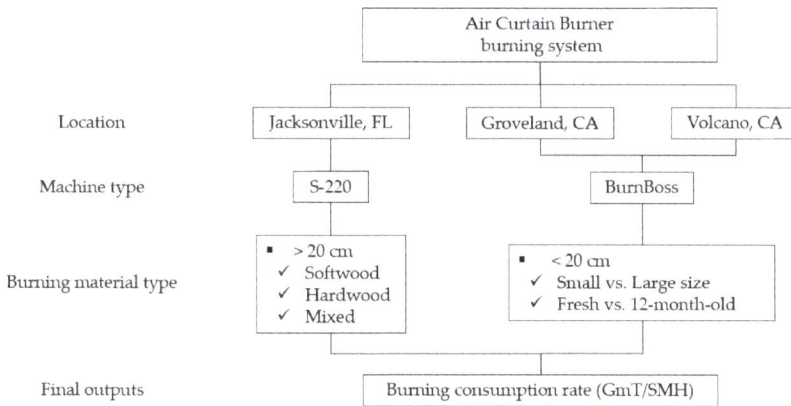

Figure 3. Integrated Air Curtain Burner burning system flow chat.

2. Materials and Methods

2.1. An ACB Burning System Set-Up and Description

During the study, we noticed that ACB operations followed five steps: putting the machine in place, first loading and kindling, air blower startup, loading (i.e., second loading), and burning it down to ash. For an initial set up of an ACB operation, these machines need to be placed on open, clear, flat ground (slope < 10%) with dry surface conditions [27]. The area should be located at least 30 m away from the outer edge of burning materials and any fire fuel [28].

Following the set-up, the operator needed to fill up a space in the ground and the inside of the FireBox with dirt to prevent the escape of smoke and embers since the FireBox is bottomless. The first loading, which is piled about 1/3 of the FireBox height consists of smaller materials (less than 10 cm in diameter) for kindling, and is then ignited with diesel fuel and a propane gas torch. At this time in the burning, greenhouse gases and smoke can temporarily be generated as the diesel fuel ignites, as there is some wait time for the materials to ignite. If the operator uses clean materials with lower moisture content (<20%) at this stage, there will be reduced emissions of CO and PM with the smoke [27,29,30]. For this reason, the materials used were not contaminated with dirt. Thus, prior to the start of air blower, a number of preparation activities should be done, including site selection, set-up, and kindle a fire. Further, first loading and kindling work can vary depending on operator and machines.

The next process for the burning is to turn on the air blower and load additional residues into the FireBox as needed to maintain combustion until the materials are burned down to ash. It should be noted that the second and following loadings should not be higher than the height of the box to retain the emissions, ember, and smoke reduction by the air curtain [19]. These processes are dependent on the type of ACB; S-type (i.e., S-220) and BurnBoss. S-220 burning operation, in which a large amount of residues were fed to a 0.5-m height below the top at a time with 25-GmT amount of residues, had a burn down time of approximately 14 h before the next experiment [31]. Thus, the S-220 or other S-types of ACB are designed to run for one day (24 h). On the other hand, when loading fuels with the BurnBoss, they should be added to 1/3 the depth of the FireBox and additional fuels added when there is enough space [27,32]. These processes were repeated until the last materials were loaded. Once the last materials were loaded, the last stage should take one or two hours to burn it down to ash. Thus, BurnBoss unit can be operated for eight hours a day.

2.2. Description of ACBs Used in the Experiments

The S 220 was a mid-sized model equipped with a 45-kW diesel engine to blow air into the FireBox. The dimensions of the FireBox used in the S-220 are 6.0-m × 1.9-m × 2.2-m (length × width × height) with 0.7-m thick steel walls filled with thermo-ceramic materials. This machine type can efficiently dispose of larger diameter (>20 cm) forest residues [26]. A potential average through-put of burning ranges from 5 to 7 GmT/h at an average fuel consumption rate of 9.5 L/h [19].

The BurnBoss was a portable prototype machine that can be moved around with a half-ton pick-up truck and a FireBox that is raised and lowered by a hydraulic lift system. This unit is only applied to off-road vehicles that do not exceed 60 km/h. The FireBox dimensions are 3.7-m (L) × 1.2-m (W) × 1.2-m (H) with 0.1-m thick thermos-ceramic material walls. It had a small (9-kW) diesel engine with a fuel consumption rate estimated at 1.1 L/h [19]. Generally, the burning consumption rate of this technology is approximately 1/2 to 1 GmT/h. The BurnBoss unit was typically loaded by hand with small (<20 cm) forest residues such as hand-pile residues and windrow along the road during the burning process.

2.3. Description of Material Types Used in the Experiments

Burning tests were conducted using two ACB machines, S-220 and BurnBoss, in three different locations. The first test was conducted over three days (4–6 August 2015) on a S-220 unit in a green waste yard located in Jacksonville, Florida. The S-220 unit was positioned at a designated location such as large green waste disposal yard or a large open area for an extended time (>6 months or year) and was used in conjunction with a rubber-tired front-end loader using a standard log grapple. The species burned for this study were loblolly pine (*Pinus taeda*), laurel oak (*Quercus laurifolia*), sand live oak (*Quercus geminate*), and myrtle oak (*Quercus myrtifolia*) trees. These materials were mainly chunk woods ranging 20 to 40 cm in diameter and 1 to 3 m in length, collected from in-forest, mill processing, and urban wood residues (Table 1). For the purpose of our study, we separated the wood fuels into three different types (softwood, hardwood, and mixed species) at an amount of 25 GmT for each burning trial test. There was no significant difference in the size and moisture content of fuel types ($p > 0.05$).

Table 1. Description of material types and weather conditions for Air Curtain Burner tests in three different locations.

Locations	Air Temperature (°C)	Relative Humidity (%)	Wind Speed (km/h)	Average Material Size in Diameter (cm)
Jacksonville, Florida (S-220)				
softwood	30	95		27
hardwood	31	87	N/A [a]	29
mix	30	85		28
Groveland, California (BurnBoss)				
small size (<10 cm)	24	38	1.8	5
large size (10 to 20 cm)	20	60	1.5	20
Volcano, California (BurnBoss installed with ember screen)				
fresh residues				
small size (<10 cm)	24	29	0.8	6
large size (10 to 20 cm)	20	38	0.5	16
12-months-old residues				
large size (10 to 20 cm)	30	36	0.3	16

[a] Data not avaliable; Note: the wind speed did not affect during ACB burning system.

A second burning test was carried out over two days (26–27 March 2016) with a BurnBoss unit in a community green waste yard area in Groveland, California (CA). Forest residues used for this test were from urban and residential fire hazard reduction and landscaping treatments, and consisted of Ponderosa pine (*Pinus ponderosa*, 80%) mixed with manzanita shrubs (*Arctostaphylos glauca*, 20%) that were less than 6-months in age. A backhoe sorted and piled approximately 10 GmT of two different fuel sizes; small-diameter fuels (<10 cm) and large-diameter fuels (10–20 cm). Larger than 10 cm fuels in diameter ranged from 1 to 2 m in length, while less than 10 cm materials were in a variety of lengths and forms. The third burning experiment was conducted in a California State Park camp ground located in Volcano, CA, USA over three days (13–15 June 2016). The BurnBoss used for this test was equipped with a cage placed on the top of the FireBox to prevent any embers from flying from the FireBox (Figure 2). An ember screen is optional to buy and helps avoid the spread of ember during the loading and burning process [19]. At this site, three different types of fresh (less than one-month-old or immediately after felling and bucking fuels) and 12-months-old Ponderosa pine (*Pinus ponderosa*) residues from commercial thinning operations were burned (Table 1). The 12-months-old residues burned were only large-diameter (10–20 cm). Fresh residues, which were from drought or insect damage, were burned immediately after felling and bucking and included small (<10 cm) and large-diameter (10–20 cm). Large fuels were generally around 1 m in length, but small materials were in a variety of lengths and forms including leaves. There was a significant difference in the moisture content of fuel depending on age ($p < 0.05$).

In both BurnBoss tests, we compared burning consumption rates with hand-pile burning (HPB) option when burning forest residues on a site. For the purpose of comparison of consumption rates, the HPB was set next to the ACB unit observed at each day of burning (Figure 4). The HPB was 3.7-m wide × 1.2-m long × 1.2-m high.

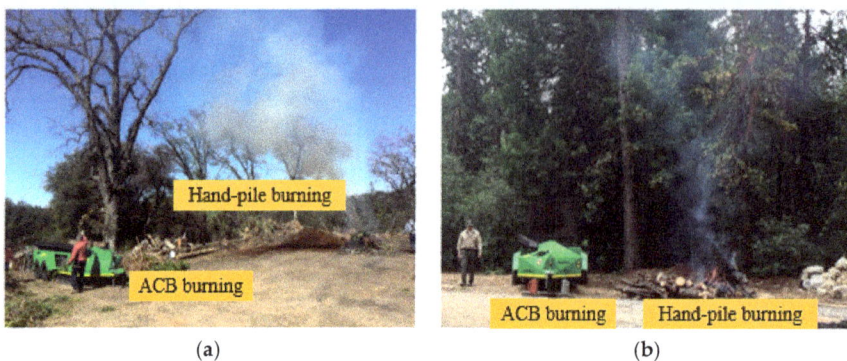

Figure 4. Comparison of the smoke occur from BurnBoss and hand-pile burning for the same types of forest residues; (**a**) Groveland and (**b**) Volcano, California. BurnBoss burnings shows little smoke and air pollutants, compared to hand-pile burnings.

2.4. Field Data Collection and Analysis

Prior to burning, the fuel material types were characterized by material species and diameter for each experiment. To measure burning consumption rates, fuel weights were measured with a PT300TM RFX portable wheel-load scale (Intercomp, Medina, MN, USA) installed on site. The weight of testing material used for this study was pre-measured and prepared by the research collaborator, following the researchers' requests. Before each burning test, moisture content was measured with a BD-2100 moisture meter (Delmhorst Instrument CO., Towaco, NJ, USA) by sampling 2 cm thick cookies of fuel samples for the greater than 10 cm slash piles. For materials less than 10 cm in diameter, we sampled portions of branches and needles to measure moisture contents. All samples were dried at 105 °C for 48 h since it was doubtful to measure a moisture content of fuels. Water content was not statistically different between BD-2100 and sampling methods ($p > 0.05$). When performing each test we also collected weather condition data such as air temperature, humidity, and wind speed using a Kestrel 3000 (Nielsen-Kellerman Co., Boothwyn, PA, USA) (Table 1).

ACBs' burning consumption rates (GmT/SMH) for each equipment and material were calculated using a time and motion study method. The burning times were recorded to the hundredths of a minute using a stopwatch. We recorded for second loading and burning time only, excluding the time prior to starting the air blower and burndown stage. This is because the preparation process was totally independent from operator control in the studied work phase. In addition, although S-220 and BurnBoss use the same principal of ACB for its burning mechanism, loading and the last stage, burndown, were different. For this reason, pure burning cycle times were recorded from air blower startup to second loading (from turned on air blower to when last fuels were loaded) during the ACB burning. The following describes how the S-220, BurnBoss, and HPB burning cycle times were measured:

- S-220 burning cycle time: For the S-220 burning test, we started recording the burning time when the air blower was turned on and stopped right after the last materials were loaded. Further, to determine the pure burning cycle time, second loading and waiting times sum up to gross-cycle times and determine to SMH.
- BurnBoss burning cycle time: For the BurnBoss burning test, we started recording the burning time when the air blower was turned on and stopped right after the last materials were loaded, producing comparable burning consumption rates (S-220 vs. BurnBoss). In addition, we continuously recorded until the materials were burned down completely to ash, because we also focused on feasibility to compare the burning consumption rates of BurnBoss with HPB.

- HPB burning cycle time: We started recording the burning time when a second loading started and completed our time measurement when the materials were burned down completely to ash. Thus, to determine the pure burning cycle time of HPB, second loading and perfectly combustion times sum up to gross-cycle times and determine to SMH. Additionally, when there were partially burned larger materials (>10 cm) in the pile, we picked them up and weighed them for evaluating the burning consumption rate of HPB.

Every test started with a "cold start" method, which means that each burn started on the bare ground (i.e., no ash from a previous fire) and required the ignition of kindling, followed by the addition of larger fuels until the fire continued on its own.

The machine rate calculation methods were used to evaluate hourly machine costs for the ACBs [33]. Fuel consumption rate, machine utilization rate, and wage were collected from the contractors (Table 2). Overhead or indirect, profit allowance cost, move-in, and transportation of residues costs were not obtained. Thus, total system costs included only the operational costs associated with supporting equipment (loader and personnel pick-up truck).

Table 2. Summary of input values and assumptions used to estimate hourly machine cost of S-220 and BurnBoss systems test in this study. The S-220 system include a loader while the BurnBoss oeprations were loaded by hand.

Cost Factors	S-220	Loader	BurnBoss [a]	Pickup Truck
Purchase price(US $)	106,000	135,000	48,900	40,000
Fuel consumption (L/h)	9.5	11.4	0.5	2.9
Utilization rate (%)	75.0	40.0	75.0	7.5
Wage (US $/h)	24.0	24.0	15.0	0.00
Hourly cost (US $/SMH)	53.9 [b]	19.2 [b]	28.5 [b]	0.7

[a] Excluding ember screen price for the machine used in Groveland and $1000 added to install ember screen for the machine used in Volcano. [b] Wage including benefits for one-man crew.

In addition, during the BurnBoss and HPB tests, we measured the flaming temperature (°C) using a Therma CAM SC640 IR camera (FLIR Systems, North Billerica, MA, USA) and Amprobe IR-750 thermometer (Amprove, Everett, WA, USA) in 10-min intervals until flame-out for each experiment. We recorded above the top of the FireBox and piles.

3. Results

The inventory data in Table 3 describe an ACB's burning consumption rates and cost. The S-220 unit was capable of burning rates of 5.7 to 6.8 GmT/SMH at a cost of US $12.8 and $10.8/GmT, respectively (Table 3). These burning operations (softwood, hardwood, and mixed species) indicated that combustion of softwood residues was 15% more efficient than both materials. During these trial tests, burning consumption rates strongly depended on the species. However, there was no significant difference in the moisture content of fuel sizes ($p > 0.05$).

In the BurnBoss tests in Groveland and Volcano, the machine's combustion rate of disposal ranged between 0.6 to 1.7 GmT/SMH at a cost between US $17.9 and $47.7/GmT, respectively (Table 3). The results indicated that there was no considerable difference in burning consumption rate between fuel size (diameter) less than 10 cm and 10–20 cm under less than 6-months age or fresh conditions. However, noticeable differences were detected by fuel age (fresh vs. 12-months-old) in Volcano. When burning 12-months-old residues after the fuel reduction operations, burning consumption rates were 70% greater compared to fresher fuels with a higher moisture content (27% and 36%). In addition, costs were high (US $47.7/GmT) when disposing of the fresh material.

Burning consumption rates were considerably different between burning options (BurnBoss vs. HPB), even though the burning experimental material properties were statistically similar ($p > 0.05$). The BurnBoss's burning consumption rate was 40–80% faster than HPB in this case study (Table 3). However, the average combustion temperature was not significantly different between the two burning options ($p > 0.05$). The smoke as indicated by pictures, is low for BurnBoss burning, which produced plumes with very low opacity only during the air blower start (Figure 4). For this reason, the ACB burning system was more an efficient and environmentally sound option to dispose of forest residues.

Table 3. Burning consumption rate, cost, maximum flame temperature of disposing forest residues using Air Curtain Burners (ACB) and hand-pile burn.

Locations	Burning Consumption Rate (GmT/SMH) [a]	Cost of Disposal (US $/GmT)	Moisture Content (%)	Ave. Flame Temperature [b] (°C)
Jacksonville, Florida (S-220)				
softwood	6.8	10.8	37	
hardwood	5.7	12.8	36	N/A [c]
mix	6.0	12.2	33	
Groveland, California (BurnBoss)				
small size (<10 cm)	0.7 (0.5 vs. HPB [d]: 0.3) [e]	40.9	26	955 (HPB [d]: 926)
large size (10 to 20 cm)	0.6 (0.5 vs. HPB [d]: 0.2) [e]	47.5	27	953 (HPB [d]: 945)
Volcano, California (BurnBoss installed with ember screen)				
fresh residues				
small size (<10 cm)	0.7 (0.3 vs. HPB [d]: 0.2) [e]	40.9	19	953 (HPB [d]: 916)
large size (10 to 20 cm)	0.6 (0.3 vs. HPB [d]: 0.1) [e]	47.7	36	957 (HPB [d]: 938)
12-months-old residues				
large size (10 to 20 cm)	1.7 (1.3 vs. HPB [d]: 0.1) [e]	17.9	17	955 (HPB [d]: 934)

[a] Green metric tons per scheduled machine hour; [b] Average temperature of combustion zone; [c] Data not avaliable; [d] Hand-pile burning method; [e] Burning cycle time was from air blower turned on to burn it down to ash.

4. Discussion and Conclusions

Utilizing forest residues for production of bioenergy and bioproducts effectively reduce the fire hazard and emissions from OPB, but collection and transportation of forest residues to power plants is often cost-prohibitive [4]. For this reason, one of the options used to reduce a huge amount of woody residues is to incinerate fuels. However, OPB has restrictions and challenges including fire hazards and emissions. The main contribution of this study was to introduce the environmental friendly forest residue disposal option of using an ACB and evaluate the economic feasibility. Therefore, this study was designed to evaluate two types of ACBs (S-220 and BurnBoss) and compare one with pile burning using a variety of forest residue types. We focused on a burning consumption rate based on the time to consume the materials loaded until the last materials are loaded. The ACB's burning consumption rates ranged between 0.6 and 6.8 GmT of forest residues/SMH at a cost US $47.5 and US $10.8/GmT, respectively. The S-220 offers a higher (85%) burning consumption rate and incurs a lower cost than the BurnBoss. The BurnBoss's burning consumption rate was 40–80% greater than HPB in this case study. In addition, BurnBoss burning produced much less smoke than the HPB.

We noticed that the S-220 unit was a good fit with a centralized operation with forest residues delivered for burning, while the BurnBoss could be frequently moved to the place where forest residues are located. For this reason, we do not suggest that one of the machines might be more effective than the other. However, it would be helpful to see the burning consumption rates and disposal costs of ACBs to select an optimal unit that meets operational needs. The burning consumption rate of the BurnBoss unit was smaller in quantity of fuels than the S-220 operation during our burning tests. A BurnBoss machine can be cost-effectively used to access the location where a small volume of fuels (e.g., hand-piled slash, small forest residues volumes along a roadside, and a few scattered drought/insect damaged trees in public parks) are piled and left for disposal. However, these options are often not carried out on harvesting sites due to the inaccessibility of forest roads such as steep, narrow, and winding conditions and tended to produce higher transportation cost of forest residues [34,35]. Particularly, in the Western U.S., a typical chip-van had limited access to harvest units due to forest roads that were constructed for a stinger-steered logging truck [35]. BurnBoss machines are easily transported with half-ton pick-up trucks [19]. For this reason, this unit was effectively used to remove forest residues that were not accessible with size reduction machines. In addition, it is effective for disposal of small volume of fuels such as hand-pile slash, paper trays or small forest residues volumes along a roadside, and a few scattered drought/insect damaged trees in public parks, which improves financial viability. On the other hand, the S-220 offers other applications and has the potential to succeed where there are large amounts of fuels on landing areas and high wildfire risk areas without local bioenergy facilities.

The burning rate or flammability was affected by species [36], moisture content or fuel ages [37,38], and airflow rate [39]. The calorific heating values (CV), which is defined as the amount of generating heat energy while a fuel is completely burned, were dependent on the species composition including differences in lignin content, hemicellulose, and density [40–42]. A softwood had a 10% greater CV because of the higher lignin content and lower hemicellulose and density than most hardwood [36,42,43]. Therefore, with S-220 burning activities, softwood burning rate was much higher than hardwood and mixed fuel tests, which explains in part the increase in CV for softwood burning relative to hardwood and mixed residues burning activities.

Moisture in fuels is one of the most important parameters in its combustion process and fresh fuels will not burn easily [44]. Dimitrakopoulos et al. [45] found that slash piles wetted could lead to a reduction in the combustion efficiency and obstruct ignition. During the BurnBoss tests, burning consumption rates were 2–3 times greater when disposing of 12-months-old material (17% moisture content) because the dry fuels require a shorter residence time to burn down to completion compared to fresh materials. Even though several studies indicated that material properties including size and volume were the primary factor driving burning rate [46], tree size did not have any effect in this case study. Thus, we noticed that dried forest residues led to shorter times for the materials to burn down completely to ash.

Previous studies have been carried out to observe the influence of airflow rate on burning rates [47,48]. If the amount of air supplied is increased, the combustion rate is greater than a critical flow rate [47–49]. Particularly, the high velocity airflow creating the air circulation supplied an oxygen-enriched environment in the FireBox that accelerated the burning process on ACBs burning system [23,26]. In addition, the high airflow tends to be attributable to a greater burning rates causing a larger high heat transfer from the flame [24,39], even though the flame temperature were not significantly different between the BurnBoss and HPB burning tests in this case study. On the other hand, OPB of fuels was limited by the supply of sufficient oxygen in natural (open-air) condition [50]. As a result, this high-speed air blower boosted and burned fuels more quickly and improved the burning consumption rates since the airflow rate was better than natural conditions.

ACB machine application can block the fugitive air pollutant emissions into the atmosphere since high velocity air created an air curtain on top of the FireBox [22,23,26]. Four previous studies, Fountainhead Engineering, Inc. [22], Zahn [24], Schapiro [26], and Miller and Lemieux (2007) found that ACB burning methods produced fewer greenhouse gases (i.e., CO), smoke (i.e., PM), and 90% less opacity than OPB when the air blower velocity was 1600 or 2000 RPM. Therefore, ACB operations were typically more efficient (e.g., reduction air pollutant emissions and smoke) within an incinerator compared with an OPB [19,23]. Further, it is expected that burning disposal methods will be adopted in forests to reduce potential environmental and fire hazards especially at the wildland-urban interfaces.

In conclusions, this was the first study to evaluate the burning consumption rates and costs of an ACB to better understand a centralized biomass disposal operation concept (S-220) and movable and mounted on a pick-up truck prototype (BurnBoss) machines. This study's findings have shown that S-220 and BurnBoss burning had considerably different burning consumption rates and disposing costs. Therefore, forest managers should consider the potential benefits and limitations of each machine to justify optimization efforts. For example, the BurnBoss is accessible to remote areas by a pick-up truck and suited for disposing of small size and volume forest residues. The ACB burning option seems to be commonly adopted in many forests to control emissions, smoke, and embers and to improve oxygen and heat supply by high velocity of airflow during the burn. Thus, this technology would be much more efficient and reduces the negative environmental and societal impact of burning forest residues. Further study is need to compare the weather condition, fuel age, amount of fuel load (i.e., second loading volume) to find the burning rates.

Acknowledgments: This project was funded by the Agricultural Research Institute, California State University: Award number 15-06-001. The authors are grateful to Woongsoon Jang and Joel Bisson for their contribution. Specially, our appreciation goes to Brian O'Connor (Air Burners, Inc., Palm City, FL, USA), Rick Whybra and Lester Scofield (PURFIRE), and Stephen Bakken (California State Park, CA, USA) for their cooperation on the operational aspects of the study.

Author Contributions: Han-Sup Han produced, designed and performed the experiments; Eunjai Lee collected and analyzed the data and wrote the article.

Conflicts of Interest: The authors declare no conflict of interest.

References and Notes

1. Bisson, J.A.; Han, S.K.; Han, H.S. Evaluating the system logistics of a centralized biomass recovery operation in Northern California. *For. Prod. J.* **2014**, *661*, 88–96. [CrossRef]
2. White, E.M. *Woody Biomass for Bioenergy and Biofuels in the United States: A Briefing Paper*; Gen. Tech. Rep. PNW-GTR-825; U.S. Department of Agriculture, Forest Service, Pacific Northwest Research Station: Portland, OR, USA, 2010; p. 45.
3. Faaij, A.P.C. Bio-energy in Europe: Changing technology choices. *Energy Policy* **2006**, *34*, 322–342. [CrossRef]
4. Tittmamm, P. The wood in the forest: Why California needs to reexamine the role of biomass in climate policy. *Calif. Agric.* **2015**, *69*, 133–137.
5. Montgomery, T.D.; Han, H.S.; Kizha, A.R. Modeling work plan logistics for centralized biomass recovery operations in mountainous terrain. *Biomass Bioenergy* **2016**, *85*, 262–270. [CrossRef]
6. Coltrin, W.R.; Han, S.-K.; Han, H.S. Costs and productivities of forest biomass harvesting operations: A literature Synthesis. In Proceedings of the Annual CoFE Council on Forest Engineering Meeting, New Bern, NC, USA, 9–12 September 2012; pp. 1–16.
7. Springsteen, B.; Christofk, T.; York, R.A.; Mason, T.; Baker, S.; Lincoln, E.; Hartsough, B.; Yoshioka, T. Forest biomass diversion in the Sierra Nevada: Energy, economics and emissions. *Calif. Agric.* **2015**, *69*, 142–149. [CrossRef]
8. Springsteen, B.; Christofk, T.; Eubanks, S.; Mason, T.; Clavin, C.; Storey, B. Emission reductions from woody biomass waste for energy as an alternative to open burning. *J. Air Waste Manag. Assoc.* **2011**, *61*, 63–68. [CrossRef] [PubMed]
9. Aurell, J.; Gullett, B.K. Emission factors from aerial and ground measurements of field and laboratory forest burns in the Southeastern U.S.: PM2.5, black and brown carbon, VOC, and PCDD/PCDF. *Environ. Sci. Technol.* **2013**, *47*, 8443–8452. [CrossRef] [PubMed]
10. Lindroos, O.; Nilsson, B.; Sowlati, T. Costs, CO_2 Emissions, and Energy balances of applying Nordic Slash Recovery Methods in British Columbia. *West. J. Appl. For.* **2011**, *26*, 30–36.
11. Miller, S.; Rhoades, C.; Schnackenberg, L. *Slash from the Past: Rehabilitating Pile Burn Scars*; Science You Can Use Bulletin Issue 15; U.S. Department of Agriculture, Rocky Mountain Research Station: Fort Collins, CO, USA, 2015.
12. Graham, R.T.; Jain, T.B.; Matthews, S. Fuel management in forests of the Inland West. In *Cumulative Watershed Effects of Fuel Management in the Western United States*; Elliot, W.J., Miller, I.S., Audin, L., Eds.; U.S. Department of Agriculture Forest Service: Fort Collins, CO, USA, 2010; Chapter 3; pp. 19–68.
13. Estrellan, C.R.; Lino, F. Toxic emissions from open burning. *Chemosphere* **2010**, *80*, 193–207. [CrossRef] [PubMed]
14. Jones, G.; Loeffler, D.; Calkin, D.; Chung, W. Forest treatment residues for thermal energy compared with disposal by onsite burning: Missions and energy return. *Biomass Bioenergy* **2010**, *34*, 737–746. [CrossRef]

15. Lemieux, P.M.; Lutes, C.C.; Santoianni, D.A. Emissions of organic air toxics from open burning: A comprehensive review. *Prog. Energy Combust. Sci.* **2004**, *30*, 1–32. [CrossRef]
16. Cetini, G. Effects of fire on properties of forest soils: A review. *Oecologia* **2005**, *143*, 1–10. [CrossRef] [PubMed]
17. Hubbert, K.; Busse, M.; Overby, S. *Effects of Pile Burning in the LTB on Soil and Water Quality*; SNPLMA 12576 Final Report; U.S. Department of Agriculture Forest Service: Flagstaff, AZ, USA, 2013; p. 66.
18. Busse, M.D.; Hubbert, K.R.; Moghaddas, E.E.Y. *Fuel Reduction Practices and Their Effects on Soil Quality*; General Technical Report PSW-GTR-241; U.S. Department of Agriculture Forest Service: Albany, CA, USA, 2014; p. 157.
19. Air Burners, Inc. Available online: http://www.airburners.com/index.html (accessed on 7 June 2017).
20. Legal Information Institute. Available online: http://www.law.cornell.edu/cfr/text (accessed on 15 June 2016).
21. Lambert, M.B. Efficiency and economy of an air curtain destructor used for slash disposal in the Northwest. Presented at American Society of Agricultural Engineers Winter Meeting, Chicago, IL, USA, 16–18 December 1972.
22. Fountainhead Engineering; Deruiter Environmental, Inc. *Final Report Describing Particulate and Carbon Monoxide Emissions from the Whitton S-127 Air Curtain Destructor*; Project #00-21; Fountainhead Engineering: Chicago, IL, USA, 2000.
23. Miller, C.A.; Lemieux, P.M. Emissions from the burning of vegetative debris in air curtain destructors. *Air Waste Manag. Assoc.* **2007**, *57*, 959–967. [CrossRef]
24. Zahn, S.M. *The Use of Air Curtain Destructors for Fuel Reduction and Disposal*; Technology & Development Program Fire Management Tech Tips 0551-1303-SDTDC; U.S. Department of Agriculture Forest Service: Albany, CA, USA, 2005; pp. 1–6.
25. Stark, D.T.; Wood, D.L.; Storer, A.J.; Stephens, S.L. Prescribed fire and mechanical thinning effects on bark beetle caused tree mortality in a mid-elevation Sierran mixed-conifer forest. *For. Ecol. Manag.* **2013**, *306*, 60–67. [CrossRef]
26. Schapiro, A.R. *The Use of Air Curtain Destructors for Fuel Reduction*; Technology & Development Program Fire Management Tech Tips 0251-1317P-SDTDC; U.S. Department of Agriculture Forest Service: Albany, CA, USA, 2002.
27. Bakken, S.R.; Forester, California State Parks. Interview in Volcano, CA, USA. Personal communication, 2016.
28. California Department of Forestry and Fire Protection (CALFIRE). Available online: http://www.fire.ca.gov (accessed on 19 November 2016).
29. Chomanee, J.; Tekasakul, S.; Tekasakul, P.; Furuuchi, M.; Otani, Y. Effects of moisture content and burning period on concentration of smoke particles and particle-bound polycyclic aromatic hydrocarbons from rubber wood combustion. *Aerosol Air Qual. Res.* **2009**, *9*, 404–411. [CrossRef]
30. Bignal, K.; Langridge, S.; Zhou, J. Release of polycyclic aromatic hydrocarbons, carbon monoxide and particulate matter form biomass combustion in a wood-fired boiler under varying boiler conditions. *Atmos. Environ.* **2008**, *42*, 8863–8871. [CrossRef]
31. O'Connor, B.; President of Air Burners Inc. Interview in Jacksonville, FL, USA. Personal communication, 2015.
32. Whybra, R.; Operator of Pur Fire. Interview in Groveland, FL, USA. Personal communication, 2016.
33. Brinker, R.; Kinard, J.; Rummer, B.; Lanford, B. Machine Rates for Selected Forest Harvesting Machines. In *Machine Rates for Selected Forest Harvesting Maines*; Auburn University: Auburn, AL, USA, 2002; p. 32.
34. Anderson, N.; Chung, W.; Loeffler, D.; Jones, J.G. A Productivity and Cost Comparison of Two Systems for Producing Biomass Fuel from Roadside Forest Treatment Residues. *For. Prod. Soc.* **2012**, *62*, 222–233. [CrossRef]
35. Han, H.-S.; Halbrook, J.; Pan, F.; Salazar, L. Economic evaluation of a roll-off trucking system removing forest biomass resulting from shaded fuelbreak treatments. *Biomass Bioenergy* **2010**, *34*, 1006–1016. [CrossRef]
36. Lowden, L.A.; Hull, T.R. Flammability behaviour of wood and a review of the methods for its reduction. *Fire Sci. Rev.* **2013**, *2*, 1–19. [CrossRef]
37. Shen, G.; Xue, M.; Wei, S.; Chen, Y.; Wang, B.; Wang, R.; Lv, Y.; Shen, H.; Li, W.; Zhang, Y.; et al. The influence of fuel moisture, charge size, burning rate and air ventilation conditions on emissions of PM, OC, EC, Parent PAHs, and their derivatives from residential wood combustion. *J. Environ. Sci.* **2013**, *25*, 1808–1816. [CrossRef]

38. Simoneit, B. Biomass burning: A review of organic tracers for smoke from incomplete combustion. *Appl. Geochem.* **2002**, *17*, 129–162. [CrossRef]

39. Regueiro, A.; Patiño, D.; Porteiro, J.; Granada, E.; Míguea, J.L. Effect of air staging ratios on the burning rate and emissions in an underfeed fixed-bed biomass combustor. *Energies* **2016**, *9*, 940. [CrossRef]

40. Moya, R.; Tenorio, C. Fuelwood characteristics and its relation with extractives chemical properties of ten fast-growth species in Cost Rico. *Biomass Bioenergy* **2013**, *56*, 14–21. [CrossRef]

41. Khider, T.O.; Elsaki, O.T. Heat value of four hardwood species from Sudan. *J. For. Prod. Ind.* **2012**, *1*, 5–9.

42. Demirbaş, A. Relationships between heating value and lignin, moisture, ash and extractive contents of biomass fuels. *Energy Expor. Exploit.* **2002**, *20*, 105–111. [CrossRef]

43. Telmo, C.; Lousada, J. Heating values of wood pellets from different species. *Biomass Bioenergy* **2011**, *35*, 2634–2639. [CrossRef]

44. Possell, M.; Bell, T.L. The influence of fuel moisture content on the combustion of Eucalyptus foliage. *Int. J. Wildland Fire* **2013**, *22*, 343–352. [CrossRef]

45. Dimitrakopoulos, A.P.; Mitsopoulos, I.D.; Gatoulas, K. Assessing ignition probability and moisture of extinction in a Mediterranean grass fuel. *Int. J. Wildland Fire* **2010**, *19*, 29–34. [CrossRef]

46. White, R.H. Fire Performance of Hardwood species. Presented at In XXI IUFRO World Congress, Kuala Lumpur, Malaysia, 7–12 August 2000.

47. Khodaei, H.; Al-Abdeli, Y.M.; Guzzomi, F.; Yeoh, G.H. An overview of processes and considerations in the modelling of fixed-bed biomass combustion. *Energy* **2015**, *88*, 946–972. [CrossRef]

48. Porteiro, J.; Patiño, D.; Moran, J.; Granada, E. Study of a fixed-bed biomass combustor: Influential parameters on ignition front propagation using parametric analysis. *Energy Fuels* **2010**, *24*, 3890–3897. [CrossRef]

49. Yang, Y.; Sharifi, V.; Swithenbank, J. Effect of air flow rate and fuel moisture on the burning behaviours of biomass and simulated municipal solid wastes in packed beds. *Fuel* **2004**, *83*, 1553–1562. [CrossRef]

50. Grendehou, S.; Koch, M.; Hockstad, L.; Pipatti, R.; Yamada, M. Incineration and Open Burning of Waste. In *Waste IPCC Guidelines for National Greenhouse Gas Inventories*; Intergovernmental Panel on Climate Change: Geneva, Switzerland, 2006; Chapter 5; pp. 77–102.

Article

How Climate Change Will Affect Forest Composition and Forest Operations in Baden-Württemberg—A GIS-Based Case Study Approach

Ferréol Berendt [1,*], Mathieu Fortin [2], Dirk Jaeger [1] and Janine Schweier [1]

[1]   Chair of Forest Operations, Albert-Ludwigs-University Freiburg, Werthmannstraße 6,
      79085 Freiburg, Germany; dirk.jaeger@foresteng.uni-freiburg.de (D.J.);
      janine.schweier@foresteng.uni-freiburg.de (J.S.)
[2]   UMR 1092, AgroParisTech/INRA, Centre de Nancy, 14 rue Girardet, Nancy CEDEX 54042, France;
      mathieu.fortin@agroparistech.fr
*   Correspondance: Ferreol.Berendt@foresteng.uni-freiburg.de; Tel.: +49-761-203-3790

Received: 6 June 2017; Accepted: 13 August 2017; Published: 16 August 2017

Abstract: In order to accommodate foreseen climate change in European forests, the following are recommended: (i) to increase the number of tree species and the structural diversity; (ii) to replace unsuitable species by native broadleaved tree species, and (iii) to apply close-to-nature silviculture. The state forest department of Baden-Württemberg (BW) currently follows the concept of Forest Development Types (FDTs). However, future climatic conditions will have an impact on these types of forest as well as timber harvesting operations. This Geographic Information System (GIS)-based analysis identified appropriate locations for main FDTs and timber harvesting and extraction methods through the use of species suitability maps, topography, and soil sensitivity data. Based on our findings, the most common FDT in the state forest of BW is expected to be coniferous-beech mixed forests with 29.0% of the total forest area, followed by beech-coniferous (20.5%) and beech-broadleaved (15.4%) mixed forests. Where access for fully mechanized systems is not possible, the main harvesting and extraction methods would be motor manual felling and cable yarding (29.1%). High proportions of large dimensioned trees will require timber extraction using forestry tractors, and these will need to be operated from tractor roads on sensitive soils (23.0%), and from skid trails on insensitive soils (18.4%).

Keywords: forest operations; timber harvesting; timber extraction; forest development types; species suitability map

---

1. Introduction

Since the beginning of the 20th century, anthropogenic greenhouse gas (GHG) emissions [1] have caused a steady rise in the mean annual temperature around the world. In Germany, the reported increase from 1881 to 2014 was 1.3 °C [2]. Besides the temperature increase, climatic simulations for Central Europe show changes in precipitation regime. While the annual precipitation may remain constant, both higher rainfall intensity [3] and more frequent droughts [4] are expected. Given the current climate change projections, diverse impacts are to be anticipated for forests, such as a northward shift of several hundred kilometers for single tree habitats [5], an altitudinal shift of 300 to 400 m [6], extended vegetation periods [7], and changes in biomass increments [8].

To improve resistance and resilience of forests to climate change, it is generally agreed that both the number of species and the structural diversity of forests should be increased [9]. Resistance and stability refers to the capacity of a system to absorb disturbances and to forestall impacts [10–12], whereas resilience is the capacity to recover and to return to the equilibrium or pre-condition state after a disturbance/perturbation [11,13,14]. Nevertheless, a major issue in forest management planning is the prediction of future forest conditions, and identification of species suitable to these future conditions. This is why projections of species distributions under both climatic and global environmental change are of great scientific and societal relevance [15]. Different approaches have been developed to identify the most suitable tree species and management strategies. These include bioclimate envelopes [16], spatio-temporal site-index predictions [7] or species distribution models [17,18]. As they are mainly focused on the suitability of individual tree species to expected future environments, they can model one species at a time in order to map its future spatial range [19]. Bolte et al. (2009) [20] mention the possibility of integrating these analyses into silvicultural concepts of forest dynamics, e.g., the Forest Development Type (FDT) approach.

The concept of FDTs was developed decades ago as a strategic approach for: (i) illustrating long-term goals for forest development in a given locality and; (ii) describing the transition of existing forest stand types into types that are well adapted to moderate climate change [21,22].

The importance of the concepts of close-to-nature or continuous-cover forestry is widely accepted [23]. Management strategies are increasingly focused on the diversification of vertical and horizontal forest structures, including a greater diversity of tree species [4]. The implementation of FDTs in Germany [24–26] follows this principle of favoring site-adapted broadleaved species. Results of the third national forest inventory showed a 7% increase in the area of broadleaved trees from 2002 to 2012, with the area increasing from 4,317,236 ha to 4,632,637 ha [27]. Increases in the area covered by broadleaved trees are also reported in Baden-Württemberg (BW). BW is a Federal state in southern Germany with forests typical for Central European conditions due to a large variation in altitudes, sites conditions, silvicultural management approaches, and stands with mixtures of broadleaved and softwood species [28]. At the moment, the most predominant tree species in BW is Norway spruce (*Picea abies* H. Karst), which covers 33.5% of the total forested land base. European beech (*Fagus sylvatica* L.) is the most common broadleaved species, covering 21.5% of the forested area [29], and it is the naturally dominant tree species [18]. Because climate change is expected to progress faster than forests can adapt [30], forest management has a particular focus on these two main tree species. The state forestry department of BW (Forst BW) intended to increase the ratio of broadleaved trees, particularly by replacing unsuitable Norway spruces with native European beeches, oaks (*Quercus robur* L. and *Quercus petraea* Liebl.), silver firs (*Abies alba* Mill.) and additional broadleaved tree species [31].

As a consequence of the tree species shift, an increase of mixed stands with high structural varieties and changing precipitation regime, changes in the degree of mechanization of felling operations will likely occur. Given the preference for motor-manual systems (chainsaws) in beech stands, this kind of operation may likely gain in popularity at the expense of single grip harvesters—which are typical in coniferous stands—in fully mechanized systems. Moreover, a greater diversity of structures in forests may favor management regimes based on natural regeneration, single-tree harvest, habitat-adapted tree species and provenances [4,32].

In addition to the altered tree species composition, future climatic conditions in BW will create additional constraints on timber harvesting operations. On frozen ground, skidders work more efficiently and cause less damage due to increased bearing capacity [33]. Because the mean annual temperature is rising, the number of days with frozen ground during the traditional logging period in winter is expected to decrease [34,35]. Moreover, the expected 35% increase in precipitation in the winter season in BW [36] will likely result in higher soil moisture content and wetter soil conditions. It is highly probable that soil moisture and the water balance will remain high during winter [37], which is not favorable to any ground-based forest operation [38]. Increased rutting, higher soil bulk densities, and lateral soil displacement are to be expected with winter operations.

In spite of all these additional constraints, it can be assumed that harvesting will still be carried out mostly during the winter season because of the increasing proportion of broadleaved trees, nature conservation aspects and work safety. Therefore, innovative timber harvesting and extraction operations that minimize soil damage caused by modern technical equipment (e.g., weight, number of axles and wheels), as well as appropriate harvesting and extraction methods (e.g., cable, horse, tethering winches) [5], need to be applied.

The key question addressed in this study is how climate change will affect future timber harvesting operations. More specifically, the research objectives were to identify, quantify and interpret expected qualitative changes in forests in BW due to climatic change ("How will future forests look?"); to describe those forest types that will potentially be the most relevant FDTs in BW; and to identify forest harvesting operations that are most adapted to expected future conditions in these FDTs.

2. Material and Methods

2.1. Concept, Tools and Data

The study partly relied on a Geographic Information System (GIS) (2015 ESRI® ArcGIS 10.3.1). Using a GIS to represent the most relevant FDTs and terrain data relevant for forest operations seemed to be the most adequate approach because (i) most site-relevant data were available in digital format, and (ii) restrictions and/or site-specific characteristics could be incorporated. This approach, which consists of combining different GIS layers with specific information, has previously been applied: to determine suitable areas for short rotation coppices [39,40]; to assess biomass potentials [41–45]; and to select appropriate timber harvesting systems [46,47]. Information about soil type, terrain slope, and stand composition as well as species suitability maps, which already include climate and site quality data, were provided by the BW Forest Research Institute (FVA BW) [48]. Additionally, data regarding soil sensitivity to traffic were also available for map units called regional site units [49] in a Microsoft Access file (2013 Microsoft ® Access ® 15.0.4857.1000) [48]. These data on the regional site units were imported into GIS. All input data were collected by forestry departments in 2010 [49], and were compiled by regional authorities and FVA BW to ensure that all data collected in BW were reported in the same format and at the same level of detail.

2.2. Species Suitability Data

The collection of data on the suitability of tree species in BW started in the 1970s [50], but the resulting "maps were originally based on expert knowledge of the site classification" [18]. The current species suitability predictions, provided by the FVA BW [48], were based on a statistical model which predicted the presence or absence of a tree species under given climatic conditions. The statistical approach is described by Hanewinkel et al. [18]. The original "presence/absence information per species [was] derived from the 'Data on Crown Condition of the systematic grid (16 × 16 km)' (Level I) from the 'International Co-operative Programme on Assessment and Monitoring of Air Pollution on Forests'" [18]. The presence/absence data were coupled with site-specific tree physiology values (mostly based on temperature and precipitation) before being statistically analyzed to identify correlations [51]. Thirty-arc-second tiles were used as spatial resolution for climate data [18]. In order to predict the species suitability to future climatic conditions, the values of explanatory variables were changed to match expected climates for the year 2050. The mean annual temperature was increased by about 2 °C, whereas the mean annual precipitation was decreased by 25 mm, mainly during the vegetation period [31]. The model generated maps of predicted probabilities of observed species. In a further step, detailed site classification information was analyzed by experts in order to provide information about the following attributes of the species [50,51]:

1. Competition strength
2. Soil protection
3. Growth performance
4. Stability

With these four attributes, the FVA BW generated a database that contained the potential of different tree species to grow under climatic conditions predicted for 2050 in each of the 5023 regional-site units. The tree suitability map had a resolution of 1:50,000 [51].

Each attribute was assigned a value on the scale from best to poorest, representing the suitability of tree species. 'Competition strength' took into account both the regeneration and the competitiveness of mature trees. 'Soil protection' considered the impact of the species on both humus and soil (e.g., the root depth). 'Stability' included biotic and abiotic dangers, and finally, 'growth performance' reflected the volume growth or market value [48].

Based on the ranking of the four attributes, an overall assessment of tree species suitability for each regional-zonal site unit was expressed using a six-category classification, with classes ranging from biologically important to inappropriate. In this study we focused on the site units that were ranked as 'biologically important', 'very suitable', 'suitable', and 'possible growth', since they represented the tree species suitable to future climatic conditions. On specific sites, some tree species were considered biologically important for humus formation, soil protection or for protective forests [48], and were therefore assessed manually. The decision matrix for assessing tree species suitability showed 135 possible combinations between the different values of the four above-mentioned attributes (Table 1). The matrix with all possible combinations for the tree species suitability classes 'very suitable', 'suitable', and 'possible growth' is shown in Table 1. For the sake of simplicity, we did not show the combinations that would lead to determining unsuitable tree species [52].

Table 1. Matrix for the assessment of tree species suitability for the classes 'very suitable', 'suitable', and 'possible growth'. The attributes 'competition strength', 'soil protection', and 'performance' are assigned values from best (1) to poorest (3), and the attribute 'stability' from best (1) to poorest (5) [52]. As an example of how the matrix works, we enclose one possible combination in red. It shows that a tree species is considered as suitable on a regional-zonal unit when the competition strength is assigned a value of 2, soil protection a value of 3, and stability and performance with values of 1 respectively.

Very suitable	Competition strength	1	2	3	1	2	1	1												
	Soil protection	1	1	1	2	2	1	1												
	Stability	1	1	1	1	1	1	2												
	Performance	1	1	1	1	1	2	1												
Suitable	Competition strength	3	1	2	2	1	1	2	3	1	2	1	1	1						
	Soil protection	2	3	3	1	2	1	1	1	2	2	3	1	1						
	Stability	1	1	1	1	1	1	2	2	2	2	2	2	3						
	Performance	1	1	1	2	2	3	1	1	1	1	1	2	1						
Possible growth	Competition strength	3	3	2	1	1	1	3	2	2	1	1	2	3	1	1	1	1	1	1
	Soil protection	3	1	2	3	2	3	2	3	1	2	3	1	1	2	3	1	1	2	2
	Stability	1	1	1	1	1	1	2	2	2	2	2	3	3	3	3	3	4	4	2
	Performance	1	2	2	2	3	3	1	1	2	2	2	1	1	1	1	2	1	1	3

2.3. Derivation of FDTs from Species Suitability Data

The concept of FDTs is currently in use in several federal states of Germany and Denmark [22,25,26,53]. It is applied in the state forests of BW, and supported by guidelines which describe 17 FDTs [24]. For a given set of climatic and stand conditions, the FDT description includes information about species distribution, rotation length, regeneration dynamics, forest management activities (tending, thinning and final cutting [28]) and timber assortments [54]. This study focused on six particular FDTs, which are described in Section 2.4. The FDTs were selected according to their current and predicted tree species: Norway spruce, beech, silver fir and oak. Altogether, these six FDTs covered 82% of the analyzed 379,215 ha forests managed by Forst BW. These public forests represented 27.6% of the forest area in BW. The area of the dataset is 424,160 ha, from which 44,945 ha were unsuitable for this study, as information was missing. The spatial extent and location of the study area is shown together with the total forest area of BW in Figure 1 [55], because for the interpretation the relevance of FDT in different areas, the remaining forest area is important. The main tree species in the state forest of BW are spruce (32.6%), beech (24.6%), silver fir (8.3%) and oaks (6.5%) [56]. This differs slightly from the overall composition of all forests in BW.

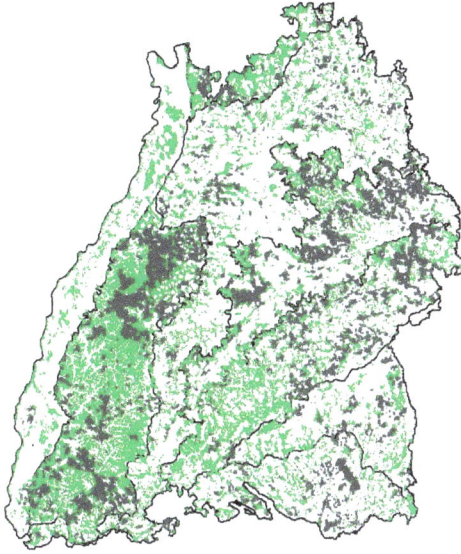

Figure 1. Spatial extent and location of the analyzed area (in grey) in comparison to whole Baden-Württemberg (BW) forest (in green) with the natural regions of third level (black lines), adapted from [55].

We followed the same assumptions as those outlined in Witt et al., Forst BW and Saar Forst [21,24,57] about the shares of species. Very suitable sites were assumed to allow the growth of a dominant tree species, meaning that the dominant tree species on these sites accounted for more than 50% of the area. The class 'possible growth' indicated that a tree species could reach proportions of between 20% and 50% of the area composition, and therefore represented an important associated species.

In addition to the species suitability, we favored conifers over broadleaved tree species on sites where both were ranked suitable. Conifers represent more than 66% of the total harvest in BW [58] and annual incomes from coniferous forests are around 100 €/ha higher than those from broadleaved forests [59]. Given the economic importance of coniferous species, it seemed reasonable to assume that forest managers would have a preference for these species whenever they are suitable. The resulting derivation of future FDTs on the tree species suitability data is represented in Figure 2.

2.4. Forest Development Types

Coniferous-beech-mixed forests (FDT1)

FDT 1 contains silver fir covering up to 60% of the total forest area. However, spruce and Douglas fir (*Pseudotsuga menziesii* Franco) may also make up high proportions on specific and suitable sites. Overall, the proportion (by area) of the dominant coniferous tree species is 50–80%, while the proportion of beech ranges between 10 and 50% [21]. Other suitable broadleaved tree species are sycamore maple (*Acer pseudoplatanus* L.), European hornbeam (*Carpinus betulus* L.), and ash (*Fraxinus excelsior* L.). Between 30–50% of the whole merchantable biomass comes from diameters of large dimension, with a diameter at breast height (DBH) greater than 50 cm [24].

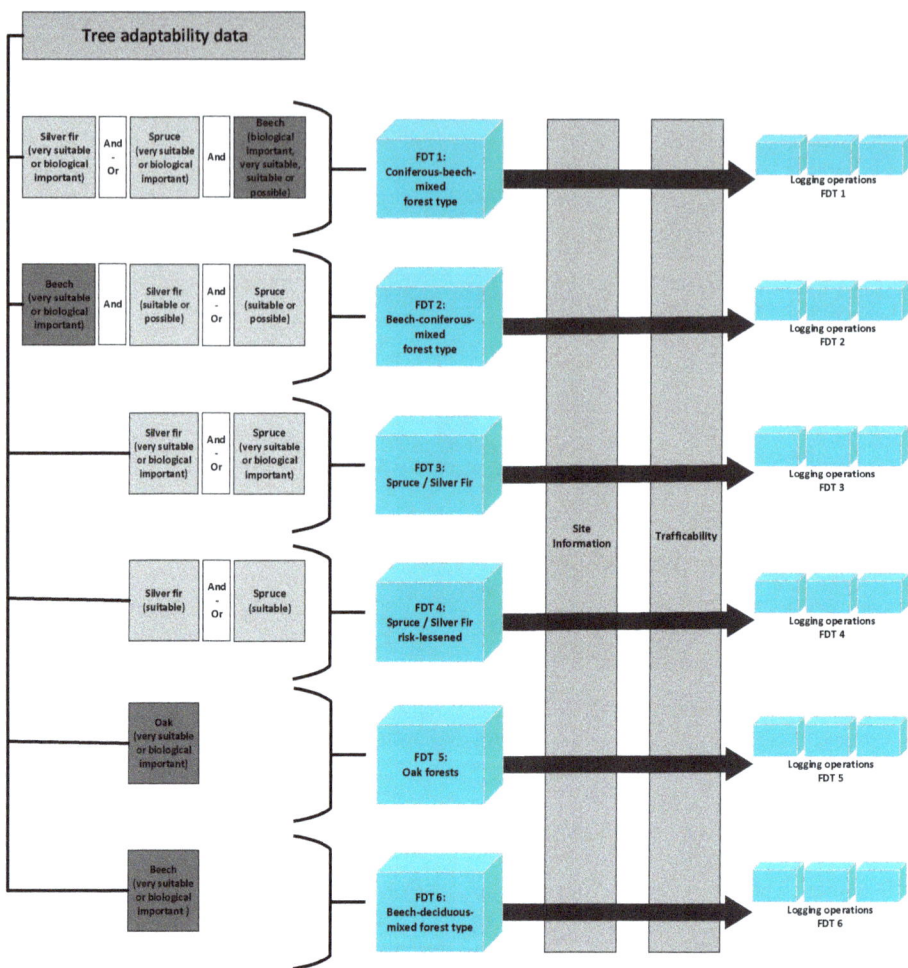

Figure 2. Flowchart of the Geographic Information System (GIS) analysis to determine Forest Development Types (FDTs) from species suitability data and timber harvesting and extraction operations from site information and trafficability data.

Beech-coniferous-mixed forests (FDT 2)

Beech-coniferous-mixed forests are dominated by beech trees (40–80% of the total area) and mixed with coniferous trees—mostly spruce or pine (*Pinus sylvestris* L.)—covering up to 40% of the total area [24,57]. In this FDT, beech is always the dominant tree species and the production goal is high quality beech timber of large dimension (DBH of 60 cm). Lower quality trees and conifers are harvested at DBH of 50 cm [24].

Reduced risk spruce forests (FDT 3)

In this FDT, the rotation lengths would be reduced to 40–60 years, during which DBHs of 40 cm can be reached [60]. The reduction of rotation length may help to limit damage by some of today's most prominent forest issues, e.g., windthrow, cambium-feeding insects, and root rot [61]. Under this new management strategy, this type of forest is expected to produce high quantities of wood for material usage, mostly lumber, because the proportion (by area) of coniferous tree species (mostly spruces), at 60–80%, is very high [24].

Silver fir forests (FDT 4)

This FDT targets high proportions of coniferous species with mainly silver fir and spruce. Commonly, wood production is oriented towards large diameters (DBH 50–80 cm) [24]. Mixture with beech is common, but beech is not dominant. Particularly when single tree selection systems—also known as Plenterwald [62]—are applied, the proportion (by area) of beech never exceeds 20% [24,57].

Oak-mixed forests (FDT 5)

Common oak and sessile oak forest types are grouped into this oak-mixed forest type. The proportion of oak species is high (at 60–90% of the total area in mixed forests) [24].

Beech-broadleaved-mixed forests (FDT 6)

This FDT has beech proportions (40–80% of the total area) similar to those of the beech-coniferous-mixed forests (FDT2). In addition to beech, sycamore maple, cherry (*Prunus avium* L.), oak and ash are the most predominant species on the sites. Coniferous tree species can be found as an admixture, representing up to 20% of the total area [24,57].

By combining the different layers of species suitability, topography and site sensitivity, it was also possible to prescribe optimal logging operations to the FDTs (Figure 2). Moreover, it was possible to assign a FDT to each site unit under the climatic conditions predicted for 2050.

2.5. Timber Harvesting and Extraction Systems

2.5.1. Soil Sensitivity

Technical terrain classification is based on three criteria: terrain slope, ground condition (bearing capacity), and ground roughness (microtopography). In our study, we focused on assessing terrain slope and soil sensitivity based on ground conditions, since microtopography data were not available for the whole study area.

Soil compaction and displacement are important aspects in forest operations [63–65]. Soil sensitivity represents the risk of irreversible soil disturbance due to machine traffic. Irreversible soil disturbances should be avoided in order to secure all the natural processes that occur in the soil, which ensure a preservation of forest ecosystems and maintain optimal productive functions in forests [66]. Our analysis was based on the soil texture for each regional site unit in the Microsoft Access file grouped into classes (Table 2) which were integrated as GIS-layers. The varying vulnerability of different soils to traffic was already classified by Wiebel [49], who grouped different soil textures into the classes (i) sensitive; (ii) insensitive, and (iii) partly sensitive [48] (Table 2).

Table 2. Risk of soil disturbance from machine traffic according to soil texture, adapted from Wiebel [49], where "−" indicates that soils are insensitive to traffic; "+" indicates that soils are sensitive to traffic; and "+/−"indicates that soils are partly sensitive to traffic.

Soil Texture	Sensitivity
Clayey	+
Loamy-clayey	+/−
Silty-loamy + clayey	+
Silty-loamy	+
Loamy	+
Loamy; sandy	+/−
Gravelly	−
Rocks	−
Varied; diverse	+/−
Organic	+/−
No data	n.a.

2.5.2. Topography

Regarding the topography, the slopes of each regional-zonal site unit were already classified by the FVA BW [48] in a GIS-layer into (a) lowlands and easy slopes (<30%); (b) medium slopes (30–45%); (c) steep slopes (>45%) and (d) others (e.g., gorges) [48,67]. The mapper overruled this classification in cases where specific site aspects may cause difficulties for logging operations on easy to medium slopes [48]. For example, both low infiltration and low bearing capacity are an indication for sites which could be overruled by the mapper.

Soil sensitivity was combined with slope classes. These two properties determine the risk of soil compaction and displacement, which are the main soil disturbances caused by vehicle traffic on forest floors [68,69]. This occurs mainly during extraction operations [70]. In addition to contributing to technical difficulties, terrain steepness also causes slippage [38] as well as erosion through runoff and soil loss [67,71] meaning that the highest soil deterioration level was experienced on slopes with inclinations above 20% [72].

Therefore, three in-stand transportation modes were identified with regard to both soil sensitivity and terrain (Figure 3): (1) Skid trails (ground-based forest operations on skid trails); (2) Tractor roads (ground-based forest operations on tractor roads) and (3) Road-based operations or cable yarding (no off-road traffic).

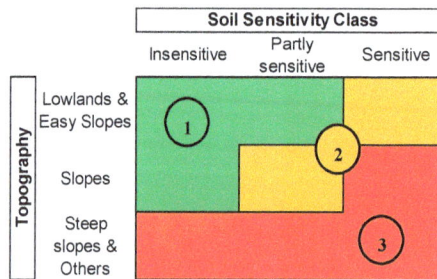

Figure 3. In-stand transportation modes with regard to soil sensitivity class and topography, focusing on timber extraction mode; where (1) indicates that traffic is possible (ground-based forest operations on skid trails); (2) indicates that low traffic is possible (ground-based forest operations on tractor roads); and (3) indicates that traffic is not possible (road-based operation/cable yarding).

2.5.3. Wood Dimensions

The DBH is a limiting factor for mechanized timber harvesting operations, depending on the type of harvester head [46]. Although harvester heads designed for diameters as large as 102 cm exist, 60% of the harvester heads on the European market have a smaller maximum felling diameter [73]. This large range in maximum felling diameters made it impossible to define a DBH limit for mechanized fellings. Kühmaier and Stampfer (2010) [46] reported a 50 cm DBH limit for mechanized fellings in softwood stands. Nevertheless, no harvester head specifically built for temperate European broadleaved tree species is on the market at this stage, and development focus is on diameters up to 35 cm [74]. Therefore, we assumed that felling would be carried out motor-manually whenever the DBH exceeded 50 cm for coniferous, and 35 cm for broadleaved tree species (Table 3).

When it comes to extraction, the choice of machine is restricted not only by terrain but also by the technical extraction mode and the volume or DBH of the trees (Table 4). For example, horses and small forestry crawlers can skid down slopes up to 45–50% [75–77], and drag volumes up to 0.6 m^3 and 1.2 m^3, respectively [78,79].

Table 3. Applied harvesting systems with regard to terrain slope and DBH.

Felling Mode	Slope	Tree Species	DBH (cm)
Chainsaw	Any	Any	any DBH
Wheeled harvester	Easy	Conifers	<50
Wheeled harvester	Easy	Broadleaves	<35
Tracked/tracked wheel harvester	Medium	Conifers	<50
Tracked/tracked wheel harvester	Medium	Broadleaves	<35

Table 4. Applied extraction systems with regard to terrain slope, log length and diameter at breast height (DBH).

Hauling Mode	Slope	Harvesting System	DBH or Volume
Skidder	Easy to medium	Tree-length/whole tree	Any
Forwarder	Easy to medium	Cut-to-length	Any
Small forestry crawler	Easy to medium	Tree-length/cut-to-length	<1.2 m^3
Horse	Easy to medium	Tree-length/cut-to-length	<0.6 m^3
Cable yarder	Any	Tree-length/whole tree	Any
Ground carriage	Easy to medium	Tree-length/cut-to-length	Any

To avoid stand damage and damage to natural regeneration, tree-length and cut-to-length operations were preferred.

2.5.4. Others Constraints

In BW, clearcuts exceeding an area of 1 ha need approval from the state forest authority [80]. Consequently, they play a minor role and were not considered in this study. Considering this, selective cuttings were assumed to be standard logging operations for final cuttings.

The cutting cycles of FDT 1, FDT 2, FDT 4 and FDT 6 were assumed to occur twice every 10 years with harvesting volumes corresponding to the volume increment [24]. FDT 3 (Reduced risk spruce forests) was assumed to be under selective logging—also known as Femelschlag—treatment, with thinning operations carried out every five years between 25 and 55 years of age, and final felling at 60, 65 and 70 years of age [60]. According to Forest Stewardship Council (FSC) standards for the region [81], the maximum cleared area during timber harvesting was set to 0.3 ha for all species, except oak and pine forests. For oak-mixed forests (FDT 5), the cleared area can be extended to 1 ha [81] (small-scale clear-cut) while still ensuring natural regeneration since oak is a species with intermediate shade tolerance.

3. Results

3.1. Forest Development Types

Conducting the GIS analysis made it possible to predict the location of the FDTs in 2050 in light of the expected climate change (Figure 4), namely a temperature increase of around 2 °C and a precipitation decrease of around 25 mm.

Coniferous-beech-mixed forests (FDT 1)

Under the above defined selection of the FDT layers, the main future forest type in BW state forests is expected to be coniferous-beech-mixed forest (FDT1) (Figure 4), with an area of 109,885 ha (29.0% of the study area). Results pertaining to the main coniferous tree species for this FDT indicated that 83,702 ha would be more favorable to silver fir (22%) than to spruce (7%).

Beech-coniferous-mixed forests (FDT 2)

At 77,736 ha (20.5%), beech coniferous-mixed forest types (Figure 4) are likely to be appropriate silvicultural options for responding to climate change.

Reduced risk spruce (FDT 3) and silver fir forests (FDT 4)

The results showed that risk-lessened spruce (Figure 4) and silver fir forests (Figure 4) will cover 21,053 ha (5.6%) and 8112 ha (2.1%) respectively. In contrast to reduced risk spruce FDT, silver fir forests (FDT4) are likely to occur mostly on sites where beech and spruce are not suitable but where silver fir is biologically important.

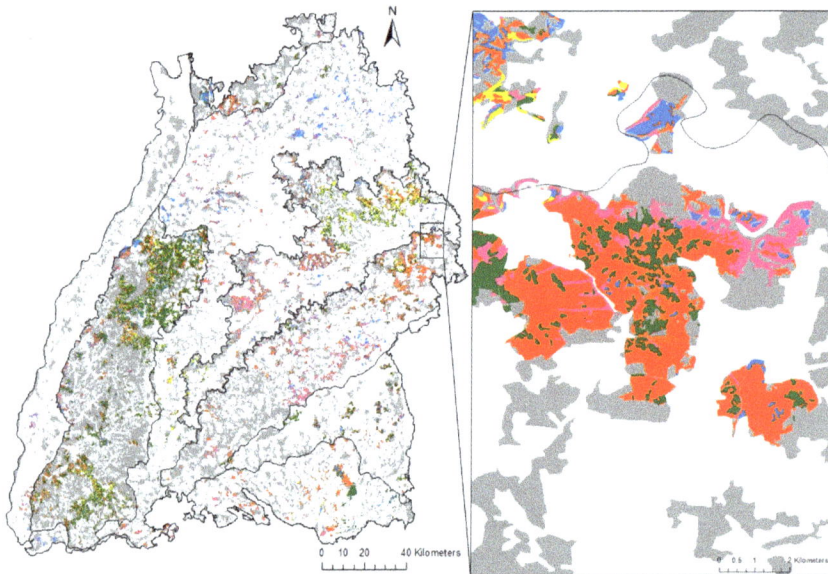

Figure 4. Location of the Forest Development Types (FDTs) in 2050 in BW state forests. In green: Coniferous-beech-mixed forests (FDT 1); in red: Beech-coniferous-mixed forests (FDT 2); in yellow: Reduced risk spruce forests (FDT3); in brown: Silver fir forests (FDT 4); in blue: Oak-mixed forests (FDT5); in pink: Beech-broadleaved forests (FDT6); in grey: Whole BW forest.

Oak-mixed forests (FDT 5)

The oak-mixed forest type includes common oak and sessile oak forest types (Figure 4) with high proportions of oak. The total area covered by this FDT will amount to 38,782 ha (10.2%).

Beech-broadleaved-mixed forests (FDT 6)

The beech-broadleaved-mixed forests (Figure 4) contain a similar proportion of beech (40–80%) to that of beech-coniferous-mixed forests. This FDT will represent an area of 58,571 ha (15.4% of the study area).

Remaining forest area

The aforementioned FDTs (1–6) are expected to cover 82.8% of the state forest of BW. The remaining 17.2% of the state forest of BW will be covered by other forest types. The main coniferous tree species may be European larch (*Larix decidua*), scots pine and Douglas fir or maple, ash, basswood and cherry as examples of broadleaved species.

3.2. Slope and Soil Sensitivity

The forest area analyzed with respect to slope classes as described in Section 2.5 is shown in Figure 5. The majority of the forest areas (55%) were located on lowlands and easy slopes. The proportion of soils sensitive to traffic was 33.5% with a majority of these located on lowlands and easy slopes (25.5%) with limited off-road traffic (Figure 5). Areas with greater terrain slope had lower proportions of soils sensitive to machine traffic.

This was supported by a GIS analysis showing that lowland sites suitable for broadleaved tree species were often more sensitive to traffic than hilly sites suitable for conifers, as shown in Table 5 for FDT 3, FDT 5, and FDT 6 (Reduced risk spruce, Oak-mixed and Beech-broadleaved-mixed FDTs, respectively).

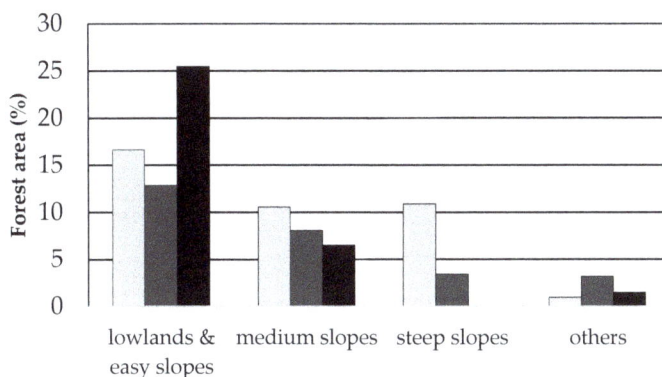

Figure 5. Distribution of the analyzed forest area with regard to topography and soil sensitivity to traffic (light grey: insensitive, dark grey: partly sensitive and black: sensitive), as a percentage of total forest area.

Table 5. Distribution of slope and soil sensitivity classes for three Forest Development Types: Risk reduced spruce forests (FDT 3), oak-mixed forests (FDT 5), and beech-broadleaved-mixed forests (FDT 6), in ha and %. The percentages are related to the area of each FDT respectively.

Slope Class	Soil Sensitivity	FDT 3 (Spruce), in ha	FDT 5 (Oak), in ha	FDT 6 (Beech), in ha
Lowlands and Easy slopes	Insensitive	6965 (33%)	3898 (10%)	2302 (4%)
	Partly Sensitive	2124 (10%)	2478 (6%)	11,670 (20%)
	Sensitive	1654 (8%)	18,012 (46%)	15,246 (26.0%)
Medium Slopes	Insensitive	3404 (16%)	5012 (13%)	2643 (4%)
	Partly Sensitive	1336 (6%)	2543 (7%)	5845 (10%)
	Sensitive	927 (4%)	2193 (6%)	5225 (9%)
Steep slopes and Others	Insensitive	2813 (14%)	3208 (8%)	9343 (16%)
	Partly Sensitive	1516 (7%)	1079 (3%)	2992 (5%)
	Sensitive Sum (ha)	314 (2%) 21,053	359 (1%) 38,782	3305 (6%) 58,571

3.3. Harvesting and Extraction Operations

Timber harvesting operations depend on varying factors such as forest type including species, tree dimension and quality, terrain slope and soil sensitivity. For the six selected FDTs, main harvesting and extraction methods (L) were determined (Figure 6 and Table 6), resulting in six different harvesting systems. Three different systems were applicable in each FDT. Overall results showed that a cable yarding system (L1) would be used on 29.1% (314,139 ha) of the total forest area. In the steep terrains, this figure would be 22.8%, and in terrain with medium slope, 6.3% (Table 6). It is very likely that the higher proportion of broadleaved trees in future forests will lead to an increased use of forestry tractors, because of both mandatory manual felling with optional cable support, and the limited trafficability of these sites. Operations will be conducted from tractor roads (23.0%) on sensitive soils and on medium slopes, whereas on insensitive soils, tractors will be operated on skid trails (18.4%) (Table 6). The combination harvester-forwarder could still be used in forests (i) with high proportions of conifer trees; (ii) with skid trail distances of 40 m when supported by chainsaw-felling (15.0%), or (iii) with skid trail distances of 20 m (4.0%) as a fully mechanized system (Table 6). On more sensitive soils, a combination of harvester and ground carriage or cable yarder is recommended (10.5%) (Table 6).

Table 6. Proportion of different felling and hauling systems (in %) applied in six selected Forest Development Types (FTDs) in Baden-Württemberg (BW). TR: tractor road, ST: skid trail.

Timber Harvesting and Extraction System	Used Abbreviation	Forest Area (%)
Chainsaw & Cable Yarder	L1	29.1
Chainsaw & Forestry tractor + Forwarder (TR)	L2	23.0
Chainsaw & Forestry tractor + Forwarder (ST)	L3	18.4
Chainsaw + Harvester & Forwarder (ST)	L4	15.0
Chainsaw + Harvester & Ground carriage/cable yarder	L5	10.5
Harvester & Forwarder (ST)	L6	4.0

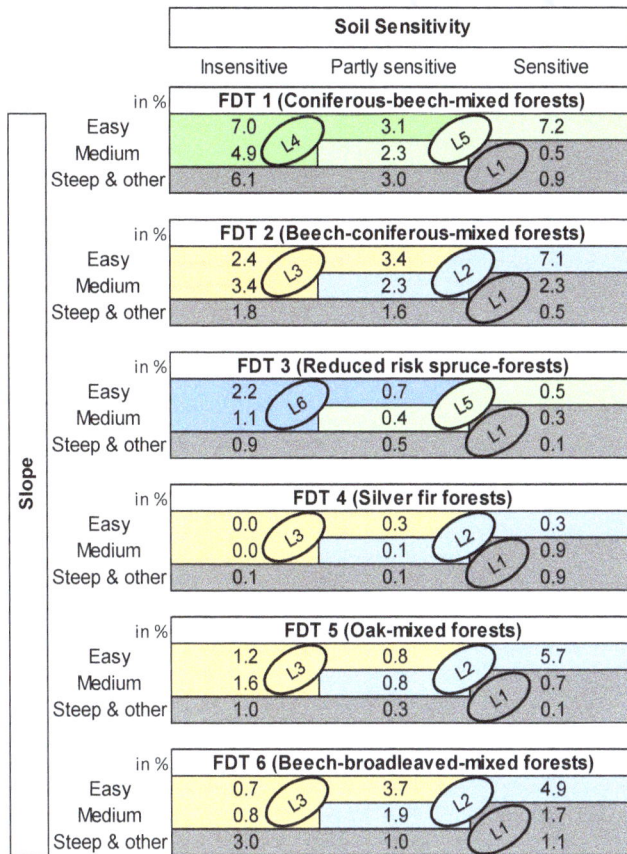

Slope	Soil Sensitivity		
	Insensitive	Partly sensitive	Sensitive
FDT 1 (Coniferous-beech-mixed forests) in %			
Easy	7.0 (L4)	3.1 (L5)	7.2
Medium	4.9	2.3	0.5
Steep & other	6.1	3.0 (L1)	0.9
FDT 2 (Beech-coniferous-mixed forests) in %			
Easy	2.4 (L3)	3.4 (L2)	7.1
Medium	3.4	2.3	2.3
Steep & other	1.8	1.6 (L1)	0.5
FDT 3 (Reduced risk spruce-forests) in %			
Easy	2.2 (L6)	0.7 (L5)	0.5
Medium	1.1	0.4	0.3
Steep & other	0.9	0.5 (L1)	0.1
FDT 4 (Silver fir forests) in %			
Easy	0.0 (L3)	0.3 (L2)	0.3
Medium	0.0	0.1	0.9
Steep & other	0.1	0.1 (L1)	0.9
FDT 5 (Oak-mixed forests) in %			
Easy	1.2 (L3)	0.8 (L2)	5.7
Medium	1.6	0.8	0.7
Steep & other	1.0	0.3 (L1)	0.1
FDT 6 (Beech-broadleaved-mixed forests) in %			
Easy	0.7 (L3)	3.7 (L2)	4.9
Medium	0.8	1.9	1.7
Steep & other	3.0	1.0 (L1)	1.1

Figure 6. Proportion (in %, relative to the total area) of the application area of different logging operations (L) regarding the Forest Development Types (FDTs), where any insensitive soil class indicates that traffic is possible (ground-based forest operations on skid trails); a partly sensitive soil class indicates that low traffic is possible (ground-based forest operations on tractor roads); and a sensitive class indicates that traffic is not possible (roadways/cable yarder forest operations) (c.f. Figure 3). L1 to L6 as described in Table 6.

4. Discussion

4.1. Advantages of Applied Methods

The use of GIS to identify Forest Development Types that will be relevant in the future turned out to be a powerful planning tool. GIS has great potential, given that data are increasingly available in digital format, and that data queries can be conducted quickly and large areas can be easily included in analyses. To the best of our knowledge, no study has addressed the future location of FDTs. Most of the assessments found in literature do not consider the actual location of future forests when estimating biomass potentials, interpreting qualitative changes in forests, or developing suitable harvesting operation strategies. By including the location, it was possible for the first time to derive harvesting systems that consider local constraints, such as soil trafficability and topography, for these forest types. We managed to quantify the expected changes in forest operations, to show trends in timber harvesting operations and to make recommendations for stakeholders.

4.2. Tree Species Composition

Climate envelope models conducted for BW show the lability of spruce forests to future climatic conditions in the lowlands [7,17]. This is one of the reasons why the proportion of spruce trees has been continuously reduced over the last 30 years as a management objective: the proportion declined from 43.5% in 1987 to 34.0% in 2012 for all forest land in BW [56]. One main policy objective of Baden-Württemberg is to achieve equal proportions of coniferous and broadleaved tree species at least in state forests [4]. Therefore, this study considers that in future forests spruce, silver fir, oaks, and beech will together represent more than 80% of the state forests of BW. Our results showed that coniferous species (spruce and silver fir) would be the dominant species in 37% of future forests whereas broadleaved species (beech and oak) would be dominant in 46%. This is clearly in line with results from Reif [4], who estimated that spruce and silver fir would grow on 39%, beech on 31% and oak on 8% of the future forest land base. Other authors have used climate envelope models to predict the future range and shifts of single tree species (mostly spruce and beech) in Europe, in Germany and in BW [6,7,17,82]. Our results showed that there would be a continuation of the trend of decreasing shares of conifers, as the proportion of forest types dominated by spruce or silver fir in state forest in BW is predicted to decline from 47.7% [83] to 37% by 2050. This situation will likely be a major issue for the wood processing industry, which mostly depends on softwood [84]. Softwood timber is of particular importance to the European construction sector [85]. The use of timber for structural products shows the greatest potential for global climate change mitigation compared to any other use of wood [86], which is mainly due to the substitution of carbon-intensive materials [30] and the long-term carbon storage of construction wood products [84].

Our results showed that the area covered by broadleaved trees will likely increase in the future. Several studies have shown the benefits of admixture of broadleaved species in conifer stands as they play a key role in forest stability and adaptation of forests to pathogens, storms and climate change. It is generally agreed that mixed stands are more resistant to biotic and abiotic disturbances [30,87] because "with an increasing number of functionally different species, the probability increases that some of these species can resist external disturbances or changing environmental conditions" [88]. Native European deciduous trees tend to be less affected by climate change than conifers [6]. In particular, the susceptibility of Norway spruce to natural hazards is much greater than that of beech, the most common tree species in BW. Moreover, a significant reduction of the financial risk can be achieved by mixing large blocks of broadleaved species with conifer stands [87]. In line with the current management strategy, beech and oaks are increasingly admixed in spruce and pine forests [30].

Non-native tree species are likely to become more important economically. Douglas fir, red oak (*Quercus rubra* L.) and Japanese larch (*Larix kaempferi* Carr.) have good adaption potential as climate conditions in their native growing regions are similar to those predicted for parts of Germany (Hickler et al., 2012). Douglas fir is often seen as a promising silvicultural option. This is supported by the German Federation of Forest Research Institutes (DVFFA) [89], whereas the German Federal Agency for Nature Protection (BfN) classified Douglas fir as an invasive species resulting in recommendations for limiting its growing area [90].

4.3. Adaption of Species Composition

The study focuses on a temperature increase of around 2 °C by 2050. Simulations and climate projections for the end of the 21st century showed that global warming will probably lie between 1.7 and 4.4 °C [91]. The higher the temperature increase, the more uncertain the predictions of tree species, FDTs and adapted forest operations are. If the temperature increase exceeds 2 °C, the adaption of existing forests will become limited [16,92]. Research on phenotypical plasticity at the single tree level, as well as on evolutionary adaption at a population level is needed.

The objective of changing forest structure and composition in order to increase the resistance and resilience of forests, is a challenge considering the long rotations under current management practices. The development towards more mixed woodland is a long-term process, and many stands first need to grow to an age where the forest can be converted [25]. The transition from current forests to the FDTs should be a fluent transition with adaptive management. The aim is to enable the FDTs to respond to change. The response option facilitates the transition to new conditions [12]. This study focused on the timber harvesting and extraction systems from future FDTs, and the transition state were not considered. The adaptive transition management has to be defined on a local scale, depending on current species composition and future FDTs. For some specific cases, Forst BW published silvicultural guidelines [24].

For private forests, the related costs and efforts could hinder the owners to apply adaptive management. Moreover, as the current income from conifer forests is much higher than those from broadleaved forests, private owners may be reluctant to adopt new management practices. Since 36% of BW forest land belongs to private owners [56], it is essential to support private owners for the success of climate change adaptation policies. The shift in the species composition within the time frame 2050–2100 will not be achieved on all forest areas, but it should be seen as a goal to initiate adaptive management no later than 2050.

4.4. Topography

Our findings showed that 55% of the forest area is located within the lowlands and easy slopes class. Previously, the third national forest inventory from 2012 (BWI^3) found that 73% of the whole forest area in BW has a terrain slope below 30% [29]. This difference may result from the mapping methodology. The guideline [48] defines that lowlands and easy slopes have a maximum inclination of 30%; however, the mapper can overrule the classification and move the topographic class "lowlands and easy slope" into the class "medium slope". Differences also occurred regarding steep slopes. According to the BWI^3, 18% of BW forest land area has a slope above 41%. In comparison, the data of our study assessed that approximately 24% of the area was on steep slope. It is obvious that the mapper over-evaluated the slopes in order to include site specificities that hamper harvesting operations. This is considered advantageous, since the resulting trafficability becomes more realistic.

4.5. Soil Sensitivity

The use of heavy machinery in forest management has significantly increased. It enhances productivity, reduces occupational health and safety risks, and lessens stand damage, but may seriously damage forest soils [64,93]. Soil compaction is a major cause of human-induced forest soil degradation [68] and "has been considered a principal form of damage associated with logging, restricting root growth and reducing productivity" [94]. Soil protection is becoming increasingly important [95], and even more soil protection will be needed in order to ensure a sustainable long-term wood supply with aggravating weather conditions [34] such as shorter frozen ground periods and higher precipitation are expected in winter, and this will have an impact on low risk traffic possibilities [96]. Conducting timber harvesting and extraction operations in late summer/early fall could offer low risk traffic opportunities [96].

Small-scaled mapping of soil types and sensitivity assists in the choice of adapted harvesting and extraction methods. Our results concerning soil sensitivity showed that around 30% of forest soils in BW could be considered sensitive to traffic, which is in line with further literature: Berleth et al. [97] even figured a proportion of 41%.

Soil water balance was indirectly included in the analysis as an input parameter for the species suitability map. However, water balance might be a useful additional layer to more precisely determine soil sensitivity as "the severity of compaction caused by forest machinery is greatly influenced by soil water content" [98].

4.6. Changes in Timber Harvesting and Extraction Systems

There is quite a variety of timber harvesting and extractions systems applied in BW. Nevertheless, the BWI^3 quantified the forest area according to forest operation conditions. According to the BWI^3, on 73.6% of the state forest area, any timber harvesting and extraction system can be used [29]. Furthermore, on 8.7% of the forest state area, only machines dedicated for steep terrain can operate and on 15.1%, no off-road traffic is possible [29]. These values do not consider tree species or tree DBH. To assess actual timber harvesting and extraction systems we had to make the following assumptions on terrain where off-road traffic was possible: (i) broadleaved trees are felled motor-manually and extracted by skidders and (ii) conifers are felled by harvesters and extracted by forwarders. Considering these tree species-based assumption, we were able to give a rough overview of current forest operations. Motor-manual systems with ground-based timber extraction would be used on 45%, harvester-forwarder systems on 40%, and motor-manual systems with cable yarder extraction on 15% of the state forest area.

Following these assumptions, the area harvested using harvesters will decrease from 40 to 30% and extraction based on forwarders even more from 40 to 19% in future. Compared to our results, the importance by area of cable yarding systems will nearly double from 15 to 29%, whereas the area with motor-manual felling and skidder extraction will remain quite constant.

4.7. Machinery

The changes in tree species composition and the growing awareness of forest soil protection may induce major technical changes for harvesting and extraction machines. The increasing number of heavy machinery, especially harvesters and forwarders, could easily lead to an over-capacity [99]. On the other hand, new technologies such as the lowland-cable-yarder, six- and eight-wheeled forestry tractors, tire width and inflation pressure, as well as cable-assist systems, help to reduce negative impacts on soil [38,100,101]. Nevertheless, such technologies (especially the lowland-cable-yarder and cable-assist systems) still need to demonstrate economic feasibility. Cable-assist systems are becoming increasingly common, but only a few studies have been published, and "the actual implementation and understanding of its limitations, is in its infancy" [102]. Therefore, and because "no European country has yet implemented specific cable-assist rules" [102], we were not able to incorporate cable-assist systems for steep slopes and sensitive soils in our study. The use of cable-assist systems could replace cable yarding operations on some sites. Therefore our estimate of cable yarding proportions could be overestimated.

The higher proportions of broadleaved trees may also change the usage of machinery on site. For safety reasons, it is useful to use a winch to support the felling process by pulling down the trees. This allows a controlled felling of the tree in the planned felling direction. Winches, small forestry crawlers, or forestry tractors equipped with winches can be used, and might become more popular for logging operations. Therefore, we recommend the use of special forestry tractors with winches. Finally, the use of harvesters for the felling and processing of both conifers and broadleaved trees may improve felling productivity, as well as work safety [103].

For future harvesting and extraction methods, it may be possible that autonomous or semi-autonomous systems will become popular. Autonomous forwarders are believed to have considerable commercial potential as they are more profitable [104,105]. However, some technical challenges are still associated with automating machines [106]. Potential is therefore seen for autonomous direct-loading systems, where a conventional harvester places processed trees directly into the bank of an autonomous forwarder [106] which could be a ground carriage system such as, for example, the ground carriage Pully developed by Konrad (Konrad Forsttechnik GmbH, Preitenegg, Austria) [107].

5. Conclusions and Outlook

In order to discuss the impact of climate change on forest structures and future harvesting operations and to map the location of the FDTs, a regional case study was performed using GIS. It was possible to prescribe distinct operations for different areas and types of forests in the state forests of the BW region. The analysis showed that, within the time horizon 2050–2100, coniferous-beech-mixed forests will probably be the main forest type in the state forest of BW, covering 29% of the total forest area. Moreover, continuous tree cover (the German "Dauerwald" concept [62]) will be applied to at least 67% of the forest area, which will certainly lead to changes in forest operations. Using trafficability classes, which are dependent on terrain slope and forest soil sensitivity to traffic, it was possible to provide recommendations regarding harvesting and extraction systems for each FDT. Data availability of current forest composition coupled with current forest management practices would be a nice asset for comparing our results to present forest operations.

Among the analyzed FDTs, 50% of the area with slopes lower than 30% has soils that are extremely sensitive to traffic. Additionally, 23% will be in terrain with slopes higher than 45% where traffic faces technical limitations. Thus, there is a strong requirement for technical developments in forest operations, especially in extraction methods such as the lowland-cable-yarders, improved forestry tractors, and forwarders to reduce ground pressure and slippage, and cable-assist systems as well as light autonomous systems. Increasing the ratio of broadleaved forests with continuous tree cover will probably increase the number of motor-manual felling operations, which in turn leads to an increased demand for manpower in the forestry sector. Also given the fact that new machines require highly specialized machine operators, it might become a challenge to acquire enough qualified forestry workers.

Future harvesting and extraction operations in more structured forests will become multifaceted through a combination of machines and manpower. These complex work systems increase the risk of accident for forestry workers when compared with fully mechanized systems which are actually linked with the lowest number of reported accidents. Enhanced and intensified work safety training and instruction guides should be developed for future timber harvesting and extraction systems.

For future research, tree-level growth models should be developed in order to improve resolution and degree of detail. This study could be used as a basis for the application of different scenarios. Through the simulation of different management strategies for each FDT and trafficability class, the prospective harvest assortments could be described with greater accuracy. Thus, in forest operations with varying harvesting intensities, assortments and machinery need to be evaluated with regard to GHG emissions.

Acknowledgments: This study was undertaken in the framework of the project "SOLVE" (Timber harvesting and transportation systems adapted to altered forest structures due to climate change), which is funded by the Federal Ministry of Food and Agriculture (BMEL) and the Federal Ministry for the Environment, Nature Conservation, Building and Nuclear Safety (BMUB) in the frame of the "Förderrichtlinie Waldklimafond" (Förderkennzeichen 28W-B-3-048-01). The UMR LERFoB is supported by a grant overseen by the French National Research Agency (ANR) as part of the "Investissements d'Avenir" program (ANR-11-LABX-0002-01, Lab of Excellence ARBRE). The article processing charge was funded by the German Research Foundation (DFG) and the University of Freiburg through the funding program Open Access Publishing.

Author Contributions: F.B., J.S. and D.J. conceived and designed the experiments; F.B. performed the experiments; F.B. and J.S. analyzed the data; M.F. contributed analysis tools; F.B. and J.S. wrote the paper with contributions from M.F. and D.J.

Conflicts of Interest: The authors declare no conflict of interest. The founding sponsors had no role in: the design of the study; the collection, analyses, or interpretation of data; the writing of the manuscript, or in the decision to publish the results.

References and Note

1. FAO. *Climate Change Guidelines for Forest Managers*; FAO Forestry Paper; FAO: Rome, Italy, 2013; Volume 172.
2. Kaspar, F.; Mächel, H. Beobachtung von Klima und Klimawandel in Mitteleuropa und Deutschland. In *Klimawandel in Deutschland*; Brasseur, G.P., Jacob, D., Schuck-Zöller, S., Eds.; Springer: Berlin/Heidelberg, Germany, 2017; pp. 17–26.
3. Kartschall, K.; Mäder, C.; Tambke, J. *Klimaänderungen, Deren Auswirkungen und was für den Klimaschutz zu Tun Ist*; Umweltbundesamt: Dessau, Germany, 2007.
4. Reif, A.; Brucker, U.; Kratzer, R.; Schmiedinger, A.; Bauhus, J. *Waldbau und Baumartenwahl in Zeiten des Klimawandels aus Sicht des Naturschutzes. Abschlussbericht eines F+E-Vorhabens im Auftrag des Bundesamtes für Naturschutz FKZ 3508 84 0200*; BfN: Bonn, Germany, 2010.

5. Hickler, T.; Bolte, A.; Harard, B. Folgen des Klimawandels für die Biodiversität in Wald und Forst. In *Klimawandel und Biodiversität—Folgen für Deutschland*; Mosbrugger, V., Brasseur, G., Schaller, M., Stribrny, B., Eds.; WBG Wiss. Buchges: Darmstadt, Germany, 2012; pp. 64–221.

6. Kölling, C.; Zimmermann, L. Die Anfälligkeiten der Wälder Deutschlands gegenüber dem Klimawandel. *Gefahrenstoffe Reinhalt. Luft* **2007**, *67*, 259–268.

7. Nothdurft, A.; Wolf, T.; Ringeler, A.; Böhner, J.; Saborowski, J. Spatio-temporal prediction of site index based on forest inventories and climate change scenarios. *For. Ecol. Manag.* **2012**, *279*, 97–111. [CrossRef]

8. Unseld, R. *Anpassungsstrategie Baden-Württemberg an die Folgen des Klimawandels. Fachgutachten für das Handlungsfeld Wald und Forstwirtschaft*; Ministerium für Umwelt, Klima und Energiewirtschaft Baden-Württemberg: Stuttgart, Germany, 2013.

9. Saha, S.; Bauhus, J. Trade-offs between climate change adaptation and mitigation objectives for forests in south-western Germany. *Forstarchiv* **2016**, *87*, 60–61.

10. Thompson, I.; Mackey, B.; McNulty, S.; Mosseler, A. *Forest Resilience, Biodiversity, and Climate Change. A Synthesis of the Biodiversity/Resilience/Stability Relationship in Forest Ecosystems*; Secretariat of the Convention on Biological Diversity: Montreal, QC, Canada, 2009.

11. Holling, C.S. Resilience and Stability of Ecological Systems. *Annu. Rev. Ecol. Syst.* **1973**, *4*, 1–23. [CrossRef]

12. Millar, C.I.; Stephenson, N.L.; Stephens, S.L. Climate change and forests of the future: Managing in the face of uncertainty. *Ecol. Appl.* **2007**, *17*, 2145–2151. [CrossRef] [PubMed]

13. Pimm, S.L. The complexity and stability of ecosystems. *Nature* **1984**, *307*, 321–326. [CrossRef]

14. Halpern, C.B. Early Successional Pathways and the Resistance and Resilience of Forest Communities. *Ecology* **1988**, *69*, 1703–1715. [CrossRef]

15. Dormann, C.F. Promising the future?: Global change projections of species distributions. *Basic Appl. Ecol.* **2007**, *8*, 387–397. [CrossRef]

16. Kölling, C. Wälder im Klimawandel: die Forstwirtschaft muss sich anpassen. In *Warnsignal Klima: Gesundheitsrisiken; Gefahren für Menschen, Tiere und Pflanzen; Wissenschaftliche Fakten; 37 Tabellen*, 2nd ed.; Lozán, J.L., Ed.; Wiss. Auswertungen: Hamburg, Germany, 2014.

17. Hanewinkel, M.; Hummel, S.; Cullmann, D.A. Modelling and economic evaluation of forest biome shifts under climate change in Southwest Germany. *For. Ecol. Manag.* **2010**, *259*, 710–719. [CrossRef]

18. Hanewinkel, M.; Cullmann, D.A.; Michiels, H.-G.; Kandler, G. Converting probabilistic tree species range shift projections into meaningful classes for management. *J. Environ. Manag.* **2014**, *134*, 153–165. [CrossRef] [PubMed]

19. Zimmermann, N.; Jandl, R.; Hanewinkel, M.; Kunstler, G.; Klling, C.; Gasparini, P.; Breznikar, A.; Meier, E.S.; Normand, S.; Ulmer, U.; et al. Potential Future Ranges of Tree Species in the Alps. In *Management Strategies to Adapt Alpine Space Forests to Climate Change Risks*; Cerbu, G.A., Ed.; InTech: Rijeka, Croatia, 2013.

20. Bolte, A.; Ammer, C.; Löf, M.; Madsen, P.; Nabuurs, G.-J.; Schall, P.; Spathelf, P.; Rock, J. Adaptive forest management in central Europe: Climate change impacts, strategies and integrative concept. *Scand. J. For. Res.* **2009**, *24*, 473–482. [CrossRef]

21. Witt, A.; Furst, C.; Frank, S.; Koschke, L.; Makeschin, F. Regionalisation of climate change sensitive forest development types for potential afforestation areas. *J. Environ. Manag.* **2013**, *127*, 48–55. [CrossRef] [PubMed]

22. Larsen, J.B.; Nielsen, A.B. Nature-based forest management—Where are we going? *For. Ecol. Manag.* **2007**, *238*, 107–117. [CrossRef]

23. Larsen, J.B. Close-to-Nature Forest Management: The Danish Approach to Sustainable Forestry. In *Sustainable Forest Management: Current Research*; García, J.M., Diez Casero, J.J., Eds.; InTech: Rijeka, Croatia, 2012.

24. ForstBW. *Richtlinie Landesweiter Waldentwicklungstypen*; Landesbetrieb Forst Baden-Württemberg, Ministerium für Ländlichen Raum und Verbraucherschutz Baden-Württemberg: Stuttgart, Germany, 2014.

25. Lower Saxony State Forest. *The LÖWE Programme. 20 Years of Long-Term Ecological Forest Development*; Lower Saxony State Forest, HenryN: Braunschweig, Germany, 2012.

26. Baumann, M. *Waldentwicklungstypen. Grundlage für Eine Dynamische Waldbaustrategie*; Pirna OT Graupa: Graupa, Germany (Staatsbetrieb Sachsenforst), 2014.

27. Schmitz, F.; Polley, H.; Henning, P. *Der Wald in Deutschland. Ausgewählte Ergebnisse der Dritten Bundeswaldinventur*; Bundesministerium für Ernährung und Landwirtschaft: Berlin, Germany, 2014.

28. Berg, S.; Schweier, J.; Brüchert, F.; Lindner, M.; Valinger, E. Economic, environmental and social impact of alternative forest management in Baden-Württemberg (Germany) and Västerbotten (Sweden). *Scand. J. For. Res.* **2014**, *29*, 485–498. [CrossRef]

29. Johann Heinrich von Thünen-Institut. Dritte Bundeswaldinventur: BWI[3]. Available online: https://bwi.info (accessed on 27 July 2017).

30. Fares, S.; Mugnozza, G.S.; Corona, P.; Palahi, M. Sustainability: Five steps for managing Europe's forests. *Nature* **2015**, *519*, 407–409. [CrossRef] [PubMed]

31. ForstBW. *Vielfältig Multifunktional Naturnah. Nachhaltigkeit im Staatswald Baden-Württemberg, Bericht*; Landesbetrieb Forst Baden-Württemberg: Stuttgart, Germany, 2014.

32. Bolte, A.; Eisenhauer, D.-R.; Ehrhart, H.-P.; Groß, J.; Hanewinkel, M.; Kölling, C.; Profft, I.; Rohde, P.; Amereller, K. Klimawandel und Forstwirtschaft: Übereinstimmungen und Unterschiede bei der Einschätzung der Anpassungsnotwendigkeiten und Anpassungsstrategien der Bundesländer. *vTI Appl. Agric. For. Res.* **2009**, *59*, 269–278.

33. Susnjar, M.; Horvat, D.; Seselj, J. Soil compaction in timber skidding in winter conditions. *Croat. J. For. Eng.* **2006**, *27*, 3–15.

34. Thees, O.; Olschewski, R. Ökonomische Überlegungen zum physikalischen Bodenschutz im Wald. In *Forum für Wissen 2013: Bodenschutz im Wald: Ziele—Konflikte—Umsetzung*; Eidgenössische Forschungsanstalt WSL: Birmensdorf, Switzerland, 2013; pp. 31–43.

35. Korn, H.; Schliep, R.; Stadler, J. *Biodiversität und Klima-Vernetzung der Akteure in Deutschland III—Ergebnisse und Dokumentation des 3. Workshops*; BfN: Bonn, Germany, 2008.

36. Gebhardt, H.; Höpker, K. *Klimawandel in Baden-Württemberg. Fakten—Folgen—Perspektiven, 3. Aktualisierte Aufl., Juni 2015*; LUBW Landesanstalt für Umwelt Messungen und Naturschutz Baden-Württemberg; Ministerium für Umwelt, Klima und Energiewirtschaft Baden-Württemberg: Stuttgart, Germany, 2015.

37. Erte, C.; Knoche, D. *Radialzuwachs von Traubeneichen und Gemeiner Kiefer bei Sich Ändernden Klimaverhältnissen*; Landeskompetenzzentrum Forst: Eberswalde, Germany, 2015.

38. Lüscher, P.; Frutig, F.; Sciacca, S. *Physikalischer Bodenschutz im Wald. Bodenschutz Beim Einsatz von Forstmaschinen*, 2nd ed.; Eidgenössische Forschungsanstalt WSL: Birmensdorf, Switzerland, 2010.

39. Aust, C.; Schweier, J.; Brodbeck, F.; Sauter, U.H.; Becker, G.; Schnitzler, J.-P. Land availability and potential biomass production with poplar and willow short rotation coppices in Germany. *GCB Bioenergy* **2014**, *6*, 521–533. [CrossRef]

40. Abolina, E.; Volk, T.A.; Lazdina, D. GIS based agricultural land availability assessment for the establishment of short rotation woody crops in Latvia. *Biomass Bioenergy* **2015**, *72*, 263–272. [CrossRef]

41. Beccali, M.; Columba, P.; D'Alberti, V.; Franzitta, V. Assessment of bioenergy potential in Sicily: A GIS-based support methodology. *Biomass Bioenergy* **2009**, *33*, 79–87. [CrossRef]

42. Viana, H.; Cohen, W.B.; Lopes, D.; Aranha, J. Assessment of forest biomass for use as energy. GIS-based analysis of geographical availability and locations of wood-fired power plants in Portugal. *Appl. Energy* **2010**, *87*, 2551–2560. [CrossRef]

43. Zambelli, P.; Lora, C.; Spinelli, R.; Tattoni, C.; Vitti, A.; Zatelli, P.; Ciolli, M. A GIS decision support system for regional forest management to assess biomass availability for renewable energy production. *Environ. Model. Softw.* **2012**, *38*, 203–213. [CrossRef]

44. López-Rodríguez, F.; Atanet, C.P.; Blázquez, F.C.; Celma, A.R. Spatial assessment of the bioenergy potential of forest residues in the western province of Spain, Caceres. *Biomass Bioenergy* **2009**, *33*, 1358–1366. [CrossRef]

45. Fernandes, U.; Costa, M. Potential of biomass residues for energy production and utilization in a region of Portugal. *Biomass Bioenergy* **2010**, *34*, 661–666. [CrossRef]

46. Kühmaier, M.; Stampfer, K. Development of a Multi-Attribute Spatial Decision SUpport System in Selecting Timber Harvesting Systems. *Croat. J. For. Eng.* **2010**, *31*, 75–88.

47. Gulci, N.; Akay, A.; Erdas, O.; Wing, M.; Sessions, J. Planning optimum logging operations through precision forestry approach. *Eur. J. For. Eng.* **2015**, *1*, 56–60.

48. FVA-BW. *Kartengrundlage. Staatswald Baden-Württemberg*; Forstliche Versuchs- und Forschungsanstalt Baden-Württemberg: Freiburg, Germany, 2016.

49. Kayser, J. *Befahrungsempfindlichkeit und Feinerschließung*; IDaMa GmbH: Freiburg, Germany, 2016.

50. Aldinger, E.; Michiels, H.-G. Baumarteneignung in der forstlichen Standortskartierung Baden-Württemberg. *AFZ DerWald* **1997**, *52*, 234–238.

51. Hanewinkel, M.; Cullmann, D.; Michiels, H.-G. Künftige Baumarteneignung für Fichte und Buche in Südwestdeutschland. *AFZ DerWald* **2010**, *19*, 30–33.

52. Michiels, H.-G. *Matrix für die Bewertung der Baumarteneignung*; Forstliche Versuchs- und Forschungsanstalt Baden-Württemberg: Freiburg, Germany, 2017.

53. MKULNV. *Wald und Waldmanagement im Klimawandel. Anpassungsstrategie für Nordrhein-Westfalen*, 2nd ed.; Ministerium für Klimaschutz, Umwelt, Landwirtschaft, Natur- und Verbraucherschutz des Landes Nordrhein-Westfalen: Düsseldorf, Germany, 2015.

54. Gadow, K.V. *Forsteinrichtung. Analyse und Entwurf der Waldentwicklung*; Universitatsverlag: Göttingen, Germany, 2005.

55. European Environment Agency (EEA) under the framework of the Copernicus programme with funding by the European Union. CORINE Land Cover: CLC-2006. Available online: http://land.copernicus.eu/pan-european/corine-land-cover/clc-2006 (accessed on 26 July 2017).

56. Kändler, G.; Cullmann, D. *Der Wald in Baden-Württemberg. Ausgewählte Ergebnisse der Dritten Bundeswaldinventur*; Forstliche Versuchs- und Forschungsanstalt Baden-Württemberg: Freiburg, Germany, 2014.

57. SaarForst. *Richtlinie für die Bewirtschaftung des Staatswaldes im Saarland*; SaarForst Landesbetrieb: Saarbrücken, Germany, 2008.

58. BMEL. Holzeinschlag in den deutschen Bundesländern nach Holzartengruppen im Jahr 2015 (in 1.000 Kubikmeter). Available online: https://de.statista.com/statistik/daten/studie/151955/umfrage/holzeinschlag-in-deutschland-nach-holzart-und-bundeslaendern/ (accessed on 21 November 2016).

59. Landtag von Baden-Württemberg. Nutzungsorientierte Baumartenzusammensetzung in der Forstwirtschaft, 2016, Drucksache 16/547. Available online: https://www.landtag-bw.de/files/live/sites/LTBW/files/dokumente/WP16/Drucksachen/0000/16_0547_D.pdf (accessed on 17 March 2017).

60. Borchers, J. *Die Fichte—Neue Zukunft für Eine Verfemte Baumart*; Waldbauerntag Werl: Werl, Germany, 2011.

61. Roberge, J.-M.; Laudon, H.; Bjorkman, C.; Ranius, T.; Sandstrom, C.; Felton, A.; Stens, A.; Nordin, A.; Granstrom, A.; Widemo, F.; et al. Socio-ecological implications of modifying rotation lengths in forestry. *Ambio* **2016**, *45* (Suppl. S2), 109–123. [CrossRef] [PubMed]

62. Pukkala, T. Plenterwald, Dauerwald, or clearcut? *For. Policy Econ.* **2016**, *62*, 125–134. [CrossRef]

63. Hartmann, M.; Niklaus, P.A.; Zimmermann, S.; Schmutz, S.; Kremer, J.; Abarenkov, K.; Luscher, P.; Widmer, F.; Frey, B. Resistance and resilience of the forest soil microbiome to logging-associated compaction. *ISME J.* **2014**, *8*, 226–244. [CrossRef] [PubMed]

64. Cambi, M.; Certini, G.; Neri, F.; Marchi, E. The impact of heavy traffic on forest soils: A review. *For. Ecol. Manag.* **2015**, *338*, 124–138. [CrossRef]

65. Labelle, E.; Jaeger, D.; Poltorak, B. Assessing the ability of hardwood and softwood brush mats to distribute applied loads. *Croat. J. For. Eng.* **2015**, *36*, 227–242.

66. Allman, M.; Jankovský, M.; Messingerová, V.; Allmanová, Z.; Ferenčík, M. Soil compaction of various Central European forest soils caused by traffic of forestry machines with various chassis. *For. Syst.* **2015**, *24*, e038. [CrossRef]

67. FVA-BW. *Richtlinie zur Feinerschliessung*; Forstliche Versuchs- und Forschungsanstalt Baden-Württemberg: Freiburg, Germany, 2003.

68. Cambi, M.; Certini, G.; Fabiano, F.; Foderi, C.; Laschi, A.; Picchio, R. Impact of wheeled and tracked tractors on soil physical properties in a mixed conifer stand. *iForest* **2016**, *9*, 89–94. [CrossRef]

69. Labelle, E.; Jaeger, D. Quantifying the use of brush matsin reducing forwarder peak loads andsurface contact pressures. *Croat. J. For. Eng.* **2012**, *33*, 249–274.

70. Kremer, J.; Frey, B.; Lüscher, P. Bodenstrukturveränderung oder Bodenschaden—Wo liegt die grenze? In *Walderschließung und Bodenschutz: Bodenverformung, Erosion, Hochwasserschutz*; Fakultät für Forst- und Umweltwissenschaften der Universität Freiburg, Forstliche Versuchs- und Forschungsanstalt Baden-Württemberg: Freiburg, Germany, 2009; pp. 39–46.

71. Solgi, A.; Najafi, A.; Sadeghi, S.H. Effects of traffic frequency and skid trail slope on surface runoff and sediment yield. *Int. J. For. Eng.* **2014**, *25*, 171–178. [CrossRef]

72. Jaafari, A.; Najafi, A.; Zenner, E.K. Ground-based skidder traffic changes chemical soil properties in a mountainous Oriental beech (Fagus orientalis Lipsky) forest in Iran. *J. Terramech.* **2014**, *55*, 39–46. [CrossRef]

73. Leszczynski, N.; Tomczak, A.; Kowalczuk, J.; Zarajcyk, J.; Wegrzyn, A.; Kocira, S.; Depo, K. The relationship between the mass of the harvester head and its maximum cutting diameter. *J. Res. Appl. Agric. Eng.* **2016**, *61*, 50–54.

74. Cacot, E.; Maire, L.; Chakroun, M.; Peuch, D.; Montagny, X.; Perrinot, C.; Bonnemazou, M. *La Mécanisation du Bucheronnage Dans le Peuplement de Feuillus. Synthèse Réalisée dans le Cadre du Projet ECOMEF Retenu à L'appel à Projet n°10 du FUI*; Institut technologique FCBA: Champs-sur-Marne, France, 2015.

75. Wirth, J.; Wolff, D. Vergleich von Pferde- und Seilschleppereinsatz beim Vorliefern von Vollbäumen. *AFZ DerWald* **2008**, *63*, 968–971.

76. Thiel, A. Imagepflege mit Pferd und Raupe: Die Landesforstverwaltung ist um eine gute Öffentlichkeitsarbeit bemüht. *Forstmasch. Profile* **2016**, *24*, 44–49.

77. Greulich, F.R.; Hanley, D.P.; McNeel, J.F.; Baumgartner, D. *A Primer for Timber Harvesting: EB 1316*; Washington State University: Pullman, WA, USA, 1999.

78. Gottlob, T. Pferd trifft Forwarder: Das Abtshagener Laubholzernteverfahren: Kombinierter Einsatz von Pferd und Tragrückeschlepper. *Starke Pferde* **2012**, *62*, 32–35.

79. Sündermann, J.; Schröder, J.; Röhe, P. *Bodenschonende Holzernte in Geschädigten Eschenbestände auf Nasstandorten. Erkenntnisse und Empfehlungen aus Fallstudien in Mecklenburg-Vorpommern*; Ministerium für Landwirtschaft, Umwelt und Verbraucherschutz Mecklenburg-Vorpommern: Schwerin, Germany, 2013.

80. *Waldgesetz für Baden-Württemberg*. Forest Act of Baden-Württemberg (Landeswaldgesetz) (*LWaldG*), 1995.

81. FSC. *Deutscher FSC-Standart 3.0. von der Mitgliedschaft am 28.6.2016 Verabschiedete Fassung*; FSC Deutschland: Freiburg, Germany, 2016.

82. Falk, W.; Hempelmann, N. Species Favourability Shift in Europe due to Climate Change: A Case Study for *Fagus sylvatica* L. and *Picea abies* (L.) Karst. Based on an Ensemble of Climate Models. *J. Climatol.* **2013**, *2013*, 1–18. [CrossRef]

83. Kändler, G. *Ergebnisse der BWI³ für den Staatswald BW*; Forstliche Versuchs- und Forschungsanstalt Baden-Württemberg: Freiburg, Germany, 2015.

84. BMEL. *Klimaschutz in der Land- und Forstwirtschaft sowie den Nachgelagerten Bereichen Ernährung und Holzverwendung*, 2nd ed.; Bundesministerium für Ernährung und Landwirtschaft: Bonn, Germany, 2016.

85. Eriksson, L.O.; Gustavsson, L.; Hänninen, R.; Kallio, M.; Lyhykäinen, H.; Pingoud, K.; Pohjola, J.; Sathre, R.; Solberg, B.; Svanaes, J.; et al. *Climate Implications of IncreasedWood Use in the Construction Sector: Towards an Integrated Modeling Framework. Arbetsrapport 257*; Swedish University of Agricultural Sciences: Umea, Sweden, 2009.

86. Lippke, B.; Wilson, J.; Meil, J.; Taylor, A. Characterizing the importance of carbon stored in wood products. *Wood Fiber Sci.* **2010**, *42*, 5–14.

87. Knoke, T.; Ammer, C.; Stimm, B.; Mosandl, R. Admixing broadleaved to coniferous tree species: A review on yield, ecological stability and economics. *Eur. J. For. Res.* **2008**, *127*, 89–101. [CrossRef]

88. Brang, P.; Spathelf, P.; Larsen, J.B.; Bauhus, J.; Bonc ina, A.; Chauvin, C.; Drossler, L.; Garcia-Guemes, C.; Heiri, C.; Kerr, G.; et al. Suitability of close-to-nature silviculture for adapting temperate European forests to climate change. *Forestry* **2014**, *87*, 492–503. [CrossRef]

89. Deutscher Forstwirtschaftsrat e.V. *Pressemitteilung 05/2015. Douglasie und Roteiche Sind Keine Invasiven Baumarten*; Deutscher Forstwirtschaftsrat: Berlin, Germany, 2015.

90. Bundesamt für Naturschutz. *BfN Kritisiert Schlussfolgerungen zur Studie zu Invasiven Baumarten. Forstliche Einstufungen zur Invasivität Sind Nicht mit Bundesnaturschutzgestz Vereinbar*; Bundesamt für Naturschutz: Bonn, Germany, 2015.

91. Schmidt, H.; Eyring, V.; Latif, M.; Rechid, D.; Sausen, R. Globale Sicht des Klimawandels. In *Klimawandel in Deutschland*; Brasseur, G.P., Jacob, D., Schuck-Zöller, S., Eds.; Springer: Berlin/Heidelberg, Germany, 2017; pp. 7–16.

92. Bolte, A.; Degen, B. Anpassung der Wälder an den Klimawandel: Optionen und Grenzen. *vTI Appl. Agric. For. Res.* **2010**, *60*, 111–118.

93. Enache, A.; Kühmaier, M.; Visser, R.; Stampfer, K. Forestry operations in the European mountains: A study of current practices and efficiency gaps. *Scand. J. For. Res.* **2015**, *31*, 412–427. [CrossRef]

94. Williamson, J.R.; Neilsen, W.A. The influence of forest site on rate and extent of soil compaction and profile disturbance of skid trails during ground-based harvesting. *Can. J. For. Res.* **2000**, *30*, 1196–1205. [CrossRef]

95.   Eisenhauer, D.-R.; Sonnemann, S. Silvicultural strategies under changing environmental conditions: Guiding principles, Target system and Forest development types. *For. Ecol. Landsc. Res. Nat. Conserv.* **2009**, *8*, 71–88.

96.   Zimmermann, L.; Hentzschel-Zimmermann, A.; Borchert, H.; Grimmeisen, W. *Das Warten auf den Bodenfrost im Wald*; 8 Bayerischer Waldbesitzertag, Freising-Weihenstephan; Bayerische Landesanstalt für Wald und Forstwirtschaft: Freising, Germany, 2011.

97.   Berleth, M.; Lelek, S.; Wolff, D. Wie sinnvoll sind weitere Gassenabstände? *AFZ DerWald* **2016**, *71*, 56–59.

98.   Labelle, E.R.; Jaeger, D. Soil Compaction Caused by Cut-to-Length Forest Operations and Possible Short-Term Natural Rehabilitation of Soil Density. *Soil Sci. Soc. Am. J.* **2011**, *75*, 2314. [CrossRef]

99.   Wippel, B.; Kastenholz, E.; Bacher-Winterhalter, M.; Storz, S.; Ebertsch, J. Praxisnahe Anhaltswerte für die mechanisierte Holzernte: Abschlussbericht. Available online: http://www.cluster-forstholz-bw.de/fileadmin/cluster/cluster_pdf/2015-10-20%20Bericht%20Praxisnahe%20Anhaltswerte.pdf (accessed on 23 February 2017).

100. Erler, J.; Duhr, M. Der Flachlandseilkran: ein Lösungsansatz zum Bodenschutz auf empfindlichen Standorten. *Der Dauerwald* **2016**, *54*, 12–20.

101. Heubaum, F. *Bodenschutz im Forstbetrieb Sachsenforst. Projekte zur Technologieerprobung*; Staatsbetrieb Sachsenforst: Pirna OT Graupa, Germany, 2015.

102. Visser, R.; Stampfer, K. Expanding Ground-based Harvesting onto steep terrain: A review. *Croat. J. For. Eng.* **2015**, *36*, 321–331.

103. Ferrari, E.; Spinelli, R.; Cavallo, E.; Magagnotti, N. Attitudes towards mechanized Cut-to-Length technology among logging contractors in Northern Italy. *Scand. J. For. Res.* **2012**, *27*, 800–806. [CrossRef]

104. Hellström, T.; Lärkeryd, P.; Nordfjell, T.; Ringdahl, O. Autonomous Forest Vehicles: Historic, envisioned, and state-of-the-art. *Int. J. For. Eng.* **2009**, *20*, 31–38.

105. Ringdahl, O. *Automation in Forestry. Development of Unmanned Forwarders*; Department of Computing Science, Umeå University: Umeå, Sweden, 2011.

106. Ringdahl, O.; Hellström, T.; Lindroos, O. Potentials of possible machine systems for directly loading logs in cut-to-length harvesting. *Can. J. For. Res.* **2012**, *42*, 970–985. [CrossRef]

107. Konrad. Ground Carriage Pully. Available online: http://www.forsttechnik.at/en/products/ground-carriage-pully.html (accessed on 7 July 2017).

Article

Can Biomass Quality Be Preserved through Tarping Comminuted Roadside Biomass Piles?

Suzanne Wetzel [1], Sylvain Volpe [2], Janet Damianopoulos [3] and Sally Krigstin [4],*

[1] Canadian Wood Fibre Centre, 580 Booth Street, Ottawa, ON K1A 0E4, Canada; suzanne.wetzel@canada.ca
[2] Scientist, FPInnovations, 570 boul. Saint-Jean, Pointe-Claire, QC H9R 3J9, Canada;
     sylvain.volpe@fpinnovations.ca
[3] Biologist, Canadian Wood Fibre Centre, 580 Booth Street, Ottawa, ON K1A 0E4, Canada;
     janet.damianopoulos@canada.ca
[4] Assistant Professor, Faculty of Forestry, University of Toronto, 33 Willcocks Street,
     Toronto, ON M5S 3B3, Canada
*   Correspondence: sally.krigstin@utoronto.ca; Tel.: +1-416-946-8507

Received: 1 August 2017; Accepted: 18 August 2017; Published: 23 August 2017

Abstract: Storage conditions play a vital role in maintaining biomass quality as a suitable bioenergy feedstock. Research has shown that biomass undergoes significant changes under different storage conditions and that these may influence its suitability for various biorefining and bioenergy opportunities. This study explores the effects of different tarp covers on the properties of stored-comminuted forest harvest residue from the Great Lakes St. Lawrence Forest. Characteristics of the biomass were evaluated upon harvesting and after one year in storage. The physical state of the different tarps used for pile coverage was monitored onsite. Results indicated that tarp material considerably affects micro-climatic conditions inside piles, yielding variation in the characteristics of stored biomass over the storage period. While plastic based tarps were easier to work with and lasted longer than paper-based tarps, the paper-based tarps were more breathable and resulted in less degradation of biomass. However, the paper-based tarps did not maintain their structural integrity for the full duration of the storage period. Moisture content of original biomass (48.99%) increased to a maximum of 65.25% under plastic cover after 1 year of storage. This negatively influenced the net heating value of the biomass, causing it to decrease from 8.58 MJ/kg to 4.06 MJ/kg. Overall, the use of covers was not considered successful in preserving the original quality of biomass but may enhance its quality for other biorefinery opportunities.

Keywords: biomass; storage; feedstock; forest; harvest residues; comminuted biomass; tarps; covering; biorefinery

---

1. Introduction

Forest biomass has gained attention as an alternative energy source to address unsustainable supplies of fossil based fuels [1]. A renewable resource, forest harvest biomass offers potential to enhance the sustainability of energy production as well as to greatly reduce greenhouse gas emissions from reliance on fossil fuels [2]. Bioenergy applications can further enhance the economic value of forest harvest operations, diversifying the forestry industry and creating new jobs in rural communities within the forestry and energy sectors [3].

Offsetting the shutdown of many Canadian pulp mills since 2005, the Canadian Forest Products Association, in cooperation with multiple partners, has introduced a "bio-pathways" program, intended to transform the Canadian forest products industry from its traditional applications to a global producer of bioenergy and bio-chemicals [4]. The market for diverse bioproducts presents opportunities to collaborate with other industries, which can further help maximize efficient use of forest resources and create new job opportunities [5]. In order to leverage these opportunities, a strong supply chain must be established to ensure the quality of the biomass used for various applications.

Establishing a strong supply chain begins with a thorough understanding of the resource. Forest harvest residue biomass is typically produced as a by-product of industrial operations and is commonly comprised of tops, branches, bark pieces and foliage. It is generated as residue either during the harvest and extraction of merchantable logs from forest sites or when extracted logs are being processed. Traditionally, residues are either left on the forest floor as material for nutrient cycling and/or erosion control, or they may be piled at roadsides and burned. With rising interest in their application as bioenergy feedstocks however, they have become valued commodities in their own right. As such, there are government sponsored innovation initiatives underway to quantify, classify and characterize feedstocks with the aim of connecting them with compatible technologies and establishing supply chains to benefit Canadian operations [6].

Properties of forest-based residues are complex due to variation in species and component composition. This complexity may be further developed during supply chain operations such as storage. Converting forest harvest residue biomass into energy or bioproducts usually requires a period of storage, which may be considered a step in the material processing. Storage is a dynamic process in which many interconnected factors affect the state of the biomass. Extended storage may cause degradation, organic matter loss and changes in energy value and fibre quality, which can reduce the overall quality of the biomass and render it ill-suited to certain applications [7]. There are multiple factors involved in storage practices that may influence the changes that biomass undergoes. These include climatic conditions, pile geometry and structure, pile size, storage time, species composition, moisture content, season of harvest, and state of the biomass [7]. However, the effects of storage are not well understood and there are many contradictory findings to date. Since much of the reliable data on suitability for biorefining applications refers to the properties of newly harvested biomass, it is important to better understand the changes that biomass undergoes in storage in order to maximize its potential as industrial feedstock.

Major processes that effect change to stored forest harvest biomass include moisture evaporation, living cell respiration, biological degradation and thermo-chemical oxidative reactions [7]. Resulting changes in the characteristics of the biomass may include moisture content, energy value, ash content and dry matter content [8], which affect its value as fuel feedstock. Moisture content is particularly important for energy production since some of the energy generated during combustion must be used for the evaporation of the water, which leaves less usable energy from feedstock. This decreases the economic value of the biomass. Another consequence of higher moisture content in stored biomass is increased biological activity due to favourable environmental conditions for bacteria and fungi growth, the presence of which can present a health and safety risk [9]. Biodegradation can, in turn, contribute to increasing the internal temperature of biomass piles, which is another major concern during storage. Elevated temperatures promote exothermic chemical reactions that can cause the piles to self-ignite which results in organic matter losses, particulate and greenhouse gas (GHG) emissions and economic peril [7].

Covering storage piles with tarps is suggested to be an inexpensive air drying technique that has been shown to be feasible and enhance biomass quality in certain operations. Covering piles ensures their protection against rain and snow, thereby preventing the introduction of extra moisture into biomass stored outdoors. Investigating the characteristics of feedstocks from silvicultural cleanings and thinnings, Nurmi and Hillebrand [10] observed that roadside drying of covered wood harvested in January resulted in lower moisture content than uncovered wood. Roser et al. [11] found similar benefits of tarping bundles of stems in different climate regions of Europe. However, applying covers to biomass has the potential to trap heat and accumulate moisture inside, which may increase the risk of degradation within biomass piles. Studies on the response of wood particles to climatic variation provide evidence that ventilation in stored biomass may result in reduced dry matter losses [12]. To address this, Afzal et al. [8] demonstrated promising results in moisture reduction of comminuted biomass using a breathable tarp cover in Eastern Canadian climatic conditions. Thus, covering storage piles with tarps, particularly those that allow for some degree of air permeability—may provide a means to lower the material's overall moisture content without significantly increasing the risk of degradation.

This paper examines the impact of two different tarp covers—moisture permeable versus moisture impermeable—on stored comminuted forest harvest residue properties in Eastern Canadian climatic conditions. Properties of interest include physical characteristics such as moisture content, bulk density and particle size, as well as thermo-chemical characteristics such as energy value, volatile matter, fixed carbon and ash content as well as elemental analysis. This study has three key objectives. Firstly, it aims to evaluate the overall biomass quality with respect to coverage during extended storage. Secondly, it aims to evaluate and compare the performance of different tarp covers by their effects on the various physical and chemical characteristics of biomass material. Finally, it aims to evaluate tarp performance by their durability and relative handling ease during tarping operations. Conclusions drawn from this study can be used to inform optimal storage strategies that maximize the desirable properties of biomass as a raw material for the purpose of direct combustion or thermal/biochemical conversion to novel forms of bio-energy and bio-materials.

2. Material and Methods

2.1. Operational Trial

JD Irving Ltd. (JDI) provided the storage site and FPInnovations (FPI) the experimental set-up for this study including pile formation and tarping operations. Samples of stored biomass were supplied to the Faculty of Forestry at the University of Toronto for characterization. Samples of biomass were obtained from piles covered with two different types of tarps as well as control piles (no tarp) at the beginning of the study and after twelve months of storage.

In September 2009, fifteen piles of comminuted forest harvest residue biomass weighing on average 200 oven dry tonnes (odt), were built in the bush at roadside near Sussex, New Brunswick. The piles were built following harvesting and flail chipping operations (bark, branches and leaves removed from stems). The piles were comprised of roughly 81% softwood species (Balsam Fir and Black Spruce) and 19% hardwood species (Poplar, Birch and Maple). Average pile dimensions were approximately 22 m × 15 m × 5 m (L × W × H). A typical pile is depicted in Figure 1 below. The flail chipping operation employs a unique silvicultural system where the harvested tree is skidded to road side and delimbed and debarked by a flail chipper. The tree enters the flail chipper and proceeds through a series of chains that remove limbs and bark from the stem. The residue material is deposited into piles at the roadside, eventually to be used as hog fuel by the mill. The merchantable stem proceeds through a chipper and is blown directly into a transport van destined for the pulp mill.

Out of fifteen piles of biomass assembled for the trial, six piles were covered with plastic-based Interwrap [13] tarps, six were covered with paper-based Walki [14] tarps and three were left uncovered to serve as control piles, providing three treatments. The Interwrap tarps are made of plastic and are 100% recyclable. The tarps used in this study are commonly used for wrapping lumber products for storage outside. The weight of the tarp used in this study is 38 kg with an average density of 70 g/m². Multiple tarps were sewn together for the purpose of the trial. Each modified tarp covered an area of 540 m². The Walki tarp are composed of 2 layers of wet strength kraft paper overlaid with a polypropylene net material to resist tearing. The Walki tarp comes in rolls 4 m wide, by 250 m long and weighs 296 kg. The average density of the tarp is 246 g/m². The manufacturer cautions that this tarp should not be used to cover material for more than a year, after which time it may start to degrade.

Figure 1. Pile of comminuted forest harvest residue at initial pile construction.

2.2. Sampling

Sampling occurred at two intervals during the course of this study. Initial sampling occurred when the piles were established in September 2009. An excavator with grapple was used to collect the samples. Five samples were taken from one pile of each treatment (control, plastic-based tarp and paper-based tarp) in September 2009. Samples were sealed in polyethylene bags and frozen at the FPI lab in Montreal and then sent to the Faculty of Forestry at the University of Toronto. Samples were returned to the freezer on arrival. These samples will be referred to as 'original' throughout the paper.

In February 2010 (after six months), an on-site evaluation was conducted to assess the integrity of the tarps. Three piles (1-control, 1-plastic-based and 1-paper-based) were designated for long-term follow-up and left untouched for one year before sampling.

For the post-storage follow-up, the remaining three piles (uncovered, plastic tarp, paper tarp) were sampled in August 2010. Four samples were taken from each pile treatment at depths of approximately 1 m, 2 m, 3 3.5 m, and 4–4.5 m. Samples were frozen and sent to the Faculty of Forestry at the University of Toronto for characterization. Testing results from original samples were used for comparison with their counterparts from the one year stored material. The average monthly temperatures duration of the trial is shown in Figure 2. The total precipitation (snow and rain) is shown in Figure 3.

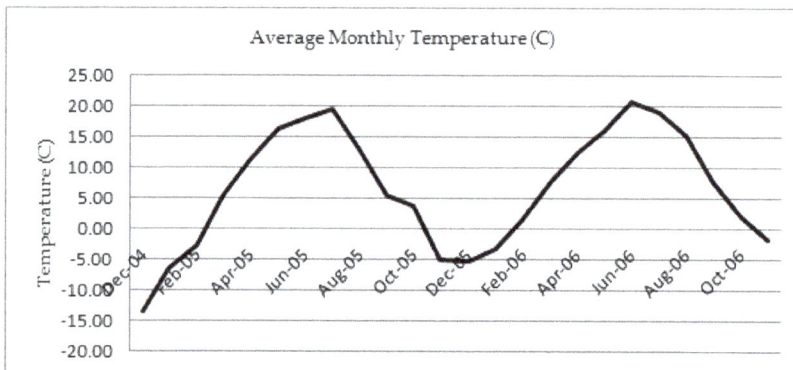

Figure 2. Fredericton, New Brunswick region average monthly temperature over one year.

68

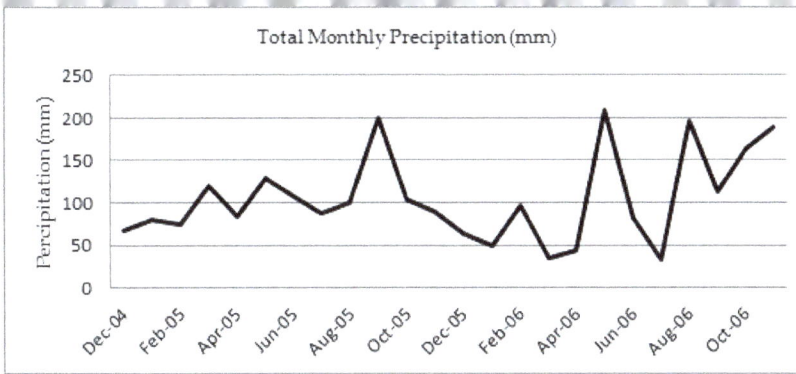

Figure 3. Fredericton, New Brunswick region total precipitation over one year.

Dataloggers were used to monitor temperature within the piles throughout the study. One datalogger was inserted into each pile. Attached to each datalogger were 4 thermocouples which were positioned at different locations within the pile. The dataloggers received a reading every 5 s for intervals of 2 h. After each 2-h interval, the dataloggers recorded the average, maximum and minimum temperature values.

2.3. Characterization

Biomass characterization included evaluation of moisture content, particle size distribution, bulk density, energy value, ash content, volatile matter percentage, fixed carbon percentage and elemental analysis. Procedures were based on standard test methods such as American Society for Testing and Materials (ASTM) and European Committee for Standardization (CEN/TS) as well as on peer-reviewed literature. Moisture content was determined on an as-received (unfrozen) basis as specified in ASTM E871-82 [15]. Three samples of 20 g each were taken from each sample bag (of approximately 5 kg) for moisture content determination. Size distribution of the biomass samples was determined first by hand sieving to separate wood chips that were too large for mechanical sieving. Air dried biomass was placed on a screen which was then shaken horizontally until the chips had been separated by size. Chips that passed through a hand sieve with screen openings smaller than 4.75 mm were then mechanically sieved using a vibrating screen. Circular openings in the screen had a width of 12.5 mm squared. Chips that passed through the screen were designated as small fractions while those that did not were designated as large fractions. The material separated into each category was weighed and calculated as a percentage of the total oven dry mass. Bulk density was determined based on air-dried volume and oven dry mass, as specified in CEN/TS 15103 [16], by pouring the biomass into a vessel of known volume and determining the mass of the biomass which filled the vessel. It must be noted that this is not a measure of the actual compressed bulk density in the pile, but rather an uncompressed bulk density giving a relative measure of the particle spacing and a proxy for particle size measurement ie.the lower the bulk density, the larger the particle size.

The Higher Heating Value (expressed in MJ/kg), which is the maximum amount of energy that can potentially be recovered on complete combustion of biomass samples, was determined by a Parr 1108 adiabatic oxygen combustion bomb calorimeter, using instrument operating instructions [17] and ASTM D2015-77 [18]. Ash content of biomass samples was assessed using a muffle furnace as per ASTM D1102-84 [19] and CEN/TS 14775 [20]. Percentages of volatiles and fixed carbon were determined by Thermo Gravimetric Analysis (TGA). Elemental analysis was conducted by the University of Toronto Department of Chemistry's Analytical Laboratory for Environmental Science Research and Training. A CHN analyzer was employed to determine carbon, hydrogen and nitrogen content. Oxygen content was then determined by subtraction from 100%. Content of each element was calculated reported on an oven dried mass basis. The detection limit of this method is 0.3%.

3. Results

3.1. Tarping Operation and Tarp Durability Evaluation

Plastic-based tarps were lighter (70 g/m² vs. 246 g/m²), more flexible and cover a greater area per sheet than the paper-based tarps. Additionally, the plastic-based tarps could be folded into an easily transportable bundle (38 kgs) relative to the paper wrap, which comes in rolls of 296 kgs. Due to their weight, the paper rolls require a roll dispenser to lift them and two workers to pull a layer of tarp over the pile. The paper-based tarps also require many layers to adequately cover a pile. On average, seven layers each 4 m wide were required, making it a physically demanding job for the workers. Using paper-based tarps also requires an excavator to follow the workers with the roll dispenser. From an operational perspective, the paper tarping operation took twice as long as the plastic tarping operation. Specific to covering large forest harvest residue piles, the tarping operation was less labour intensive and simpler when employing the plastic-based tarps as compared to the paper-based tarps, Figure 4.

(a) (b)

Figure 4. Tarping operation using, (**a**) plastic-based tarp and (**b**) paper-based tarp.

After six months in use, a field evaluation of the two types of tarps revealed superior performance of the plastic tarp to the paper tarp. After six months of exposure to sun, rain, wind and snow, the plastic-based tarps retained their waterproof properties and remained intact all around the piles, as shown in Figure 5. The paper-based tarp showed signs of severe degradation and in some locations had completely lost their protective value, as shown in Figure 6. The degradation of the paper-based tarp on the top of the pile might be explained by Buggelen's wet lens theory [21]. According to Buggelen, a pile of biomass will expel hot/moist air through its top like a chimney as it draws in fresh air from its sides. As noted in this trial, the paper-based tarp on the sides of the piles, where the temperature is not expected to dramatically increase, and there is less snow and rain accumulation, remained in good condition through the six month period, while the tarp at the top of the pile was thoroughly degraded.

Figure 5. Plastic-based tarps after six months.

Figure 6. Paper-based tarp after six months.

3.2. Biomass Characterization

A comparison of the characteristics from biomass samples taken at the time of pile formation with those taken after long-term (1-year) storage was performed, as shown in Table 1. The moisture content and bulk density for the original material is an average of four samples from each of three random piles with three replicates per sample ($n = 36$). The post-storage moisture contents are calculated from one pile at five different depths and three replicates ($n = 15$). Post-storage values were averaged from the different strata samples within a treatment. To assess the differences across time and between tarp treatment, a two-way analysis of variance followed by Tukey's Honestly Significant Difference (HSD) test using R statistical computing software [22] was used. The results are summarized in Table 1.

Table 1. Properties of comminuted biomass including original material and samples after 1 year storage with no tarp, plastic-based tarp and paper-based tarp.

Parameter	Original Material	Post-Storage Control	Post-Storage Plastic Tarp	Post-Storage Paper Tarp
Pile Moisture Content (ar)%	48.99 [a] (2.35) [b] a 36 [c]	61.77 (9.39) b 15	65.25 (5.06) b 15	52.40 (13.87) a 15
Bulk Density (od-kg/m^3)	112.78 (26.94) bc 36	89.78 (9.10) ab 9	124.49 (17.53) c 9	73.25 (6.72) a 9
Large Particle Fraction (% od mass)	60.65 (8.81) a 3	56.58 (8.26) a 9	42.94 (7.88) b 9	57.42 (7.32) a 9
Small Particle Fraction (% od mass)	39.35 (8.81) a 3	43.42 (8.26) a 9	57.06 (7.88) b 9	42.58 (7.32) a 9
Proximate Analysis				
Volatiles % Oven Dried	71.1 (4.85) b 12	73.03 (2.40) b 6	59.54 (4.98) a 4	76.38 (3.04) b 4
Fixed carbon % Oven Dried	16.47 (2.13) a 12	21.78 (2.64) b 6	21.92 (0.56) b 4	20.14 (3.57) ab 4
Ash % Oven Dried	8.94 (5.08) a 12	5.18 (1.08) a 6	18.54 (5.52) b 4	3.48 (0.53) a 4
Higher Heating Value (MJ/kg-od)	19.51 (1.08) ab 12	20.35 (0.39) b 6	18.37 (1.05) a 6	20.43 (0.21) b 6
Net Heating Value (MJ/kg)	8.58 (0.75) c 12	4.31 (0.54) a 6	4.06 (0.53) a 6	5.81 (0.45) b 6
Energy Density (MJ/m^3)	967.6	386.9	505.4	425.6
Carbon (%)	49.92 (2.23) a 9	51.6 (0.6) a 3	48.6 (0.9) b 3	51.3 (0.7) a 3
Hydrogen (%)	5.70 (0.26) a 9	5.6 (0.1) ab 3	4.86 (0.1) c 3	5.46 (0.06) b 3
Oxygen (%)	43.74 (2.25) a 9	42.2 (0.6) a 3	46.0 (0.6) b 3	43.3 (0.8) a 3
Nitrogen (%)	0.64 (0.09) a 9	0.6 (0.04) a 3	0.61 (0.03) a 3	0.57 (0.06) a 3

[a] Numbers in table are mean values; [b] Values in parentheses are the standard deviations; [c] Values below means and standard deviations denote sample size; Different letters across rows denote a significant difference according to Tukey's HSD at the 95% confidence interval.

3.3. Physical Characteristics

3.3.1. Moisture Content

After one year of storage, both the untarped control pile and the plastic-based tarp pile experienced a significant increase in moisture content ($p < 0.01$) as compared to the original material. While the paper-based samples also underwent an increase in moisture content, this change was not significant, as shown in Table 1 and Figure 7. Tarping of piles ensures protection against rain and snow ingress to the pile, which should reduce the moisture content over the storage period. However, reduced air circulation and moisture evaporation brought about by an impermeable tarp actually traps moist warm air in the pile environment, increasing biological degradation and moisture content [23]. In general, it is a well established fact that moisture permeable tarps or sheets help to lower the pile moisture content. On the other hand, impermeable covers resulted in higher moisture content in the pile because surface evaporation of moisture is difficult and the majority of moisture is trapped under the covering.

The effectiveness of the paper tarp cannot be properly surmised from the trial because of its degradation and loss of function in a six-month period. This may have contributed to the high variability in moisture contents in this pile because some areas remained covered and other areas were completely exposed to the environment. The paper-based tarps were thoroughly degraded on the top-middle section of the pile where heat would have concentrated, however they were relatively intact after six months around the edges of the pile. This may have helped stave off excess moisture later in the storage period. The pile's configuration may have also opened up a vent for the moisture to leave. A study by Jirjis [24] cited a ventilation tunnel as the reason for a woodchip pile losing both substantial heat and moisture content while in storage. Potentially, using the paper-based cover with an exposed area at the top centre of the pile to permit moisture and heat transfer away from the biomass to the atmosphere may be worth exploring.

One of the primary quality aspects of biomass when used as feedstock to industrial boilers is the consistency of the moisture content. It can be seen in this trial that the original material has a much lower variability ($sd = 2.35$) as compared to the material stored in piles for 1 year, as shown in Figure 7. It has been noted in previous works that moisture will migrate in a pile and produce areas of relatively drier and wetter biomass; in other words, there was found to be a significant interaction effect of storage time and location within a pile on biomass moisture [25]. Our finding suggests that prolonged storage of biomass, under any of the three storage regimes, reduces the quality of the biomass as a feedstock for industrial boilers because of the increase in variability of the moisture in the fuel.

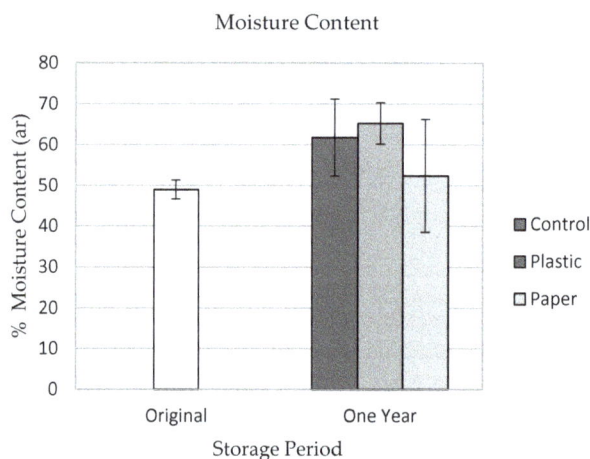

Figure 7. Change in moisture content of comminuted biomass after a storage period of one year.

3.3.2. Particle Size Distribution

The average particle size distribution of the biomass did not change after the 1-year storage, with the exception of the material in the plastic tarped pile, as shown in Figure 8. In this pile, there was a small reduction in the proportion of larger particles as compared to the small particles ($p < 0.05$). At the end of the storage period, plastic-based samples had a significantly lower percentage of large fractions and significantly higher percentage of small fractions than the original material and the uncovered control samples ($p < 0.05$) as well as the paper-based samples ($p < 0.01$). These results provide evidence that the impermeable plastic-based tarp facilitated a greater degradation of the biomass material. These findings are supported by the higher bulk density of the material in the plastic-tarp pile. The bulk density of this material is significantly higher than material in either the control pile or the paper-tarped pile (Table 1). Higher bulk density suggests a reduced internal airflow which will result in higher internal temperature [25], increasing degradation (both microbial and chemical), increasing moisture generation and ultimately greater decomposition and material loss. Relating observations from our analysis, the plastic-tarped pile had higher moisture content, higher bulk density and the highest proportion of small particles as compared to the other storage regimes, thus suggesting that the plastic-based tarp is not ideal for maintaining the original biomass quality.

Particle Size Distribution of Original Material and Post 1 Year Storage

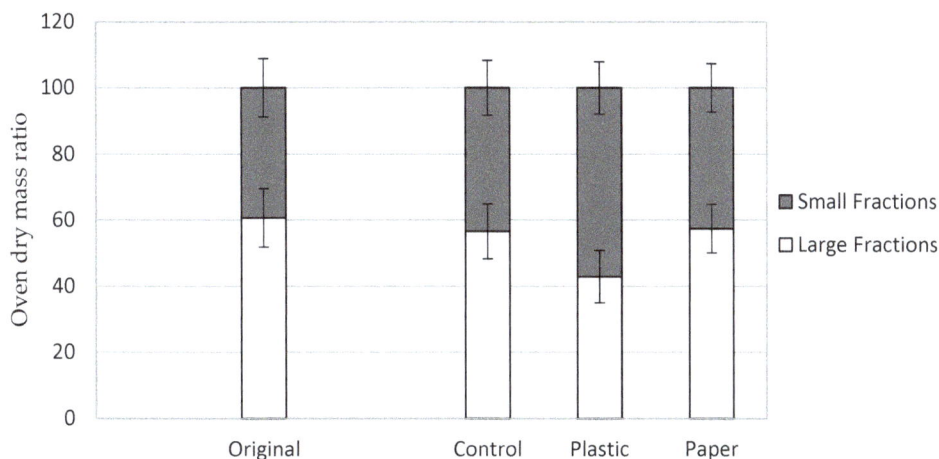

Figure 8. Change in particle size distribution (% o.d. mass ratio) of comminuted biomass after a storage period of one year.

3.4. Thermochemical Characteristics

3.4.1. Heating Values

The higher heating value of biomass stored for 1 year under the various tarp regimes, were not significantly different to the original biomass material (Table 1). However, the net heating values of the 1-year old stored biomass were significantly lower than the original material. This can be explained by the higher moisture in all of the post storage samples as compared to the original material. Moisture is generated within the stored biomass from the biological breakdown of the lignocellulosic material. Aerobic and anaerobic metabolic processes generate carbon dioxide, water and energy on consumption of organic matter. Previous research has shown a link between increased moisture content in stored woodchip piles and decreased net heating value [25,26], which aligns with our findings.

From an energy density (MJ/m^3) perspective, the original material is at least twice as high as any of the post storage materials, as shown in Table 1. The higher energy density of the fresh material signifies that one load of fresh material will be equivalent to at least two loads of any of the post storage materials. This finding suggests that the cost of transportation to the energy generating facility, per unit of energy of the post storage material, will be double that of the original material.

Proximate analysis shows that the post-storage material under the plastic-tarp was significantly different to the original biomass. It possessed a lower volatile content, a higher fixed carbon content and a much greater inorganic (ash) content. These changes support the hypothesis that the plastic tarp storage regime caused degradation of the material over the extended storage period. This is contrary to findings reported by Feist et al. 1971, that plastic covers could deter decomposition by limiting the availability of [27] oxygen to the aerobic micro-organisms. The plastic tarp piles bore the greatest temperature change, increasing a maximum of 11 °C compared to a 4 °C increase in the uncovered control piles. Internal pile temperature readings of the plastic tarped piles were in excess of 60 °C, with the highest at 76 °C. Ernston et al. [28] note that degradation happens in successive stages in which low molecular weight sugars and other easily degradable compounds are first metabolized followed by cellulose and finally lignin, if more specialized fungal species are present. The mould fungi typically found in forest residue chips may consume cellulose and hemicellulose whereas white-rot fungi, which consume lignin, are rare in forest residue chips [29]. The higher fixed carbon proportion in the proximate analysis of the stored woods is a good indicator of the selective loss of the carbohydrate versus lignin through microbial activity.

The ash content of all of the biomass tested in this study was higher than has generally been reported for other forest residue biomass. Acquah et al. [26] found ash contents of less than 5% for biomass collected from an industrial biomass harvesting site. The biomass in that study was comprised of a substantial amount of white wood from undersized and undesirable species in addition to tops/branches and bark. The material collected in this study is far "dirtier" in that it consists entirely of bark, branches and foliage. It is apparent from our results that the bark is contaminated with soil and grit from the skidding and handling operations, creating the unusually high ash content. Examination of ash content of forest harvest residues by other researchers has found that ash percentage increased over storage time due to degradation and loss of organic fraction [26].

The material in the plastic-tarped pile had a very high proportion of ash as compared to any of the other biomass materials. In fact, 18% is abnormally high and suggests that the samples were likely highly contaminated with inorganics such as soil and grit. The standard deviation for this sample is also relatively high, suggesting that the grit and soil were more predominate in certain parts of the pile than others. Our findings show that the plastic-tarp covered material is the only material that had significantly higher ash than the original material, indicating an apparent loss of organic matter. The fixed carbon content of the plastic-tarped material was higher as compared to the original material, suggesting that it may be a preferred feedstock for torrefied pellets, where pyrolyzed organic matter (char) is the desired outcome.

3.4.2. Ultimate Analysis

The carbon–hydrogen–oxygen content of the biomass materials is quite consistent across all samples. The plastic-tarp biomass is the only material that deviates significantly from the original biomass. The molecular formula (derived from the percent composition) for the original biomass is $C_{4.2}H_{5.7}O_{2.7}$ or $(C_{4.2}H_{0.3}(H_2O)_{2.7})$, while the molecular formula of the plastic-tarp is $C_{4.1}H_{4.9}O_{2.9}$ or $(C_{4.1}O_{0.45}(H_2O)_{2.45})$, clearly showing that the stored material has become more oxygenated and has a lower water of constitution. So, while the plastic tarped biomass has been degraded from its original composition, it may lend itself better to biorefining options such as pyrolysis or torrefaction. The lower water of constitution and higher fixed carbon content offer an interesting alternative to the original biomass for this application.

4. Conclusions

The quality characteristics of "fresh" forest biomass are commonly accepted as the standard measure of biomass when biomass is being considered for industrial applications. This study illustrates that long-term storage of forest biomass can significantly alter its chemical, physical and thermal properties and lead to higher costs over the entire bioenergy supply chain. Although the loss of dry matter was not accounted for in this study, increased costs to deliver stored biomass can be a consequence of degradation, resulting in lower net energy density and ensuing higher transport costs and GHG emissions. In addition, the tarping operation added extra cost and GHG emissions to the supply chain through materials, labour and extra machinery time.

The results of this study did not support the efficiency of using covers to preserve the quality of comminuted biomass piles stored at the roadside. Biomass covered with plastic-based tarps experienced accelerated degradation and increased moisture content over uncovered control piles due to an apparent lack of ventilation. Biomass covered with paper-based tarps was not much better off than uncovered biomass after six months in storage, at which point the tarps had lost their protective properties. The paper-based tarp did not retain its structural integrity past six months. As this rapid degradation made a proper comparison between tarps impossible, it would be worthwhile to investigate the efficiency of this tarp over a shorter storage period or to conduct a similar study using a different type of ventilated cover that may better withstand harsh weather conditions.

This study underscores the importance of continuing to evaluate residues in order to help handling practices reduce decomposition and preserve quality. Although they did not prove effective tools for the purpose of this study, in the greater context of the supply chain, covers have great potential to promote desirable biomass qualities. For instance, if a supplier wishes to change pile characteristics before end-use, a cover could be employed as an incubator for pre-treatment. If the end-use requires a biochar or torrefied pellet, then higher degradation may be the favourable condition. In addition, accelerating degradation can result in a material that is more amenable to enzymatic treatment as it is already partially broken down. At present, the supply chain is largely uncontrolled; it is based on the availability of residues rather than the demand for their specific applications. This highlights the need to develop standards that will help optimize biomass supply based on specific user-demand to ultimately improve bioenergy applications in Canada.

Acknowledgments: This work would not have been possible without the significant contributions of Gifty Ewurama Acquah and Gireesh Gupta. A The authors also gratefully acknowledge the ongoing support of Natural Resources Canada's Program of Energy and Research Development (PERD) to this work.

Author Contributions: Suzanne Wetzel, Sally Krigstin and Sylvain Volpe conceived and designed the experiments; Sylvain Volpe and colleagues from FPInnovations contributed materials; Sally Krigstin and students from the University of Toronto as well as Sylvain Volpe and colleagues from FPInnovations performed the experiments; Sally Krigstin and Janet Damianopoulos ran the statistical software; Sally Krigstin wrote the paper.

Conflicts of Interest: The authors declare no conflict of interest. The founding sponsors had no role in the design of the study; in the collection, analyses, or interpretation of data; in the writing of the manuscript, and in the decision to publish the results

References

1. Ontario Ministry of Natural Resources. Available online: https://www.ontario.ca/document/forest-biofibre-allocation-and-use (accessed on 28 September 2016).
2. Hall, J.P. Sustainable production of forest biomass for energy. *For. Chron.* **2002**, *78*, 391–396. [CrossRef]
3. Ministry of Forests, Lands and Natural Resource Operations, British Columbia. Available online: https://www.for.gov.bc.ca/hfd/pubs/docs/tr/tr081.htm (accessed on 29 September 2016).
4. Forest Products Association of Canada. Available online: http://www.fpac.ca/forest-industry-innovation/construction/ (accessed on 22 November 2016).
5. Duchesne, L.C.; Wetzel, S. The bioeconomy and the forestry sector: Changing markets and new opportunities. *For. Chron.* **2003**, *79*, 860–864. [CrossRef]

6. Natural Resources Canada. Available online: http://www.nrcan.gc.ca/forests/industry/products-applications/13359 (accessed on 28 September 2016).

7. Krigstin, S.; Wetzel, S. A review of mechanisms responsible for changes to stored woody biomass fuels. *Fuel* **2016**, *175*, 75–86. [CrossRef]

8. Afzal, M.T.; Bedane, A.H.; Sokhansanj, S.; Mahmood, W. Storage of comminuted and uncomminuted forest biomass and its effect on fuel quality. *BioResources* **2009**, *5*, 55–69.

9. Sebastian, A.; Madsen, A.M.; Martensson, L.; Pomorska, D.; Larsson, L. Assessment of microbial exposure risks from handling of biofuel wood chips and straw-effect of outdoor storage. *Ann. Agric. Environ. Med.* **2006**, *13*, 139–145. [PubMed]

10. Nurmi, J.; Hillebrand, K. The characteristics of whole-tree fuel stocks from silvicultural cleanings and thinnings. *Biomass Bioenergy* **2007**, *31*, 381–392. [CrossRef]

11. Roser, D.; Mola-Yudego, B.; Sikanen, L.; Prinz, R.; Gritten, D.; Emer, B.; Väätäinen, K.; Erkkilä, A. Natural drying treatments during seasonal storage of wood for bioenergy in different European locations. *Biomass Bioenergy* **2011**, *35*, 4238–4247. [CrossRef]

12. Suadicani, K.; Heding, N. Wood preparation, storage, and drying. *Biomass Bioenergy* **1992**, *2*, 149–156. [CrossRef]

13. WeatherPro® Lumber Wrap & Protective Packaging. Available online: http://www.interwrap.com/Protective-Packaging/WeatherPro.html (accessed on 21 August 2017).

14. Walki, 2010. Walki®Biomass Cover. Available online: http://www.walki.com/material/attachments/walki.com/brochures/53sD8gFOD/Walki-BiomassCover_gb.pdf (accessed on 21 August 2017).

15. American Society for Testing and Materials (ASTM). *Standard Test Method for Moisture Analysis of Particulate Wood Fuels*; ASTM E871–82; ASTM: West Conshohocken, PA, USA, 2006.

16. European Committee for Standardization (CEN). *Solid Biofuels, Methods for the Determination of Bulk Density*; CEN/TS 15103; CEN: Brussels, Belgium, 2005.

17. Parr Instrument Company. *Operating Instructions No. 205M for the Parr 1108 Adiabatic Oxygen Combustion Bomb Calorimeter*; Parr Instrument Company: Moline, IL, USA, 2008.

18. American Society for Testing and Materials (ASTM). *Gross Calorific Value of Solid Fuel by the Adiabatic Bomb Calorimeter*; ASTM D2015–77; ASTM: West Conshohocken, PA, USA, 1983.

19. American Society for Testing and Materials (ASTM). *Standard Test Method for Ash in Wood*; ASTM D1102–84; ASTM: West Conshohocken, PA, USA, 2007.

20. European Committee for Standardization (CEN). *Solid Biofuels, Method for the Determination of Ash Content*; CEN/TS 14775; CEN: Brussels, Belgium, 2004.

21. Buggelen, R. Outside storage of wood chips. *Biocycle* **1999**, *40*, 32–34.

22. Maechler, M.; Rousseeuw, P.; Struyf, A.; Hubert, M.; Hornik, K. Cluster Analysis Basics and Extensions. R package. R package version 1.14.4. 2013.

23. Nurmi, J.; Hillebrand, K. Storage alternative affect fuelwood properties of Norway spruce logging residue. *N. Z. For. Sci. J.* **2002**, *31*, 289–297.

24. Jirjis, R. Storage and drying of wood fuel. *Biomass Bioenergy* **1995**, *9*, 181–190. [CrossRef]

25. Casal, M.D.; Gil, M.V.; Pevida, C.; Rubiera, F.; Pis, J.J. Influence of storage time on the quality and combustion behaviour of pine woodchips. *Energy* **2010**, *35*, 3066–3071. [CrossRef]

26. Acquah, G.E.; Krigstin, S.G.; Wetzel, S.; Cooper, P.; Cormier, D. Heterogeneity of forest harvest residue from Eastern Ontario Biomass Harvests. *For. Prod. J.* **2016**, *66*, 164–175. [CrossRef]

27. Feist, W.C.; Springer, E.L.; Hajny, G.J. Encasing wood chip piles in plastic membranes to control chip deterioration. *Tappi J.* **1971**, *54*, 1140–1142.

28. Ernston, M.-L.; Jirjis, R.; Rasmuson, A. Experimental determination of the degradation rate for some forest residue fuel components at different temperatures and oxygen concentrations. *Scand. J. For. Res.* **1991**, *6*, 271–287. [CrossRef]

29. Jirjis, R. Enumeration and distribution of fungi in stored fuel chip piles. *Mater. Org.* **1989**, *24*, 27–38.

Article

Forestry Best Management Practices Relationships with Aquatic and Riparian Fauna: A Review

Brooke M. Warrington [1], W. Michael Aust [1,*], Scott M. Barrett [1], W. Mark Ford [2], C. Andrew Dolloff [3], Erik B. Schilling [4], T. Bently Wigley [5] and M. Chad Bolding [1]

[1]   Department of Forest Resources and Environmental Conservation, Virginia Tech, 228 Cheatham Hall, 310 West Campus Dr., Blacksburg, VA 24061, USA; bmwarring@gmail.com (B.M.W.); sbarrett@vt.edu (S.M.B.); bolding@vt.edu (M.C.B.)

[2]   U.S. Geological Survey, Virginia Cooperative Fish and Wildlife Research Unit, Virginia Tech, Blacksburg, VA 24061, USA; wmford@vt.edu

[3]   USDA Forest Service, Southern Research, Station Forest Watershed Science, Virginia Tech, Blacksburg, VA 24061, USA; adolloff@fs.fed.us

[4]   National Council for Air and Stream Improvement, Inc., 104 East Bruce St., Aubrey, TX 76227, USA; eschilling@ncasi.org

[5]   National Council for Air and Stream Improvement, Inc., PO Box 340317, Clemson, SC 29634, USA; bwigley@ncasi.org

*   Correspondence: waust@vt.edu; Tel.: +1-540-231-4523

Received: 17 August 2017; Accepted: 1 September 2017; Published: 7 September 2017

Abstract: Forestry best management practices (BMPs) were developed to minimize water pollution from forestry operations by primarily addressing sediment and sediment transport, which is the leading source of pollution from silviculture. Implementation of water quality BMPs may also benefit riparian and aquatic wildlife, although wildlife benefits were not driving forces for BMP development. Therefore, we reviewed literature regarding potential contributions of sediment-reducing BMPs to conservation of riparian and aquatic wildlife, while realizing that BMPs also minimize thermal, nutrient, and chemical pollution. We reached five important conclusions: (1) a significant body of research confirms that forestry BMPs contribute to the protection of water quality and riparian forest structure; (2) data-specific relationships between forestry BMPs and reviewed species are limited; (3) forestry BMPs for forest road construction and maintenance, skid trails, stream crossings, and streamside management zones (SMZs) are important particularly for protection of water quality and aquatic species; (4) stream crossings should be carefully selected and installed to minimize sediment inputs and stream channel alterations; and (5) SMZs promote retention of older-age riparian habitat with benefits extending from water bodies to surrounding uplands. Overall, BMPs developed for protection of water quality should benefit a variety of riparian and aquatic species that are sensitive to changes in water quality or forest structure.

Keywords: best management practices; forest operations; riparian species; silviculture; wildlife

---

1. Introduction

Forestry best management practices (BMPs) were developed and implemented to protect physical and chemical aspects of water quality relative to the Clean Water Act of 1972 [1–6]. Prior to development and implementation of forestry BMPs, adverse impacts from forest operations to aquatic environments included increases in water temperature, deposition of fine sediment and increases in concentrations of nutrients and other chemicals, altered loading of coarse and fine organic matter in streams as well as disruption in stream channel form [7,8]. BMP guidelines were developed in the 1970s and refined over time as new information and practices were developed [1,5]. Today forestry BMPs are widely adopted, implemented, and studied [9]. Furthermore, reviews of forestry BMP research conclude that properly applied forestry BMPs protect water quality and critical habitat [1,2,10,11]. Specifically, forestry BMPs address potential impacts of sedimentation, temperature change, and changes in chemical regimes by significantly reducing or eliminating sediment, nutrient, and other pollution inputs [1–3,6,8,10–12]. Following widespread BMP implementation in the United States, water quality impacts from forestry operations have been reduced by over 90% from operations in the pre-BMP era [2].

The water quality protections that forestry BMPs afford are believed to have positive effects on riparian and aquatic species, yet information about the specific effects that forestry BMPs have on wildlife, biodiversity, and other ecological functions have not been fully examined or synthesized. Over the past two decades, forest management research projects have incorporated surveys of species abundance, biological diversity, and measures of nutrient cycling/food chain interactions into monitoring protocols to more fully assess BMP effectiveness [13–15]. Overall, such projects have concluded that forestry BMPs conserve portions of affected forested ecosystems and provide protection for a variety of species. For example, Lockaby et al. [13] examined forest harvesting in Southern bottomland hardwoods and concluded that there would be few lasting effects on ecosystem processes "as long as best management practices are followed." The overall goal of our review is to document how forestry BMP implementation affects aquatic and riparian species.

2. Materials and Methods

Because the numbers of species that might be affected by sediment reductions due to BMP implementation are too numerous to summarize concisely, we focused our literature search on 322 faunal species included in a recent multi-species status assessment conducted by the U.S. Fish and Wildlife Service [16]. This list included all categories of vertebrate (fish, amphibians, reptiles, mammals, and birds) in the region as well as a broad cross-section of invertebrates (Table 1) that are potentially affected by sediment. These riparian and aquatic species were being considered for potential classification as threatened/endangered status in the southeastern United States. Throughout this manuscript, we elected to use the phrase "riparian and aquatic species," as some of these species migrate between the riparian and aquatic environments during different periods of the life cycle, with different activities, or as the riparian zones experience overbank flooding. Our initial search revealed that specific information relating the impact of BMPs on targeted individual species was limited due to the lack of natural history information and species-specific knowledge of habitat associations. Thus, we developed this review of BMP effect literature on information from species in the same genus or with similar ecologies, rather than restricting our search to individual species. In effect, this expanded the scope of the search to an evaluation of potential sediment effects on aquatic and riparian species in the region, not just those originally considered by the U.S. Fish and Wildlife Service. We examined research data regarding assessments of BMP effectiveness for protecting riparian and aquatic fauna during forest operations and supplemented this information with information regarding habitat relations, life histories, and home ranges. We used *Google Scholar* to access information and used combinations of keywords including species names, species groupings, species life histories, forestry best management practices, riparian forests, forest operations, water quality, and silviculture. We generally restricted the geographical setting to the southeastern U.S. However, we often expanded the locations of the search to fill gaps. Articles produced after the introduction of BMPs were targeted. The search provided over 300 peer-reviewed journal articles, theses/dissertations, and government publications, although we are reporting on only the most relevant and non-duplicative for the sake of brevity.

Peer-reviewed articles were given precedence, but other sources of information (theses, dissertations, government/technical publications) were used to help fill information gaps. We focused on studies performed in the southeastern United States and published after the passage of the Federal Water Pollution Control Act of 1972. We summarized information on the effects of BMPs by geographic location on faunal species (or genera) with the specific intent of developing overarching, comprehensive conclusions.

Table 1. Taxonomic groups and the number of species listed in the recent status assessment and the focus of the literature review.

Taxonomic Group	Number of Species
Crayfish	83
Fish	48
Mussels	48
Snails	44
Beetles	18
Amphibians	15
Dragonflies	14
Reptiles	13
Caddisflies	9
Stoneflies	8
Amphipods	6
Mammals	4
Butterflies	4
Birds	3
Isopods	2
Fairy shrimp	1
Moths	1
Springfly	1

3. Results and Discussion

3.1. BMP Implementation and Benefits to Riparian Ecosystems

The overwhelming consensus of the literature was that forestry BMP acceptance and implementation levels have increased dramatically since the 1970's, reaching over 90% in 2015, and that forestry BMPs are effective for reduction of nonpoint source pollution. Numerous reviews from regions across the U.S. have been conducted on forestry BMP effectiveness, and all have concluded that forestry BMPs are effective, e.g., [1–4,6,10,11,17]. Recently, Cristan et al. [9] conducted a survey of forestry BMP implementation across the United States and found that overall BMP implementation levels in the Southeast were slightly over 92%, which was an increase of approximately 8% since a previous study by the Southern Group of State Foresters in 2012. The combination of BMP effectiveness with widespread BMP acceptance and implementation levels provides compelling evidence that BMP programs protect water quality.

3.2. Sediment

Sediment is the most commonly identified nonpoint source pollutant associated with most designations of stream impairment in the United States [11,18]. Although forest operations are relatively minor contributors of sediment, accounting for about 7% of stream sediment impairment [18], sedimentation is the primary water quality concern associated with forest operations [1,4,5,19,20]. Unfortunately, studies of the specific relationships between the ameliorating effects of forest BMPs for sediment reduction on aquatic/riparian species are very limited. However, there is a significant body of work demonstrating the deleterious impacts of excess sediment on aquatic species.

Species that depend on aquatic respiration may experience lethal or sub-lethal respiratory impairment from the precipitation of suspended sediment on gills [21–24]. Sub-lethal effects of elevated sediment potentially include reduced immunity to disease, depressed growth rates, and impaired feeding and reproduction [22,23,25–33].

Sedimentation also has the potential to transform benthic substrate by homogenizing benthic environments, reducing habitat complexity and structural diversity and eliminating important microhabitats [21,30,32–36]. Dissolved oxygen levels also can be reduced by sediment loading [32,35–38]. Ultimately, high fine sediment loads can alter community composition and disrupt trophic level interactions [21,24,27,33,39–43]. Much of this information was developed from studies of sediment derived from land uses other than forestry or from sites without BMPs. These worst-case examples nonetheless emphasize the negative effects of sediment and suggest that any reduction in the rate or amount of sediment delivered to water bodies, as is the purpose of forestry BMPs, should be beneficial to aquatic species [1,2,13,15,44–47].

3.3. BMP Effectiveness

The literature also indicates that BMPs developed for forest roads, skid trails, stream crossings, and streamside management zones (SMZs) have greater potential to reduce sedimentation and also benefit riparian and aquatic species. Of all silvicultural activities, roads and skid trails have the largest potential to contribute excess sediment [48,49]; poorly designed and maintained roads and skid trails have been shown to increase soil erosion, regardless of harvest intensity [44,45,47]. Forestry BMPs for road and skid trails have consistently been shown to be effective for limiting sediment inputs [49–58]. Effective sediment control from forest roads involves appropriate design and template selection, minimization of road grade (particularly at stream crossing approaches), use of water control and retention structures, and achieving adequate cover or surfacing for travel surfaces [48].

Road and skid trail stream crossing approaches surfaced with grass, slash, or gravel are examples of highly effective BMPs that can slow or prevent sediment inputs to aquatic environments [59–63]. Changes in macroinvertebrate functional feeding groups and assemblages in streams associated with harvest treatments and forest road construction and maintenance activities implemented with BMPs are often indistinguishable from natural variations [64].

Forestry BMPs also include a number of water control and diversion methods to prevent sedimentation, particularly on roads and skid trails. Water turnouts and water bars can reduce stream water sedimentation by diverting water flow away from streams and into riparian filter strips, thereby reducing runoff velocity and allowing sediment to be deposited on land [11,58,61]. While stream crossings with a greater area will have greater erosion potential, water turnout and wing ditch BMPs will decrease the stream crossing approach length, thereby reducing the amount of potential sediment inputs [62]. In the Virginia Piedmont, Brown et al. [61] found that appropriate spacing of water control structures can reduce sediment loss. A continuous berm along the edge of a forest road in the Coastal Plain of North Carolina reduced sediment loss by an average of 99% [65]. Maximizing water bar surface roughness and increasing water bar frequency are also effective measures in reducing sediment delivery to streams [66]. Lang et al. [58] found that ditch BMPs can be used to effectively reduce sediment contributions from road ditches.

Streamside management zones and riparian buffers have consistently been shown to reduce sedimentation in aquatic environments [53,54,67]. On the Appalachian Plateau region of eastern Kentucky, Arthur et al. [68] found that in watersheds where BMPs (including riparian buffers) were applied, water yield, sediment flux, and nutrient inputs were similar to non-harvested watershed sites. Lakel et al. [53] found that SMZs trapped approximately 89–97% of watershed erosion before reaching streams. In the Georgia Piedmont, Ward and Jackson [69] found that SMZs were effective for trapping sediment from overland flow, averaging 81% efficiency [11]. Although SMZs applied to harvest units in the highly erodible bluff hills of the Gulf Coastal Plain in Mississippi were not effective in reducing overland flow, they reduced total suspended solid (TSS) concentrations due to the preservation of riparian characteristics [70]. Variability in results among studies may be attributed to differences in topography, geology/soils, and hydrology, but the results of this particular study were attributed to the formation of gullies that breached the SMZ [70]. BMP efficacy is dependent on site-specific characteristics (i.e., site history, disturbance or logging history, topography, slope, climate, etc.); studies consistently have shown the capacity for forestry BMPs to effectively prevent excessive sedimentation of aquatic environments [1–4,10,11,17].

3.4. Stream Crossings

In the absence of BMPs, both permanent [71] and temporary stream crossings have the potential to cause long-lasting effects on aquatic and riparian environments and organisms if not properly designed or maintained [72–75]. However, application of BMP technology at stream crossings can help substrate heterogeneity and stream flow regimes, and retain streambank integrity [72,76–78], all of which are important environmental characteristics for maintaining and conserving healthy aquatic and terrestrial riparian wildlife [71,79].

Stream crossing location and design are important aspects of forestry BMPs. Improper selection of stream crossings can cause changes in hydrology and sediment inputs, which in turn may influence population dynamics and in-stream and terrestrial habitat [72]. Maintaining stream channel morphology helps maintain ecological communities and populations of riparian species [72]. To safeguard aerial, terrestrial, and aquatic riparian wildlife, stream crossing designs need to consider river morphology, depth, velocity, stream flow, and scouring potential [80–83]. These considerations are vital to maintaining healthy wildlife populations; poorly designed stream crossings increase sedimentation, cause issues with aquatic organism passage, influence population dynamics, and contribute to loss of suitable habitat [72,76]. Legacy stream crossings installed prior to the era of BMPs may require replacement or remediation [57]. Although bridges, which usually can be installed to preserve natural channel shape, are a preferred stream crossing type, particularly over larger streams, other factors such as traffic requirements, structural loading capabilities, site features, and economic feasibility often favor use of other stream crossing types [48,83,84].

Culverts typically are the structure of choice for crossings in smaller streams. Culverts that are improperly sized for the watershed can have reduced capacity to pass water and sediment or accommodate fish passage [83]. Water velocity at the exit from culverts that are undersized for the watershed can exceed the capacity of the channel and cause greater downstream scour [85]. Increased velocity and potential culvert suspension could result in habitat and population fragmentation of aquatic species by creating a barrier to aquatic organism movement upstream [73,76,83,86]. Proper culvert size selection is a very basic BMP application [48]. By considering the benefits and potential impacts of stream crossing selection options at a particular site, land managers can select an appropriate stream crossing that will minimize potential ecological and environmental impacts [76,84].

Potentially negative effects of culverts are also contingent on other design parameters. With careful planning and knowledge of local options, selection of stream crossing types that allow adequate organism passage can be both cost-effective and ecologically compatible. Some designs, such as open-bottom culverts, can mimic natural channel shape and substrate and preserve natural hydrological attributes [76–78,83]. Although research on the effects of stream crossings on riparian and aquatic wildlife other than fish [86,87] are relatively limited [71,77,78], culvert BMPs and culvert design have been shown to influence mussels [88], crayfish [77,78], snails [89–92], and aquatic insects [93–96]. The complexity of culvert material, design, and placement and potential effects on a variety of aquatic organisms warrants further investigation. Current BMP effectiveness research suggests that minimizing the numbers of crossings, placement of stream crossings at sites that minimize channel disturbances, use of appropriately sized and installed culverts, and disconnecting sources of erosion from stream crossing approaches will benefit sediment-sensitive aquatic organisms [48,72,83].

3.5. Streamside Management Zones

SMZs, also known as riparian buffers, forest buffers, and filter strips, are a particularly important type of BMP because they provide a zone for water quality protection between managed lands and the stream [97–99]. For example, SMZs promote sediment and nutrient trapping by slowing and often preventing entry into aquatic systems [53,100]. Appropriately managed and designed SMZs moderate light infiltration, dampen or minimize aquatic and terrestrial temperature gradients, slow nutrient flow, maintain hypoheic and hydrologic function, and preserve riparian vegetative composition and structure [2,15,101–106].

However, SMZs also provide forest habitat for a variety of species and maintain structures important to faunal communities in the presence of forestry operations [27,40,47,107,108]. SMZs provide coarse woody debris and detritus inputs that serve as food and habitat structure for aquatic species [109–112] and critical microhabitat for riparian organisms for nesting, roosting, feeding, or breeding. Implementation requirements and subsequent success of an SMZ depends on the aquatic species of concern, and to what extent managers need to protect riparian and aquatic ecosystems from disturbance [47].

As different species of wildlife have different sensitivities to environmental changes, they also have differing requirements for optimal environments, including canopy composition, width of the riparian buffer, and patch length along the stream [34]. SMZs can be designed to address these habitat requirements. Alterations to aquatic and terrestrial temperature gradients could impact aquatic biota, restricting movement, limiting biological functions, and altering habitat suitability [20,47,113,114]. SMZs provide shade and relatively stable canopy composition and streambank stability; therefore, these riparian buffers help preserve terrestrial and aquatic temperature regimes in riparian areas, safeguarding and maintaining wildlife populations [2,15,102,114–118].

Canopy cover influences light and temperature regimes, which are important to many vertebrate species but also to other sensitive fauna such as dragonflies (Odonata) and moths and butterflies (Lepidoptera) [116,119–121]. Species preferences for light infiltration, solar radiation, and temperature regimes differ [47,122]. Although some species are sensitive to changes in these parameters, others may benefit from the manipulation and alteration of riparian vegetation [123]. Riparian zones can be managed to favor the life history for a particular species, but specific information for many threatened and endangered species is lacking and some species have conflicting life history needs. Thus, additional research is needed to ascertain how forestry riparian zones can be managed to best address various species' habitat requirements.

SMZs also provide a source for coarse woody debris, snags, tree cavities, and rotting logs [39,112]. These structures create diverse habitat and microtopography, benefiting a variety of riparian species [112,124] and species within the surrounding landscapes where these structures may be less abundant. The benefits of coarse woody debris in aquatic ecosystems depend on the amount and size of such debris [2,39,112]. Because coarse woody debris can increase habitat diversity, it is an essential component of eastern stream ecosystems, and is required by fish populations [39] and some turtles [124].

Woody debris is typically generated during harvesting operations, and riparian buffers can be managed to provide woody debris input into aquatic ecosystems [39,110–112,125,126]. Although additional nutrient inputs sometimes can enhance fish habitat in headwater streams, in some cases, these inputs could stimulate downstream eutrophication [2]. Fresh slash and debris inputs may elevate water temperatures and decrease dissolved oxygen in still or very slow-flowing water [2], although predictions are challenging due to complexity in natural ecosystems [2,127]. However, natural input of woody debris provided by SMZs should benefit aquatic and riparian wildlife by providing microhabitat and providing allochthonous organic matter to aquatic ecosystems [39,112].

States developed recommendations for SMZ widths primarily to address the goal of water quality protection, and most studies have found that these widths were adequate for protection against thermal, sediment, and nutrient pollution [53]. However, recommendations for SMZ widths typically were not designed specifically to meet objectives related to terrestrial wildlife species associated with riparian ecosystems. Appropriate riparian buffer characteristics for meeting objectives related to these species likely vary depending on site-specific vegetative, hydrologic, and geographic characteristics, adjacent forest structural conditions, and harvesting practices used in adjacent stands [128,129]. However, the literature generally indicates that habitat conditions for many riparian and aquatic species are positively affected by implementation of SMZs.

Because BMPs protect water quality and in-stream structure, and provide heterogeneity of vegetation structure in riparian zones, implementation of BMPs has been shown to benefit aquatic biota and their habitat [130]. For example, from 2006 to 2010, DaSilva et al. [131] studied stream metabolic rates upstream and downstream from a loblolly pine (*Pinus taeda*) stand that was harvested with Louisiana's current BMPs. They quantified rates of net ecosystem productivity (NEP), gross primary productivity (GPP), community respiration (CR), and the GPP/CR ratio. No calculated metabolic rate was significantly changed by the timber harvest. Thus, the authors concluded that "timber harvests of similar intensity with Louisiana's current BMPs may not significantly impact stream biological conditions".

Bioassessment is a common technique for assessing biological integrity of streams [132] and has been used to characterize macroinvertebrates' response to timber harvests with forestry BMPs. Harvesting practices without properly implemented BMPs can negatively influence stream macroinvertebrate populations [133]; however, multiple studies in the Southeast have reported little to no change in aquatic macroinvertebrate community diversity following timber harvesting with BMPs [100,134–140]. Changes in invertebrate communities, when they do occur, generally reflect a shift from allochthonous to autochthonous food resources in streams draining harvested watersheds that is relatively short-lived (<5 years) due to rapid vegetation regrowth [7]. We briefly summarize results from several studies in the Southeast that have used bioassessment methods to study macroinvertebrate responses to BMPs.

Adams et al. [134] studied whether forestry BMPs effectively reduced harvesting impacts on stream habitat and macroinvertebrates in five physiographic regions in South Carolina. They found that most sites with BMPs scored high on rapid bioassessment protocols III (RBPs) established by the Environmental Protection Agency to assess stream health. Thus, the authors concluded that BMPs were effective in protecting macroinvertebrate assemblages.

Kedzierski and Smock [135] examined macroinvertebrate production and macrophyte growth in harvested and non-harvested sections of a low-gradient, sand-bottomed blackwater stream in the Virginia Coastal Plain. A section of the catchment had been clearcut three years prior to sampling and no additional harvesting occurred in the upstream area of the catchment. Macroinvertebrate production was higher in the stream reach of the harvested tract (103 g m^{-2}) than in the reach of the non-harvested tract stream (41 g m^{-2}). Production in the stream of the harvested tract was dominated by collector-filterers living on macrophytes as well as collector-gatherers. Other macroinvertebrate functional feeding groups showed little response to harvesting.

Vowell [136] evaluated Florida's BMPs for protecting aquatic ecosystems during intensive forestry operations that included clearcutting, mechanical site preparation, and machine planting. Sample streams were selected across Florida's major ecoregions. Stream condition index (SCI) bioassessments were conducted at points along each stream, above and below the treatment area. No significant difference in the SCI was observed between the reference and treatment stream segments. Vowell concluded that the proper implementation of forestry BMPs provides effective protection of aquatic resources.

Williams et al. [137] used stream survey data to evaluate timber harvesting influences on physical stream features and macroinvertebrate assemblages in three drainage basins in the Ouachita Mountains of Arkansas. Variability in macroinvertebrate assemblages was largely explained by drainage basin differences and year of sampling. The interaction between timber harvesting and drainage basins suggested that differences in physical stream features were important for determining the effects of logging within individual basins. Furthermore, harvesting did not influence diversity of macroinvertebrates in these small headwater streams. The authors suggest that natural variability in hydrology and in-stream physical features were the primary drivers of assemblage differences and not the effects of harvesting.

Carroll et al. [100] evaluated effectiveness of SMZs to protect water quality, aquatic habitat, and macroinvertebrate communities in low-order streams in north central Mississippi. Three replications of three SMZ treatments (clearcutting with no SMZ, clearcutting with an SMZ, and unharvested reference) were evaluated using response variables that included water quality, mineral soil exposure, and net soil deposition or erosion. One year following harvest, no differences in response variables between harvested sites with stream SMZs and reference streams were observed. Streams in harvested sites without SMZs had significantly higher stream temperatures and declining habitat stability ratings, but increased macroinvertebrate density compared to reference streams.

Vowell and Frydenborg [138] evaluated effectiveness of Florida's forestry BMPs for herbicide applications using methods similar to those used by Vowell [136]. Following a pretreatment assessment, study streams were re-sampled one and two years following herbicide applications to forests adjacent to streams. No significant differences in the Stream Condition Index were observed between reference and test portions of the streams that could be attributed to practices that included chemical applications.

Grippo and McCord [141] used bioassessment of benthic macroinvertebrates to evaluate effectiveness of Arkansas' silvicultural BMPs in protecting the water quality and biological integrity of streams adjacent to harvest areas. They found few significant differences in water quality or biological variables that could be associated with silviculture. Differences between upstream and downstream sites, when noted, were present before as well as after timber harvest. Differences in relative abundance variables (e.g., percent EPT) were typically location-specific and unrelated to silviculture activities.

Griswold et al. [139] conducted pre- and post-harvest sampling of benthic macroinvertebrates from four first-order streams draining the Dry Creek watershed in southwestern Georgia. They found differences in community structure between pre- and post-harvest periods, but responses of macroinvertebrates to harvest treatment and SMZ thinning were subtle. Relative abundance and total taxa all increased in the control and treatment sites after harvest, suggesting communities may have responded to increased streamflow due to increased rainfall during the study period. Overall, the macroinvertebrate communities appear to have been more strongly influenced by environmental factors (e.g., stream flow, water chemistry, and canopy cover) than by SMZ thinning and harvesting of adjacent stands.

McCord et al. [140] examined macroinvertebrate assemblages in six Arkansas low-order streams following harvesting with implementation of BMPs. Stream samples were collected above and below harvested tracts. BMP implementation rates on the harvested tracts ranged between 89% and 100%. Deficiencies in BMPs were generally limited to poorly designed erosion controls; however, no evidence of sedimentation was observed in any of the study stream reaches. Harvesting did not reduce taxonomic richness but did significantly influence several relative abundance metrics. Overall, Arkansas' forestry BMPs were effective in protecting water quality and biological integrity in five of the six study stream reaches examined.

Simpson et al. [142] used a Before-After-Control-Impact study design to assess effectiveness of Texas forestry BMPs for protecting water quality and biological integrity of four streams on intensively managed silvicultural sites in east Texas. Biological and physiochemical monitoring (both grab samples and stormwater samples) was conducted above and below treatment areas. The physiochemical data showed no statistically significant difference as a result of treatment. Following treatment, the biological data showed a shift in habitat quality at two sites and for fish at another site compared to the reference. Although change was detected, the treatment sites generally showed improved conditions for Aquatic Life Use Index (fish) and the Habitat Quality Index. Treatment had no negative effect on water quality and biology.

In addition to protecting water quality and biological integrity of aquatic ecosystems, BMPs also benefit riparian ecosystems. For example, SMZs in the southeastern U.S. provide habitat for species associated with mature deciduous forests and may provide travel corridors for some species, e.g., [123,143–146]. In intensively managed forest landscapes, SMZs promote spatial heterogeneity and enhance landscape conservation value, e.g., [147,148].

4. Conclusions

This literature review indicates that forestry operations in the pre-BMP era had the potential to negatively affect water quality and aquatic and riparian species, and that current forestry BMPs can help protect water quality and habitat conditions for a variety of riparian and aquatic wildlife during forestry operations. Although there are relatively few direct evaluations of the specific effects of forestry BMPs on individual aquatic or riparian species, the effects of forestry BMPs are likely beneficial for the following reasons.

Riparian and aquatic species in general benefit from reduction of anthropogenic pollutants. Forestry BMPs, which were specifically designed to limit sediment, nutrient, and other pollutant entry into streams, help protect habitat for many riparian and aquatic species. State forestry agencies report that forestry BMP implementation levels across the United States are above 90%, and within the Southeastern region overall implementation rates are about 92% [11]. Forestry BMPs specifically target roads, skid trails, and stream crossings, as these forest operations have greater potential to cause water quality problems if BMPs are not applied. Forestry BMPs have been shown in numerous research investigations and comprehensive reviews to protect water quality from sediment, nutrient, and chemical pollution. Thus, a wide variety of species that are negatively influenced by such pollutants should benefit from BMPs that protect water quality. The BMPs recommended by several of the southeastern states have been shown to protect water quality, and multiple studies have reported little to no change in aquatic macroinvertebrate community diversity following timber harvesting with BMPs. Therefore, BMPs should benefit individual aquatic and riparian species that are negatively influenced by sediment and other pollutants.

Stream crossings receive particular attention in forestry BMP guidelines in the southeastern United States for several reasons. Research indicates that stream crossings with inadequate or no BMPs are likely to provide direct connectivity of sediment generated from road systems to hydrologic networks. As a result, BMPs for stream crossings were developed to minimize effects on stream water quality and stream dependent organisms. Therefore, appropriate stream crossings, adherence to installation recommendations, and other properly implemented water quality BMPs will help protect aquatic and riparian species.

SMZs are an especially important type of BMP because managed riparian buffers provide habitat and water quality benefits for both riparian and aquatic organisms. Riparian buffers protect the stream from thermal pollution, which can negatively affect a host of species. Riparian buffers also provide leaf litter and woody debris, both of which are critically important to aquatic food chains, stream habitat, and stream structure and morphology. Riparian forests are zones where sediment, nutrients, and chemicals can be trapped and transformed by physical, soil and plant processes. Finally, SMZs provide habitats for species associated with riparian forests and potentially provide refugia for species affected by adjacent forest management activities. Current state SMZ recommendations or requirements maintain water quality and greatly reduce potential risks to aquatic and riparian species during forest management.

Although implementation of BMPs has been shown to benefit aquatic macroinvertebrate communities, information about the direct effects of forestry BMPs for forest operations on many individual aquatic and riparian faunal species is limited. Therefore, additional research investigating the responses of aquatic and riparian species and communities to modern forestry practices that include implementation of BMPs, is warranted.

Acknowledgments: This project received financial and logistical support from the National Council for Air and Stream Improvement (NCASI), the Department of Forest Resources and Environmental Conservation (FREC) at Virginia Polytechnic Institute and State University, the Virginia SHARP Logger Program, and the McIntire-Stennis Program of the National Institute of Food and Agriculture, US Department of Agriculture. Any use of trade, firm, or product names is for descriptive purposes only and does not imply endorsement by the U.S. Government.

Author Contributions: Brooke M. Warrington conducted the literature review and drafted the original document as part of her Master of Forestry program at Virginia Tech. W. Michael Aust was a co-principal investigator on the grant, served as the co-chair of the graduate committee, and served as the collaborator and editor for various versions of the document. Scott M. Barrett served as the co-chair of the graduate committee, and edited various versions of the document. W. Mark Ford was a co-principal investigator on the grant, served as a member of the graduate committee, and edited various versions of the document. C. Andrew Dolloff was a co-principal investigator on the grant, served as a member of the graduate committee, and edited various versions of the document. Erik B. Schilling was an NCASI collaborator, contributed additional literature to the document, and edited various versions of the document. T. Bently Wigley was an NCASI collaborator, contributed additional literature to the document, and edited various versions of the document. M. Chad Bolding was a co-principal investigator on the grant and served as an editor for various versions of the document.

Conflicts of Interest: The authors declare no conflict of interest.

References

1. Aust, W.M.; Blinn, C.R. Forestry best management practices for timber harvesting and site preparation in the eastern United States: An overview of water quality and productivity research during the past 20 years (1982–2002). *Water Air Soil Pollut. Focus* **2004**, *4*, 5–36. [CrossRef]

2. Ice, G. History of innovative best management practice development and its role in addressing water quality limited waterbodies. *J. Environ. Eng.* **2004**, *130*, 684–689. [CrossRef]

3. Shepard, J.P. Water quality protection in bioenergy production: The US system of forestry Best Management Practices. *Biomass Bioenergy* **2006**, *30*, 378–384. [CrossRef]

4. Edwards, P.J.; Williard, K.W.J. Efficiencies of forestry best management practices for reducing sediment and nutrient losses in the eastern United States. *J. For.* **2010**, *108*, 245–249.

5. Cristan, R.; Aust, W.M.; Bolding, M.C.; Barrett, S.M. Status of state forestry best management practices for the southeastern United States. In *Proceedings of the 18th Biennial Southern Silvicultural Research Conference, Knoxville, TN, USA, 2–5 March 2015*; South. Res. Sta. Gen. Tech. Rep. SRS-212; U.S. Department of Agriculture Forest Service: Asheville, NC, USA.

6. Edwards, P.J.; Wood, F.; Quinlivan, R.L. *Effectiveness of Best Management Practices that Have Application to Forest Roads: A Literature Synthesis*; North. Res. Sta.: Gen. Tech. Rep. NRS-163; U.S. Department of Agriculture Forest Service: Newtown Square, PA, USA, 2016; 171p.

7. Webster, J.R.; Golladay, S.W.; Benfield, E.F.; Meyer, J.L.; Swank, W.T.; Wallace, J.B. Catchment disturbance and stream response: An overview of stream research at Coweeta Hydrologic Laboratory. *River Conserv. Manag.* **1992**, *15*, 232–253.

8. Fortino, K.; Hershey, A.E.; Goodman, K.J. Utility of biological monitoring for detection of timber harvest effects on streams and evaluation of best management practices: A review. *J. N. Am. Benthol. Soc.* **2004**, *23*, 634–646. [CrossRef]

9. Cristan, R.; Aust, W.M.; Bolding, M.C.; Barrett, S.M.; Munsell, J.F.; Schilling, E. National status of state developed and implemented forestry best management practices in the United States. *For. Ecol. Manag.* **2017**. [CrossRef]

10. Anderson, C.J.; Lockaby, B.G. The effectiveness of forestry best management practices for sediment control in the southeastern United States: A literature review. *South. J. Appl. For.* **2011**, *35*, 170–177.

11. Cristan, R.; Aust, W.M.; Bolding, M.C.; Barrett, S.M.; Munsell, J.F.; Schilling, E. Effectiveness of forestry best management practices in the United States: Literature review. *For. Ecol. Manag.* **2016**, *360*, 133–151. [CrossRef]

12. Lakel, W.A.; Aust, W.M.; Dolloff, C.A. Seeing the trees along the streamside: Forested streamside management zones are one of the more commonly recommended forestry best management practices for the protection of water quality. *J. Soil Water Conserv.* **2006**, *61*, 22A–29A.

13. Lockaby, B.G.; Jones, R.H.; Clawson, R.G.; Meadows, J.S.; Stanturf, J.A.; Thornton, F.C. Influences of harvesting on functions of floodplain forests associated with low-order, blackwater streams. *For. Ecol. Manag.* **1997**, *90*, 217–224. [CrossRef]

14. Wigley, T.B.; Roberts, T.H. Landscape-level effects of forest management on faunal diversity in bottomland hardwoods. *For. Ecol. Manag.* **1997**, *90*, 141–154. [CrossRef]

15. Quinn, J.M.; Boothroyd, I.K.G.; Smith, B.J. Riparian buffers mitigate effects of pine plantation logging on New Zealand streams: 2. Invertebrate communities. *For. Ecol. Manag.* **2004**, *191*, 129–146. [CrossRef]

16. USDI Fish and Wildlife Service. *Endangered and Threatened Wildlife and Plants; Partial 90-Day Finding on a Petition to List 404 Species in the Southeastern United States as Endangered or Threatened With Critical Habitat, Proposed Rule*; Federal Register No. 187; USDI Fish and Wildlife Service: Washington, DC, USA, 27 September 2011; pp. 59836–59862.

17. Schilling, E.; Ice, G. *Assessing the Effectiveness of Contemporary Forestry Best Management Practices (BMPs): Focus on Roads*; Special Report No. 12-01; National Council for Air and Stream Improvement: Research Triangle Park, NC, USA, 2012.

18. U.S. Environmental Protection Agency. *National Water Quality List: 2000 Report to Congress*; U.S. Environmental Protection Agency: Washington, DC, USA, 2002.

19. Binkley, D.; Brown, T.C. *Management Impacts on Water Quality of Forests and Rangelands*; Rocky Mtn. For. & Range Exp. Sta.; U.S. Department of Agriculture Forest Service: Fort Collins, CO, USA, 1993.

20. Beschta, R.L. Long-term patterns of sediment production following road construction and logging in the Oregon Coast Range. *Water Resour. Res.* **1978**, *14*, 1011–1016. [CrossRef]

21. Newcombe, C.P.; Macdonald, D.D. Effects of suspended sediments on aquatic ecosystems. *N. Am. J. Fish. Manag.* **1991**, *11*, 72–82. [CrossRef]

22. Sutherland, A.; Meyer, J. Effects of increased suspended sediment on growth rate and gill condition of two southern Appalachian minnows. *Environ. Biol. Fishes* **2007**, *80*, 389–403. [CrossRef]

23. Kefford, B.J.; Zalizniak, L.; Dunlop, J.E.; Nugegoda, D.; Choy, S.C. How are macroinvertebrates of slow flowing lotic systems directly affected by suspended and deposited sediments? *Environ. Pollut.* **2010**, *158*, 543–550. [CrossRef] [PubMed]

24. Wood, S.L.; Richardson, J.S. Impact of sediment and nutrient inputs on growth and survival of tadpoles of the western toad. *Freshw. Biol.* **2009**, *54*, 1120–1134. [CrossRef]

25. Nalepa, T.F. Status and trends of the Lake Ontario macrobenthos. *Can. J. Fish. Aquat. Sci.* **1991**, *48*, 1558–1567. [CrossRef]

26. Needham, J.G.; Minter, J.; Westfall, J.; May, M.L. *Dragonflies of North America*; Scientific Publishers: Gainesville, FL, USA, 2000.

27. St-Onge, I.; Magnan, P. Impact of logging and natural fires on fish communities of Laurentian Shield lakes. *Can. J. Fish. Aquat. Sci.* **2000**, *57*, 165–174. [CrossRef]

28. Anthony, J.L.; Downing, J.A. Exploitation trajectory of a declining fauna: A century of freshwater mussel fisheries in North America. *Can. J. Fish. Aquat. Sci.* **2001**, *58*, 2071–2090. [CrossRef]

29. Broekhuizen, N.; Parkyn, S.; Miller, D. Fine sediment effects on feeding and growth in the invertebrate grazers *Potamopyrgus antipodarum* (Gastropoda, Hydrobiidae) and *Deleatidium* spp. (Ephemeroptera, Leptophlebiidae). *Hydrobiologia* **2001**, *457*, 125–132. [CrossRef]

30. Herrig, J.; Shute, P. Aquatic Animals and their Hhabitats. In *Southern Forest Resource Assessment*; Wear, D.N., Carter, D.R., Prestemon, J., Eds.; South. Res. Sta. Gen. Tech. Rep. SRS-53; U.S. Department of Agriculture Forest Service: Asheville, NC, USA, 2002; pp. 537–580.

31. Berger, C. *Wild Guide: Dragonflies*; Stackpole Books: Mechanicsburg, PA, USA, 2004; ISBN 0-8117-2971-0.

32. Watters, T.G.; Hoggarth, M.A.; Stansbery, D.H. *The Freshwater Mussels of Ohio*; The Ohio University Press: Columbus, OH, USA, 2009; ISBN 978-0-8142-1105-2.

33. Thorp, J.H.; Rogers, C.D. *Field Guide to Freshwater Invertebrates of North America*; Elsevier: San Diego, CA, USA, 2011; ISBN 978-0-12-381426-5.

34. Jones, E.B.D.; Helfman, G.S.; Harper, J.O.; Bolstad, P.V. Effects of riparian forest removal on fish assemblages in Southern Appalachian streams. *Conserv. Biol.* **1999**, *13*, 1454–1465. [CrossRef]

35. Ames, T., Jr. *Caddisflies: A Guide to Eastern Species for Anglers and Other Naturalists*; Stackpole Books: Mechanicsburg, PA, USA, 2009; ISBN 978-1-5718-8210-3.

36. Johnson, P.D. *Sustaining America's Aquatic Biodiversity: Freshwater Snail Biodiversity and Conservation*; VCE Pub. 420-530; Virginia Cooperative Extension: Blacksburg, VA, USA, 2009.

37. Kreutzweiser, D.; Capell, S.; Good, K. Effects of fine sediment inputs from a logging road on stream insect communities: A large-scale experimental approach in a Canadian headwater stream. *Aquat. Ecol.* **2005**, *39*, 55–66. [CrossRef]

38. Kondratieff, B.C. *Smokies Needlefly*; South Carolina Department of Natural Resources: Columbia, SC, USA, 2005.

39. Richards, C.; Hollingsworth, B. Managing Riparian Areas for Fish. In *Riparian Management in Forests of the Continental Eastern United States*; Verry, E.S., Hornbeck, J.W., Dolloff, C.A., Eds.; Lewis Publishers: New York, NY, USA, 2000; pp. 157–168. ISBN 978-1-56-670501-1.

40. Nislow, K.H.; Lowe, W.H. Influences of logging history and riparian forest characteristics on macroinvertebrates and brook trout (*Salvelinus fontinalis*) in headwater streams (New Hampshire, USA). *Freshw. Biol.* **2006**, *51*, 388–397. [CrossRef]

41. Moseley, K.R.; Ford, W.M.; Edwards, J.W.; Schuler, T.M. Long-term partial cutting impacts on *Desmognathus* salamander abundance in West Virginia headwater streams. *For. Ecol. Manag.* **2008**, *254*, 300–307. [CrossRef]

42. Moseley, K.R.; Ford, W.M.; Edwards, J.W. Local and landscape scale factors influencing edge effects on woodland salamanders. *Environ. Monit. Assess.* **2009**, *151*, 425–435. [CrossRef] [PubMed]

43. Williams, J.D.; Bogan, A.E.; Garner, J.T. *Freshwater Mussels of Alabama and the Mobile Basin in Georgia, Mississippi and Tennessee*; The University of Alabama Press: Tuscaloosa, AL, USA, 2008; ISBN 978-0-8173-1613-6.

44. Patric, J.H. Soil erosion in the eastern forest. *J. For.* **1976**, *74*, 671–677.

45. Patric, J.H. Harvesting effects on soil and water in the eastern hardwood forest. *South. J. Appl. For.* **1978**, *2*, 66–73.

46. Newbold, J.D.; Erman, D.C.; Roby, K.B. Effects of logging on macroinvertebrates in streams with and without buffer strip. *Can. J. Fish. Aquat. Sci.* **1980**, *37*, 1076–1085. [CrossRef]

47. Crow, T.R.; Baker, M.E.; Barnes, B.V. Diversity in Riparian Landscapes. In *Riparian Management in Forests of the Continental Eastern United States*; Verry, E.S., Hornbeck, J.W., Dolloff, C.A., Eds.; Lewis Publishers: Boca Raton, FL, USA, 1999; p. 402. ISBN 978-1-5667-0501-1.

48. Aust, W.M.; Bolding, M.C.; Barrett, S.B. Best management practices for low-volume roads in the Piedmont region: Summary and implications of research. *J. Transp. Rev. Board* **2015**, *2472*, 51–55. [CrossRef]

49. Sosa-Perez, G.; MacDonald, L.H. Reductions in road sediment production and road-stream connectivity from two decommissioning treatments. *For. Ecol. Manag.* **2017**, *398*, 116–129. [CrossRef]

50. Swift, L.W. Forest road design to minimize erosion in the Southern Appalachians. In *Proceedings of Forestry and Water Quality: A Mid-South Symposium*; Blackman, B.G., Ed.; University of Arkansas: Monticello, Little Rock, AR, USA, 1985; pp. 141–151.

51. Grace, J.M., III. Control of sediment export from the forest road prism. *Am. Soc. Agric. Biol. Eng.* **2002**, *45*, 1127–1132.

52. Wear, D.N.; Greis, J.G. Southern forest resource assessment: Summary of findings. *J. For.* **2002**, *100*, 6–14.

53. Lakel, W.A.; Aust, W.M.; Bolding, M.C.; Dolloff, C.A.; Keyser, P.; Feldt, R. Sediment trapping by streamside management zones of various widths after forest harvest and site preparation. *For. Sci.* **2010**, *56*, 541–551.

54. Clinton, B. Stream water responses to timber harvest: Riparian buffer width effectiveness. *For. Ecol. Manag.* **2011**, *261*, 979–988. [CrossRef]

55. Sawyers, B.C.; Bolding, M.C.; Aust, W.M.; Lakel, W.A. Effectiveness and implementation costs of overland skid trail closure techniques in the Virginia Piedmont. *J. Soil Water Conserv.* **2012**, *67*, 300–310. [CrossRef]

56. Wade, C.R.; Bolding, M.C.; Aust, W.M.; Lakel, W.A. Comparison of five erosion control techniques for bladed skid trails in Virginia. *South. J. Appl. For.* **2012**, *36*, 191–197. [CrossRef]

57. Brown, K.R.; McGuire, K.J.; Aust, W.M.; Hession, W.C.; Dolloff, C.A. The effect of increasing gravel cover on forest roads for reduced sediment delivery to stream crossings. *Hydrol. Proc.* **2015**, *29*, 1129–1140. [CrossRef]

58. Lang, A.J.; Aust, W.M.; Bolding, M.C.; McGuire, K.; Schilling, E.B. Comparing sediment trap data with erosion models for evaluation of haul road stream crossing approaches. *Trans. Am. Soc. Agric. Biol. Eng.* **2017**, *60*, 393–408.

59. Kochenderfer, J.N.; Helvey, J.D. Using gravel to reduce soil losses from minimum-standard forest roads. *J. Soil Water Conserv.* **1987**, *42*, 46–50.

60. Wade, C.R.; Bolding, M.C.; Aust, W.M.; Lakel, W.A.; Schilling, E.B. Comparing sediment trap data with the USLE-Forest, RUSLE2, and WEPP-road erosion models for evaluation of bladed skid trail BMPs. *Trans. ASABE* **2012**, *55*, 403–414. [CrossRef]

61. Brown, K.R.; Aust, W.M.; McGuire, K.J. Sediment delivery from bare and graveled forest road stream crossing approaches in the Virginia Piedmont. *For. Ecol. Manag.* **2013**, *310*, 836–846. [CrossRef]

62. Wear, L.R.; Aust, W.M.; Bolding, M.C.; Strahm, B.D.; Dolloff, C.A. Effectiveness of best management practices for sediment reduction at operational forest stream crossings. *For. Ecol. Manag.* **2013**, *289*, 551–561. [CrossRef]

63. Vinson, J.A.; Barrett, S.M.; Aust, W.M.; Bolding, M.C. Evaluation of bladed skid trail closure methods in the ridge and valley region. *For. Sci.* **2017**, *63*, 432–440. [CrossRef]

64. Gravelle, J.A.; Link, T.E.; Broglio, J.R.; Braatne, J.H. Effects of timber harvest on aquatic macroinvertebrate community composition in a northern Idaho watershed. *For. Sci.* **2009**, *55*, 352–366.

65. Appelboom, T.; Chescheir, G.; Skaggs, R.; Hesterberg, D. Management practices for sediment reduction from forest roads in the coastal plains. *Trans. ASAE* **2002**, *45*, 337–344. [CrossRef]

66. Litschert, S.E.; MacDonald, L.H. Frequency and characteristics of sediment delivery pathways from forest harvest units to streams. *For. Ecol. Manag.* **2009**, *259*, 143–150. [CrossRef]

67. Clayton, J.L.; Stihler, C.W.; Wallace, J.L. Status of and potential impacts to the freshwater bivalves (Unionidae) in Patterson Creek, West Virginia. *Northeast. Nat.* **2001**, *8*, 179–188. [CrossRef]

68. Arthur, M.A.; Coltharp, G.B.; Brown, D.L. Effects of best management practices on forest streamwater quality in eastern Kentucky. *J. Am. Water Resour. Assoc.* **1998**, *34*, 481–495. [CrossRef]

69. Ward, J.M.; Jackson, C.R. Sediment trapping within forestry streamside management zones: Georgia Piedmont, USA. *J. Am. Water Resour. Assoc.* **2004**, *40*, 1421–1431. [CrossRef]

70. Keim, R.F.; Schoenholtz, S.H. Functions and effectiveness of silvicultural streamside management zones in loessial bluff forests. *For. Ecol. Manag.* **1999**, *118*, 197–209. [CrossRef]

71. Levine, J.F.; Bogan, A.E.; Pollock, K.H.; Devine, H.A.; Gustafson, L.L.; Eads, C.B.; Russell, P.P.; Anderson, E.F. *Final Report: Distribution of Freshwater Mussel Populations in Relationship to Crossing Structures*; North Carolina State University: Raleigh, NC, USA, 2003.

72. Gibson, R.J.; Haedrich, R.L.; Wernerheim, C.M. Loss of fish habitat as a consequence of inappropriately constructed stream crossings. *Fisheries* **2005**, *30*, 10–17. [CrossRef]

73. Park, D.; Sullivan, M.; Bayne, E.; Scrimgeour, G. Landscape-level stream fragmentation caused by hanging culverts along roads in Alberta's boreal forest. *Can. J. For. Res.* **2008**, *38*, 566–575. [CrossRef]

74. Aust, W.M.; Carroll, M.B.; Bolding, M.C.; Dolloff, C.A. Operational forest stream crossings effects on water quality in the Virginia Piedmont. *South. J. Appl. For.* **2011**, *35*, 123–130.

75. Nolan, L.; Aust, W.M.; Barrett, S.M.; Bolding, M.C.; Brown, K.; McGuire, K. Estimating costs and effectiveness of upgrades in forestry best management practices for stream crossings. *Water* **2015**, *7*, 6946–6966. [CrossRef]

76. Warren, M.L.; Pardew, M.G. Road crossings as barriers to small-stream fish movement. *Trans. Am. Fish. Soc.* **1998**, *127*, 637–644. [CrossRef]

77. Foster, H.R.; Keller, T.A. Flow in culverts as a potential mechanism of stream fragmentation for native and nonindigenous crayfish species. *J. N. Am. Benthol. Soc.* **2011**, *30*, 1129–1137. [CrossRef]

78. Louca, V.; Ream, H.M.; Findlay, J.D.; Latham, D.; Lucas, M.C. Do culverts impact the movements of the endangered white-clawed crayfish? *Knowl. Manag. Aquat. Ecosyst.* **2014**, *414*, 14. [CrossRef]

79. Poole, K.E.; Downing, J.A. Relationship of declining mussel biodiversity to stream-reach and watershed characteristics in an agricultural landscape. *J. N. Am. Benthol. Soc.* **2004**, *23*, 114–125. [CrossRef]

80. Diamond, J.M.; Bressler, D.W.; Serveiss, V.B. Assessing relationships between human land uses and the decline of native mussels, fish, and macroinvertebrates in the Clinch and Powell River watershed, USA. *Environ. Toxicol. Chem.* **2002**, *21*, 1147–1155. [CrossRef] [PubMed]

81. Merrill, M.A. The Effects of Culverts and Bridges on Stream Geomorphology. Master's Thesis, North Carolina State University, Raleigh, NC, USA, 2005.

82. Bambarger, A.R. Freshwater Mussel Communities of the Florida Parishes, Louisiana: The Importance of Spatial Scale. Master's Thesis, Louisiana State University, Baton Rouge, LA, USA, 2006.

83. Diebel, M.W.; Fedora, M.; Cogswell, S.; O'Hanley, J.R. Effects of road crossings on habitat connectivity for stream-resident fish. *River Res. Appl.* **2015**, *31*, 1251–1261. [CrossRef]

84. Levine, J.F.; Eads, C.B.; Cope, W.G.; Humphries, L.F.; Bringolf, R.B.; Lazaro, P.R.; Shea, D.; Pluym, J.V.; Eggleston, D.; Merril, M.A.; et al. *Final Report: A Comparison of the Impacts of Culverts Versus Bridges on Stream Habitat and Aquatic Fauna*; North Carolina State University: Raleigh, NC, USA, 2007.

85. Jensen, K.M. Velocity Reduction Factors in Near Boundary Flow and the Effect on Fish Passage through Culverts. Master's Thesis, Brigham Young University, Provo, UT, USA, 2014.

86. Kemp, P.S.; O'Hanley, J.R. Procedures for evaluating and prioritising the removal of fish passage barriers: A synthesis. *Fish. Manag. Ecol.* **2010**, *17*, 297–322. [CrossRef]

87. Hotchkiss, R.H.; Frei, C.M. *Design for Fish Passage at Roadway-Stream Crossings: Synthesis Report*; Federal Highway Administration: McLean, VA, USA, 2007.

88. Vaughan, D.M. *Potential Impact of Road-Stream Crossings (Culverts) on the Upstream Passage of Aquatic Macroinvertebrates*; Report to the USDA Forest Service; The Xerces Society: Portland, OR, USA, 2002.

89. Rivera, C.J.R. *Obstruction of the Upstream Migration of the Invasive Snail Cipangopaludina chinensis by High Water Currents*; Summer UNDERC Project (BIOS 35502: Practicum in Field Biology); University of Notre Dame: Notre Dame, IN, USA, 2008.

90. Jackson, S.D. Ecological considerations in the design of river and stream crossings. In *International Conference on Ecology and Transportation*; University of Massachusetts Amherst: Amherst, MA, USA, 2003; pp. 24–29.

91. Resh, V.H. Stream crossings and the conservation of diadromous invertebrates in South Pacific island streams. *Aquat. Conserv.* **2005**, *15*, 313–317. [CrossRef]

92. Clennon, J.A.; King, C.H.; Muchiri, E.M.; Kitron, U. Hydrological modelling of snail dispersal patterns in Msambweni, Kenya, and potential resurgence of *Schistosoma haematobium* transmission. *Parasitology* **2007**, *134*, 683–693. [CrossRef] [PubMed]

93. Blakely, T.; Harding, J.; McIntosh, A. *Impacts of Urbanisation in Okeover Stream, Christchurch (Report)*; Freshwater Ecology Research Group, University of Canterbury: Christchurch, New Zealand, 2003; p. 25.

94. Harding, J.; Neumegen, R.; van den Braak, I. Where have all the caddis gone? The role of culverts, and spiders. In Proceedings of the American Geophysical Union Spring Meeting, New Orleans, LA, USA, 23–27 May 2005. Abstract Number NB14C-01.

95. Blakely, T.J.; Harding, J.S.; McIntosh, A.R.; Winterbourn, M.J. Barriers to the recovery of aquatic insect communities in urban streams. *Freshw. Biol.* **2006**, *51*, 1634–1645. [CrossRef]

96. Smith, R.F.; Alexander, L.C.; Lamp, W.O. Dispersal by terrestrial stages of stream insects in urban watersheds: A synthesis of current knowledge. *J. N. Am. Benthol. Soc.* **2009**, *28*, 1022–1037. [CrossRef]

97. Lowrance, R.; Altier, L.S.; Williams, R.G.; Inamdar, S.P.; Sheridan, J.M.; Bosch, D.D.; Hubbard, R.K.; Thomas, D.L. REMM: The riparian ecosystem management model. *J. Soil Water Conserv.* **2000**, *55*, 27–34.

98. Lee, K.H.; Isenhart, T.M.; Schultz, R.C. Sediment and nutrient removal in an established multi-species riparian buffer. *J. Soil Water Conserv.* **2003**, *58*, 1–8.

99. Newbold, J.D.; Herbert, S.; Sweeney, B.W.; Kiry, P.; Alberts, S.J. Water quality functions of a 15-year-old riparian forest buffer system. *J. Am. Water Resour. Assoc.* **2010**, *46*, 299–310. [CrossRef]

100. Carroll, G.D.; Schoenholtz, S.H.; Young, B.W.; Dibble, E.D. Effectiveness of forestry streamside management zones in the sand-clay hills of Mississippi: Early indications. *Water Air Soil Pollut. Focus* **2004**, *4*, 275–296. [CrossRef]

101. Zokaites, C. *Living on Karst: A Reference Guide for Landowners in Limestone Regions*; Cave Conservancy of the Virginias, Virginia Department of Conservation and Recreation: Richmond, VA, USA, 1997.

102. Kiffney, P.M.; Richardson, J.S.; Bull, J.P. Responses of periphyton and insects to experimental manipulation of riparian buffer width along forest streams. *J. Appl. Ecol.* **2003**, *40*, 1060–1076. [CrossRef]

103. Summerville, K.S.; Dupont, M.M.; Johnson, A.V.; Krehbiel, R.L. Spatial structure of forest lepidopteran communities in oak hickory forests of Indiana. *Environ. Entomol.* **2008**, *37*, 1224–1230. [CrossRef] [PubMed]

104. Summerville, K.S.; Courard-Hauri, D.; Dupont, M.M. The legacy of timber harvest: Do patterns of species dominance suggest recovery of lepidopteran communities in managed hardwood stands? *For. Ecol. Manag.* **2009**, *259*, 8–13. [CrossRef]

105. Fong, D.W. Management of subterranean fauna in karst. In *Karst Management*; van Beynen, P.E., Ed.; Springer: Dordrecht, The Netherlands, 2011; pp. 201–224. ISBN 978-94-007-1206-5.

106. Summerville, K.S.; Saunders, M.R.; Lane, J.L. The Lepidoptera as predictable communities of herbivores: A test of niche assembly using the moth communities of Morgan-Monroe State Forest. In *The Hardwood Ecosystem Experiment: A Framework for Studying Responses to Forest Management*; Swihart, R.K., Saunders, M.R., Kalb, R.A., Haulton, G.S., Michler, C.H., Eds.; U.S. Department of Agriculture, Forest Service, Northern Research Station: Newtown Square, PA, USA, 2013; pp. 237–252.

107. Dickson, J.G.; Williamson, J.H. Small Mammals in Streamside Management Zones in Pine Plantations. In *Proceedings of the Symposium on Management of Amphibians, Reptiles, and Small Mammals in North America, Flagstaff, AZ, USA, 19–21 July 1988*; Rocky Mtn. For. & Range Exp. Sta.: Gen. Tech. Rep. RM-166; U.S. Department of Agriculture Forest Service: Fort Collins, CO, USA; pp. 375–378.

108. Miller, D.A.; Thill, R.E.; Melchiors, M.A.; Wigley, T.B.; Tappe, P.A. Small mammal communities of streamside management zones in intensively managed pine forests of Arkansas. *For. Ecol. Manag.* **2004**, *203*, 381–393. [CrossRef]

109. Sweeney, B.W. Effects of Streamside Vegetation on Macroinvertebrate Communities of White Clay Creek in Eastern North America. *Proc. Acad. Nat. Sci. Phila.* **1993**, *144*, 291–340.

110. Flebbe, P.A.; Dolloff, C.A. Trout use of woody debris and habitat in Appalachian wilderness streams of North Carolina. *N. Am. J. Fish. Manag.* **1995**, *15*, 579–590. [CrossRef]

111. Hilderbrand, R.H.; Lemly, A.D.; Dolloff, C.A.; Harpster, K.L. Effects of large woody debris placement on stream channels and benthic macroinvertebrates. *Can. J. Fish. Aquat. Sci.* **1997**, *54*, 931–939. [CrossRef]

112. Dolloff, C.A.; Webster, J.R. Particulate organic contributions from forests and streams: Debris isn't so bad. In *Riparian Management in Forests of the Continental Eastern United States*; Verry, E.S., Hornbeck, J.W., Dolloff, C.A., Eds.; Lewis Publishers: Boca Raton, FL, USA, 2000; pp. 125–138. ISBN 978-1-5667-0501-1.

113. Holtby, L.B. Effects of logging on stream temperatures in Carnation Creek, British Columbia, and associated impacts on the coho salmon (*Oncorhynchus kisutch*). *Can. J. Fish. Aquat. Sci.* **1988**, *45*, 502–515. [CrossRef]

114. Hickey, M.B.C.; Doran, B. A review of the efficiency of buffer strips for the maintenance and enhancement of riparian ecosystems. *Water Qual. Res. J. Can.* **2004**, *39*, 311–317.

115. Verry, E.S.; Dolloff, C.A.; Manning, M.E. Riparian ecotone: A functional definition and delineation for resource assessment. *Water Air Soil Pollut. Focus* **2004**, *4*, 67–94. [CrossRef]

116. Hamer, K.C.; Hill, J.K.; Benedick, S.; Mustaffa, N.; Sherratt, T.N.; Maryati, M.; Chey, V.K. Ecology of butterflies in natural and selectively logged forests of northern Borneo: The importance of habitat heterogeneity. *J. Appl. Ecol.* **2003**, *40*, 150–162. [CrossRef]

117. Remsburg, A.J.; Olson, A.C.; Samways, M.J. Shade alone reduces adult dragonfly (Odonata: Libellulidae) abundance. *J. Insect Behav.* **2008**, *21*, 460–468. [CrossRef]

118. Myers, J.H.; Cory, J.S. Population cycles in forest lepidoptera revisited. *Annu. Rev. Ecol. Evol. Syst.* **2013**, *44*, 565–592. [CrossRef]

119. Swift, L.W.; Messer, J.B. Forest cuttings raise temperatures of small streams in the southern Appalachians. *J. Soil Water Conserv.* **1971**, *26*, 111–116.

120. Samways, M.J.; Taylor, S. Impacts of invasive alien plants on Red-Listed South African dragonflies (Odonata). *S. Afr. J. Sci.* **2004**, *100*, 78–80.

121. Janisch, J.E.; Wondzell, S.M.; Ehinger, W.J. Headwater stream temperature: Interpreting response after logging, with and without riparian buffers, Washington, USA. *For. Ecol. Manag.* **2012**, *270*, 302–313. [CrossRef]

122. Ford, W.M.; Chapman, B.R.; Menzel, M.A.; Odum, R.H. Stand age and habitat influences on salamanders in Appalachian cover hardwood forests. *For. Ecol. Manag.* **2002**, *155*, 131–141. [CrossRef]

123. Rudolph, D.C.; Dickson, J.G. Streamside zone width and amphibian and reptile abundance. *Southwest. Nat.* **1990**, *35*, 472–476. [CrossRef]

124. Sterrett, S.C.; Smith, L.L.; Schweitzer, S.H.; Maerz, J.C. An assessment of two methods for sampling river turtle assemblages. *Herpetol. Conserv. Biol.* **2010**, *5*, 490.

125. Bisson, P.A.; Bilby, R.E.; Bryant, M.D.; Dolloff, C.A.; Grette, G.; House, R.A.; Murphy, M.L.; Koski, K.V.; Sedell, J.R. Large Woody Debris in Forested Streams in the Pacific Northwest: Past, Present, and Future. In *Proceedings of the Symposium on Streamside Management: Forestry and Fishery Interactions, Seattle, DC, USA, 1987*; University of Washington: Seattle, DC, USA, 1987.

126. Hairston-Strang, A.B.; Adams, P.W. Potential large woody debris sources in riparian buffers after harvesting in Oregon, USA. *For. Ecol. Manag.* **1998**, *112*, 67–77. [CrossRef]

127. Hartman, G.; Scrivener, J.; Miles, M. Impacts of logging in Carnation Creek, a high-energy coastal stream in British Columbia, and their implication for restoring fish habitat. *Can. J. Fish. Aquat. Sci.* **1996**, *53*, 237–251. [CrossRef]

128. Foley, D.H. Short-Term Response of Herpetofauna to Timber Harvesting in Conjunction with Streamside-Management Zones in Seasonally-Flooded Bottomland-Hardwood Forests of Southeast Texas. Master's Thesis, Texas A&M University, College Station, TX, USA, 1994.

129. DeMaynadier, P.G.; Hunter, M.L., Jr. The relationship between forest management and amphibian ecology: A review of the North American literature. *Environ. Rev.* **1995**, *3*, 230–261. [CrossRef]

130. Broadmeadow, S.; Nisbet, T.R. The effects of riparian forest management on the freshwater environment: A literature review of best management practice. *Hydrol. Earth Syst. Sci. Dis.* **2004**, *8*, 286–305. [CrossRef]

131. DaSilva, A.; Xu, Y.J.; Ice, G.; Beebe, J.; Stich, R. Effects of timber harvesting with best management practices on ecosystem metabolism of a low gradient stream on the United States Gulf Coastal Plain. *Water* **2013**, *5*, 747–766. [CrossRef]

132. Hutchens, J.J.; Batzer, D.P.; Reese, E. Bioassessment of silvicultural impacts in streams and wetlands of the eastern United States. *Water Air Soil Pollut. Focus* **2004**, *4*, 37–53. [CrossRef]

133. Gurtz, M.E.; Wallace, J.B. Substrate-mediated response of stream invertebrates to disturbance. *Ecology* **1984**, *65*, 1556–1569. [CrossRef]

134. Adams, T.O.; Hook, D.D.; Floyd, M.A. Effectiveness monitoring of silvicultural best management practices in South Carolina. *South. J. Appl. For.* **1995**, *19*, 170–176.

135. Kedzierski, W.M.; Smock, L.A. Effects of logging on macroinvertebrate production in a sand-bottomed, low-gradient stream. *Freshw. Biol.* **2001**, *46*, 821–833. [CrossRef]

136. Vowell, J.L. Using stream bioassessment to monitor best management practice effectiveness. *For. Ecol. Manag.* **2001**, *143*, 237–244. [CrossRef]

137. Williams, L.R.; Taylor, C.M.; Warren, M.L., Jr.; Clingenpeel, J.A. Large-scale effects of timber harvesting on stream systems in the Ouachita Mountains, Arkansas, USA. *Environ. Manag.* **2002**, *29*, 76–87. [CrossRef]

138. Vowell, J.L.; Frydenborg, R.B. A biological assessment of best management practice effectiveness during intensive silviculture and forest chemical application. *Water Air Soil Pollut. Focus* **2004**, *4*, 297–307. [CrossRef]

139. Griswold, M.W.; Winn, R.T.; Crisman, T.L.; White, W.R. *Dry Creek Long-Term Watershed Study: Assessment of Immediate Response of Aquatic Macroinvertebrates to Watershed Level Harvesting and Thinning of Streamside Management Zones*; U.S. Department of Agriculture, Forest Service, Southern Research Station: Asheville, NC, USA, 2006; pp. 392–395.

140. McCord, S.B.; Grippo, R.S.; Eagle, D.M. Effects of silviculture using best management practices on stream macroinvertebrate communities in three ecoregions of Arkansas, USA. *Water Air Soil Pollut.* **2007**, *184*, 299–311. [CrossRef]

141. Grippo, R.S.; McCord, S.B. *Bioassessment of Silviculture Best Management Practices in Arkansas*; Arkansas State University College of Science and Mathematics: Jonesboro, AR, USA, 2006.

142. Simpson, H.; Work, D.; Harrington, S. *Evaluating the Effectiveness of Texas Forestry Best Management Practices: Results from the Texas Silvicultural BMP Effectiveness Monitoring Project 2003–2007*; Texas Forest Service: Lufkin, TX, USA, 2008.

143. Machtans, C.S.; Villard, M.-A.; Hannon, S.J. Use of riparian buffer strips as movement corridors by forest birds. *Conserv. Biol.* **1996**, *10*, 1366–1379. [CrossRef]

144. Lindenmayer, D.B.; Hobbs, R.J. Fauna conservation in Australian plantation forests–A review. *Biol. Conserv.* **2004**, *119*, 151–168. [CrossRef]

145. Shirley, S.M.; Smith, J.N. Bird community structure across riparian buffer strips of varying width in a coastal temperate forest. *Biol. Conserv.* **2005**, *125*, 475–489. [CrossRef]

146. Perkins, D.W.; Hunter, M.L., Jr. Effects of riparian timber management on amphibians in Maine. *J. Wildl. Manag.* **2006**, *70*, 657–670. [CrossRef]

147. Lindenmayer, D.B.; Manning, A.D.; Smith, P.L.; Possingham, H.P.; Fischer, J.; Oliver, I.; McCarthy, M.A. The focal-species approach and landscape restoration: A critique. *Conserv. Biol.* **2002**, *16*, 338–345. [CrossRef]

148. Fischer, J.; Lindenmayer, D.B.; Manning, A.D. Biodiversity, ecosystem function, and resilience: Ten guiding principles for commodity production landscapes. *Front. Ecol. Environ.* **2006**, *4*, 80–86. [CrossRef]

Article

Characterizing Rigging Crew Proximity to Hazards on Cable Logging Operations Using GNSS-RF: Effect of GNSS Positioning Error on Worker Safety Status

Ann M. Wempe * and Robert F. Keefe

Forest Operations Research Lab, College of Natural Resources, University of Idaho, 875 Perimeter Drive, Moscow, ID 83844-1133, USA; robk@uidaho.edu
* Correspondence: awempe@uidaho.edu; Tel.: +1-208-885-6695

Received: 9 August 2017; Accepted: 18 September 2017; Published: 23 September 2017

Abstract: Logging continues to rank among the most lethal occupations in the United States. Though the hazards associated with fatalities are well-documented and safe distances from hazards is a common theme in safety education, positional relationships between workers and hazards have not been quantified previously. Using GNSS-RF (Global Navigation Satellite System-Radio Frequency) transponders that allow real-time monitoring of personnel, we collected positioning data for rigging crew workers and three common cable logging hazards: a log loader, skyline carriage, and snag. We summarized distances between all ground workers and each hazard on three active operations and estimated the proportion of time crew occupied higher-risk areas, as represented by geofences. We then assessed the extent to which positioning error associated with different stand conditions affected perceived worker safety status by applying error sampled in a separate, controlled field experiment to the operational data. Root mean squared error was estimated at 11.08 m in mature stands and 3.37 m in clearcuts. Simulated error expected for mature stands altered safety status in six of nine treatment combinations, whereas error expected for clearcuts affected only one. Our results show that canopy-associated GNSS error affects real-time geofence safety applications when using single-constellation American Global Positioning System transponders.

Keywords: GNSS-RF; GPS; GNSS positioning error; logging safety; geofences

---

1. Introduction

Ground workers on cable logging operations work in close proximity to multiple, moving hazards, including highly active heavy equipment, raw materials, and other objects that are swung, dragged, dropped, and dislodged on steep slopes. Proximity to these hazards creates potentially injurious situations for cable logging workers [1–4]. Unlike mechanized, ground-based operations in which employees are generally working within enclosed machine cabs, cable operations rely on ground crew who work unprotected alongside equipment and other hazards in a dynamic environment. Hand fallers and members of the rigging crew face increased risk of injury from hazards such as falling limbs or falling live (green) and dead trees, as well as rolling logs and rocks on steep slopes [1,4,5]. Although United States Occupational Safety and Health Administration (OSHA) and state-level regulations require felling of standing dead trees (snags) within active logging areas [6] (1910.266(h)(1)(vi)), snags may still be present on the periphery of units and during initial work periods prior to felling.

The dangerous nature of logging work is reflected in the industry's high fatality injury rates, as published annually by the Bureau of Labor Statistics (BLS). The BLS' 2015 Census of Fatal Occupational Injuries reported 132.7 logger deaths for every 100,000 full-time employees, which was the highest rate of any profession in the United States in 2015 [7]. The rate increased 20% from 2014, when logging also ranked as the most fatal occupation [8]. Lefort et al., who characterized logger injuries in the late 1980s and early 1990s, noted that mechanization of the logging industry had reduced the total number of workplace accidents, but had triggered an increase in injury severity [3]. They tattributed this trend to the changing nature of exposure; ground crew are now working closer to the landing where they face impacts from moving logs and machinery. In 2015, the BLS identified trees, logs, or limbs as the primary source of fatal injury in 41 of 80 total occupational deaths in the logging industry, while 14 deaths were attributed to machinery [9]. Consistent with reports by the Bureau of Labor Statistics, an analysis of Worker's Compensation claims in West Virginia indicated objects, primarily trees, snags, or logs, striking crew members accounted for 47% of injuries, more than any other cause [1]. Similarly, according to claims records from eight southern states in 1997, falling trees or limbs and moving logs caused the most accidents (28% of injuries), followed by equipment, including skidders, feller-bunchers, dozers, and loaders (23%) [4].

GNSS-RF (Global Navigation Satellite System-Radio Frequency) transponders have the potential to reduce the incidence of injuries and fatalities on logging operations by improving situational awareness. GNSS-RF units determine their coordinates from one or more navigation satellite systems, including the United States' Global Positioning System (GPS), Russia's Global Navigation Satellite System (GLONASS), China's BeiDou, or Europe's Galileo. They then transmit those coordinates to other units locally by data transfer using radio frequency transmission. Used in conjunction with mobile devices such as phones or handheld tablets, or onboard computers, GNSS-RF technology can provide a real-time, systemic visualization of all the interacting components of logging operations, supplementing voice communications used conventionally on two-way radios and signal horns such as Talkie-Tooters (Rothenbuhler Engineering, Sedro Woolley, WA, USA). With knowledge of ground crew positions in relation to potential hazards, machine operators could make more informed decisions based on the known locations of workers displayed on maps on mobile devices and, in some cases, supplement the use of conventional audible communication with visual or audible alerts indicating worker presence in work zones delineated by geofences [10–12].

GNSS has been utilized widely in forestry for decades. GNSS is integrated into Geographic Information Systems (GIS) to map ownerships and delineate stand boundaries, forest road locations, and other features on timber sales [13]. Mobile positioning devices have been installed on harvesting machines to track movement over the course of harvest operations and assess soil impacts and performance [14,15]. GNSS is increasingly being used in place of traditional, observational methods to characterize productive cycle-times of forest machines [16–18]. Harvesters have been fitted with GNSS devices to collect tree positioning data [19,20]. Development of GNSS paired with RF for real-time positioning is emerging quickly in forestry and has a variety of potential uses including operational and wildland fire logistics, real-time optimization, and safety [10–12,16].

Situational awareness can be augmented further by combining GNSS-RF positioning with virtual boundaries known as geofences, which delineate hazardous areas, silvicultural treatments, or work zones on timber sales [10,11]. Geofences provide a means by which to monitor the current locations of people, equipment, or other resources relative to spatial boundaries and can be programmed to alert users of crossing events [21]. They have been successfully integrated into various industries to help resolve positional monitoring and restriction needs [21–26] and have potential applications in logging to alert machine operators about ground-worker proximity [10–12].

To improve logging safety, geofence boundaries need to account for dynamic positional relationships between workers and hazards. As ground workers move throughout cable corridors, spatial proximity of people to one more pieces of equipment, snags, skyline rigging, and harvested resources are in constant flux. OSHA, which establishes guidelines and regulations for safe practices on logging operations in the United States, does not provide explicit safe distance recommendations for most logging equipment. Rather, it relies upon workers to interpret safe proximity in situational context. OSHA regulation 1910.266(f)(2)(vii) states that a "machine shall be operated at such a distance from employees and other machines such that operation will not create a hazard for an employee" [6]. Oregon OSHA Division 7 (2009), as well as common industrial safety awareness campaigns, advise workers to stay "in the clear", which generally is translated as a distance equivalent to the length of a tree or log being transported to the landing [27]. However, if loggers frequently occupy areas less than one tree length from a hazard, the geofence associated with that hazard may need to be smaller than the recommended safety distance in order for operators to discern between normal activity and higher risk situations, or early warning signals may need to be deployed. Knowledge of positional relationships will also help define GNSS accuracy needs. If ground crew generally work within 5–10 m of a hazard, positioning errors greater than 5 m may be detrimental to safety; whereas lower accuracies may still be useful in improving general awareness if workers already avoid proximity to hazardous areas. The use of mobile geofences, which can move with hazards, introduces additional considerations, such as geofence alert accuracies associated with the geometry of multiple moving components [12].

Although the sources of occupational injuries and fatalities are well-documented for logging and use of geofences for logging safety applications has been studied in designed experiments, spatial analysis of the actual positional relationships between workers and some common hazards on active operations has not been quantified or summarized previously. In fact, despite the widespread attention to spatial proximity in safety training as well as state and federal regulations in forestry, there has been virtually no prior analysis of actual positional movements among ground workers of the sort that is now possible using GNSS-RF technology. In this paper, we characterized the real-time positions of ground crew workers and three common situational hazards during active cable operations using coordinates collected by Raveon Atlas PT GNSS-RF devices (Raveon Technologies Corp, San Diego, CA, USA), which feature a VHF data modem combined with a 12-channel GNSS receiver that receives position information from a single constellation, the American NAVSTAR GPS system. It is important to note that the devices do not receive positional information from GLONASS, BeiDou, or Galileo, as some other current GNSS-RF devices do. We summarized safe worker–hazard distances by calculating the amount of time, in one second increments, that workers occupied zones outside ("safe") and inside ("unsafe") circular geofence boundaries assigned to each hazard. Because forest overstory is known to impact GNSS accuracy [28–30], we also conducted a designed experiment on the University of Idaho Experimental Forest to quantify canopy impacts on receiver accuracy in both mature and recently clearcut stands. These conditions correspond to the early stages of harvesting operations (canopy intact), transitioning to later stages (canopy removed) that result during typical clearcut operations in the northwestern United States. We then used simulation to re-analyze our operational data, in order to evaluate the extent to which canopy-induced error, as determined in the earlier designed experiment, affected the GNSS-characterized safety status of ground workers over the course of active operations.

Our specific objectives were to determine whether the proportion of unsafe time, defined as time spent inside one or more hazard geofences, differed by (1) hazard type (loader, carriage, snag); (2) timber sale; or (3) GNSS environment (observed, mature, clearcut).

2. Materials and Methods

2.1. Controlled Experiment

2.1.1. Data Collection

To estimate the impact of GNSS error on operational positioning data collected at logging operations, we first calculated Atlas PT error in a controlled experiment on the University of Idaho Experimental Forest (UIEF) in Princeton, Idaho (USA). The UIEF encompasses canopy features and slopes representative of north Idaho mixed-conifer forests, ranging in age from recent clearcuts to mature stands approximately 90 years old. Eight stands were selected for sampling in the Flat Creek and East Hatter units of the UIEF, all located at mid-elevation (approximately 915 m) on the north slope of Moscow Mountain in the Palouse Range, in the vicinity of 46.8413° latitude, −116.7734° longitude. Four stands were clearcut harvested within 5 years prior to the experiment, which took place in October and November of 2016. The other four stands were over-mature, with most trees approximately 80–90 years old, having regenerated after railroad logging in the early 20th century. Based upon plot inventories completed in each stand following sampling, tree heights ranged from 3.2 to 40.3 m in mature sites (mean of 18.2 m), and diameters at breast height (DBH) ranged from 13 to 89 cm (mean of 31 cm). Stands were comprised of ponderosa pine (*Pinus ponderosa*), grand fir (*Abies grandis*), western larch (*Larix occidentalis*), western white pine (*Pinus monticola*), Douglas-fir (*Pseudotsuga menziesii*), and western red cedar (*Thuja plicata*). Table 1 shows stand characteristics measured at time of sampling (azimuth, slope) and during inventory (height, DBH), as well as sampling conditions, including satellite availability and constellation. The number of in-view satellites was recorded every second by each of four Atlas PT transponders during sampling and then averaged across all devices. A Differential GPS (DGPS) GNSS receiver, the Arrow 100 made by EOS Positioning Systems (Terrebonne, QC, Canada), collected position dilution of precision (PDOP) values at each Atlas PT location. Low PDOP values (less than 4) indicate lower GNSS positioning error and are a function of satellite constellation orientation.

Table 1. Controlled experiment: University of Idaho Experimental Forest stand characteristics and sampling conditions.

Stand	Cover	Mean Height (m)	Mean DBH (cm)	Azimith (°)	Slope (%)	Date	Mean Satellites	Mean PDOP
260	Mature	23.52	33	95	37	10/12/16	7	1.6
58	Clearcut	NA	NA	352	18	10/17/16	8	1.3
531	Clearcut	NA	NA	165	35	10/19/16	8	1.4
290	Mature	14.6	25	35	5	10/19/16	7	1.5
345	Clearcut	NA	NA	130	8	10/24/16	10	1.3
139	Mature	16.7	31	347	43	10/24/16	6	1.6
524	Mature	17.3	31	27	14	11/10/16	6	1.7
262	Clearcut	NA	NA	205	2	11/17/16	8	1.2

At each stand, we collected GNSS positioning data using four Atlas PT transponder units. We sampled for thirty minutes at a transmission frequency of one second, allowing for a potential of 1800 total observations per unit per site (actual signal transmission efficiency ranged from 0.739 to 0.997). Each unit was fastened to a wooden post using plastic zip ties, such that the base of the radio antenna was positioned at a height of one meter above the ground. The Atlas PTs were arranged in a triangular plot as shown in Figure 1, with a centrally located unit ("A") positioned 25 m in slope-distance from unit "B", 50 m from "C", and 75 m from "D". The orientation (azimuth) of unit B from A was selected randomly prior to sampling, and orientations for C and D were measured using a Suunto azimuth compass at 120° (±0.5°) from unit B's orientation.

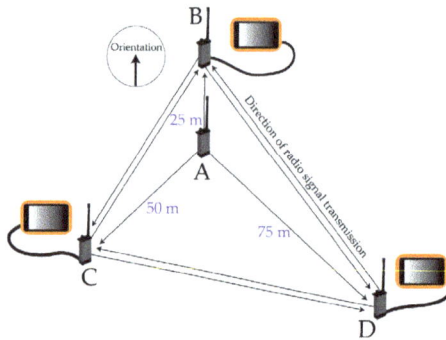

Figure 1. Design of the controlled experiment.

During sampling, each Atlas PT transmitted its location coordinates to each other unit once per second via radio frequency. Units B, C, and D were connected to Dell Venue Pro 8 5855 (Dell Inc., Round Rock, TX,

USA) tablets equipped with Raveon RavTrack PC real-time tracking software (Version 6.5, 2015, Raveon Technologies Corp, San Diego, CA, USA) and Microsoft Access™ (Version 16.0, 2016, Microsoft, Redmond, WA, USA). Tablets automatically logged transmissions to an MS Access database for subsequent analysis. The same Atlas PT and tablet were used for each position (A, B, C, D) at every stand.

2.1.2. Estimation of Error

To determine the positioning error associated with Atlas PTs, we compared the recorded (observed) coordinates to reference coordinates determined in real-time using the Arrow 100. The Atlas PTs receive single frequency (L1) signals from the United States GPS system and are capable of static horizontal accuracies of less than 2.5 m 50% of the time and less than 5 m 90% of the time. The Arrow 100 (also single frequency) is a multi-constellation receiver utilizing GLONASS and BeiDou in addition to GPS. Differential correction with a Satellite Based Augmentation System (SBAS), which is the Wide Area Augmentation System (WAAS) in the United States, enables it to achieve accuracies less than 60 cm. After the four Atlas PTs were situated in the arrangement described above, the Arrow 100 was placed at each Atlas PT position where it recorded coordinates and position dilution of precision (PDOP) values (see Table 1). After sampling, we projected all data to the Universal Transverse Mercator (UTM) projection, which has units in meters, using ArcGIS 10.3 software (Version 10.3.1, 2015, Esri, Redlands, CA, USA). We then calculated the horizontal error for each observation (each 1 second-interval transmission) as the hypotenuse distance between the two sets of UTM easting (denoted UTM_e) and UTM northing (denoted UTM_n) points (actual—observed), as shown in Equation (1). Observed coordinates were retrieved from unit B's transmission log of all four Atlas PT positions over the sample period.

$$Error = e = \sqrt{(act.UTM_e - obs.UTM_e)^2 + (act.UTM_n - obs.UTM_n)^2} \tag{1}$$

We determined if this error varied by stand, cover, or individual transponder unit using an Analysis of Variance (ANOVA), and then we identified significant sources of variation among stands and transponders using a Bonferroni multiple comparison test.

Error was summarized for each unit and each stand as the root mean square error (RMSE), which is a measure of the difference between predicted (based on the Arrow 100) and observed (based on the Atlas PT) values.

$$RMSE = \sqrt{\frac{\sum e^2}{n}} \tag{2}$$

where e represents error as calculated in Equation (1), and n represents the sample size (number of 1-second transmissions). All calculations and statistical analyses for the study were completed using R open source statistical computing software (Version 3.3.1, 2016, The R Foundation, Vienna, Austria) and are presented in the Results section [31].

2.2. Operational Sampling

2.2.1. Data Collection

GNSS positioning data were collected using Atlas PT units at three active cable logging operations in north Idaho (see Figure 2) on slopes ranging from 40–65%. All logging activities were conducted by professional, certified logging contractors on regular, operational timber sales at three ownerships: Idaho Department of Lands state endowment land (John Lewis Pole, or JLP), Potlatch Corp (Wash Trap South, or WTS) and the University of Idaho Experimental Forest (Upper Hatter, or UH). All operations were rigged for uphill yarding using motorized carriages. The state and industrial operations had swing yarders (Linkbelt 90 and Skagit GT-4, respectively) and the contractor working on the UIEF had a custom excaliner constructed on a John Deere carrier. Ground crew were responsible for setting chokers (the hooker, in regional terminology) and unhooking chokers when logs reached the landing (the chaser). The WTS and UH timber sales were clearcut operations, while the state JLP timber sale was a cedar pole harvest. Under Idaho law, cedar poles are required to be removed prior to other harvesting on cable operations when more than 10 poles per acre are present.

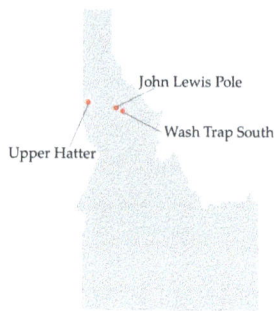

Figure 2. Idaho state map, with logging operation sites shown in red.

Excluding road (corridor) repositioning and equipment repairs, at least ten hours of positioning data were collected during regular operations at each harvest over the course of three days per site, with sampling occurring between mid-August and December 2016. Worker Atlas PT units were placed in radio pouches for protection and then distributed to ground crew, who wore the radio pouches on their belts. Machine-mounted units were attached to the external, metal grating covering cab windows, where the unit antennas (GPS and radio) were unobstructed. Skyline carriage units were secured to the carriage top using zip ties, such that antennas were exposed to the sky. At each site, an Atlas PT placed uphill on the yarder cab was connected to a Dell tablet for data collection and real-time visualization of unit positions with Raveon RavTrack software using the methods as in the controlled field experiment. In addition to collecting fluid GNSS coordinates of mobile hazards with the Atlas PT units, we also identified a snag or danger tree within the harvest unit and recorded its location with a Garmin 64 handheld GPS (Garmin Ltd., Olathe, KS, USA). Thus, our raw data comprised of GNSS positioning for two to three ground workers and three types of operational hazards: machinery (the loader), equipment (the carriage), and a stationary, environmental hazard (the snag) (see Figure 3). After sampling, coordinates from the Atlas PTs and Garmin were converted to UTM for all analyses.

Figure 3. Typical cable logging operation with (**A**) chaser; (**B**) yarder; (**C**) loader; (**D**) carriage; (**E**) snag; and (**F**) hooker (note: images are not drawn to scale). Yellow ellipses highlight areas with increased risk of injury associated with the three types of hazards shown. Of the three hazards, snag GNSS coordinates were fixed (static), while loader and carriage locations were dynamic.

2.2.2. Summarizing Worker Proximity to Hazards

We defined worker safety in terms of worker position relative to circular geofences surrounding each of the three hazards. Conceptually, the area inside each geofence was assumed to represent a work area with higher risk of fatal or non-fatal traumatic injury due to the potential for being struck by the hazard. While work inside these areas is necessary on partially mechanized operations, the presence of ground workers within geofenced areas require increased situational awareness and caution on the part of both ground workers and operators to avoid accidents. Areas beyond geofence borders encompass safe work areas, where the risk of injury from striking hazards is generally lower. Thus, at any given time, a worker was positioned either inside or outside the hazard geofences of 0–3 hazards and was classified as either safe or unsafe relative to each. Similarly, at any time, a given hazard such as the skyline carriage might have as many as three rigging crew workers in close proximity.

Using the statistical programming environment, R, we created geofences centered at each hazard's coordinates, as recorded by the associated Atlas PT (or Garmin, in the case of the snag). The carriage geofence was assigned a radius of 30 m, approximating one tree length, to encompass the risk of being struck by both the carriage itself, swinging choker cables, or logs being yarded by the carriage. The loader geofence also had a radius of 30 m (one tree length). The snag's geofence radius of 60 m represented two tree lengths, which is the standard recommended safe working distance from danger trees published by OSHA [6] (1910.266(h)(1)(vi)).

For each one second time stamp in the operational data, we calculated the distance of all ground crew from each of the three hazards. For the purposes of spatial analysis, and in order to summarize 6–7 entities moving dynamically in time and space, we grouped proximities in 5 m increments from 0–350 m. We then summed the frequency at which workers occupied each proximity zone at each of the three timber sales, as well as the proportion of time spent in safe and unsafe zones, internal and external to the geofence associated with each hazard. To simplify analysis, ground workers were analyzed as a group rather than individually. Rigging crew workers in the region regularly alternate roles, switching among, for example, hooking and chasing, and thus summarizing across all work tasks allowed workers to retain the same Atlas PT units without stopping to switch. Also, for the purposes of analysis, the locations of the workers and hazards reported by Atlas PT GNSS positioning were considered observed coordinates with an expected degree of error comparable to the error evaluated previously in the controlled experiment.

We used the Marascuillo Procedure for comparing multiple proportions to test the null hypotheses that the proportion of observed worker presence in unsafe areas did not differ by (1) timber sale or (2) hazard type. The Marascuillo Procedure compares the test statistic (see Equation (3)) to a critical value (Equation (4)) calculated for each pair of proportions in a way that accounts for degrees of freedom when comparing multiple proportions simultaneously.

$$value = |p_i - p_j| \tag{3}$$

$$critical\ range = r_{ij} = \sqrt{X^2_{1-\alpha,\ k-1}} \sqrt{\frac{p_i(1-p_i)}{n_i} + \frac{p_j(1-p_j)}{n_j}}, \tag{4}$$

where $X^2_{1-\alpha,k-1}$ is the chi-square distribution with a confidence interval of $1 - \alpha$ (α is the significance level) and degrees of freedom equal to $1 - k$, (k equals the number of populations). p_i represents the proportion for sample i, p_j represents the proportion for sample j, and n represents the sample size. If the value from Equation (3) is greater than the critical range, then the two compared proportions are significantly different.

2.2.3. Simulation of GNSS Error

A simulation script was written in the R language in order to assess the effect of horizontal positioning error on the safety status of individuals. For each one second time stamp in the operational data, a one second observation was selected at random from one of the four mature or clearcut plots in the controlled experiment described previously. We applied an error adjustment to the operational data based upon the UTM easting and UTM northing differences, as well as azimuth (in degrees) from the actual (Arrow 100) and observed (Atlas PT) coordinates. Thus, we assumed for the purposes of analysis that each worker location in the operational data was uncorrected, and then shifted each coordinate individually by a distance and direction corresponding to either mature canopy or clearcut error accuracy from the controlled experiment. To simplify analysis, we assumed that hazard locations were true coordinates; thus, they were not adjusted during simulation. 500 iterations of the simulation script were processed. After resampling and application of error adjustments to worker positions, inter-point distances from each of the three jobsite hazards were again summarized in zones of 5 m increments, and the proportions of safe and unsafe status were determined. Since adjustments to the operational data were sampled from individual GNSS errors recorded on multiple different sites and dates, simulated data do not represent the true location of each worker at a given time. Rather, we utilized simulation to provide an indication of the degree of impact to be expected from positioning error in relation to a fixed point (the geofence) in each GNSS environment (observed, mature, or clearcut).

We determined whether GNSS positioning error would impact definitions of workers as safe or unsafe based on proportions of time spent inside geofenced hazard zones. Using the Marascuillo Procedure, we tested the null hypothesis that unsafe proportions were equal for observed, mature, and clearcut data for each hazard at each site, where observed data represented worker positions as recorded by the Atlas PTs, mature data represented simulated worker positions accounting for GNSS error associated with canopy, and clearcut data represented simulated worker positions accounting for GNSS error under un-obstructed conditions. The Marascuillo Procedure was performed for each iteration, and the mean value was compared to the mean critical range to determine if proportions differed significantly.

3. Results

3.1. Controlled Experiment: Estimating Atlas PT Positioning Error

Plot-level Root Mean Squared Error (RMSE) calculated for all four Atlas PT units within each stand (Equation (2)) ranged from 2.64 m to 4.09 m in clearcuts, with the best accuracy achieved in Stand 531. By contrast, RMSE ranged from 8.56 m to 14.34 m in mature stands, with the lowest accuracy occurring in Stand 524 (see Table 2 for RMSE by stand and unit). The RMSE of all mature stands combined was 11.08 m, while the overall RMSE of clearcuts was 3.37 m. With the exception of one unit (B in Stand 58), RMSE values in clearcut conditions are under the 5 m accuracy expected for Atlas PTs 90% of the time. However, none of the devices in mature stands achieved this level of accuracy.

Table 2. Root mean square error (RMSE) of Atlas PT GNSS horizontal positioning error, in meters, at each unit position (A, B, C, and D) and across all units (last column).

Cover	Stand	Unit RMSE (m)				
		A	B	C	D	All Units
MATURE	260	13.07	8.29	12.52	4.79	10.34
	290	11.68	6.91	7.48	7.28	8.56
	139	11.23	7.68	12.78	7.90	10.14
	524	18.17	11.16	11.59	15.30	14.34
CLEARCUT	58	4.33	5.36	2.98	1.46	3.81
	531	2.42	1.67	3.16	3.01	2.64
	345	1.38	1.49	3.09	3.66	2.67
	262	3.96	3.69	4.52	4.07	4.09

Actual error calculated for each second of sampling (Equation (1)) varied significantly by stand (F-statistic = 4735, p-value < 2×10^{-16}), cover (F-statistic = 25,390, p-value < 2×10^{-16}), and individual transponder unit (F-statistic = 337.3, p-value < 2×10^{-16}). The Bonferroni multiple comparison test comparing all stands indicated that only two stands did not differ significantly from one another (clearcut units 345 and 531, with p = 1.00). Multiple comparison indicated that all transponder units differed significantly from one other (p-values less than 2×10^{-16}) except for units B and D (p = 0.089). Figure 4 illustrates actual error variation across stands of different cover types.

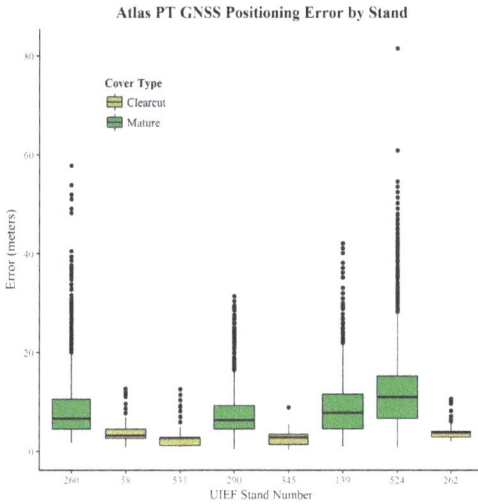

Figure 4. Boxplot comparing Atlas PT actual GNSS error in the controlled experiment across eight stands: four mature (green) and four clearcut (tan).

Figure 5 illustrates the distribution of Atlas PT GNSS positions collected over each 30-min sampling period compared to the single coordinates recorded by the EOS Arrow 100 at each Atlas PT location. The largest actual error observed for an Atlas PT in a mature stand was 81.5 m, and the largest error observed in a clearcut stand was 12.6 m.

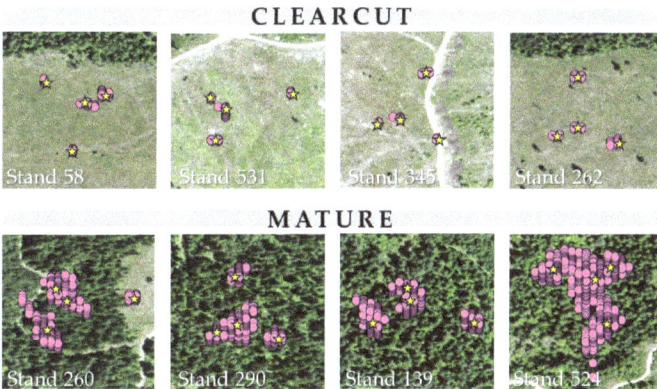

Figure 5. Visual comparison of Atlas PT coordinates (purple dots) and Arrow 100 coordinates (yellow stars) in clearcut versus mature stands of the controlled experiment. Scale is 1:1500.

Figure 6 illustrates the distribution of worker–hazard distances for each of the three hazards at each harvesting operation. Bars show frequency of ground worker presence within distance zones in increments of 5 m, ranging from 0 to 350 m from the specified hazard. John Lewis Pole plots represent three ground workers (the chaser, bucker, and hooker), while Wash Trap South and Upper Hatter encompass positioning data of two workers (the chaser and hooker). Each of three days is overlaid for a given site and hazard, except for the John Lewis Pole Carriage, which included two days of sampling, and Upper Hatter Loader, which included one day. Distances are based on GNSS coordinates collected every one second (s); thus, a frequency of 6000 corresponds to 6000 s (100 min) spent inside a given proximity interval. The proportion of time in which ground crew occupied zones defined as unsafe due to increased risk of injury or fatality is summarized in Table 3. Unsafe zones were defined as distances between 0–30 m for the loader and carriage and 0–60 m for the snag. Observed (Obs.) values are based upon GNSS positions collected by Atlas PTs during sampling and subsequent calculations of distances from hazards in R. Mature (Mat.) and clearcut (Clear.) values represent mean values for the GNSS environment simulated with and without mature forest overstory across 500 iterations applied to observed data based on sampled horizontal positioning error as measured in the controlled experiment. Proportions cover three sampling days at each of three sites: John Lewis Pole (JLP), Wash Trap South (WTS), and Upper Hatter (UH). Sample sizes are indicated in parentheses below each set of proportions. Differences in sample sizes reflect missing GNSS coordinates for a hazard, either due to positioning or transmission error, or because the equipment designated as a hazard was not in operation for a portion of the sampling period. Proportions shown in Table 3 were used in the Marascuillo Procedure analysis.

Across all days and all sites, ground workers spent a combined 18.5 hours (h) within 30 m of the loader geofence (34.6% of their time), 21.4 h within 30 m of the carriage (38.7% of time), and 32.3 h within 60 m of the snag (46.7%). It is important to note that our results represent collective ground worker positioning data, pooling the chaser and hooker (as well as a bucker for John Lewis Pole). Chasers generally work close to the landing while hookers work varying distances along the cable corridor, so proximity to landing hazards such as the loader would be expected to differ for the two workers.

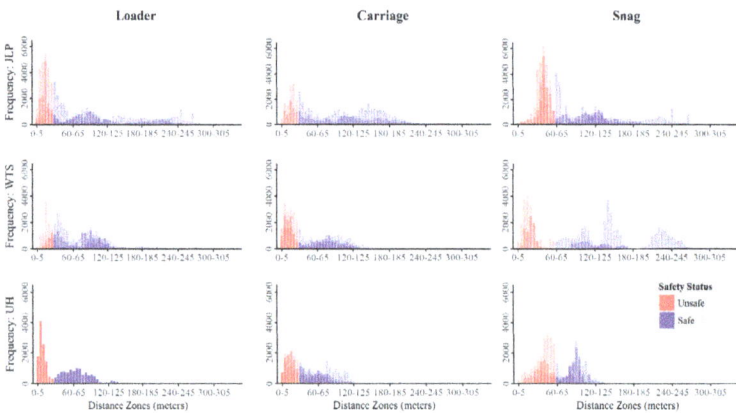

Figure 6. Distribution of ground worker distances from each of three hazards (loader, carriage, and snag) across three logging operations (John Lewis Pole, Wash Trap South, and Upper Hatter). Vertical bars are in 5 m increments. Red bars indicate location inside of the hazard's geofence (Safe); blue bars indicate location outside of the geofence (Unsafe). Shades of color represent different sampling dates.

Table 3. Proportion of instances (1-second intervals) when ground workers occupied unsafe zones associated with each hazard based on observed, mature, and clearcut data, with sample size, n, shown in parentheses.

Site	Loader			Carriage			Snag		
	Obs.	Mat.	Clear.	Obs.	Mat.	Clear.	Obs.	Mat.	Clear.
JLP	0.404	0.382	0.406	0.252	0.238	0.257	0.503	0.497	0.512
		$(n = 100{,}886)$			$(n = 72{,}251)$			$(n = 110{,}495)$	
WTS	0.226	0.220	0.225	0.477	0.449	0.477	0.363	0.361	0.362
		$(n = 66{,}872)$			$(n = 70{,}430)$			$(n = 76{,}926)$	
UH	0.492	0.487	0.491	0.449	0.423	0.449	0.528	0.498	0.523
		$(n = 21{,}742)$			$(n = 56{,}490)$			$(n = 61{,}826)$	

Results from the Marascuillo Procedure comparing proportions among hazards and sites are shown in Table 4. For each pair of comparisons, the table shows the value (Equation (3)) representing the Marascuillo test statistic and the critical range (Equation (4)). If the value exceeds the critical range ("yes"), the difference in the two compared proportions is significant. The Marascuillo Procedure compared 36 total proportions but only the 18 tests of interest in our study are shown. They include comparing (1) each hazard across all three sites: Loader (Tests 3, 4, and 12), Carriage (6, 7, and 14), and Snag (8, 9, and 15) and (2) each site for all three hazard types: John Lewis Pole (1, 2, and 5), Wash Trap South (10, 11, and 13), and Upper Hatter (16, 17, and 18). All 18 hazard and site comparisons of interest were significant, indicating that the proportion of worker presence inside geofence boundaries varied by hazard type and site.

Table 4. Results of the Marascuillo Procedure, comparing unsafe proportions across hazards and sites (alpha = 0.05).

Test	Compared Proportions	Value (Test Statistic)	Critical Range	Value > Critical Range?
1	JLP_L-JLP_C	0.152	0.009	yes
2	JLP_L-JLP_S	0.099	0.008	yes
3	JLP_L-WTS_L	0.178	0.009	yes
4	JLP_L-UH_L	0.088	0.015	yes
5	JLP_C-JLP_S	0.251	0.009	yes
6	JLP_C-WTS_C	0.225	0.01	yes
7	JLP_C-UH_C	0.197	0.01	yes
8	JLP_S-WTS_S	0.14	0.009	yes
9	JLP_S-UH_S	0.025	0.01	yes
10	WTS_L-WTS_C	0.251	0.01	yes
11	WTS_L-WTS_S	0.137	0.009	yes
12	WTS_L-UH_L	0.266	0.015	yes
13	WTS_C-WTS_S	0.114	0.01	yes
14	WTS_C-UH_C	0.028	0.011	yes
15	WTS_S-UH_S	0.165	0.01	yes
16	UH_L-UH_C	0.043	0.016	yes
17	UH_L-UH_S	0.036	0.016	yes
18	UH_C-UH_S	0.079	0.011	yes

Results of the Marascuillo Procedure comparing unsafe proportions between observed, operational data with and without simulated canopy and clearcut error effects are shown in Table 5. The table summarizes the results for nine separate tests, each with three comparisons in which observed (Obs.), mature (Mat.), and clearcut (Clear.) proportions (see Table 3) were compared for a single hazard at each timber harvest. Observed proportions differed significantly from mature proportions for six of nine hazards: JLP Loader, JLP Carriage, JLP Snag, WTS Carriage, UH Carriage, and UH Snag, but differed significantly for only one clearcut proportion (JLP Snag). Mature and clearcut proportions differed significantly from each other for six of nine hazards: JLP Loader, JLP Carriage, JLP Snag, WTS Carriage, UH Carriage, and UH Snag.

Table 5. Results of the Marascuillo Procedure, comparing unsafe proportions of observed, mature, and clearcut data (alpha = 0.05).

Test	Compared Proportions	Mean Value (Test Statistic)	Mean Critical Range	Value > Critical Range?
JLP Loader	Obs.-Mat.	0.022	0.005	yes
	Obs.-Clear.	0.002	0.005	no
	Mat.-Clear.	0.024	0.005	yes
JLP Carriage	Obs.-Mat.	0.014	0.006	yes
	Obs.-Clear.	0.005	0.006	no
	Mat.-Clear.	0.019	0.006	yes
JLP Snag	Obs.-Mat.	0.007	0.005	yes
	Obs.-Clear.	0.009	0.005	yes
	Mat.-Clear.	0.016	0.005	yes
WTS Loader	Obs.-Mat.	0.006	0.006	no
	Obs.-Clear.	0.001	0.006	no
	Mat.-Clear.	0.005	0.006	no
WTS Carriage	Obs.-Mat.	0.027	0.007	yes
	Obs.-Clear.	0.000	0.007	no
	Mat.-Clear.	0.027	0.007	yes
WTS Snag	Obs.-Mat.	0.002	0.006	no
	Obs.-Clear.	0.000	0.006	no
	Mat.-Clear.	0.002	0.006	no
UH Loader	Obs.-Mat.	0.005	0.012	no
	Obs.-Clear.	0.001	0.012	no
	Mat.-Clear.	0.004	0.012	no
UH Carriage	Obs.-Mat.	0.026	0.007	yes
	Obs.-Clear.	0.001	0.007	no
	Mat.-Clear.	0.027	0.007	yes
UH Snag	Obs.-Mat.	0.030	0.007	yes
	Obs.-Clear.	0.005	0.007	no
	Mat.-Clear.	0.024	0.007	yes

4. Discussion

Our results showed clearly that the nature of positional relationships was complex and varied both between sites and between hazard types in each treatment comparison tested. Distinct, multi-modal patterns of worker proximity to hazards were evident, and the locations of peak distances where workers tended to spend more time varied by day. Although we did not formally test differences among the three days sampled at each site, it was evident graphically when overlaying the distributions of proximity (Figure 6) that distinct patterns of spatial proximity exist and change over time. These trends likely correspond to, for example, hookers gradually working further from the loader as they set chokers and yard materials to the landing from further down the hill, or gradually working either closer to or further away from snag hazards identified adjacent to the harvest units.

A more nuanced analysis of individual worker positions relative to multiple hazards, such as studying hooker or chaser movements separately, could help to better quantify the spatial and temporal nature of positional relationships during normal work. However, we felt that our analysis reflected the reality of cable logging, in which multiple hazards are present simultaneously for any given worker, often in different directions. For example, a member of the rigging crew setting chokers near the top of the hill may be at risk of impact from rolling logs inadvertently bumped by the loader at the log deck. At the same time, he or she may also be at risk of being hit by a rotating log attached to a choker as the carriage begins to laterally yard logs toward the skyline if not sufficiently 'in the clear' at a safe distance horizontally across the hillslope from the carriage (and log). Simultaneously, a snag on the perimeter of the corridor could fall if dislodged by the log being yarded or a cable under tension. Although we focused on three possible hazards, the reality of cable logging is that multiple, other concerns are also present, including the processor swinging logs, pinch points caused as swing yarders, loaders, and processors rotate adjacent to the cut slope of the logging road, possible chain shot from the processor, and loose boulders in the corridor that may become dislodged.

Results of our controlled experiment on the UI Experimental Forest showed that the positioning accuracy of the GNSS-RF transponders used in our study was greatly affected by canopy. RMSE for the Atlas PT GNSS receivers in clearcuts was 3.37 m; whereas in mature, 90-year old mixed conifer stands, the RMSE was 11.08 m. The error observed in our study represents a function of variables affecting positioning accuracy, including Atlas PT receiver quality, the satellite geometry for the specific times and dates of sampling (see Table 1 for PDOP values), and multipath effects unique to the individual environments of each site. Improved GNSS receivers may demonstrate higher accuracies, even in mature stands. GNSS error associated with forest canopy has been well-documented though [28–30], so the observed variation in positioning accuracy by cover type is consistent with past studies.

When simulation was used to evaluate the relative importance of variable GNSS accuracy on worker safety status during active logging, results clearly showed that canopy-induced error did significantly affect the safety status, as defined using geofences. It is important to note that simulated canopy and clearcut error impacts on worker safety status were based on resampling from positioning data obtained at different locations, dates, and times than the operational sampling, so our results serve as an approximate estimation of canopy effects; actual error observed at active logging operations may differ due to topography or other factors. Further, error estimates based on static positioning in the controlled experiment were likely more conservative than error associated with dynamic positioning during active logging operations [32,33]. A further caveat we wish to highlight is that the statistical method used in our analysis to evaluate differences among sites and hazards, the Marascuillo Procedure, does not formally account for potential correlation that exists between adjacent location sample points in time and space. To the extent possible, we addressed this issue through the use of an analytical script that involved randomized resampling from our experimental data. For subsequent analysis, development of an analytical method that incorporates hierarchical modeling, including both fixed and random effects, into the procedure may help address impacts of possible correlated data structures associated with real-time GNSS.

Applications of real-time positioning for logging safety need to account for the reality that both mature and clearcut conditions, and associated impacts on GNSS accuracy, occur over the course of most conventional harvesting operations in the northwestern U.S. When a harvest unit has been felled in its entirety and the rigging crew is working in the open, more accurate positioning is possible. However, higher errors should be expected for GNSS-RF applications related to manual fallers or feller-bunchers, or in partial harvesting operations, such as the John Lewis Pole cedar pole harvest. Lower accuracies attributed to canopy cover may also be compounded by terrain effects, which can reduce satellite fix rates in forested areas, particularly in valleys [28,29,34]. For example, GNSS accuracies of devices associated with the rigging crew could vary between hookers working downhill and chasers working closer to ridgelines.

If sub-meter accuracy is desired under canopies, similar to precision forestry applications that require accurate marking of skid trails or individual trees [35], ground based augmentation systems (GBAS) may be necessary. GBAS determine the degree of error and transmit corrections to rover units which can then re-calculate their positions accordingly [19]. Haughlin et al. recently achieved 0.94-m accuracy on a harvester using RTK (real-time kinematic) correction, compared to 7 m with GNSS alone [20]. It is also important to note that the Atlas PT transponders used in this study relied on only the United States' NAVSTAR GPS constellation for position determination. Many current GNSS devices, including even consumer-grade handheld units for recreational use, are multi-constellation devices that determine position using not only GPS, but also the Russian Global Navigation Satellite System (GLONASS). Emerging devices will soon also utilize European (Galileo) and Chinese (BeiDou) navigational satellite systems as well. It is likely that newer GNSS-RF transponders capable of multi-constellation positioning will have higher accuracy in forested, mountainous locations where the number of trackable satellites may be diminished. However, use of multi-constellation sensors will not eliminate the multipath error endemic to highly reflective environments such as forests [32]. Similarly, although GBAS can greatly reduce GNSS positioning errors under canopies, differential correction cannot account for multipath effects. Even DGPS receivers will demonstrate higher errors in mature stands than in clearcuts.

According to our distance-based definitions of safe and hazardous work areas, the rigging crews we evaluated spent, on average, over one-third of their work day in unsafe conditions associated with the loader and carriage and nearly half of the day near snags. The simulated proportions of time spent in unsafe zones based on expected mature stand error varied significantly from observed proportions for six of nine tests; thus, using the technology evaluated in our study, accuracy errors associated with GNSS-RF devices under the canopy do impact perceptions of safety on logging operations, even when using basic, dichotomous definitions based on presence inside or outside a geofence. Devices with greater accuracy capabilities, at least through multi-constellation GNSS processing, and preferably RTK or other improved localization, are recommended for fine-resolution applications such as worker positioning around the landing. Proportions of safe and unsafe time differed significantly between observed and clearcut data in only one test, indicating that the higher accuracies achievable in clearcut conditions enable greater reliability in geofence alerts.

Use of GNSS-RF technology for safety applications on logging operations should be proportional to accuracy limitations. Given the large GNSS error observed under mature forest canopy in our designed experiment, single-constellation GNSS-RF radios such as the Raveon Atlas PT should only be deployed for very coarse monitoring of worker locations to improve general situational awareness and communication in forested environments; no operator decisions should be made based on observed, transmitted locations indicating the proximity of workers to jobsite hazards. That said, our operational sampling results offer a glimpse into the novel sorts of analyses that are becoming possible with real-time, networked positioning solutions in operational forestry. There is tremendous potential for improving both the safety and efficiency of logging through analysis of the high resolution spatial and temporal data that results from deployment of GNSS-RF and similar location-based services in production forestry.

Future research on GNSS-RF use for logging safety may wish to consider both vertical and horizontal positioning to better account for overhead hazards, such as the carriage, and to better specify inter-element distances on steep slopes. Future studies may also address how current positioning devices and systems can be adapted specifically for forestry applications, such as improvements to the user interface that allow loggers to utilize the technology easily and effectively with little distraction to normal work flow. This could entail display and sound settings or possible integration with other forms of data acquisition. For instance, Light Detecting and Ranging (LiDAR) information collected on snag locations could be synchronized with GNSS data to note worker proximity to snags or other environmental hazards [36]. Safety applications could also incorporate a more fluid warning system, such as through a series of proximity alerts that indicate increasing levels of danger associated with proximity to one or more hazards.

5. Conclusions

Atlas PT GNSS-RF positioning accuracy using only the NAVSTAR GPS system was more than three times greater in clearcut harvest units than under mature forest canopies. Error associated with mature overstory significantly affected the perception of worker safe or unsafe proximity to situational hazards. Ground workers spend approximately one-third of their time within areas of increased risk adjacent to mobile hazards such as the loader and carriage. Multi-constellation GNSS processing technology or other methods to improve localization accuracy are needed to provide the level of positioning detail necessary to avoid accidents with these fast-moving, dynamic hazards. In clearcut conditions, where errors are generally under 4 m, differential correction or other improved localization may be less critical but still recommended, especially for positioning at the landing and along the chute below the yarder.

Acknowledgments: This project was funded by the U.S. Centers for Disease Control and Prevention (CDC) National Institute for Occupational Safety and Health (NIOSH) grant number 5 U01 OH010841. The authors would like to thank Potlatch Corporation and the Idaho Department of Lands for permitting us to conduct operational sampling on their timber sales and for the logging contractors who supported our research by wearing transponders or putting them on equipment, as well as agreeing to allow us to observe their daily operations. We would also like to thank Molly Rard for assisting in set up of the controlled experiment, Andrew Naughton and Kevin Cannon for completing plot inventories on the UIEF, and Eloise Zimbelman, Darko Veljkovic, and Sarah Parkinson for aiding in operational sampling.

Author Contributions: Ann Wempe and Robert Keefe conceived and designed the experiments, performed the experiments, analyzed the data, and wrote the paper.

Conflicts of Interest: The authors declare no conflict of interest.

References

1. Bell, J.L.; Helmkamp, J.C. Non-fatal injuries in the West Virginia logging industry: Using workers' compensation claims to assess risk from 1995 through 2001. *Am. J. Ind. Med.* **2003**, *44*, 502–509. [CrossRef] [PubMed]
2. Bordas, R.M.; Davis, G.A.; Hopkins, B.L.; Thomas, R.E.; Rummer, R.B. Documentation of hazards and safety perceptions for mechanized logging operations in East Central Alabama. *J. Agric. Saf. Health* **2001**, *7*, 113–123. [CrossRef] [PubMed]
3. Lefort, A.J.; Pine, J.C.; Marx, B.D. Characteristics of injuries in the logging industry of Louisiana, USA: 1986–1998. *Int. J. For. Eng.* **2003**, *14*, 75–89.
4. Shaffer, R.M.; Milburn, J.S. Injuries on feller-buncher/grapple skidder logging operations in the Southeastern United States. *For. Prod. J.* **1999**, *49*, 24–26.
5. Sygnatur, E.F. Logging is Perilous Work. *Compens. Work. Cond.* **1998**, *3*, 1–7.
6. Occupational Safety and Health Administration. Regulations (Standards-29 CFR), 1910.266 2014. Available online: http://www.osha.gov/pls/oshaweb/owadisp.show_document?p_table=STANDARDS&p_id=9862 (accessed on 3 May 2017).
7. Bureau of Labor Statistics 2015 Census of Fatal Occupational Injuries. Available online: https://www.bls.gov/iif/oshwc/cfoi/cfch0014.pdf (accessed on 25 April 2017).
8. Bureau of Labor Statistics 2014 Census of Fatal Occupational Injuries. Available online: www.bls.gov/iif/oshwc/cfoi/cfch0013.pdf (accessed on 17 October 2016).
9. Bureau of Labor Statistics Occupational Injuries/Illnesses and Fatal Injuries Profiles. Available online: https://data.bls.gov/gqt/InitialPage (accessed on 24 April 2017).
10. Keefe, R.F.; Eitel, J.U.H.; Smith, A.M.S.; Tinkham, W.T. Applications of multi-transmitter GPS-VHF in forest operations. In Proceedings of the 47th International Symposium on Forestry Mechanization and 5th International Forest Engineering Conference, Gerardmer, France, 23–26 September 2014.
11. Grayson, L.M.; Keefe, R.F.; Tinkham, W.T.; Eitel, J.U.H.; Saralecos, J.D.; Smith, A.M.S.; Zimbelman, E.G. Accuracy of WAAS-enabled GPS-RF warning signals when crossing a terrestrial geofence. *Sensors* **2016**, *16*, 912. [CrossRef] [PubMed]
12. Zimbelman, E.G.; Keefe, R.F.; Strand, E.K.; Kolden, C.A.; Wempe, A.M. Hazards in motion: Development of mobile geofences for use in logging safety. *Sensors* **2017**, *17*, 822. [CrossRef] [PubMed]
13. Bettinger, P.; Wing, M.G. *Geographic Information Systems: Applications in Forestry and Natural Resource Management*; McGraw-Hill, Higher Education: New York, NY, USA, 2004; p. 230. ISBN 0072562420.
14. Carter, E.A.; McDonald, T.P.; Torbert, J.L. Application of GPS technology to monitor traffic intensity and soil impacts in a forest harvest operation. In Proceedings of the Tenth Biennial Southern Silvicultural Research Conference, Shreveport, LA, USA, 16–18 February 1999.
15. Taylor, S.E.; McDonald, T.P.; Veal, M.W.; Grift, T.E. Using GPS to evaluate productivity and performance of forest machine systems. In Proceedings of the First International Precision Forestry Symposium, Seattle, WA, USA, 17–20 June 2001.
16. Becker, R.M.; Keefe, R.K.; Anderson, N.M. Use of real-time GNSS-RF data to characterize the swing movements of forestry equipment. *Forests* **2017**, *8*, 44. [CrossRef]
17. McDonald, T.P.; Fulton, J.P. Automated time study of skidders using global positioning system data. *Comput. Electron. Agric.* **2005**, *48*, 19–37. [CrossRef]

18. Strandgard, M.; Mitchell, R. Automated time study of forwarders using GPS and a vibration sensor. *Croat. J. For. Eng. J. Theory Appl. For. Eng.* **2015**, *36*, 175–184.

19. Kaartinen, H.; Hyyppä, J.; Vastaranta, M.; Kukko, A.; Jaakkola, A.; Yu, X.; Pyörälä, J.; Liang, X.; Liu, J.; Wang, Y.; et al. Accuracy of kinematic positioning using Global Satellite Navigation Systems under forest canopies. *Forests* **2015**, *6*, 3218–3236. [CrossRef]

20. Hauglin, M.; Hansen, E.H.; Næsset, E.; Busterud, B.E.; Gjevestad, J.G.O.; Gobakken, T. Accurate single-tree positions from a harvester: A test of two global satellite-based positioning systems. *Scand. J. For. Res.* **2017**. [CrossRef]

21. Reclus, F.; Drouard, K. Geofencing for fleet & freight management. In Proceedings of the 9th International Conference on Intelligent Transport Systems Telecommunications, Lille, France, 20–22 October 2009; pp. 353–356. [CrossRef]

22. Anderson, D.M. Virtual fencing—Past, present and future. *Rangel. J.* **2007**, *29*, 65–78. [CrossRef]

23. Butler, Z.; Corke, P.; Peterson, R.; Rus, D. From robots to animals: Virtual fences for controlling cattle. *Int. J. Robot. Res.* **2006**, *25*, 485–508. [CrossRef]

24. Marsh, R.E. Fenceless Animal Control System Using GPS Location Information. U.S. Patent 5868100 A, 9 February 1999.

25. Sheppard, J.K.; McGann, A.; Lanzone, M.; Swaisgood, R.R. An autonomous GPS geofence alert system to curtail avian fatalities at wind farms. *Anim. Biotelem.* **2015**, *3*, 1–8. [CrossRef]

26. Umstatter, C. The evolution of virtual fences: A review. *Comput. Electron. Agric.* **2011**, *75*, 10–22. [CrossRef]

27. Oregon Occupational Safety and Health Division. Oregon Administrative Rules, Chapter 437, Division 7 Forest Activities 2003. Available online: http://osha.oregon.gov/OSHARules/div7/div7.pdf (accessed on 3 May 2017).

28. Deckert, C.; Bolstad, P.V. Forest canopy, terrain, and distance effects on Global Positioning System point accuracy. *Photogramm. Eng. Remote Sens.* **1996**, *62*, 317–321.

29. Liu, C.J.; Brantigan, R. Using differential GPS for forest traverse surveys. *Can. J. For. Res.* **1995**, *25*, 1795–1805. [CrossRef]

30. Rempel, R.S.; Rodgers, A.R. Effects of differential correction on accuracy of a GPS animal location system. *J. Wildl. Manag.* **1997**, *61*, 525–530. [CrossRef]

31. R Core Team. *R: A Language and Environment for Statistical Computing*; R Foundation for Statistical Computing: Vienna, Austria, 2016. Available online: https://www.R-project.org/ (accessed on 30 December 2016).

32. Parkinson, B.W.; Enge, P. Differential GPS. In *Progress in Astronautics and Aeronautics: Global Positioning System: Theory and Applications*; Parkinson, B.W., Spilker, J.J., Eds.; American Institute of Aeronautics and Astronautics, Inc.: Washington, DC, USA, 1994; Volume 2, pp. 3–49. ISBN 1600864201.

33. Wang, L.; Li, Z.; Zhao, J.; Zhou, K.; Wang, Z.; Yuan, H. Smart device-supported BDS/GNSS real-time kinematic positioning for sub-meter-level accuracy in urban location-based services. *Sensors* **2016**, *16*, 2201. [CrossRef] [PubMed]

34. D'Eon, R.G.; Serrouya, R.; Smith, G.; Kochanny, C.O. GPS Radiotelemetry error and bias in mountainous terrain. *Wildl. Soc. Bull.* **2002**, *30*, 430–439.

35. Blum, R.; Bischof, R.; Sauter, U.H.; Foeller, J. Tests of reception of the combination of GPS and GLONASS signals under and above forest canopy in the Black Forest, Germany, using choke rings antennas. *Int. J. For. Eng.* **2015**, *27*, 2–14. [CrossRef]

36. Wing, B.M.; Ritchie, M.W.; Boston, K.; Cohen, W.B.; Olsen, M.J. Individual snag detection using neighborhood attribute filtered airborne lidar data. *Remote Sens. Environ.* **2015**, *163*, 165–179. [CrossRef]

Article

The Effect of Customer–Contractor Alignment in Forest Harvesting Services on Contractor Profitability and the Risk for Relationship Breakdown

Mattias Eriksson [1,*], Luc LeBel [2] and Ola Lindroos [3]

[1]   SCA Forest Products, Skepparplatsen 1, SE-851 88 Sundsvall, Sweden
[2]   Département des Sciences du Bois et de la Forêt, Faculté de Foresterie et de Géomatique,
      2405 rue de la Terrasse, Université Laval, Québec, QC G1V 0A6, Canada; luc.lebel@sbf.ulaval.ca
[3]   Department of Forest Biomaterials and Technology, Swedish University of Agricultural Sciences,
      Skogsmarksgränd, SE-901 83 Umeå, Sweden; ola.lindroos@slu.se
*   Correspondence: mattias.eriksson.skog@sca.com; Tel.: +46-60-193-174

Received: 23 August 2017; Accepted: 16 September 2017; Published: 25 September 2017

Abstract: In forest operations, the interface between forest companies and harvesting contractors is of special importance, considering that it is the first link in the forest industry's supply chains. Supply operations account for a significant share of the final costs of wood products (up to 50%). This study investigates the effect of customer–contractor alignment on contractors' profit margins and on the risk for business relationship breakdown. Alignment is empirically measured for a Swedish forest company and 74 of its harvesting contractors, who were monitored during a four-year period. Two measures of alignment are employed: (1) the customer-perceived value of the contractors' services; and (2) the contractors' perceived alignment with the forest company expectations. Results indicate that the two measures of alignment are largely independent from each other, and that customer-perceived value affects both contractor profitability and the risk of relationship breakdown. Conflict between the two parties and lack of trust for the customer were found to be common complaints among contractors who ceased working for the studied forest company. Consequently, customer–contractor alignment should be considered a key objective by contractors who strive for business success, and also by forest companies who wish to improve their supply chain performance.

Keywords: forest harvesting contractor; supply chain alignment; customer-perceived value; contractor profitability; business relationship; supply risk

---

1. Introduction

1.1. Background

Forests are major sources of renewable products, annually contributing some 3.5 billion cubic meters of wood to the world economy and thus making the forest industry a potentially important part of the bioeconomy of the future [1] (p. 329). However, this potential development depends on the forest industry managing to stay competitive against, for instance, the concrete, plastic, and petroleum industries. Although it benefits from its status as a renewable product, wood must compensate for inherent disadvantages such as its geographical dispersion of production, acute sensitivity to weather and climate, and public scrutiny of its forest management. Competition in today's markets often stands between supply chains of interlinked companies rather than between individual firms. To stay competitive in such an environment requires purposeful actions in regard to production technology and operations management. In literature discussing management of supply chains, alignment of a company with its suppliers and distribution channels is often identified as a key success factor (e.g., [2,3]). The notion of alignment may be considered from several different perspectives. In this study, alignment between two firms in a supply chain is considered as a consequence of two factors: (a) the supplying firm's performance in relation to its customer's requirements; and (b) the degree to which the configuration of the relationship corresponds to the supplying firm's preferences. Ensuring that all parties in a supply chain align towards a common goal is no simple task, and limited information and guidelines seem to be available to production managers. In the forest industry, the outsourcing of forest operations has become common practice all over the world (e.g., [4]). Consequently, many forest companies rely on the services of small or medium-sized harvesting contractors (e.g., [5,6]). Despite the long tradition in the forest industry of working with contracted harvesting resources, large variations in contractor performance have been found regarding, for instance, contractor profitability (e.g., [7–10]), technical efficiency [11–13], and various aspects of the service delivered to the customer [14,15].

For forest companies who rely on the services of a number of contractors, such variation in performance means that some contractors are more in alignment with the companies' service needs than others [14], which ultimately may lead to difficulties for the forest companies in fulfilling their own downstream requirements. Eriksson et al. [14] addressed this issue by analyzing harvesting contractor performance from a customer's point of view. They suggested four generic approaches to contractor alignment: incentives, supplier development, use of power advantage, and active sourcing. Further, they developed a process that can be used to choose a strategy for aligning the performance of a company's contractor fleet with the company's needs. The process is designed to provide contractor-specific blends of the four generic approaches to help forest companies to maximize the effect of their efforts to improve contractor performance. The sole purpose of this process is contractor performance alignment from the customer's perspective, which of course overlooks the contractors' perspectives on the relationship. This may be a limitation of the process, since the risk for poor contractor performance is not the only risk associated with outsourced harvesting operations. The most obvious risk for a company may be that one or several of its suppliers decide to take their business elsewhere or terminate their enterprise, both of which could cause serious supply shortages if the company is unable to regain its suppliers or find replacements. To avoid such a situation, it may benefit companies better to align with their contractors in order to ensure that the contractors commit to their customer. To achieve such commitment, the alignment of performance may need to be supplemented by alignment of other dimensions such as strategy, culture, leadership style [16], pricing, reliability, responsiveness, and quality management [17].

In the Nordic countries, harvesting contractors perform fully mechanized cut-to-length thinning and final felling operations, by use of harvesters and forwarders. In a series of articles in the magazine of the Swedish Forest Machine Owners Association (e.g., [18–21]), the business environment for harvesting contractors provided by large forest companies is highlighted in numerous statements as a major threat to the contractors' prosperity and ultimately to the whole forest industry. Such statements, paired with reports of the low profitability of many contractors (e.g., [7–9]), stalling or declining productivity [22,23], and rising costs ([24] and [1] (p. 249)), give the impression of serious structural problems in the forest sector. Researchers interested in contractor performance have reported large variations (e.g., [9,11,15,25]), even among contractors working within the same business context [14,26]. For the most part, researchers have focused on contractor-specific factors as explanatory variables. Such factors include, for instance, leadership and processes [27], knowledge, performance measurement [25], and the number of customers [11,28,29]. However, the suggestion from practitioners (such as the Swedish Forest Machine Owners Association) that the business environment is a key determinant of contractor performance has largely been overlooked by researchers, despite indications that good relations with a single customer is often common among high-performing contractors [11,28,29]. Further, few, if any, studies have taken a longitudinal perspective to the identification of the factors that lead to sustained benefits. All in all, little is known about what harvesting contractors can do to maximize their performance, or how customers of harvesting services can reduce their risk of contractor-related problems. Below, the two perspectives on the customer–contractor relationship are described and the concepts of alignment, contractor performance, and supply risk are outlined.

1.2. Two Perspectives on the Customer–Contractor Relationship

The relationship between harvesting contractors and their customers, like any relationship between two independent sides, may be looked upon from two perspectives where each side has their own unique expectations and perceptions. From the customer's perspective, low cost has historically been the most frequently used measure in evaluating contractor performance [6]. Nowadays, customers of harvesting services also consider other more qualitative measures, with high contractor performance across a range of service aspects being necessary to fulfill customer requirements [14,30]. The switch from a cost-only focus in forest harvest contracting to a situation where many aspects of the service are valued is similar to the broader understanding of customer value that now prevails in a more general business context. For instance, Grönroos [31] suggests that a customer's perception of the value of a service—apart from its monetary cost—depends on the perceived value of the core solution (i.e., the physical result of the service), the perceived value of various additional services associated with the core solution, and the perceived cost of upholding the business relationship (e.g., the cost of supervision).

From the contractors' perspective, the ability to operate efficiently is to some extent determined by the working environment [11]. The working environment may differ between customers of harvesting services depending on, for instance, the customers' different needs of flexibility [15,32]. Further, preferences on how the harvesting service should be performed may differ considerably between service buyers [26]. This suggests that contractors may utilize variations in potential customers' preferences and operating environments to gain benefits by securing contracts with the customer whose needs the contractor can meet at the lowest cost.

1.3. Alignment

The notion of alignment has been used to address a variety of issues related to a company's internal operation, such as the relationship between a company's functional strategies to its holistic business strategy [33,34]. In supply chains involving multiple organizations, misalignments between supply chain partners may harm the overall effectiveness [2], and may be measured in several dimensions such as, for instance, cost reduction efforts, pricing, reliability, responsiveness, and quality management [17]. Eriksson et al. [14] assumed the customer's perceived value of a harvesting contractor's services as a measure of alignment, and suggested a process that forest products companies can use to actively manage the performance of their contractor fleets. However, taking only one perspective into account (as did Eriksson et al. [14]) is bound to give a somewhat incomplete description of the involved parties' alignment. Accordingly, it may be necessary to measure, in addition to the customer's perceived value of a service, the degree to which the contractors perceive alignment with their customer in order to achieve a complete picture of the alignment of studied relationships. Such a contractor-centered perspective is taken by Mäkinen [28], who in addition to considering the financial success of contractors also measured their capacity utilization and satisfaction with their customer. He found that the financially most successful contractor group also was the most satisfied with their customer. They also had the highest capacity utilization. This led him to conclude that the key success factor appeared to be working for a customer who provided a beneficial operating environment. This may be the clearest example from the literature that describes a situation where the customer's alignment with the contractor gives mutual benefits for both parties. In Mäkinen's study [28], the benefit for the contractor was financial success, and for the customer in the form of high contractor capacity utilization (and supposedly a competitive price).

To get a more complete picture of customer–contractor alignment, this study will address the notion from the previously described two perspectives: the customer's and the contractor's.

Customer-perceived value [31] is used to reflect the customer's perspective on relationship alignment, whereas the contractor perspective is represented by contractor perceptions on the business and operating environments offered by their customer.

1.4. Contractor Performance

The term 'firm performance' comprises a range of evaluative factors and assessments, but all measurements of a firm's performance need to somehow support the firm's strategic objectives [35]. Among harvesting contractors, a variety of reasons have been identified behind the foundation or takeover of harvesting firms, and their strategic objectives also differ markedly [36], thus making a comprehensive and yet homogenous definition of contractor performance a challenge. However, profitable operations allow contractors to stay in business and to acquire resources that can be used to reach other objectives—personal or professional; it is this which prompts the use of profitability as the primary measure of contractor performance. Focusing on this criterion has the advantage that it simplifies the measure of contractor performance and enables direct comparisons of performance between contractors; this essential simplicity explains its popularity in studies of harvesting contractors (e.g., [9,25]). Profitability may be measured with a range of key ratios, usually based on a firm's earnings put in relation to either the firm's turnover, or some proportion of the firm's assets. Net profit margin (i.e., the share of the turnover that harvesting contractors have left after all expenses and taxes have been deducted) may be the most commonly used key ratio in analyses of firm profitability. For Finnish and Swedish conditions, published surveys of contractors' finances have indicated an average net profit margin for harvesting contractors of around 4–7% with some annual fluctuations, and with a considerable share of the contractors reporting losses [7,9]. The profit margin key ratio gives an indication of the contractor firm's ability to price its services and control its costs, and ultimately illustrates how effective the firm is in creating excess resources.

1.5. Supply Risk

Outsourcing in itself introduces some risk to a firm's supply, since by definition it involves the firm giving up some of its control in exchange for some potential benefits [37]. In the worst case scenario, this loss of control may lead to higher procurement costs, interrupted downstream operations, or an inability to meet demand, if suppliers fail to perform as expected. To mitigate such risks, a frequently used method is to use several sources for a certain product or service, so that unexpected disruptions at a single supplier will have a limited effect on the total supply [38]. In the forest industry, a common setting is one where large forest companies employ several small harvesting contractors (e.g., [39]). Consequently, any disruptions in a contractor's service need to be of significant magnitude to cause much of a problem for the average customer of harvesting services. On the other hand, a current trend in Sweden is to minimize roundwood stocks, so if one or more contractors in a certain area completely stop production, even large forest companies may face serious difficulties in meeting mill demand, or in fulfilling harvesting agreements with private landowners. Thus, such disruptions may cause costly problems and possibly lost business opportunities. In this context, it is a benefit for forest companies to have stable, or at least predictable, relationships with their contractors. To address these issues, this study evaluates supply risk by considering the risk for breakdown of the customer–contractor relationship, more specifically by following the relationship status over several years to see whether a breakup has occurred or not.

1.6. Study Objectives

As shown above, poor contractor performance within the forest industry is likely to cause supply disturbances, harvesting operations out of alignment with downstream requirements, and poor contractor profitability. Hypothetically, the degree of such problems can be related to the degree of alignment between harvesting contractors and their customers. However, these issues have previously not been experimentally demonstrated. Therefore, the overall aim of this study was to improve our knowledge of the effect of the customer-contractor relationship on contractor performance and the supply risk for the customer of harvesting services. The specific objectives of this study were to

(1) define and apply criteria for measurement of alignment from both the customers' and the contractors' perspectives
(2) analyze the relationship between customer–contractor alignment and contractor profitability
(3) analyze the effect of both customer–contractor alignment and contractor profitability on the risk for relationship breakdown over time.

This study analyzes customer–contractor alignment from two perspectives:

(1) Customer perceptions of the received value from contractor services.
(2) Contractor perceptions of how well customers' business and operational environments align with the contractors' preferences.

The main hypothesis of the study is that customer–contractor alignment can foster both a lasting relationship and performance levels that meet customer expectations on high quality services as well as contractor expectations on high profits.

2. Materials and Methods

This study builds on previous research by Eriksson and Lindroos [26] and Eriksson et al. [14], with the parts used here being summarized in Section 3.1. In that previous work, alignment was studied from the customer perspective in a case of a large Swedish forest industry company and 74 of its harvesting contractors employed on a long-term basis during the period 2006–2009. The current study supplements the previously collected data with the contractors' perspectives on their alignment with the customer, and with data on their profitability during the same time period. Further, this study also adds a longitudinal dimension by tracking the development of the customer–contractor relationship and the contractors' profitability in 2011–2013 (ca. 4 years after the initial data collection) to measure how the more recent situation relates to the historic relationship with their customer. The study can be classified as using mixed methods, since use of qualitative methods (interviews and questionnaires) were mixed with the use of quantitative methods (e.g., collection of production data). Full coverage of contractors and customer representatives were aimed for in all data collections.

2.1. Measurement of Customer-Perceived Value of Harvesting Services

The customer-perceived value of a service was defined by Grönroos [31] as a composite of four generic attributes: the core solution, additional services, price, and relationship costs. Eriksson and Lindroos [26] adapted this framework for the study of forest harvesting services, identifying several sub-attributes. Furthermore, they measured the customer-perceived value of the harvesting services from each of the contractors included in this study, using survey items for sub-attributes associated with three generic attributes: core solution; additional services; and relationship costs. These items assessed the customer's satisfaction with contractors' services on 10-graded Likert scales. Further, the long-term ability of the contractors to maintain a competitive service price was assessed by identifying and collecting data for some key performance indicators reflecting harvesting and forwarding efficiency. These indicators (Table 1) were used by the customer company to monitor operations and are described more in detail in Eriksson et al. [14]. The data on the four measures collected in the previous studies [14,26] were used as input on customer-perceived value in this study.
Prior to performing analyses, all variables were standardized (mean value = 0, standard deviation (SD) = 1) to allow for easier interpretation of figures and comparison of regression coefficients. In addition, a composite index of customer-perceived value of harvesting services (CPV index) was constructed by calculating a standardized (mean value = 0, SD = 1) mean value of the standardized scores for the attribute-specific measures for each contractor.

Table 1. Schematic view of the measures of customer-perceived value as used in this study.

Name of Measure in This Study	Associated Generic Attribute [31]	Associated Sub-Attributes [26]	Type of Data	Method of Calculation
Core solution	Core solution	Log quality Thinning quality Environmental considerations	Survey of customer perceptions	Mean of sub-attribute scores
Additional services	Additional services	Flexibility Delivery performance Management Collaboration	Survey of customer perceptions	Mean of sub-attribute scores
Operational efficiency	Price	Harvester productivity Harvester utilization rate Forwarder productivity Forwarder utilization rate	Machine follow-up data	Equation (1) (see below)
Relationship costs	Relationship costs	Daily communication Business relationship	Survey of customer perceptions	Mean of sub-attribute scores

Equation (1) describes the calculation of the composite measure of operational efficiency used in subsequent analyses. A multiplicative format was chosen to reflect the interdependence between harvesters and forwarders in the studied harvesting system.

$$\text{Operational efficiency} = \text{Relative harvester productivity} \times \text{Harvester utilization rate} \times \text{Relative forwarder productivity} \times \text{Forwarder utilization rate} \tag{1}$$

2.2. Measurement of Contractor-Perceived Alignment

Information on the contractors' perceptions of their alignment with the customer was collected through a questionnaire that all participating contractors were asked to complete. It was designed to provide a reliable measure of how well the relationship with the customer corresponded to contractor preferences. Five items were developed and pre-tested on two contractors who were not part of the final sample to ensure their validity. The final questionnaire comprised five statements describing their perceived alignment with the client in 2009, which was the end of the period covered by the data presented in Section 3.1. Contractors were asked to indicate the extent to which they agreed. The following items were used (translated from Swedish):

(1) The company was the preferred client of my services five years ago.
(2) My preferred way of working was compatible with the company's way of working five years ago.
(3) My values were compatible with the company's values five years ago.
(4) Having the company as a client helped me fulfill my business objectives five years ago.
(5) I had a good business relationship with the company five years ago.

Answers were given on a 1–10 Likert scale, where 1 meant that the contractor totally disagreed, and 10 meant that the contractor totally agreed with the statement. Finally, all contractors were asked to tell without any formal restrictions their reasons for working with the client and, if applicable, their reasons for not working for the client anymore.

All information was collected by telephone in the summer of 2014. At least three attempts were made to reach contractors that did not initially respond. Fifty-five contractors were reached in this manner. Of the remaining 19 contractors, 18 contractors could not be reached, of which one had passed away; seven did not respond to any contact attempts despite still being in operation; while the remaining ten had dissolved their companies and it proved impossible to obtain valid contact information. In addition to these 18 contractors, one contractor declined to participate in the study despite assurances of strict anonymity. A total response rate of 76% was therefore obtained.

In addition, a composite index of contractor-perceived customer alignment (Alignment index) was constructed by calculating the standardized (mean value = 0, SD = 1) mean value of the scores for the above described items for each responding contractor.

114

2.3. Records of Customer–Contractor Relationship Status

Four groups of contractors were formed to analyze changes in relationship status. The groups were formed according to their relationship status with the case company at the end of 2013: the first group comprised contractors still employed by the company; the second was of bankrupted or dissolved contractor firms; the third was of firms who had left the case company to start business relationships with other customers; and finally the fourth group comprised both the contractors from the second and third groups, i.e., all contractors who were no longer employed by the company regardless of the reasons. Further, a binary variable was constructed where 1 was assigned to contractors that still worked for the customer on a long-term basis at the end of 2013, and 0 was given to all other contractors. Contractor firms in which the owner had retired but transferred the firm to a family member were considered to be the same entity.

2.4. Records of Contractor Profitability

All incorporated businesses are by law required to annually file their financial reports at the Swedish Companies Registration Office, where the reports are publicly available for a fee. In this study, the contractors' net profit margin was selected as a measure of contractor profitability and collected from the annual financial reports of the studied contractors. Yearly net profit margins were aggregated to three-year averages for each contractor to reduce the effect of extraordinary events—such as machine investments or sales—on contractor financial key ratios. Net profit margins were recorded for two three-year aggregates. The first period covered the three fiscal years for each contractor that best corresponded to Eriksson and Lindroos's [26] study period (2006–2009), and the second aggregate covered the latest three financial statements from each contractor that were publicly available in June 2014. It was not possible to obtain financial information for the exact same periods from each contractor since starting dates for their fiscal years differ.

2.5. Analyses

Due to the aforementioned non-responses, data on the contractors' perceived alignment were missing for 24% of the contractors comprised in the initial dataset. Data imputation was used to compensate for the consequent lack of data in the variable alignment index. This imputation was made according to a model in which non-respondents were assigned alignment index values based on whether they were still employed by the customer or not. The rationale behind the model was derived from the significant differences in alignment index between contractor. The customer-perceived value index did not differ noticeably between respondents and non-respondents, but a fairly large—albeit statistically insignificant—difference in net profit margin of 3% was noted to the respondents' advantage. Consequently, the applied imputation model should be sufficient to avoid problematic effects in the subsequent analyses due to non-response errors.

A one-way ANOVA was used to test for pairwise comparisons between contractor groups, and Pearson's test was used to analyze correlations between variables. Ordinary least squares regression was used to analyze factors affecting contractor profitability, and binary logistic regression was used to analyze factors affecting the risk for relationship breakdown. The critical significance level was set to 0.05 in all analyses, but parameters with higher p-values are sometimes presented to show tendencies in the data. In the regression analyses, all independent variables were entered in standardized format (mean = 0, SD = 1) to allow for easy comparisons between variables, whereas the dependent variables were kept in their original form.

3. Results

Results are presented using the layout applied in the methods section: (1) contractor-perceived customer alignment, (2) customer-perceived value, (3) risk for relationship breakdown, and (4) contractor profitability.

3.1. Contractor-Perceived Customer Alignment

Most of the contractors who responded to the survey in this study indicated that their customer was well aligned with their needs (see Table 2). In contrast to the general picture, however, a few contractors perceived severe misalignment, as indicated by scores at the very low end of the item scales. All items were highly correlated to each other (Pearson's correlations between 0.43 and 0.66 significant at the p-value < 0.001 level), indicating a reliable composite measure.

Table 2. Contractors' perceptions of their customer in 2009 (1 = completely disagree, 10 = completely agree). n= 55 for all items. SD = standard deviation and Q = quartile.

	Item	Mean	SD	Min	Q1	Median	Q3	Max
1.	The company was the preferred client of my services.	7.2	2.3	1.0	6.0	8.0	8.0	10.0
2.	My preferred way of working was compatible with the company's way of working.	7.1	2.0	2.0	6.0	7.0	8.0	10.0
3.	My values were compatible with the company's values.	6.8	2.1	1.0	6.0	7.0	8.0	10.0
4.	To have the company as a client helped me fulfill my business objectives.	6.4	2.5	1.0	5.0	7.0	8.0	10.0
5.	had a good business relationship with the company.	7.5	2.5	1.0	7.0	8.0	9.0	10.0
	Alignment index [1]	0.0	1.0	−3.2	−0.64	0.21	0.72	2.0

[1] Standardized mean of standardized item values (mean value = 0, SD = 1).

3.2. Customer-Perceived Value of Harvesting Services

In the data collected by Eriksson and Lindroos [26], the customer expressed a relatively high appreciation for the contractors' services (Table 3), at least for the three attributes measured on a Likert scale. Common to all customer-perceived value attributes is that they cover large ranges, indicative of significant potential improvements in the supply chain that could be achieved by a closer alignment of contractor performance with customer requirements.

Table 3. Statistics for customer-perceived value for the period 2007–2009. $n = 74$ for all attributes.

Attribute	Mean	SD	Min	Q1	Median	Q3	Max
Core solution [1]	7.3	1.2	4.7	6.5	7.3	8.0	10.0
Additional services [1]	6.5	1.4	2.9	5.8	6.6	7.7	8.8
Relationship costs [1]	7.4	1.3	1.3	6.7	7.5	8.2	10.0
Operational efficiency [2]	0.75	0.19	0.40	0.63	0.73	0.83	1.33
CPV Index [3]	0	1.0	−2.6	−0.51	0.038	0.71	2.0

[1] Measured on a 1–10 Likert scale. [2] Index as described in Equation (1). [3] Standardized mean of standardized attribute values (mean value = 0, SD = 1). CPV: customer-perceived value of harvesting services.

The two alignment measures were plotted against each other (Figure 1). This allowed for the identification of two groups of contractors that stand out from the rest: group A, consisting of four contractors that expressed a very low perceived alignment with the customer, and group B, consisting of five contractors for which the customer perceived a very low value from their services. When the contractors in group A were asked why they perceived such misalignment, all but one (who declined to give a more detailed answer) described a very low level of trust in their client, frequently associating their customer with attributes such as having no respect or trust, unprofessional behavior, and dishonesty. These responses are clearly indicating a relationship with a certain level of conflict rather than alignment. For group B, the two contractors with the lowest perceived alignment had a similar (although not quite as severe) opinion of their customer, whereas the remaining three contractors gave no such indication of dissatisfaction. Ten of the interviewed contractors stopped working for the customer after the first study period. The reasons for this cessation varied, but three aspects were identified in the answers provided by the interviewees. First, seven out of the ten contractors who stopped working for the customer described a severely deteriorating relationship where they ultimately had lost trust in the customer's officials. It was not possible to clearly establish the root cause of these conflicts, but it is worth noting that many of the contractors who had quit performed rather poorly from the customer's perspective. In comparison, those contractors who did not quit performed noticeably above average. Second, four contractors indicated low price as the main reason why they decided to look elsewhere for better options. Three out of these contractors were also included in the first group. Finally, one of the interviewed contractors had been forced to file for bankruptcy (which also was the case for an additional three contractors in the non-responding group).

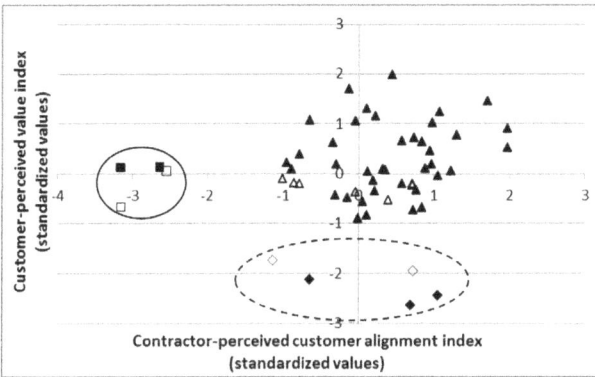

Figure 1. Customer–contractor alignment from two different perspectives for 55 responding contractors. Filled (black) symbols represent contractors that still worked for the customer, whereas unfilled symbols represent contractors that ceased to work for the customer. Group A (squares, and circled by solid line) include low perceived alignment by contractor, and Group B (diamonds, and circled by a dotted line) include low performing contractors from the client's perspective.

3.3. Risk for Relationship Breakdown

The status of the relationships between the customer and its contractors was monitored from the end of 2009 to the end of 2013 (i.e., if the contractors were still employed by the customer or not). During this time span, about 30% of the contractors stopped working for the company (Table 4). All contractors were divided into two groups according to the status of their relationship at the end of 2013. Contractors who had quit working for the customer were further segregated into two categories based on their current business status (i.e., if they were still in operation or not). Pairwise comparisons between groups revealed that contractors who continued to work with the customer showed a significantly higher contractor-perceived customer alignment index as well as customer-perceived value index than contractors who had quit (Table 4). Unsurprisingly, low profitability was a typical characteristic of contractor firms that later went bankrupt or were liquidated. Furthermore, contractors who decided to end the relationship for other reasons (mostly to start working for another customer) were on average less profitable than those who decided to stay. In the second period, no difference in profitability could be observed between contractor groups, indicating that the contractors who abandoned the case company for other customers actually managed to find customers whose business proposals were more in alignment with their own preferences.

Table 4. Relation between relationship status and customer–contractor alignment; customer-perceived value of harvesting services; and contractor profitability.

Variable	Still Employed ($n = 52$)	Dissolved Relationship		
		Total ($n = 22$)	Bankruptcy or Liquidation ($n = 12$)	Other Reasons ($n = 10$)
Alignment index [1]	0.23 [A]	−0.64 [B]	−0.43	−0.89
CPV index	0.23 [A]	−0.45 [B]	−0.42	−0.49
Average net profit margin				
Fiscal years 2007–2009	0.05 [A]	−0.03 [B]	−0.05	0.00
Fiscal years 2011–2013	0.05 [A]	0.05 [B]	-	0.05

[1] Deviating number of respondents: fifty-five in total; forty-four still employed, three in bankruptcy or liquidated, and eight who quit for other reasons. Note: Within rows, different superscript letters (A and B) indicate significant ($p < 0.05$) differences (one-way ANOVA) between still employed contractors and those who no longer worked for the customer. Group differences between the still employed contractors and the two sub-groups according to the reason for the ending of the customer–contractor relationship were not analyzed because of limited sizes of the sub-groups.

Binary logistic regression was used to analyze what led to the observed events among the contractors. For the full dataset, net profit margin and perceived value from additional services (which in this case mostly corresponds to the customer's perception of how well the contractor runs his business) both significantly increased the chance of a sustained relationship, whereas perceived value from the core solution rather counter-intuitively increased the risk of a dissolved relationship (Table 5, model a). Although there was not a significant statistical relationship ($p = 0.104$), there was still some limited evidence that the contractors' perceived alignment with the customer, as measured by the alignment index variable, had a tendency to affect the chance of a sustained relationship. Indeed, most interviewees who had taken their business elsewhere reported that they did so because of dissatisfaction, or even conflict, with the customer. After calculation of the probabilities for a continued relationship and classifying probabilities over 50% as cases where a continued relationship was likely, model a managed to predict the relationship status after the study period correctly for 80% of all contractors. However, the hit ratio was unevenly distributed, with 90% of continuously employed contractors and 55% of contractors who quit classified correctly according to model a. This is indicative of the complexity of modelling relationship breakdown.

An analysis was also made for a subset of the data comprising all contractors who currently are active, either as contractors with the case company or for some other customer (Table 5, model b). In this analysis, the same pattern emerged as with the full dataset, but with one distinction: net profit margin did not affect the chance of a sustained relationship for this group. Consequently, there is no evidence that contractors who took their business elsewhere did so or were forced to do so because of low profitability. This result may suggest that other factors are considered when a contractor decides to change customer, or the customer decides to terminate a contract. Model b managed to predict the relationship status correctly for 87% of all the surviving contractors. The hit ratio was unevenly distributed, with 98% of continuously employed contractors and 30% of contractors who quit classified correctly according to model b, which further underlines the difficulty of accurately predicting relationship breakdown. For this group of contractors, the probability of an average contractor quitting was only 8%.

Table 5. Binary logistic regression models with relationship status as dependent variable (continued relationship = 1; terminated relationship = 0, with a continued relationship as reference event).

Model	Variable	Parameter Estimate	Standard Error	Wald	p-Value	Odds Ratio	Deviance p-Value
a	Full model	-	-	-	-	-	0.865
	Intercept	1.511	0.406	3.72	<0.001	-	-
	Additional services	1.270	0.460	2.76	0.006	3.56	-
	Net profit margin	1.086	0.443	2.45	0.014	2.96	-
	Core solution	−0.913	0.456	−2.00	0.045	0.40	-
	Alignment index	0.597	0.367	1.63	0.104	1.82	-
b [1]	Full model	-	-	-	-	-	0.992
	Intercept	2.445	0.609	4.01	<0.001	-	-
	Additional services	1.729	0.711	2.43	0.015	5.63	-
	Core solution	−1.494	0.685	−2.18	0.029	0.22	-
	Alignment index	0.733	0.429	1.71	0.087	2.08	-

[1] Model developed on a subset of the data for which n = 62 contractors who were still in business after the second study period.

None of the contractors whom the customer perceived to deliver above-average value and who simultaneously made profits, which represented 42% of the total sample, quit working for the customer during the studied period (see Figure 2). However, 51% of the remaining contractors (i.e., 30% of the total number of contractors) who either made losses, were perceived to deliver below-average value, or both, had quit working for the customer, indicating the importance of achieving mutual benefit to sustain a business relationship.

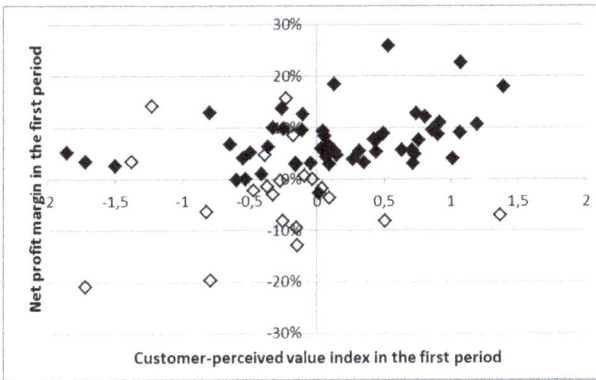

Figure 2. Net profit margin and customer-perceived value index in the first period (ca. 2006–2009). Filled (black) symbols represent contractors that still worked for the customer, whereas unfilled symbols represent contractors that ceased to work for the customer.

3.4. Contractor Profitability and Its Predictors

The average contractor in the study was profitable during the first study period. Profitability ranged from −21 to +23% and more than a quarter of the contractors were losing money, indicating a difficult situation for many of them (Table 6). Average profit increased in the second study period. However, by the end of the second period, 12 of the contractors had been liquidated or gone bankrupt, reducing the sample from 74 to 62 contractors. Of these 62 contractors, nine had switched to other customers and were still active (constituting the majority of category "other reasons" in Table 4).

Table 6. Contractors' profitability for each studied period.

Period	n			Average Net Profit Margin (%)				
		Mean	SD	Min	Q1	Median	Q3	Max
1 (fiscal years 2007–2009)	74	3	8	−21	−1	2	7	23
2 (fiscal years 2011–2013)	62	5	5	−8	1	3	8	21

In this study, customer-perceived value and its attributes, as well as the contractor-perceived customer alignment index, was hypothesized to relate to the contractor's net profit margin. Many of these variables were found to correlate to each other, but the alignment index did not correlate significantly to other variables except for the customer-perceived value attribute of operational efficiency (Table 7). The customer-perceived value index, however, showed a significant positive correlation to net profit margin in both periods, identifying customer satisfaction as a key objective for contractors who pursue financial objectives. The attributes of customer-perceived value all correlated significantly to the customer-perceived value index (0.43–0.66), and most of them also correlated to each other (0.28–0.57), with the exception of operational efficiency which only correlated significantly to additional services (0.46). Judging by the correlation of net profit margin between periods, profitable contractors tend to stay profitable. Notably, all significant correlations in the matrix are positive.

Table 7. Pearson's correlations between the numerical variables in the study.

Variable	Alignment Index [1]	Net Profit Margin, Period 1	Net Profit Margin, Period 2 [2]
Net profit margin, period 1	0.16	-	-
Net profit margin, period 2 [2]	0.027	0.45 *	-
CPV index	0.13	0.32 *	0.29 *
Core solution	−0.13	−0.01	0.04
Additional services	0.19	0.33 *	0.26 *
Operational efficiency	0.30 *	0.34 *	0.32 *
Business relationship	0.00	0.26 *	0.19 †

[†] $p = 0.05$–0.1, [*] $p < 0.05$. Note: $n = 74$, except for [1] where $n = 55$ respondents and [2] where $n = 62$ surviving firms.

As could have been expected from the observed correlations (Table 7), the customer-perceived value index had a significant effect on contractor net profit margin in the first period at the $p < 0.05$ level, whereas the significance level needed to be relaxed somewhat to detect any effect of the alignment index ($p = 0.09$; Table 8, model A). Replacing the customer-perceived value index with its attributes revealed that the only attribute with a significant effect on contractor profitability was operational efficiency (Table 8, model B). This is most likely linked to the nature of the predominant business model in this case, in which the vast majority of the work is paid by piece rates, which intrinsically means that more efficient contractors will have better cash flows. In this more detailed model (B), no tendency was observed for the alignment index, indicating that the tendency observed in model A may have been caused by the correlation between operational efficiency and the alignment index (see Table 7) rather than by the alignment index on its own merits. The explanatory power of both models is low (R^2 of 11.6 and 12.5%, respectively), something which is possibly related to a high frequency of extreme observations in the net profit margin variable (see Figure 2).

In the second period, the customer-perceived value index had a significant effect on net profit margin, but when replaced by its attributes, only operational efficiency remained significant. However, when entering the net profit margin from the first period as a variable in the regression, all other variables lost their explanatory power—indicating past profitability as the key determinant of present profitability (Table 8, model C), at least for cases when no information on other aspects of present performance is available. Focusing only on contractors who continued to work for the customer, both the magnitude of the effect and the explanatory power of past profitability increased substantially (Table 8, model D), for which the parameter increased from 0.32 to 0.57, and R^2 from 13 to 34% compared to the model (C) developed from the full dataset. This indicates that a contractor's profit level is likely to be sustained as long as the contractor retains the relationship with the same customer. Notably, the average profitability for contractors who went to work for other customers was 5% percent units higher in the second period than in the first (see Table 4), even though the difference was not significant. This suggests that the change of customer may have worked as a reset for the profitability for this group by giving them a fresh start in a new environment, possibly an environment more in alignment with their requirements.

Table 8. Models explaining observed variation in contractor net profit margin.

Period	Model	Variable	Parameter Estimate	Standard Error	p-Value	R^2-adj	RMSE
1	A [1]	Full model	-	-	0.005	0.118	0.0720
		Intercept	0.0256	0.008	0.003	-	-
		CPV index	0.0234	0.009	0.011	-	-
		Alignment index	0.0142	0.009	0.102	-	-
	B [2]	Full model	-	-	0.003	0.125	0.0717
		Intercept	0.0259	0.008	0.003	-	-
		Operational efficiency	0.0225	0.009	0.011	-	-
		Relationship costs	0.0147	0.009	0.093	-	-
2	C [1]	Full model	-	-	0.002	0.133	0.0501
		Intercept	0.0328	0.008	<0.01	-	-
		Profit margin period 1	0.3173	0.099	0.002	-	-
	D [3]	Full model	-	-	0.002	0.344	0.0444
		Intercept	0.0182	0.008	0.027	-	-
		Profit margin period 1	0.5674	0.108	<0.001	-	-

[1] Full dataset. [2] Full dataset, CPV index replaced with its attributes. [3] Subset of the data comprising 52 contractors who continued to work for the customer. R^2-adj = adjusted level of explained variance; RMSE = root mean square error.

To summarize, customer–contractor alignment was considered from both the customer's and the contractors' perspectives. The two variables, customer-perceived value index and contractor-perceived alignment index, were developed and measured to enable analysis of alignment from both perspectives and were found to be uncorrelated to each other (Table 7). However, contractors' operational efficiency was found to relate positively to the contractor-perceived alignment index (Table 7), suggesting that high operational efficiency may benefit both sides of the relationship. This is supported by the finding that the customer-perceived value index and its attribute operational efficiency were both significantly and positively related to current contractor profitability. To predict future profitability, past profitability was found to be the best predictor, especially for contractors who stayed in the relationship with the customer (Table 8). Contractors who quit working for the customer were significantly less aligned to the customer from both the customer's and the contractors' perspectives (Table 4), and stood out from other contractors by reporting, for instance, conflict and lack of trust with the customer's officials. High contractor net profit margin and high customer-perceived value from additional services decreased the likelihood of relationship breakup, whereas high customer-perceived value from the core solution increased the risk (Table 5).

4. Discussion

Customer-perceived value of contractors' services and contractor-perceived alignment with the customer were hypothesized to affect both contractors' profitability and the probability of a retained customer relationship. A positive relationship would indicate that a contractor who manages to engage with a customer that offers a business environment in alignment with the contractors' needs, and who manages to meet that customer's expectations, would be in a position of competitive advantage compared to his or her peers. The study did indeed find partial support for this hypothesis, but neither customer-perceived value (or its attributes) nor the alignment index managed to explain much variance in profitability. An interpretation of this is that the primary source of competitive advantage for many harvesting contractors may be found in internal factors such as effective entrepreneurship or leadership rather than in their relationship to the customer, a tentative conclusion which concurs with findings from earlier studies (e.g., [25,27,29]). However, the differences between contractors who decided to terminate their relationship and contractors who continued to work for the customer indicate that a change of customer may contribute to resetting profitability for contractors who switch to a different business or operating environment. This indicates that a change of customer may be an attractive move for contractors in some specific situations, for instance, when a contractor's production resources better align with some other customer's operating environment. In 1994, Anderson et al. [40] reported that present customer satisfaction and past profitability both had positive effects on present profitability in a cross-industrial study of 77 Swedish companies, corresponding well with the results of our study. This study, in addition to considering a customer-perceived value index, also developed and analyzed measures associated with its attributes: core solution, additional services, price, and relationship costs. This indicated operational efficiency as the primary driver of profitability. Seemingly, operational efficiency is the alignment 'focal point' of this case, since it is also specified by the customer as a key attribute of the harvesting service [26], and was the only customer-perceived value attribute that correlated significantly to contractor-perceived customer alignment (see Table 7). Many of the interviewed contractors also perceived high expectations from the customer on maintaining a high pace in production and to constantly improve efficiency; these demands seem to have alienated some, primarily less efficient, contractors, and in extreme cases even led to open conflict. The commonly adopted piece rate pricing policy in the forest industry suggests that operational efficiency may be considered a general focal point for relationship alignment in forest harvesting services. Hypothetically, however, this may vary between companies.

High perceived value from additional services safeguarded against relationship breakdown (see Table 5). This is likely caused by the fact that one of the main points of outsourcing is to reduce the need for first level supervision by letting the contractor take on this role in exchange for a profit. Since many of the items used to quantify additional services were associated with how well a contractor runs the business (see [26], for details), high performance in additional services indicates that this transfer of managerial responsibility is working well. If a contractor is perceived to fail in managing the crew or crews, the customer may face a situation where it needs to allocate resources to the close supervision of the contractor to avoid receiving sub-standard services; this may lead the customer to look for other options to free up these resources.

The results from this study showed, somewhat counter-intuitively, an increased risk of a terminated relationship associated with high customer-perceived value from the core solution (i.e., in this case, perceived log quality, thinning quality, and environmental considerations, see [26]). This finding resembles the results from a previous study [41], in which trucking contractors' service level to the customer was found to have a negative effect on contractor profitability, indicating low or no return on some core attributes of the service. A possible explanation is if the customer is, for instance, interested in quality being above a certain threshold level rather than maximized service quality [42]. Hypothetically, a contractor in such a case—one who is providing services far above the threshold—would have the incentive to change customer if the potential new customer is also offering a premium for quality. Relationships that end as a result of such recognition may not be entirely negative when considering the forest industry as a whole. Indeed, it contributes to relocating contractor resources and capabilities to supply chains where they are better appreciated, and presumably yield better benefits.

From the forest company's perspective, the benefit that contractors get from finding the right customer underlines the importance of providing a business environment suited for contractors that they wish to keep in service. In this study, the fairly high scores that the majority of contractors gave to the items measuring their perceived alignment with the customer (see Table 2) indicate that the customer company provides a reasonably attractive environment for most of its contractors.

Further, the positive relationship between customer-perceived value and contractor profitability shows that the company provides incentives for the contractors to fulfill the company's needs. However, on a more detailed level, most attributes of customer-perceived value did not have any effect on contractor profitability. This indicates low incentives for contractors to focus on improvement of important aspects of their services, such as log quality, thinning quality, and environmental considerations, all associated with the attribute 'core solution'. Consequently, the customer would be likely to benefit from adjusting its incentives program so that the contractors gain appreciably by performing well in the most important service attributes to the customer [14]. The chance for a sustained relationship was found to relate positively to contractor profitability and high perceived value from additional services (i.e., in this case, contractor management, collaboration, flexibility, and delivery performance, see [26] for details) whereas high perceived value from the core solution posed a risk for relationship breakup. Consequently, these variables could be used by the company to monitor the level of risk for flight or defaults in its contractor fleet and to evaluate, for instance, its pricing of harvesting services. In addition, interviews with contractors who had ceased working for the company identified lack of trust and conflicts with company officials as important causes of relationship breakdown. This indicates that the contractors' perceived level of trust or conflict with the customer may lead to contractor flight, and consequently is a driver of supply risk for the customer, underlining the importance of good contractor relations for the customer.

This study used a limited sample to investigate customer–contractor alignment and its effects on contractor success in the forest harvesting business. Inherently, this will, of course, make some of the results case specific or at least somewhat biased due to the specific conditions of the sample. For instance, when analyzing the performance of contractors operating as third party service-supplier, Erlandsson et al. [15] have suggested adding the forest owner's perspective to the here addressed costumer and contractor perspectives. Häggström and Lindroos [43] suggest a wider approach for performance in forest operations in general, as a product of complex interactions between human, technology, organization, and environment. Hence, there is growing support for a need of a more holistic view on performance in general, and especially for contractors. Moreover, cases of large forest companies relying on fleets of small harvesting contractors can be found in the forest industry all over the world. This, together with findings from the cited literature supporting the results, indicates that some generalizations can be made from our study. However, some special caution is advised when interpreting the results relating to specifics of the group of contractors who quit working for their customer. The response rate for this group was comparatively low, which may have caused errors due to possible non-response bias.

Naturally there are also limitations and possibilities for improvements in the applied methods and studied aspects of customer–contractor alignment. As indicated above, the performance within forest operations is dependent on numerous, entangled factors and actors. The scope could thus be broadened to cover aspects omitted in the current study. For instance, strategies and expectations for the innovation processes required for continuously improved performances both influence and are influenced by the relationships between costumer and contractor [44,45]. Thus, alignment within, for instance, those aspects could also be included in future analyses.

The use of self-reported data enables gathering of perceptual data that otherwise are difficult to come by, but the method also has limitations [46]. To mitigate the possible methodological shortcomings with self-reported data, a combination of various empirical and self-reported sources were used. For instance, the perceptions reported by customers were compared with the perceptions reported by contractors (e.g., Figure 1). Thus, the absolute values of the respective perception indexes might have been influenced by, for instance, same-source bias, but the influence on the comparisons of (standardized) indexes from different sources should be low. Another possible methodological weakness of the study is that it is in part based on retrospective interviews and surveys with contractors, making faulty recollections a potential source of error. Such response errors have been shown to relate negatively to the saliency of the recalled event [47]. Since the relationship with the forest

123

company certainly represents a salient part of the contractors' professional life at the time of the study, the likelihood of significant effects due to response errors should be low.

5. Conclusions

The objectives of this study were to define and operationalize alignment from both the customers' and the contractors' perspectives, and to empirically analyze its effect on contractor profitability and the development of the customer–contractor relationship over time. Measures of customer-perceived value of contractor services, and contractor-perceived alignment to the customer were developed and applied in a study of a Swedish forest company and its employed contractors.

Findings from this study demonstrate effects of previous contractor profitability, and the customer's perceived value of the contractors' services on contractor profitability. Further, we demonstrate that contractor profitability reduces the risk of relationship breakup, whereas attributes of the customer's perceived value of contractor services have mixed effects. Contractors who decided to change customer or liquidate their firm during the studied period reported conflict and lack of trust as important reasons for ending the relationship with the customer, which indicates an interesting area for future research.

Contractors in the forest harvesting business should seek out the customer that offers business opportunities that are in close alignment with their capabilities and resources to optimize their chances of success. When contractors are employed by the best possible customer, results from this study indicate that the best strategy is to focus on internal efficiency and to be perceived as a promising contractor by the customer. For contractors working on a piece rate basis, this primarily means focusing on operational efficiency to maintain a high cash flow and competitively price their services. Forest companies, on the other hand, would benefit from applying a process for management of its contractor fleet, such as the one suggested by Eriksson et al. [14], to improve alignment of its contractors. Further, companies may benefit from monitoring the profitability of their contractors and use it as an indicator of their contractor-related supply risk.

Results from this study show that customer–contractor alignment does affect outcomes for both contractors and their customers. Consequently, alignment between supply chain partners should be a priority for both practitioners and researchers interested in improving forest operations from a supply chain perspective. If successful, such efforts would contribute to simultaneously improve the value of contractor services to downstream actors and provide contractors with better business opportunities.

Acknowledgments: We would like to thank Pelle Gemmel and Magnus Larsson for helpful comments on earlier versions of the manuscript. This study was conducted as part of the Swedish Forest Industry Research School on Technology (FIRST) program.

Author Contributions: M.E. conceived and designed the study, collected and analyzed the data, and wrote the paper. L.L. and O.L. conceived and designed the study and contributed to the writing of the paper.

Conflicts of Interest: The authors declare no conflict of interest.

References

1. Anon. *Skogsstatistisk Årsbok 2014 (Swedish Statistical Yearbook of Forestry 2014)*; Skogsstyrelsen: Jonkoping, Sweden, 2014; pp. 249, 329. (In Swedish and English)
2. Lee, H.L. The triple—A supply chain. *Harv. Bus. Rev.* **2004**, *82*, 102–112. [PubMed]
3. Cavinato, J.L.; Flynn, A.E.; Kauffmann, R.G. *The Supply Management Handbook*, 7th ed.; McGraw-Hill: New York, NY, USA, 2006.
4. Häggström, C. Human Factors in Mechanized Cut-to-Length Forest Operations. Ph.D. Thesis, Swedish University for Agricultural Sciences, Acta Universitatis Agriculturae Sueciae, Umeå, Sweden, 2015.
5. St-Jean, E.; LeBel, L. The Influence of decisional autonomy on performance and strategic choices—The case of subcontracting SMEs in logging operations. In *Global Perspectives on Sustainable Forest Management*; Okia, C.A., Ed.; InTech: Rijeka, Croatia, 2012; pp. 59–74.

6. Ager, B. *Skogsbrukets Rationalisering och Humanisering 1900–2011 och Framåt (The Rationalization and Humanization of Forestry 1900–2011 and Onwards)*; Arbetsrapport 378; Swedish University for Agricultural Sciences: Umeå, Sweden, 2012. (In Swedish with English summary)

7. Berg, S. Skogsentreprenadföretagens Lönsamhet (Logging Contractors' Profitability). Master's Thesis, Swedish University for Agricultural Sciences, Uppsala, Sweden, 2009. (In Swedish with English summary)

8. Soirinsuo, J.; Mäkinen, P. Importance of the financial situation for the growth of a forest machine entrepreneur. *Scand. J. For. Res.* **2009**, *24*, 264–272. [CrossRef]

9. Penttinen, M.; Mikkola, J.; Rummukainen, A. *Profitability of Wood Harvesting Enterprises*; Working Papers of the Finnish Forest Research Institute; Finnish Forest Research Institute: Helsinki, Finland, 2009; Volume 126.

10. Kelly, M.C.; Germain, R.H. Is it efficient to single-handedly run a multi-machine harvesting operation? A case study from the Northeast United States? *Int. J. For. Eng.* **2016**, *27*, 140–150. [CrossRef]

11. LeBel, L.; Stuart, W.B. Technical efficiency evaluation of logging contractors using a nonparametric model. *J. For. Eng.* **1998**, *9*, 15–24.

12. Carter, D.R.; Cubbage, F.W. Stochastic frontier estimation and sources of technical efficiency in southern timber harvesting. *For. Sci.* **1995**, *41*, 576–593.

13. Conrad, J.-L., IV; Vokoun, M.M.; Prisley, S.P.; Bolding, C. Barriers to logging production and efficiency in Wisconsin. *Int. J. For. Eng.* **2017**, *28*, 57–65. [CrossRef]

14. Eriksson, M.; LeBel, L.; Lindroos, O. Management of outsourced forest harvesting operations for better customer-contractor alignment. *For. Policy Econ.* **2015**, *53*, 45–55. [CrossRef]

15. Erlandsson, E.; Lidestav, G.; Fjeld, D. Measuring quality perception and satisfaction for wood harvesting services with a triad perspective. *Int. J. For. Eng.* **2017**, *28*, 18–33. [CrossRef]

16. Chorn, N.H. The "alignment" theory: Creating strategic fit. *Manag. Decis.* **1991**, *29*, 20–24. [CrossRef]

17. Siguaw, J.A.; Simpson, P.M. Toward assessing supplier value: Usage and importance of supplier selection, retention, and value-added criteria. *J. Mark. Channels* **2004**, *11*, 3–31. [CrossRef]

18. Anon. Entreprenadföretagare om Skogsindustrin: Omodern och Oproffsig (Harvesting contractor about the forest industry: Out of date and unprofessional). *Skogsentreprenören* **2012**, *1*, 16. (In Swedish)

19. Sandström, U. Är entreprenörsstrategin i det svenska skogsbruket på väg att totalhaverera? (Is the contracting strategy of Swedish forest industry completely failing?) *Skogsentreprenören* **2012**, *2*, 10–11. (In Swedish)

20. Sandström, U. Ris och ros för nyckeltalen (Cheers and boos for the key ratios). *Skogsentreprenören* **2012**, *1*, 10–11. (In Swedish)

21. Sandström, U. Det är dags att stoppa nu (It is time to stop now)! *Skogsentreprenören* **2013**, *1*, 13. (In Swedish)

22. Brunberg, T. *Produktiviteten vid drivning från 2008 till 2011 (Productivity in Logging 2008 to 2011)*; Resultat No 2; Skogforsk: Uppsala, Sweden, 2012. (In Swedish with a summary in English)

23. Anon. *Finnish Statistical Yearbook of Forestry 2010*; Finnish Forest Research Institute: Helsinki, Finland, 2012; p. 472.

24. Brunberg, T. *Skogsbrukets kostnader och intäkter 2011—Kraftig ökning av skogsbrukskostnaden (Forestry Costs and Revenues 2011—Major Increase in Costs)*; Resultat No 5; Skogforsk: Uppsala, Sweden, 2012. (In Swedish with a summary in English)

25. Cacot, E.; Emeyriat, R.; Bouvet, A.; Helou, T.E. Tools and analysis of key success factors for mechanized forest contractors specializing in mechanized harvesting in the Aquitaine region. In Proceedings of the FORMEC 2010, Padova, Italy, 11–14 July 2010.

26. Eriksson, M.; Lindroos, O. Customer-perceived value in forest harvesting operations. In Proceedings of the 34th Council of Forest Engineering, Quebec City, QC, Canada, 12–15 June 2011.

27. Norin, K.; Thorsén, Å. *Skogsbrukets "bästa" entreprenadföretag—Deras starka sidor och vad de vill förbättra (The Top Logging Contractors—Their Strengths and Aims)*; Resultat No 9; Skogforsk: Uppsala, Sweden, 1998. (In Swedish with a summary in English)

28. Mäkinen, P. Success factors for forest machine entrepreneurs. *J. For. Eng.* **1997**, *8*, 27–35.

29. Norin, K.; Karlsson, A. *Så arbetar en vinnare—Djupintervjuer med tio lönsamma skogsentreprenörer (What Makes a Winner?—Searching Interviews with Ten Successful Forestry Contractor Businesses)*; Resultat No 21; Skogforsk: Uppsala, Sweden, 2010. (In Swedish with a summary in English)

30. Ulaga, W.; Eggert, A. Value-based differentiation in business relationships: Gaining and sustaining key supplier status. *J. Mark.* **2006**, *70*, 119–136. [CrossRef]

31. Grönroos, C. Value-driven relational marketing: From products to resources and competencies. *J. Mark. Manag.* **1997**, *13*, 407–419. [CrossRef]
32. Erlandsson, E. The impact of industrial context on procurement, management and development of harvesting services: A comparison of two Swedish forest owners associations. *Forests* **2013**, *4*, 1171–1198. [CrossRef]
33. Nollet, J.; Ponce, S.; Campbell, M. About "strategy" and "strategies" in supply management. *J. Purch. Supply Manag.* **2005**, *11*, 129–140. [CrossRef]
34. Cousins, P.D. The alignment of appropriate firm and supply strategies for competitive advantage. *Int. J. Oper. Prod. Manag.* **2005**, *25*, 403–428. [CrossRef]
35. Tangen, S. Performance measurement: From philosophy to practice. *Int. J. Prod. Perform. Manag.* **2004**, *53*, 726–737. [CrossRef]
36. Drolet, S.; LeBel, L. Forest harvesting entrepreneurs, perception of their business status and its influence on performance evaluation. *For. Policy Econ.* **2010**, *12*, 287–298. [CrossRef]
37. Zsidisin, G.A. Managerial perceptions of supply risk. *J. Supply Chain Manag.* **2003**, *39*, 14–25. [CrossRef]
38. Chopra, S.; Meindl, P. *Supply Chain Management: Strategy, Planning, and Operation*, 5th ed.; Pearson: London, UK, 2013.
39. Häggström, C.; Kawasaki, A.; Lidestav, G. Profiles of forestry contractors and development of the forestry-contracting sector in Sweden. *Scand. J. For. Res.* **2013**, *28*, 395–404. [CrossRef]
40. Anderson, E.W.; Fornell, C.; Lehmann, D.R. Customer satisfaction, market share, and profitability: Findings from Sweden. *J. Mark.* **1994**, *58*, 53–66. [CrossRef]
41. Lindström, J. Kartläggning av Ruttplaneringsprocesser för Rundvirkestransportörer (Mapping of the Vehicle Routing Processes in Timber Transport). Master's Thesis, Arbetsrapport 285, Swedish University of Agricultural Sciences, Uppsala, Sweden. (In Swedish with a summary in English)
42. Yang, C.-C. The refined Kano's model and its application. *Total Qual. Manag.* **2005**, *16*, 1127–1137. [CrossRef]
43. Häggström, C.; Lindroos, O. Human, technology, organization and environment—A human factors perspective on performance in forest harvesting. *Int. J. For. Eng.* **2016**, *27*, 67–78. [CrossRef]
44. Rametsteiner, E.; Weiss, G. Innovation and innovation policy in forestry: Linking innovation process with systems models. *For. Policy Econ.* **2006**, *8*, 691–703. [CrossRef]
45. Štěrbová, M.; Loučanová, E.; Paluš, H.; L'ubomír, I.; Šálka, J. Innovation strategy in Slovak forest contractor firms—A SWOT analysis. *Forests* **2016**, *7*, 118. [CrossRef]
46. Podsakoff, P.M.; Organ, D.W. Self-reports in organizational research: Problems and prospects. *J. Manag.* **1986**, *12*, 531–544. [CrossRef]
47. Mathiowetz, N.A.; Duncan, G.J. Out of work, out of mind: Response errors in retrospective reports of unemployment. *J. Bus. Econ. Stat.* **1988**, *6*, 221–229.

Article

A Robust Productivity Model for Grapple Yarding in Fast-Growing Tree Plantations

Riaan Engelbrecht [1], Andrew McEwan [2] and Raffaele Spinelli [3,4,*

[1]   Power Brite Sdn Bhd, Parkcity Commerce Square, Bintulu 97000, Sarawak, Malaysia; rengelbrecht10@gmail.com

[2]   Forest Engineering Dept., Nelson Mandela Metropolitan University, Private Bag X6531, George 6530, South Africa; Andrew@cmo.co.za

[3]   Istituto per la Valorizzazione del Legno e delle Specie Arboree, Consiglio Nazionale delle Ricerche, Via Madonna del Piano 10, Sesto Fiorentino 50019, Italy

[4]   Australian Forest Operations Research Alliance, University of the Sunshine Coast, Locked Bag 4, Maroochydore DC, QLD 4558, Australia

*   Correspondence: spinelli@ivalsa.cnr.it; Tel.: +39-335-5429-798

Received: 23 September 2017; Accepted: 13 October 2017; Published: 17 October 2017

Abstract: New techniques have recently appeared that can extend the advantages of grapple yarding to fast-growing plantations. The most promising technique consists of an excavator-base un-guyed yarder equipped with new radio-controlled grapple carriages, fed by another excavator stationed on the cut-over. This system is very productive, avoids in-stand traffic, and removes operators from positions of high risk. This paper presents the results of a long-term study conducted on 12 different teams equipped with the new technology, operating in the fast-growing black wattle (*Acacia mangium* Willd) plantations of Sarawak, Malaysia. Data were collected continuously for almost 8 months and represented 555 shifts, or over 55,000 cycles—each recorded individually. Production, utilization, and machine availability were estimated, respectively at: 63 m^3 per productive machine hour (excluding all delays), 63% and 93%. Regression analysis of experimental data yielded a strong productivity forecast model that was highly significant, accounted for 50% of the total variability in the dataset and was validated with a non-significant error estimated at less than 1%. The figures reported in this study are especially robust, because they were obtained from a long-term study that covered multiple teams and accumulated an exceptionally large number of observations.

Keywords: productivity; logging; steep terrain; cable logging; *Acacia mangium*

---

1. Introduction

Compared with natural forests, tree plantations offer many benefits, including fast growth, rationalized management and pre-defined product target [1]. Industrial tree plantations are characterized by very high yields, which result from the use of selected genotypes under favorable soil and climate conditions [2]. The fastest growing tree plantations are found in the Southern Hemisphere, and yield up to 40 m^3 of solid wood per hectare per year [3]. As a result, many forest companies in Europe and North America have transferred their capital away from domestic forest ventures [4] and towards highly-productive plantations in South America, South Africa, and Australasia [5].

Plantation forestry has an enormous potential for the efficient supply of fiber and renewable fuel [6], and may already account for almost half of the global production of industrial wood [7]. In fact, experts estimate that the importance of plantations will continue to grow [8], and that by 2050 75% of the global fiber supply will be sourced from dedicated plantations established on farm land [9]. In that regard, plantation forestry offers better environmental performances than conventional agriculture and animal husbandry [10], and is widely acknowledged as a sustainable source of wood fiber [11]. Despite concerns about potential social impacts [12], plantation forestry does not seem to imply a loss of jobs compared with traditional agriculture [13], and it may create high quality jobs in underdeveloped regions [14].

Ultimately, tree plantations contribute to economic and social development, and help in offsetting the increased CO_2 emission [15] resulting from the rapid development of countries in the Southern Hemisphere [16]. Of course, the environmental and social benefits of plantation forestry are only realized if proper management techniques are adopted [17]. That is especially critical for harvesting, which has the highest potential for impact. Safe, low-impact harvesting should be adopted in all cases, making use of modern equipment that is specifically designed for this purpose. Today, forestry equipment manufacturers offer new machines capable of matching the strictest specifications for environmental performance and safe operation [18]. In particular, operator stations are always enclosed in protected cabs, which mitigate the consequences of accidents and are designed to minimize the risk of long-term occupational disease [19].

Complete mechanization of harvesting work is especially difficult on steep terrain, where excessive slope gradient prevents stand access to forest vehicles [20]. Under these conditions, operators generally resort to cable yarding or to building a dense network of skid trails [21]. Unfortunately, neither solution can guarantee low-impact and safe operation at the same time. Cable logging configures as a low-impact harvesting technique, but it requires that operators are on the ground, exposed to serious hazard [22]. On the other hand, opening trail after trail certainly cannot be defined as low-impact harvesting: in fact, the building of skid trails on steep slopes is a main contributor to site impact, especially in tropical areas [23].

There is an urgent need to find cost-effective, low-impact and safe technologies for steep terrain harvesting, as flat land is monopolized by farming and rural development, and plantations can only expand to hill country [24]. For instance, experts predict that by 2020 the volume of wood harvested in New Zealand with cable technology will be larger than the volume harvested with conventional ground-based equipment [25]. Similarly, loggers in different places such as France [26] and the Pacific Northwest [27] must now tackle the challenging terrain left behind in the recent past because it was too steep for low-cost mechanized harvesting with ground-based technology.

A possible solution is offered by grapple yarders, which offer the typical environmental benefits of cable extraction while dispensing with manual assistants placed in risk areas [28]. With these machines, a mechanically-operated grapple clutches the load, removing the need for one or more operators to station at the loading site and manually hitch the loads with chain or wire slings [29]. Grapple yarders are not a new development, and their use has been documented since at least the early 1960s [30]. In fact, they were designed for logging old-growth, where loads generally consisted of one large piece. Landing the grapple on such a big target was relatively easy, and grapple yarders have remained popular in the Pacific Northwest for as long as big tree logging has been widely accepted [31]. However, plantation forestry offers much smaller trees, and optimum load size is only achieved when multiple trees are assembled into the same load, which favors choker slings over grapples [32]. Under these conditions, grapple yarders are potentially less productive because they move smaller loads or spend more time to accumulate a large enough load [33]. Fortunately, a new work technique seems able to recover the handicap and release the full potential of grapple yarding: this consists of detaching a mid-size excavator to feed the grapple with pre-bunched loads, which dramatically boosts grapple yarder production [34]. In this configuration, the grapple yarder becomes a very effective system for logging steep terrain plantations, and may compete successfully with most other options [35]. However, this system is so new that very little information is available on its performance, despite the large potential and widespread interest. To-date, all available studies on the subject consist of short-term case studies that cannot be generalized and are not capable of producing an accurate representation of downtime, as required for long term production estimates [36].

Therefore, the goal of this study is to produce a general model for the productivity of innovative grapple yarding techniques in plantation forestry, based on a large number of observations, and conducted over extended periods on multiple teams. Only in that way can one produce reliable productivity estimates for this technique, which may integrate the inherent variability introduced by terrain, machine type, individual team proficiency, and seasonal fluctuations. In particular, the study aims at (1) determining reference values for net productivity and utilization, (2) categorizing downtime, (3) discriminating between equipment options, and (4) modeling net productivity as a function of relevant independent variables. Such knowledge will allow accurate operation planning, which is crucial to precision management.

2. Materials and Methods

2.1. Site and Equipment

The opportunity for this study was provided by the massive introduction of the new system to the large plantation established and managed by the Grand Perfect Sdn Bhd consortium 45 km east of Bintulu, in Sarawak, Malaysia (Figure 1).

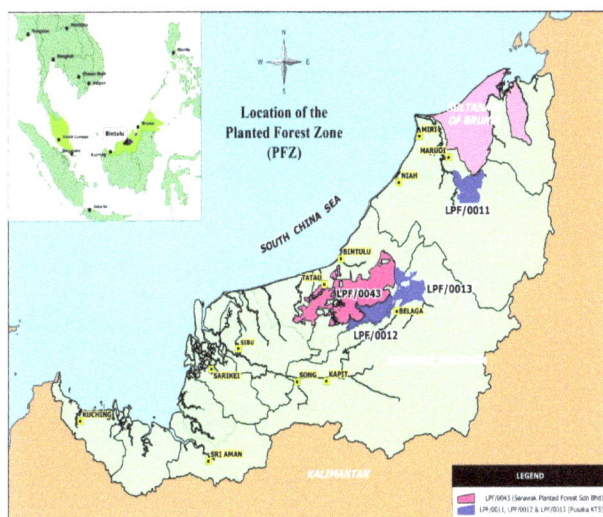

Figure 1. Location of the study sites.

The Sarawak State Government tasked the consortium with planting 150,000 hectares of hill country with black wattle (*Acacia mangium* Willd.), in order to supply a new Kraft Pulp Mill being built in the area [37]. When the time for harvesting approached, the search was launched for a system that could guarantee high productivity, safe operation, and low impact. In particular, managers wanted to avoid the use of bulldozers and other ground-based equipment, which are especially detrimental to the red-yellow podzolic soils characterizing Sarawak forest land [38]. Furthermore, the new system had to be simple enough for operation by relatively inexperienced labor, since the rapid industrialization of Borneo is drawing specialized workers away from rural areas and into factories, as occurs in many other developing countries [39].

Within three years, the consortium commissioned 12 complete yarding operations, each consisting of four machines: an excavator-based grapple yarder for extracting whole-tree loads to the landing edge, a second excavator placed near the yarder tasked with moving loads from in front of the yarder to the stack, a third excavator stationed on the slope for feeding pre-bunched loads to the grapple carriage, and finally a fourth excavator used as a mobile tailhold, turned on only during line change—this last one being an old machine valuable for its weight more than for its power. The excavator-based yarder was equipped with a double drum hydraulic winch set, a tower extension bolted onto the boom stick, and an innovative remote-controlled grapple carriage [40]. The yarder was set up in a semi-live shotgun skyline configuration (Figure 2).

Figure 2. One of the excavator-based un-guyed grapple yarders used in the study (shotgun configuration).

The use of an excavator as the base for the grapple yarder offers significant benefits—especially global availability, ease of operation, versatility, relatively low cost, and robustness [41]. Further, the large mass of the excavator base and the use of the boom as an outrigger allow the guylines to be dispensed with [42]. In turn, un-guyed yarders offer the main advantage of quick repositioning, which allows cost-effective operation in short corridors and reduction of corridor spacing, to the benefit of minimizing the need for time-consuming and potentially high-impact lateral yarding [43]. This is a crucial asset when extracting timber from short steep slopes and waterlogged gullies, frequently encountered in the Grand Perfect plantations. Furthermore, quick repositioning and capacity to swing incoming loads to the side facilitate better use of landing space, enabling operation directly from the forest road, as the yarder can easily move aside once the available stacking space is full [44].

Field work for this research was conducted between 4 October 2016 and 25 May 2017. In order to integrate all possible variability, the study covered all four plantation districts, namely: Anap/Tatau, Tubau, Kakus, and Kemena (Figure 3). For the same reason, the study included all 12 machine teams, which represented two slightly different set-ups (Table 1).

Table 1. Main characteristics of the excavator-based un-guyed yarders on test.

Operation	Type	Heavy	Medium
Teams	#	1 to 7	8 to 12
Base machine	Make	Kobelco	Doosan
Base machine	Model	SK330	DX340 LCA
Base machine	kW	209	184
Winch	Make	Alpine	Alpine
Winch	Model	MDWS 12	MDWS 10
Mainline pull	kN	120	100
Mainline speed	$m\,s^{-1}$	9.0	7.3
Carriage	Make	Alpine	Alpine
Carriage	Model	Hydraulic	Hydraulic
Grapple open	mm	1480	1900
Grapple area	m^2	1.71	3.48

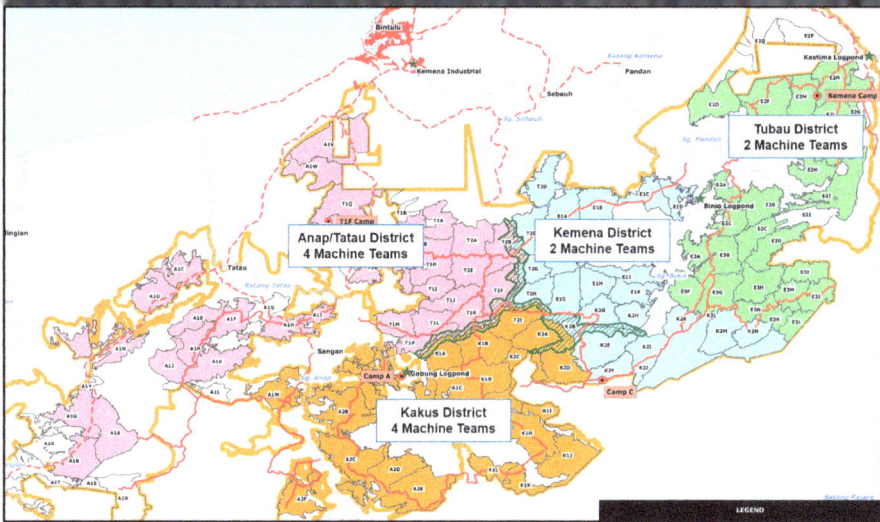

Figure 3. Position of the study compartments and of the 12 machine teams involved with the study.

The compartments were all monocultural black wattle, planted between 1997 and 2003 at a 3 m × 3 m final spacing. Treatment was a clearcut at the end of rotation, without any previous thinning. At the time of cut, most compartments were aged between 10 and 15 years. The pre-harvest inventory determined the following stand characteristics: mean diameter at breast height = 18.7 ± 5.2 cm; total height = 19.5 ± 2.8 m; stocking = 225 ± 140 m^3 ha^{-1}. In fact, rotation age was longer than originally planned, but growth stagnated after the first 7–8 years and therefore conditions can be considered representative of normal black wattle plantations, and more in general of hardwood plantations in tropical and sub-tropical countries.

All compartments were felled motor-manually with chainsaws. After felling, trees were pre-bunched with a medium-size excavator, specifically modified for the purpose by mounting suitable guarding, a log grapple, and a fixed heel.

2.2. Data Collection

A data collection sheet was designed for collection of the following data on a daily basis: machine and operator ID, begin and end of shift (h:min), duration of any delays longer than 5 min (min:ss), description of the delay, duration of each extraction cycle (min:ss), number of pieces in each load (n°), and line length (m). On the same form, a separate page was designed for introducing the total length and the diameter at the top and the butt ends of all pieces in a sample load, which was to be measured four times a day at regular intervals (08:30 h, 10:30 h, 13:30 h, and 17:00 h). For the purpose of the measurements, each team supervisor was equipped with forms, a digital stopwatch, laser range-finder, inclinometer, caliper, and measuring tape. Five research days were spent to train yarder teams to record information accurately, and data sheets were regularly collected and checked by the operation manager.

A vector ruggedness measure (VRM) was determined for each compartment, using a digital elevation model (DEM) acquired from 2-m resolution LiDAR measurements. The VRM is able to estimate the heterogeneity of terrain features independently of slope values, and is considered able to differentiate smooth, steep hillsides from broken terrain with variable gradient and aspect [45].

Eventually, the complete dataset included 555 shifts, for a total of 1346 lines, 54,624 valid cycles and 3517 h of worksite time (excluding 239 h of study delays, which were removed from the dataset).

The study covered the extraction of over 125,000 m^3 or over 240,000 stems, with total length between 10 and 30 m.

2.3. Data Analysis

Data were analyzed both at the shift level and at the cycle level. Descriptive statistics were used for reporting the main results of the study. Then, individual variables were tested for compliance with the main statistical assumptions for a parametric test. In particular, linearity and normality were checked by observing residual plots and distribution histograms, respectively. Equality of variance was checked with Levene's test. Normal (or normalized) data was tested through the analysis of covariance (ANCOVA), with the aim of determining the significance and the strength of all relevant effects—especially machine type, team, distance, and piece size. Eventual differences were pinned on the specific treatments using Tukey–Kramer's test. Non-normal data were analyzed using non-parametric techniques, and treatments were separated using Scheffe's test, which is particularly robust to violations of the normality assumption [46]. The significance of any differences between distributions was checked using classic x^2 (chi-square) analysis.

Time and productivity data were also analyzed with multiple regression techniques in order to estimate significant relationships between these variables and other relevant variables. The effect of categorical data was introduced by generating suitable indicator variables [47]. Validation is a prerequisite of production models derived from time study data [48], and it was conducted according to the same procedure recently used by Adebayo et al. [49] and Spinelli et al. [50] for similar modeling studies. The dataset was partitioned at random into two subsets: the first subset, containing 95% of the observation number was used to calculate appropriate productivity relationships through regression analysis; the second subset, with the remaining 5% of the observations (reserved data), was used to validate the regressions obtained above. To this purpose, the models were used to predict the reserved data, then a paired t-test was used to check if there were any significant differences between predicted and observed figures.

All statistical analyses were conducted with the SAS Statview 5.01 software package (SAS Institute Inc., Cary, NC, USA), for $\alpha < 0.05$.

3. Results

Daily production averaged 226 m^3 for a mean shift duration equal to 6.3 h (Table 2). Mean piece and load size were 0.5 m^3 and 2.4 m^3, respectively. Teams commonly worked 2.4 lines per shift, with a mean length of 103 m and a maximum length of 250 m. That resulted in a mean extraction density of 0.9 m^3 per m of line.

Utilization averaged 63%, with small differences between teams (Table 3). Only team 7 emerged with a significantly higher utilization rate than the others, reaching 80%. While the erratic nature of delays made it difficult to find significant differences between the impact of different delay types, a x^2 analysis of the frequency of delay events showed that team 7 had significantly lower occurrences of mechanical delay and line change events, pointing at higher mechanical reliability and better organization of the harvest area as the possible reasons for the better performance of team 7. In fact, mechanical availability was high for all machines and never went below 89%. As for line change, it is possible that team 7 performed line changes when the machines were idle for some other reasons, so that line change time would be covered by some other delay event category. In all cases, the incidence of personnel delays was relatively high, which was due to the inclusion of the main lunch break within shift time (Figure 4).

Table 2. Main results obtained from the shift level study of the grapple yarder teams.

		Mean per Shift	SD	Min	Max
Lines	#	2.4	1.3	0	7
Cycles	#	97	43	12	233
Delays	events	2.8	4.4	0	36
Mechanical delays	events	0.4	0.7	0	5
Personnel delays	events	0.6	0.6	0	3
Operational delays	events	1.8	3.1	0	19
Work time	h	3.7	1.5	0.5	9.8
Line change time	h	0.9	1.2	0.0	6.6
Delay time	h	1.7	3.0	0.0	22.1
Mechanical delays	h	0.3	1.0	0.0	6.1
Personnel delays	h	0.9	0.9	0.0	3.7
Operational delays	h	0.5	0.7	0.0	7.0
Production	m^3	226	103	22	579
Piece volume	m^3	0.52	0.13	0.18	1.23
Load size	pieces	4.6	0.9	3.1	8.6
Load volume	m^3	2.4	0.6	1.1	4.5
Yarding distance	m	103	40	10	251
Stacking distance	m	20	12	10	60
Productivity	$m^3\ PMH^{-1}$	63	20	15	133
Productivity	$m^3\ SMH^{-1}$	39	17	8	117

Notes: m^3 = cubic meters solid volume over bark; SD = Standard deviation; Stacking distance = the distance between the wood stacks and the yarder chute, covered by the stacking unit when removing incoming loads from under the yarder; PMH = productive machine hours, excluding delay and line change time; SMH = scheduled machine hours, including delays and line change time.

Table 3. Machine utilization and distribution of the number of delay events by grapple yarder team.

Team	Utilization	Cycles	Mechanical Delay	Personnel Delay	Operational Delay	Line Changes
#	%	n°	Events	Events	Events	Events
1	58 [b]	4144	18	23	55	**101**
2	60 [b]	4315	24	22	**132**	26
3	63 [b]	6465	26	24	41	53
4	62 [b]	6028	30	43	95	57
5	68 [b]	5205	14	38	41	67
6	60 [b]	929	5	7	20	10
7	80 [a]	5032	3	22	37	<u>5</u>
8	60 [b]	8732	17	49	111	133
9	57 [b]	5706	53	44	**153**	90
10	63 [b]	3390	17	24	**138**	47
11	64 [b]	2497	13	26	**96**	11
12	57 [b]	1425	9	10	40	12

Notes: Utilization = Productive time/Worksite time; different superscript letters on the utilization figures denote a significant difference for $\alpha < 0.05$; figures in bold represent a significantly higher frequency of occurrence for the event type indicated in the column; underlined figures represent a significantly lower frequency of occurrence for the event type indicated the column.

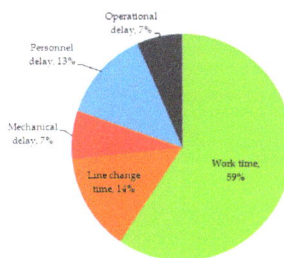

Figure 4. Breakdown of worksite time by main work activities for the 12 grapple yarder teams in the study.

Slope gradient varied from 23% to 49%, with a mean value of 33%. VRM ranged between 1 (even terrain) and 3 (moderately rugged terrain), with the average value at 2 (slightly rugged terrain). Terrain characteristics did not seem to have any strong effects on yarder performance, although there was some evidence that extraction distance increased with slope gradient and decreased with terrain ruggedness. This might be related to the practice of moving all trees to the bottom of the slope during pre-bunching in steep sites, and to the difficulty in getting enough deflection on long lines in rugged terrain. Data analysis also suggested that the frequency of line changes increased with slope, but the time required for the change decreased with it, possibly because an increase in the number of lines for the same harvest implied that the distance between two adjacent lines was shorter, and that made line change easier and faster. In any case, these relationships were quite weak, which made it difficult to estimate reliable factors for the effects of slope gradient and terrain ruggedness on line length, line change frequency, and line change duration.

The analysis of covariance showed that net work productivity in m^3 per productive machine hour (PMH) was affected by piece size, yarder type (heavy or medium), and by the interaction of piece size with yarder type. All these effects were significant at the <0.0001 level, for η^2 values of 10%, 1%, and 1%, respectively.

The analysis of data indicated that the two yarder types (heavy and medium) were deployed on different sites, with the heavy yarder handling larger piece sizes than the medium yarder (Table 4). On the other hand, the latter was able to accumulate a significantly larger number of pieces per load, resulting in a larger load volume—although that took a longer time. The capacity of medium yarders to collect more pieces per load was the result of these machines being equipped with wider-opening grapples than were installed on heavy yarders. Heavy yarders worked longer shifts and had higher utilization, although the differences were small. Line change was faster for the heavy yarders, likely because they worked on shorter distances. However, heavy yarders changed lines more often and the incidence of line change time over total time was the same for both yarder types. Heavy yarders experienced fewer but longer delay events, with the same outcome of no difference between yarder types for the incidence of downtime. As expected, productivity was significantly higher for heavy yarders, as the combined result of stronger winches, faster line speed, larger piece size, and shorter extraction distance. The productive edge of heavy yarders was 25% and 17%, depending on whether delays were included or not.

Table 4. Comparison between medium and heavy grapple yarder types.

		Medium Yarder		Heavy Yarder		p-Value	Test
		Mean	SD	Mean	SD		
Piece volume	m^3	0.47	0.11	0.55	0.13	<0.0001	t-test
Load size	pieces	5.6	0.9	4.2	0.5	<0.0001	MW
Load volume	m^3	2.61	0.59	2.28	0.59	<0.0001	t-test
Yarding distance	m	123	48	93	30	<0.0001	MW
Stacking distance	m	22	12	24	8	0.0005	MW
Work time	h shift^{-1}	3.42	1.41	3.87	1.57	0.0011	t-test
Total time	h shift^{-1}	6.13	2.48	6.42	2.41	0.0495	MW
Utilization	%	60.1	29.1	64.1	33.0	0.0065	MW
Delay time	%	27.4	21.5	23.9	17.1	0.1317	MW
Line change time	%	12.4	17.0	12.0	14.7	0.2427	MW
Line change time	h line^{-1}	1.16	1.18	0.78	0.83	0.0016	MW
Delay events	# shift^{-1}	3.4	3.0	2.4	2.0	0.0005	MW
Delay events	h event^{-1}	0.67	0.65	0.91	0.75	<0.0001	MW
Productivity	m^3 PMH^{-1}	56	18	66	21	<0.0001	MW
Productivity	m^3 SMH^{-1}	33	15	42	17	<0.0001	MW

Notes: m^3 = cubic meters solid volume over bark; SD = Standard deviation; p-Value = results of the unpaired comparison test; t-test = unpaired comparison test is a classic student t-test because data comply with the statistical assumptions; MW = unpaired comparison test is a non-parametric Mann–Whitney test, because data do not comply with the statistical assumptions; PMH = productive machine hours, excluding delay and line change time; SMH = scheduled machine hours, including delays and line change time.

The net productivity of teams 1, 4, 6, and 8 was between 20% and 30% higher than the grand average, and the difference was significant. While these differences could partly depend on better working conditions, they hinted at the effect of operator skills on operation performance. Therefore, these teams were marked as "top teams", and the group was tested as an independent variable in multiple regression analysis, in order to check if the effect of team choice was significant in addition to the effects of work conditions.

The model was highly significant, and could explain half of the variability in the data pool (Table 5). The relationships described by the model were all logical: productivity increased with piece volume and the number of pieces in a load, and decreased with both line length and stacking distance. Productivity was lower for the medium yarders, and the difference between the two yarder types increased with piece size, with heavy yarders performing increasingly better with larger trees, also because they were fitted with smaller grapples and may have encountered more difficulty when trying to accumulate many pieces in a single load. Top teams were more productive than the average, but their margin eroded with piece size, indicating that with a large piece size most operators can achieve a high productivity, and that skills are truly tested with small pieces rather than with large ones. The model was successfully validated. Reserved data were predicted with an error of 0.3%, and the difference between actual and predicted productivity figures was not significant ($p = 0.6943$).

Table 5. Regression model for predicting net grapple yarder productivity as a function of significant independent variables.

$P = a + b$ Vol $+ c$ N$^\circ$ $+ d$ Line $+ e$ Stack $+ f$ Medium Vol $+ g$ Top Vol $+ h$ Top				
Adjusted R^2 = 0.501; n = 42,927; F = 6157.6; $p < 0.0001$				
	Coeff	SE	t-Value	p-Value
a	−50.515	1.163	−43.4	<0.0001
b	132.724	2.068	64.2	<0.0001
c	14.222	0.108	131.6	<0.0001
d	−0.127	0.003	−42.2	<0.0001
e	−0.124	0.012	−10.3	<0.0001
f	−25.143	0.778	−32.3	<0.0001
g	−36.148	2.662	−13.6	<0.0001
h	29.559	1.427	20.7	<0.0001

Where: P = net productivity in m^3 PMH^{-1}, excluding delays; n = number of valid observations; SE = Standard error; Vol = average piece volume, m^3 over bark; N$^\circ$ = number of pieces per load; Line = Line length in m; Stack = stacking distance in m; Medium = indicator variable for medium yarder: if yarder is medium = 1, if heavy = 0; Top = indicator variable for top team: if top team = 1, if not = 0.

The model was used to estimate a break-even piece size between the two yarder types. The calculation was conducted under the following hypotheses: line length = 100 m, stacking distance = 20 m, use of a standard team (not a top team), and a number of pieces per load that was found to be stable at 4.1 for the heavy yarder, and varied according to the following equation for the medium yarder: n$^\circ$ of pieces = 6.88 – 2.73 piece volume in m^3. The graph suggests that heavy yarders should be preferably deployed on those sites where mean piece volume is larger than 0.6 m^3 (Figure 5). Of course, this is true for medium yarders with a larger grapple than installed on heavy yarders, and it is likely that the results will change if heavy yarders will also be equipped with a wide-opening grapple.

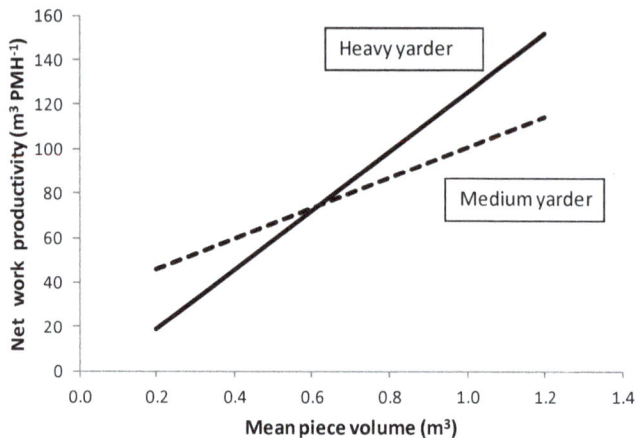

Figure 5. Net productivity of the two excavator-based grapple yarder types as a function of piece volume, estimated using the regression models in Table 5.

4. Discussion

Firstly, it is important to state the limitations of the study upfront, namely: (a) the observational character of the experiment, (b) the recording of data directly by operators, and (c) the inaccuracy inherent to some of the record types.

The observational character of the experiment is the inevitable consequence of conducting a long-term study of commercial operations, which was necessary to reflect realistic work conditions and to accumulate a very large number of observations. This study would have not been sustainable if experimental conditions had to be controlled. In fact, the very large number of observations gathered with this study is very likely to counteract the shortcomings of an unbalanced dataset, since treatment balance is especially critical with small datasets, while large datasets are less sensitive to imbalance, and in fact they are often imbalanced without much prejudice to the quality of results [51].

The recording of data directly by operators is likely to result in some loss of accuracy, because operators cannot be as skillful and motivated as professional researchers [52]. Unfortunately, it would have been too expensive to hire 12 researchers for over 6 months, and the only alternative to investing operators with data collection would have been that of resorting to automatic data collection [53]. However, automatic data collection systems cannot collect all necessary data, and still need operator input, although much reduced [54,55]. In order to limit the risk for gross inaccuracies, operators were carefully trained and supervised, and the data collection routine was simplified as much as possible [56].

The simplification of the data collection process actually brings the discussion to the third limitation, consisting of the use of inherently inaccurate indicators. That is the case of piece volume and line length. Adopting the mean piece volume of four loads and applying that figure to over 90 loads is likely to result in some error, and so does the adoption of total line length as the reference distance for all turns extracted on that given line. On the other hand, both errors are likely contained, because of the relatively homogeneous stem size characterizing most plantation compartments, and because of the general practice of moving stems downhill and towards the tail anchor during pre-bunching.

Without accepting these limitations, it would have been very difficult to study multiple teams and collect a very large number of observations. While work study theory recommends that models are built using large data pools obtained by sampling many different operators [57], the reality is that most of the available forest engineering bibliography consists of case-studies [58], with few notable exceptions [50,59]. This study fulfils both canonical requirements for producing a reliable model, and that is confirmed by the results of the statistical analysis. While data imbalance and various errors may have weakened the accuracy of the results, the study is still able to determine the effect of the most important variables that affect productivity, and offers a regression model that is highly significant, accounts for 50% of the total variability in the dataset, and is validated with a non-significant error estimated at less than 1%.

The reference figures presented in this study substantiate recent claims about the very high productivity of the new grapple yarding system. Such encouraging results confirm the inherent potential of the new technology, but depend at least in part on the short extraction distance and the benefits of pre-bunching, which has been shown to increase yarder productivity by more than 30% [33,34]. On the other hand, piece size is much smaller than reported in New Zealand or US studies, and offers a better representation of black wattle and fast-growing hardwood plantations in general [60].

The productivity levels recorded in this study compare quite well with the figures reported in other previous studies of similar grapple yarders, swing yarders, and excavator-based un-guyed yarders (Table 6). The closest match is the study by Amishev and Evanson [34], reporting a productivity of 86 m^3 PMH^{-1}. When the model developed in this study is used to calculate yarder productivity under the same line length and piece size conditions, the result is 83 and 98 m^3 PMH^{-1} for the medium and the heavy yarders, respectively. This may support the general validity of our productivity model.

Table 6. Comparison of results between this study and previous grapple and swing yarder studies.

Yarder Type	Carriage Type	Line Length m	Piece m^3	Productivity m^3 PMH^{-1}	Operation Type	Country	Cycles n	Reference
Alpine MDWS	Grapple	103	0.52	63	Clearcut	Malaysia	54,624	this study
Madill 124	Grapple	100	0.81	58	Clearcut	Australia	184	[33]
Thunderbird 6355	Grapple	160	0.85	86	Clearcut	New Zealand	123	[34]
Thunderbird 255	Slings	233	1.52	39	Clearcut	New Zealand	165	[61]
Madill 122	Slings	267	0.71	44	Clearcut	USA	70	[62]
Timbco T425	Slings	80	0.55	15	Thinning	USA	218	[63]
CAT 315 L	Slings	80	1.43	30	Thinning	USA	237	[63]
Doosan DX 210W	Slings	120	0.28	11	Clearcut	Norway	149	[44]
Modified JCB	Slings	130	0.35	17	Clearcut	Ireland	90	[42]

Notes: Piece = piece size; m^3 = cubic meters solid volume over bark; PMH = productive machine hours, excluding delay and line change time.

This model also includes the effect of operator proficiency, indicating that particularly skillful teams can outperform average teams by 10% to 30%, depending on piece size and yarder type. Such figures are fully compatible with those reported for Finnish harvester operators by Ovaskainen et al. [64] and Karha et al. [65], who also indicated a bracket between 20% and 40%. While it may be difficult to rate operators correctly [66], knowledge of possible variations is important when assessing between-team variability for planning purposes [67].

In that regard, it is also important to remark that the team that clearly emerged above all the others for minimum downtime was not among the teams that were rated as top teams for their high net productivity and fast work pace. This simple fact suggests that dexterity and good time management represent different skills, which may not be concurrent within the same team. In fact, any future upgrades should target time management and utilization rather than net productivity, since the latter is already in line with any of the documented figures or well above them, whereas utilization and shift duration are still relatively small and could be improved.

The model can also discriminate between yarder types, indicating preferential conditions for yarder type selection. Of course, the differences found with the model are only valid for the specific machine configurations covered with the study, and may vary if specific improvements were made on the yarders. That is specifically the case of grapple selection: the better performance of medium yarders with smaller trees is due at least in part to the wider opening grapple they received, and therefore one may improve the small-tree performance of heavy yarders by fitting them with a larger grapple.

However, the model cannot account for the effect of slope gradient and ruggedness, but that might be due to the relatively small field of variation for both factors rather than to poor model characteristics. In fact, both slope gradient and terrain ruggedness are somewhat moderate, which is the condition for effective pre-bunching. On steeper slopes, one may resort to winch-assist technology for pre-bunching, as it has already been done in other cases [35]. While terrain conditions would be suitable for the introduction of a complete winch-assist ground-based harvesting system, the need for minimizing soil impacts suggests sticking with the current grapple yarder system. In this case, a valid compromise could be reached by introducing a winch-assist feller-buncher to replace manual felling and separate pre-bunching. This measure may dramatically increase productivity and worker safety. However, winch-assist ground-based harvesting systems require highly specialized personnel, which may represent a limiting factor in developing economies [68].

5. Conclusions

The productivity models estimated in this study are especially robust, because they were obtained from a long-term study that covered multiple teams and accumulated an exceptionally large number of observations, as prescribed by the classic canons of model development. The figures and the model offered in this paper fill an urgent knowledge need, as the new grapple yarding technique is becoming increasingly popular and attracts growing attention. While gained specifically on one yarder make and plantation type, the information in this study can be extended to other similar machines and plantations, because the sheer volume of data allows cautious generalization.

Acknowledgments: Special thanks are due to the Grand Perfect Pusaka Sdn Bhd that allowed access to their operations, as part of a farseeing strategy to pursue innovation and excellence through research. We also thank all harvesting supervisors and managers at Power Brite Sdn Bhd for their commitment and dedication during the lengthy data collection process, Peter Lai Soon for helping with training yarder operators for the study, and Julia Jita for her kind assistance with data filing. This project has been partly supported by the Bio Based Industries Joint Undertaking under the European Union's Horizon 2020 research and innovation program under grant agreement No. 720757 Tech4Effect.

Author Contributions: Riaan Engelbrecht, Andrew McEwan and Raffaele Spinelli conceived and designed the experiment; Riaan Engelbrecht collected the data in cooperation with Andrew McEwan and Raffaele Spinelli; Raffaele Spinelli and Andrew McEwan processed the data, in cooperation with Riaan Engelbrecht; Riaan Engelbrecht, Andrew McEwan and Raffaele Spinelli wrote the paper.

Conflicts of Interest: The authors declare no conflict of interest.

References

1. Campinhos, E., Jr. Sustainable plantations of high-yield shape Eucalyptus trees for production of fiber: The Aracruz case. *New For.* **1999**, *17*, 129–143. [CrossRef]
2. Stape, J.; Binkley, D.; Ryan, M.; Fonseca, S.; Loos, R.; Takahashi, E.; Silva, C.; Silva, S.; Hakamada, R.; Ferreira, J.; et al. The Brazil Eucalyptus Potential Productivity Project: Influence of water, nutrients and stand uniformity on wood production. *For. Ecol. Manag.* **2010**, *259*, 1684–1694. [CrossRef]
3. Siry, J.; Cubbage, F.; Ahmed, M. Sustainable forest management: Global trends and opportunities. *For. Policy Econ.* **2005**, *7*, 551–561. [CrossRef]
4. Lonnstedt, L.; Sedjo, R. Forestland ownership changes in the United States and Sweden. *For. Policy Econ.* **2012**, *14*, 19–27. [CrossRef]
5. Laaksonen-Craig, S. Foreign direct investment in the forest sector: Implications for sustainable forest management in developed and developing countries. *For. Policy Econ.* **2004**, *6*, 359–370. [CrossRef]
6. Sedjo, R. The potential of high-yield plantation forestry for meeting timber needs. *New For.* **1999**, *17*, 339–360. [CrossRef]
7. Food and Agriculture Organization (FAO). Global Forest Resources Assessment 2015. Available online: http://www.fao.org/3/a-i4793e.pdf (accessed on 16 October 2017).

8. Ragauskas, A.; Williams, C.; Davison, B.; Britovsek, G.; Cairney, J.; Eckert, C.; Frederick, W., Jr.; Hallett, J.; Leak, D.; Liotta, C.; et al. The path forward for biofuels and biomaterials. *Science* **2006**, *311*, 484–489. [CrossRef] [PubMed]

9. Sohngen, B.; Mendelsohn, R.; Sedjo, R. Forest management, conservation, and global timber markets. *Am. J. Agric. Econ.* **1999**, *81*, 1–13. [CrossRef]

10. Bremer, L.; Farley, A. Does plantation forestry restore biodiversity or create green deserts? A synthesis of the effects of land-use transitions on plant species richness. *Biodivers. Conserv.* **2010**, *19*, 3893–3915. [CrossRef]

11. Berndes, G.; Hoogwijk, M.; Van den Broek, R. The contribution of biomass in the future global energy supply: A review of 17 studies. *Biomass Bioenergy* **2003**, *25*, 1–28. [CrossRef]

12. Charnley, S. Industrial plantation forestry. Do local communities benefit? *J. Sustain. For.* **2005**, *21*, 35–57. [CrossRef]

13. Paul, K.; Reeson, A.; Polglase, P.; Ritson, P. Economic and employment implications of a carbon market for industrial plantation forestry. *Land Use Policy* **2013**, *30*, 528–540. [CrossRef]

14. Landry, J.; Chirwa, P. Analysis of the potential socio-economic impact of establishing plantation forestry on rural communities in Sanga district, Niassa province, Mozambique. *Land Use Policy* **2011**, *28*, 542–551. [CrossRef]

15. Machado, R.; Conceição, S.; Leite, H.; de Souza, A.; Wolff, E. Evaluation of forest growth and carbon stock in forestry projects by system dynamics. *J. Clean. Prod.* **2015**, *96*, 520–530. [CrossRef]

16. Rochedo, P.; Costa, I.; Império, M.; Hoffmann, B.; Merschmann, P.; Oliveira, C.; Szklo, A.; Schaeffer, R. Carbon capture potential and costs in Brazil. *J. Clean. Prod.* **2016**, *131*, 280–295. [CrossRef]

17. Food and Agriculture Organization (FAO). Responsible Management of Planted Forests: Voluntary Guidelines. Available online: http://www.fao.org/docrep/009/j9256e/J9256E03.htm (accessed on 16 October 2017).

18. Spinelli, R.; Magagnotti, N. The effects of introducing modern technology on the financial, labour and energy performance of forest operations in the Italian Alps. *For. Policy Econ.* **2011**, *13*, 520–524. [CrossRef]

19. Visser, R.; Stampfer, K. Expanding ground-based harvesting onto steep terrain: A review. *Croat. J. For. Eng.* **2015**, *36*, 321–331.

20. Visser, R.; Berkett, H. Effect of terrain steepness on machine slope when harvesting. *Int. J. For. Eng.* **2015**, *26*, 1–9. [CrossRef]

21. Food and Agriculture Organization (FAO). Logging and Transport in Steep Terrain. Available online: http://www.fao.org/docrep/016/ap015e/ap015e00.pdf (accessed on 16 October 2017).

22. Montorselli, N.; Lombardini, C.; Magagnotti, N.; Marchi, E.; Neri, F.; Picchi, G.; Spinelli, R. Relating safety, productivity and company type for motor-manual logging operations in the Italian Alps. *Accid. Anal. Prev.* **2010**, *42*, 2013–2017. [CrossRef] [PubMed]

23. Pinard, M.; Barker, M.; Tay, J. Soil disturbance and post-logging forest recovery on bulldozer paths in Sabah, Malaysia. *For. Ecol. Manag.* **2000**, *130*, 213–225. [CrossRef]

24. Nahuelhual, L.; Carmona, A.; Lara, A.; Echeverría, C.; González, M. Land-cover change to forest plantations: Proximate causes and implications for the landscape in south-central Chile. *Landsc. Urban Plan.* **2012**, *107*, 12–20. [CrossRef]

25. Raymond, K. Innovation to increase profitability of steep terrain harvesting in New Zealand. *N. Z. J. For.* **2012**, *57*, 19–23.

26. Spinelli, R.; Visser, R.; Riond, C.; Magagnotti, N. A survey of logging contract rates in the southern European Alps. *Small-Scale For.* **2017**, *16*, 179–193. [CrossRef]

27. FPInnovations Steep Slope Harvesting Initiative. Available online: http://www.coastforest.org/fpinnovations-steep-slope-harvesting-initiative/ (accessed on 13 October 2017).

28. WorkSafe, B.C. *Grapple Yarder and Supersnorkel Handbook*; Workers' Compensation Board: Richmond, BC, Canada, 1992; 196p.

29. Howard, A. Production equations for grapple yarding in Coastal British Columbia. *West. J. Appl. For.* **1991**, *6*, 7–10.

30. Studier, D.; Binkley, V. *Cable Logging Systems. Division of Timber Management*; USDA Forest Service: Portland, OR, USA, 1974; 190p.

31. De Souza, A. A Study of Production and Ergonomic Factors in Grapple Yarding Operations Using an Electronic Data Logger System. Ph.D. Thesis, University of British Columbia, Vancouver, BC, Canada, 1983.

32. Helton, J. A Comparison of Grapple Yarding and Choker Yarding in British Columbia. Master's Thesis, University of New Brunswick, Fredericton, NB, Canada, 1985.

33. Acuna, M.; Skinnell, J.; Evanson, T.; Mitchell, R. Bunching with a self-levelling feller-buncher on steep terrain for efficient yarder extraction. *Croat. J. For. Eng.* **2011**, *32*, 521–531.

34. Amishev, D.; Evanson, T. Innovative methods for steep terrain harvesting. In Proceedings of the FORMEC 2010 Conference "Forest Engineering: Meeting the Needs of the Society and the Environment", Padova, Italy, 11–14 July 2010.

35. Visser, R.; Raymond, K.; Harrill, H. Mechanizing steep terrain harvesting operations. *N. Z. J. For.* **2014**, *59*, 3–8. [CrossRef]

36. Cavalli, R. Prospects of research on cable logging in forest engineering community. *Croat. J. For. Eng.* **2012**, *33*, 339–356.

37. Chua, A. More forest plantations. *Sarawak Tribune*, 2 December 1996.

38. Pinard, M.; Putz, F.; Tay, J.; Sullivan, T. Creating timber harvesting guidelines for a reduced-impact logging project in Malaysia. *J. For.* **1995**, *93*, 41–45.

39. Hoffmann, S.; Jaeger, D.; Lingenfelder, M.; Schoenherr, S. Analyzing the efficiency of start-up cable yarding crew in southern china under new forest management perspectives. *Forests* **2016**, *7*, 33. [CrossRef]

40. Alpass, P. Alpine shovel yarders go international. *S. Afr. For. Mag.* **2010**, *10*, 14–15.

41. Torgersen, H.; Lisland, T. Excavator-based cable logging and processing system: A Norwegian case study. *Int. J. For. Eng.* **2002**, *13*, 11–16.

42. Devlin, G.; Klvač, R. How technology can improve the efficiency of excavator-base cable harvesting for potential biomass extraction—A woody productivity resource and cost analysis for Ireland. *Energies* **2014**, *7*, 8374–8395. [CrossRef]

43. Talbot, B.; Ottaviani-Aalmo, G.; Stampfer, K. Productivity analysis of an un-guyed integrated yarder-processor with running skyline. *Croat. J. For. Eng.* **2014**, *35*, 201–210.

44. Talbot, B.; Stampfer, K.; Visser, R. Machine function integration and its effect on the performance of a timber yarding and processing operation. *Biosyst. Eng.* **2015**, *135*, 10–20. [CrossRef]

45. Sappington, J.; Longshore, K.; Thompson, D. Quantifying landscape ruggedness for animal habitat analysis: A case study using bighorn sheep in the Mojave Desert. *J. Wildl. Manag.* **2007**, *71*, 1419–1426. [CrossRef]

46. Statistical Analysis System (SAS). *StatView Reference*; SAS Publishing: Cary, NC, USA, 1999; 528p.

47. Olsen, E.; Hossain, M.; Miller, M. *Statistical Comparison of Methods Used in Harvesting Work Studies*; Oregon State University, Forest Research Laboratory: Corvallis, OR, USA, 1998; 31p.

48. Howard, A. Validating forest harvesting production equations. *Trans. ASAE* **1992**, *35*, 1683–1687. [CrossRef]

49. Adebayo, A.; Han, H.S.; Johnson, L. Productivity and cost of cut-to-length and whole-tree harvesting in a mixed-conifer stand. *For. Prod. J.* **2007**, *57*, 59–69.

50. Spinelli, R.; Hartsough, B.; Magagnotti, N. Productivity standards for harvesters and processors in Italy. *For. Prod. J.* **2010**, *60*, 226–235. [CrossRef]

51. Payne, R. General balance, large data sets and extensions to unbalanced treatment structures. *Comput. Stat. Data Anal.* **2003**, *44*, 297–304. [CrossRef]

52. Nuutinen, Y.; Väätäinen, K.; Heinonen, J.; Asikainen, A.; Röser, D. The accuracy of manually recorded time study data for harvester operation shown via simulator screen. *Silv. Fenn.* **2008**, *42*, 63–72. [CrossRef]

53. Manner, J.; Nordfjell, T.; Lindroos, O. Automatic load level follow-up of forwarders' fuel and time consumption. *Int. J. For. Eng.* **2016**, *27*, 151–160. [CrossRef]

54. Holzleitner, F.; Kanzian, C.; Höller, N. Monitoring the chipping and transportation of wood fuels with a fleet management system. *Silv. Fenn.* **2013**, *47*, 11. [CrossRef]

55. Spinelli, R.; Magagnotti, N.; Pari, L.; De Francesco, F. A comparison of tractor-trailer units and high-speed forwarders used in Alpine forestry. *Scand. J. For. Res.* **2015**, *30*, 470–477. [CrossRef]

56. Spinelli, R.; Laina-Relaño, R.; Magagnotti, N.; Tolosana, E. Determining observer and method effects on the accuracy of elemental time studies in forest operations. *Balt. For.* **2013**, *19*, 301–306.

57. Harstela, P. Work studies in forestry. *Silv. Carelica* **1991**, *18*, 41.

58. Lindroos, O.; Cavalli, R. Cable yarding productivity models: A systematic review over the period 2000–2011. *Int. J. For. Eng.* **2016**, *27*, 79–94. [CrossRef]

59. Eriksson, M.; Lindroos, O. Productivity of harvesters and forwarders in CTL operations in northern Sweden based on large follow-up datasets. *Int. J. For. Eng.* **2014**, *25*, 179–200. [CrossRef]

60. Spinelli, R.; Ward, S.; Owende, P. A harvest and transport cost model for Eucalyptus spp. fast-growing short rotation plantations. *Biomass Bioenergy* **2009**, *33*, 1265–1270. [CrossRef]

61. A mechanized swing yarder operation in New Zealand. Available online: http://fgr.nz/documents/download/4592 (accessed on 16 October 2017).

62. Madill 122 Interlock swing yarder. Available online: http://fgr.nz/documents/download/4844 (accessed on 16 October 2017).

63. Largo, S.; Han, H.S.; Johnson, L. Productivity and cost evaluation for non-guyline yarders in northern Idaho. In Proceedings of the COFE Conference "Machines and People: The Interface", Hot Springs, AK, USA, 27–30 April 2004.

64. Ovaskainen, H.; Uusitalo, J.; Väätäinen, K. Characteristics and significance of a harvester operator's working technique in thinnings. *Int. J. For. Eng.* **2004**, *15*, 67–77.

65. Kärhä, K.; Rönkö, E.; Gunne, S. Productivity and cutting costs of thinning harvesters. *Int. J. For. Eng.* **2004**, *15*, 43–55.

66. Purfürst, T.; Lindroos, O. The correlation between long-term productivity and short-term performance ratings of harvester operators. *Croat. J. For. Eng.* **2011**, *32*, 509–519.

67. Mola-Yudego, B.; Picchi, G.; Röser, D.; Spinelli, R. Assessing chipper productivity and operator effects in forest biomass operations. *Silv. Fenn.* **2010**, *49*, 14. [CrossRef]

68. Raymond, K. Innovative harvesting solutions: A step change harvesting research programme. *N. Z. J. For.* **2010**, *55*, 4–9.

Article

Impacts of Early Thinning of a *Eucalyptus globulus* Labill. Pulplog Plantation in Western Australia on Economic Profitability and Harvester Productivity

Mauricio Acuna [1],*, Martin Strandgard [2], John Wiedemann [3] and Rick Mitchell [4]

[1] Australian Forest Operations Research Alliance (AFORA), University of the Sunshine Coast, Private bag 12, Hobart 7001, TAS, Australia
[2] Australian Forest Operations Research Alliance (AFORA), University of the Sunshine Coast, 500 Yarra Boulevard, Richmond 3121, VIC, Australia; mstrandg@usc.edu.au
[3] WA Plantation Resources (WAPRES), P.O. Box 444, Manjimup 6258, WA, Australia; john.wiedemann@wapres.com.au
[4] Australian Forest Operations Research Alliance (AFORA), University of the Sunshine Coast, 35 Shorts Place, Albany 6330, WA, Australia; rmitchel@usc.edu.au
* Correspondence: macuna@usc.edu.au; Tel.: +61-(03)-6237-5623

Received: 13 September 2017; Accepted: 30 October 2017; Published: 1 November 2017

Abstract: The impact of the manipulation of plantation stocking density on individual tree size can affect final harvest costs and machine productivity. This paper investigated the impact of four early-age thinning treatments applied to a *Eucalyptus globulus* Labill. pulplog plantation in south-west Western Australia on economic profitability and harvester productivity. Eighteen sample plots were randomly laid out in the study area. The nominal 700, 500, and 400 stems per hectare (sph) plots were thinned to waste 3.2 years after establishment while the nominal 1000 sph (UTH) plots were left unthinned. The economic analysis showed that all thinning treatments resulted in a lower Land Expectation Value (LEV) and net financial loss over the full rotation at their theoretical optimal rotation age when compared with the unthinned control treatment. Tree growth and form were positively impacted by thinning. However, associated reductions in harvesting costs were less than the value losses resulting from reduced per hectare yield.

Keywords: land expectation value; thinning; *Eucalyptus globulus*; stocking density; harvesting productivity; Australia

---

1. Introduction

Over 900,000 hectares of eucalypt plantations (>50% of which is *Eucalyptus globulus* Labill.) have been established in Australia, primarily since 1990 and principally as a source of export chiplogs [1]. These plantations have typically been planted at a stocking density of approximately 1000–1250 stems per hectare (sph) with early weed control and fertiliser application and a planned rotation length of ten years.

A key silvicultural tool used by plantation managers to achieve their objectives is the manipulation of plantation stocking density through initial stocking or thinning. A large number of studies across both coniferous (e.g., *Pinus radiata* D.Don [2], *Pinus sylvestris* L. [3], *Picea mariana* (Mill.) BSP, *Picea glauca* (Moench) Voss, and *Pinus resinosa* Sol. ex Aiton [4]) and hardwood (e.g., *Eucalyptus nitens* H. Deane & Maiden [5], *Eucalyptus grandis* W. Hill ex Maiden [6], *Populus deltoides* W. Bartram ex Marshall [7]) species have concluded that increasing initial stocking density increases total wood volume per hectare and decreases mean tree volume and diameter, although the effect of stocking density on tree traits tends to reduce with increasing age [8]. Mean tree height has also been found to decrease as stocking density increases, but the effect is only apparent at extremely high stocking densities [9]. Accordingly, higher stocking densities are generally favoured for pulpwood plantations, where maximum wood volume is the main objective, and lower stocking densities are generally favoured for sawlog plantations where large individual tree sizes are required [10]. Lower initial stocking densities also allow machine access for management activities [11]. However, for species such as eucalypts where planting stock and establishment costs are relatively high, the additional volume production from higher stocking densities needs to be balanced against higher initial costs [5,8]; hence, initial stocking densities of eucalypt plantations are generally between 800 and 2000 sph for both sawlog and pulpwood regimes [12].

Thinning to remove small or defective trees (thinning from below) is common practice in sawlog plantations to improve average stem form and accelerate the growth of retained stems [13,14]. Thinning is also used to control pests and diseases [15,16] and reduce stand water use [17,18]. Planting stands at higher stocking densities followed by thinning suppresses the growth of lower branches—and, hence, knot sizes—and provides more trees from which to select final crop trees [5]. However, growth of lower branches can increase post-thinning [19], machinery used in thinning operations can damage retained stems [11,20], and debris left on site from thinning operations can increase the risk and severity of fires [21,22] and pest infestations [23]. Delaying the age at which thinning is performed can increase thinning wood volumes and, hence, returns, but can expose retained stems to wind damage (as they are more slender than if thinned early) [11], reduce the size of the final crop trees [24], and increase the risk of drought-induced mortality [25]. Heavy early thinning to final crop stocking can result in lost wood production due to the retained trees never fully utilising the site during the rotation [26].

Plantation stocking density impacts on individual tree sizes can affect final harvest costs and productivity as tree size has been shown in numerous previous studies to be the main driver of mechanised harvesting costs and productivity (e.g., [27–30]). For example, [31] found that pre-commercial thinning in their study reduced final harvest costs by over 25% as a result of significant improvements in harvester and forwarder productivity caused by larger post-thinning mean tree sizes. The relationship between machine productivity or harvest cost and tree size is non-linear with a significant decrease in productivity and increase in harvest cost per cubic metre being observed for trees with a volume less than approximately 0.2 m^3 [28,29]. Wood losses due to stem breakage can also increase as stem size decreases when stems are being processed by harvesters [27,32] or chippers [33]. However, some of the harvesting gains from increased tree size at reduced stocking densities can be lost through increased branch thickness reducing harvester productivity [34].

The impacts of plantation stocking density on final harvest costs need to be considered in the context of the total costs and returns for a rotation. A simulation of the impact of pre-commercial thinning on final harvest costs and returns from *Pinus banksiana* Lamb. stands found that the additional costs associated with thinning were more than offset by the reduction in final harvest costs resulting from larger and fewer trees [35].

A number of financial decision tools, including Net Present Value (NPV) and Internal Rate of Return (IRR), have been used to compare the value of alternative forestry investments or to compare potential forestry investments with non-forestry options. Deficiencies in the use of NPV and IRR in the context of forestry investments were highlighted by [36], who advocated the use of the classical solution: Faustmann's Land Expectation Value (LEV) [37]. LEV is the present value of an infinite series of rotations, assuming that future stand growth and prices are known, with the optimal rotation age being that at which the present value is maximised. The original work by Faustmann has been extended to allow for additional factors including stochastic factors, such as risk from fire [21] or changes in timber prices [38], and provision of non-timber forest services or amenities such as recreation and flood control [39].

The reported study investigated the effect of a range of stocking densities on standing tree and harvesting traits in a *Eucalyptus globulus* plantation in south-west Western Australia (WA). The study aimed to (i) quantify the effect of stocking density on standing tree attributes, including diameter at breast height over bark (DBHOB), tree height, tree volume, and tree form attributes (branches and forks); (ii) quantify the effect of stocking density on harvesting performance including machine hourly

143

productivity and cost per cubic metre; (iii) conduct an economic analysis and determine optimal rotation ages of a range of stocking densities assuming an *Eucalyptus globulus* plantation with infinite rotations; and (iv) conduct a sensitivity analysis to identify which factors had the greatest impact on LEV.

2. Materials and Methods

2.1. Stocking and Harvesting Trial

The study was established on a property owned by Western Australian Plantation Resources (WAPRES) near Greenbushes, Western Australia (33°48′002745.9″ S, 116°04′38.7″ E). Mean annual rainfall recorded at the nearest weather station was ~850 mm, predominantly falling in winter.

The study site was planted in early July 1999 with *Eucalyptus globulus* at 1000 sph (spaced 5.0 m between rows and 1.9 m between trees). A stocking trial was then established on the site in September 2002 (3.2 years after establishment) by thinning stems to waste to investigate stocking density impacts on tree growth and stand production. Chemical weed control was undertaken at four months and at one year after planting. Prior to the trial establishment, the plantation suffered damage from parrots. These parrots can attack the dominant tree leader which can result in poor stem form, particularly forks [40].

Eighteen sample plots were randomly laid out across the study site. Fifteen plots were thinned to 700, 500, or 400 sph. The "1000" sph (the unthinned control, hereafter named "UTH") and 700 sph treatments were replicated three times, whereas the 500 and 400 sph treatments were replicated six times (completely randomised design). The thinning-to-waste treatments (700, 500, and 400 sph) prioritised removal of trees of poorer form (mainly resulting from parrot damage). Thinning did not completely eliminate forking in the treated plots as some further forking (including parrot damage) occurred post-treatment.

The harvesting trial was conducted during the final felling of the site in January 2009 (at 9.5 years of age). Harvesting was carried out by an experienced operator using a tracked excavator-based Cat 322L harvester equipped with a 20-inch Waratah HTH620 head. The study site terrain was firm and even, with slopes ranging from flat to a gentle side slope (average 6°). Harvesting focused on the production of logs with a nominal length of 5.2 m and a minimum small-end diameter of 50 mm.

Harvester work elements (Table 1) were recorded for each tree during the harvesting of the 18 plots. Each treatment plot was 35 m long × 30 m (six rows) wide. The harvester–processor worked along strips consisting of three rows at a time. All work elements for each tree, plot, and treatment were accurately timed and manually recorded from a safe distance during felling using a Hanhart 2656 1/100 minute digital stopwatch. Work elements included brushing or clearing, felling, moving, processing, stacking or bunching, and travel time. The harvesting operation was also recorded using a handheld digital video recorder, and a second camera mounted in the harvester cabin to allow post-harvest data validation.

Table 1. Harvesting work elements recorded during the study.

Work Element	Description
Felling	Begins when crane starts to engage the tree and ends when processing commences
Processing	Debarking, delimbing, and bucking (i.e., cross-cutting) of logs. Commences when tree is horizontal or feed rollers commence turning
Brushing or clearing	Removal or movement of slash, undergrowth, or unmerchantable trees
Moving	Not associated with felling and processing, harvester moving within a pass (3 harvested rows per pass)
Travelling	Movement between passes or bays
Delay	Any interruption that causes the harvester to cease working during a shift

2.2. Yield and Harvest Productivity Modelling

The diameter at breast height over bark (DBHOB), tree height, and survival were measured in each treatment plot six times during the trial (at ages 3.2, 3.4, 4.3, 5.4, 7.6, and 9.5 years). Plot and treatment results included: mean DBHOB, basal area, under-bark volume, mean tree height, stocking, and survival. The DBHOB and height measurements were used to estimate the total under-bark volume of all stems using a taper function developed by the trial site owner WAPRES for *Eucalyptus globulus* plantations. Under-bark volume estimated using this taper function has previously been found to accurately predict recovered volume in operational plantations of similar site quality and tree size. At age 9.5, each tree was also subjectively assessed to determine the expected impact on harvester productivity of three major form criteria (branchiness, forking, and sweep) using a 2-class coding system where Class 1 meant no anticipated impact of form factor on harvesting/processing productivity and Class 2 meant a possible or expected impact of form factor on harvesting/processing productivity. The volume increment was used to develop a yield model, which was used to generate mean annual increment (MAI) curves from ages 5 to 12 (i.e., beyond harvest age) and as an input into the economic analysis of the four stocking treatments.

The time and motion study data was used to develop a general harvesting productivity model. The stepwise regression subroutine implemented in the GLMSELECT procedure of the software SAS/STAT® was used to develop this model, starting from a maximal model which included all independent variables: thinning treatment (indicator), tree size (continuous), branchiness (binary), and forking (binary). Variables with no statistically significant effect ($p > 0.05$) were then excluded one by one from the model. Productivity and tree size were transformed to their natural logarithm to homogenise the variance of the dependent variable to improve fit. Harvesting productivity was predicted in cubic metre solid per productive machine hour excluding all delays (PMH). PMH is defined as the portion of shift time that is spent producing output.

An analysis of variance (ANOVA) test, implemented in the ANOVA procedure of the software SAS/STAT®, was run on plot-level data collected at the time of harvest to test for statistically significant differences between each treatment for the variates DBHOB, tree height, basal area, and harvest productivity. A Shapiro–Wilk normality test and an equal variance test were performed (significance level 0.05), and the Holm–Sidak test [41] was used to perform pairwise multiple comparisons between treatments. This test is a variation of the original method presented by [42] which, although it does not compute intervals, has more power than the Bonferroni method for multiple comparisons.

2.3. Economic Analysis

The financially optimal stocking treatment was determined by calculating the Land Expectation Value (LEV) for each treatment and selecting the treatment that maximised LEV. LEV is the present value per unit area of the projected costs and revenues from an infinite series of identical even-age forest rotations, starting initially from bare land [43,44]. It is the primary approach for assessing and selecting management options for even-aged stands when the objective is to maximise the financial return from growing timber. In its simplest formulation, the LEV calculation assumes for all rotations that each rotation is of equal length, with the same sequence of events within each rotation, and the same net revenue associated with each event within the rotation.

The mathematical formulation to calculate LEV (Equation (1)) is based on the Faustmann formula [37]. The types of costs and revenues included in the calculation of LEV were (1) stand establishment cost; (2) annual leasing cost; (3) miscellaneous costs that occur in the middle of the rotation (thinning to waste) and end of the rotation (harvesting and transportation costs); and (4) net revenue for the sale of the wood at the end of the rotation.

$$LEV = \frac{\left[-E + \sum_{t-1}^{R-1} \frac{I_t}{(1+r)^t} + \frac{A[(1+r)^R-1]}{r(1+r)^R} + \frac{PY}{(1+r)^R} - \frac{HY}{(1+r)^R} \right](1+r)^R}{(1+r)^R - 1} \tag{1}$$

where

LEV = Land Expectation Value per hectare,

R = rotation length (years),

E = stand establishment cost per hectare,
A = annual land leasing cost per hectare,
I_t = thinning cost per hectare occurring after plantation establishment and before the final harvest,
Y = expected pulplog yield (m³) per hectare at the end of the rotation,
P = mill gate price of pulplogs per m³,
H = harvesting and transportation cost per m³, and
r = real interest rate.

Expenditure and revenue figures that approximate real values at the time of the study (all costs expressed in Australian dollars) were used to calculate the LEV for each thinning (stocking) alternative used. Estimated values for establishment cost (year 0), annual land leasing cost, and thinning cost (year 3.2) were $1450 ha^{-1}, $500 ha^{-1}, and $400 ha^{-1}, respectively. A constant value was used for the thinning cost because it consisted mainly of a fixed cost component. Testing of lower thinning cost values for the 700 sph and 500 sph treatments did not change the relative LEV order. Harvesting costs per m³ were calculated from the results obtained with the harvester–processor productivity model developed in the study, and an assumed hourly cost of $220 PMH^{-1}. Transport cost was estimated at $10 m^{-3}. A mill gate price for the logs of $75 m^{-3} was used to determine revenue per unit area. An interest rate of 7%, as is commonly used in forestry projects in Australia, was used for the analysis.

For each treatment, the theoretical rotation age that maximised LEV was calculated using the What'sBest® (Lindo Systems Inc., Chicago, IL, USA) solver package for MS-Excel. A sensitivity analysis was conducted on several key parameters to determine their impact on LEV and rotation age. These parameters included establishment cost, annual leasing cost, thinning cost, yield per hectare, harvest productivity, mill gate price, and interest rate. Supplementary tornado charts were constructed to compare the relative importance of each parameter and to assess their impact on LEV.

3. Results

3.1. Tree and Stand Factors at Time of Harvest

Tree growth and stand production within each stocking treatment at time of harvest (age 9.5) are summarised in Table 2. Thinning at age 3.2 clearly had a significant impact on tree growth. Average DBHOB increased by 45% and average tree height increased by 19% when moving from the UTH treatment to the 400 sph treatment. Despite the increased tree growth in the thinned plots, overall stand production (tonnes per hectare) at time of harvest was less than that in the UTH plots. The average final merchantable yield for the 400, 500, and 700 sph treatments was consistently approximately 7–8% less than that for the UTH treatment.

Table 2. Tree and stand factors at time of harvest. Values within a row sharing a letter were not significantly different at $p = 0.05$. UTH corresponds to the unthinned control treatment.

Tree and Stand Factors	Target Stocking (Trees ha^{-1})			
	UTH	700	500	400
Number of treatment plots	3	3	6	6
Actual merchantable stocking, trees ha^{-1}	978	637	489	393
Mean tree diameter (DBHOB), mm	167 a	205 b	226 b	253 c
Mean tree height, m	19.8 a	21.4 ab	22.2 ab	23.7 b
Mean standing tree volume, m³ tree^{-1}	0.233 a	0.286 ab	0.366 b	0.464 c
Stem form (Forking), % of trees				
Class 1	62	62	77	77
Class 2	38	38	23	23
Merchantable yield, tonnes ha^{-1} *	194.6	179.9	178.1	180.2
Differential		−8%	−8%	−7%

* The weight to volume conversion ratio was 1:1 based on samples of logs measured and weighted to confirm standard average ratio.

As the thinning treatment at age 3.2 targeted poor-form stems, it had a positive impact on the overall stand form within each treatment (Table 2). In the UTH and 700 sph treatments, 38% of the stems had major forks compared with only 23% in the 500 and 400 sph treatments. As few Class 2 sweep and branchiness trees were observed in the study, they were excluded from the analysis.

The ANOVA test found statistically significant differences between treatments for the DBHOB, Tree height, and Tree volume variates (Table 2). Figure 1 shows a combined histogram/box plot chart for the variate Tree volume, which had the biggest impact on harvest productivity.

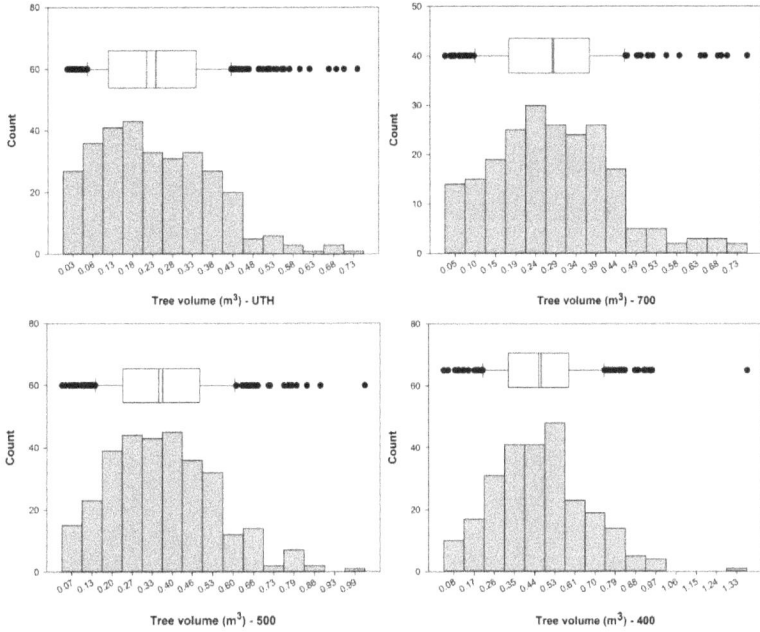

Figure 1. Combined histogram/box-plot charts for variate "Tree volume" by stocking treatment. The vertical red line in the box plot corresponds to the mean value.

With the exception of the 700 and 500 sph treatments, multiple comparisons for the variate DBHOB were statistically significant between the treatments. The comparison between the 400 sph and UTH treatments for the variate Tree height was the only comparison found to be statistically significant. The comparisons between the UTH and 700 sph treatments and between the 700 and 500 sph treatments were not significantly different for the variate Tree volume. All other comparisons for this variate were significantly different. A greater proportion of the trees in the UTH and 700 sph treatments were concentrated in the lower part of the tree volume range compared with those in the 500 and 400 sph treatments. Seventy-five percent of the trees had a tree volume greater than 0.11, 0.18, 0.23, and 0.32 m^3, in the UTH, 700, 500, and 400 sph treatments, respectively (Figure 1).

3.2. Yield and MAI Curves

Standing volume curves for each treatment from age 5 and projected to age 12 are shown in Figure 2. Across the age range, the volume per hectare of the unthinned treatment (UTH) was 9.7%, 10.4%, and 17.9% greater than those of the 700, 500, 400 sph treatments, respectively. The volume per hectare of the 400 sph treatment equalled that of the 500 and 700 sph treatments at around age 11, and of the UTH treatment at around age 12. There was little difference between the volume per hectare of the 700 and 500 sph treatments from ages 5 to 12. A similar trend occurred for the MAI values. The MAI of the 400 sph treatment increased at a greater rate than that of the other treatments until year 10.5 when it equalled those of the 700 and 500 sph treatments. It equalled the MAI of the

UTH stocking at about age 12. Maximum MAI values (biological rotation age) were reached at ages 9.5, 10.0, 9.6, and 11.3, for the UTH, 700, 500, and 400 sph treatments, respectively (Figure 3).

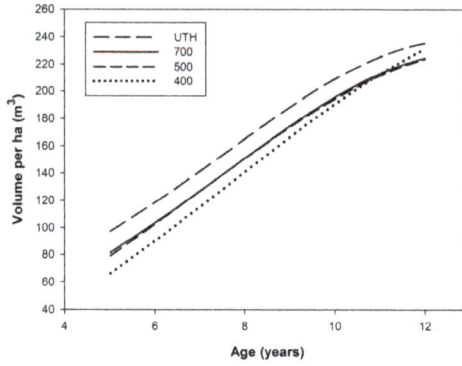

Figure 2. Standing volume curves by stocking treatment.

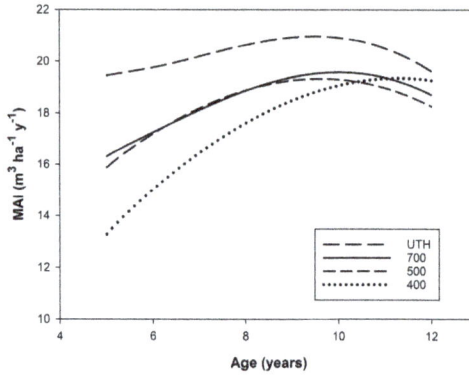

Figure 3. Mean annual increment (MAI) curves by stocking treatment.

3.3. Harvest Producivity Study and Modelling

One thousand and forty-eight trees (cycles) were timed for the harvester–processor. The mean time per tree for each work element in a full cycle is presented in Table 3. As expected, the most time-consuming work elements were processing and felling which, on average, accounted for 78.6% and 16.3% of the total cycle time, respectively.

Table 3. Summary of harvester–processor time study. Values within a row sharing a letter were not significantly different at $p = 0.05$.

Work Element	UTH		700		500		400	
	Mean time per cycle, sec.	% of cycle time	Mean time per cycle, sec.	% of cycle time	Mean time per cycle, sec.	% of cycle time	Mean time per cycle, sec.	% of cycle time
Felling	16.1	17.6	14.4	16.9	14.3	14.8	15.7	15.8
Processing	71.3 a	78.1	66.5 a	78.2	77.3 b	79.9	77.8 b	78.2
Brushing or cleaning	0.8	0.8	0.12	0.1	0.26	0.3	0.20	0.2
Moving	3.0 a	3.3	3.8 ab	4.5	4.7 bc	4.9	5.6 c	5.6

Table 3. *Cont.*

	UTH		700		500		400	
Travelling	0.2	0.2	0.2	0.2	0.2	0.2	0.2	0.2
Total	91.4 a	100.0	85.0 a	100.0	96.8 b	100.0	99.5 b	100.0

Results show that both harvester–processor cycle times and processing times were significantly greater for the 500 and 400 sph treatments than for the UTH and 700 sph treatments. This was mainly the result of increases in the proportion of forks and tree size as presented in Table 2. Moving time was also significantly greater at lower stockings as the harvester–processor had to travel further to reach each harvested tree. The harvester–processor's productivity regression model (Equation (2)) includes coefficients and statistically significant variables ($p < 0.05$) (tree volume and forking) selected using the stepwise procedure. The corresponding productivity equation is presented in Equation (3). The regression model's adjusted $r^2 = 0.85$. The variables Tree Volume and Forking explained 78% and 7% of the variation in productivity, respectively.

$$ln(Productivity) = 3.848 - 0.301 * Forking + 0.668 * ln(TreeVolume) \tag{2}$$

$$Productivity = \exp^{(3.848-0.301*Forking+0.668*ln(TreeVolume))} \tag{3}$$

Harvester–processor productivity increased with increasing tree volume, though the rate of increase was less for trees with forks (Figure 4). Using the productivity model (Equation (3)), the yield per ha equations (Figure 2), and the actual merchantable stocking values (Table 2), the mean predicted harvester–processor productivity values for the UTH, 700, 500, and 400 sph treatments at the rotation age (9.5 years) were 14.5, 18.3, 22.8, and 25.9 m^3 PMH^{-1}, respectively. The impact of forks on harvester–processor productivity became more prominent as tree volume increased.

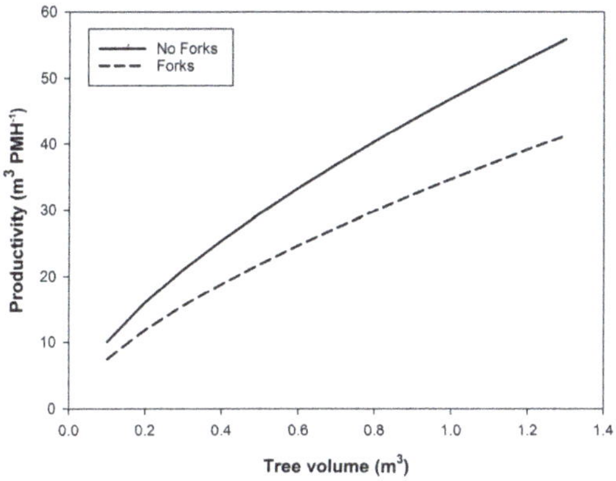

Figure 4. Harvesting productivity as a function of tree volume with and without forks. PMH is the productive machine hour excluding delays.

3.4. LEV by Stocking Treatment

The LEV peaked at ages 9.6, 10.0, 9.8, and 10.8 (theoretical optimal rotation ages), for the UTH, 700, 500, and 400 sph treatments, respectively (Figure 5). For all treatments, the theoretical optimal rotation age exceeded the trial rotation length (age 9.5). The 700 sph treatment was found to have a substantially lower LEV due mainly to it having a lower harvesting productivity (higher harvesting cost per cubic metre) compared with the 500 and 400 sph treatments, and a lower revenue compared with that of the UTH treatment due to the 700 sph treatment having the second-lowest yield per hectare and the second-highest rotation length.

149

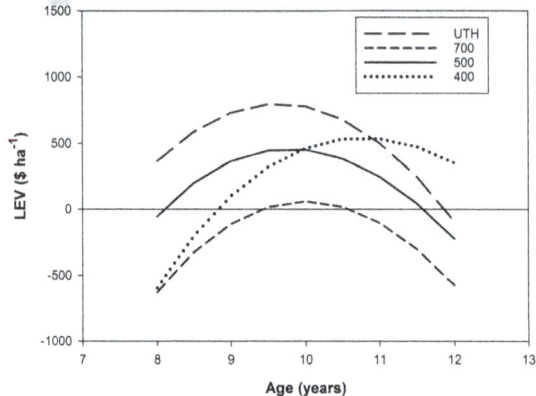

Figure 5. Land expectation values (LEVs) by stocking treatment.

At their theoretical optimal rotation ages, the yields per hectare for the UTH, 700, 500, and 400 sph treatments were 202.2 m^3 (LEV = 799.8 $ ha^{-1}), 195.7 m^3 (LEV = 58.8 $ ha^{-1}), 190.2 m^3 (LEV = 458.8 $ ha^{-1}), and 207.7 m^3 (LEV = 542.8 $ ha^{-1}), respectively. Therefore, in comparison with the UTH treatment, net losses of 3.6 $ t^{-1}, 1.54 $ t^{-1}, and 1.3 $ t^{-1}, were obtained for the 700, 500, and 400 sph treatments, respectively. To match the LEV of the UTH treatment, the yield per hectare of the 700, 500, and 400 sph treatments, would need to increase to 207.4 m^3 ha^{-1} (6.0%), 195.4 m^3 ha^{-1} (2.7%), and 212.9 m^3 ha^{-1} (2.5%), respectively.

3.5. Sensitivity Analysis on LEV

Figure 6 presents a tornado chart showing the sensitivity analysis results for the 400 sph treatment for a number of factors impacting LEV (and the corresponding lowest and highest ranges). The trend was very similar for the other treatments. Variation in mill gate price and yield per ha had the highest impact on LEV. Unfavourable conditions, such as higher thinning costs, establishment costs, and leasing costs, as well as lower yields per hectare, harvesting productivities, and lower mill gate prices, negatively impacted LEV and postponed the optimal rotation age. As expected, increasing interest rates resulted in an earlier rotation age (11.0 and 10.6 years for interest rates of 6% and 8%, respectively).

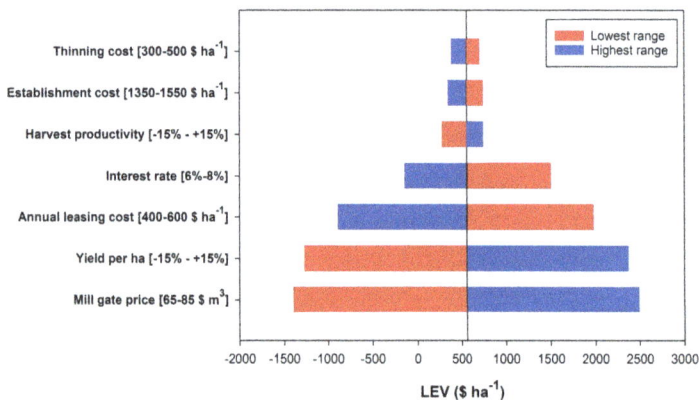

Figure 6. Tornado chart showing the effect of various factors on LEV (400 sph treatment). The lowest and highest range evaluated for each factor are shown in parentheses.

4. Discussion

As would be expected based on previous studies (e.g., [2–6]), the reduction in inter-tree competition in the thinned treatments increased tree growth (particularly DBHOB and tree volume) relative to the tree growth in the unthinned treatment. Stocking reduction had less impact on tree height, with the only significant difference recorded being between the UTH and 400 sph treatments. Although the individual tree volumes in the thinned plots were larger than those in the unthinned plots, the total merchantable volume in the thinned plots was consistently 7–8% less relative to that in the unthinned plots. This supported the finding of [26] that heavy early thinning to final stocking may prevent trees from fully occupying the site over the course of the rotation. However, when the volume growth of the trial plots was projected past the rotation age, the total volume of the 400 sph treatment plots was predicted to equal that of the unthinned plots at around age 12.

As has been found in numerous previous studies (e.g., [27–30]), the main driver of harvester productivity was found to be tree size, with productivity increasing as tree size increased. Tree size accounted for 78% of the variation in harvester productivity. Forking was the major stem form defect recorded in the current trial and accounted for 7% of the variation in harvester productivity. The proportion of forking in the current trial was much greater than that reported in recent trials in other Australian *Eucalyptus globulus* plantations [45,46]. The forking was likely to have mainly resulted from parrot damage recorded at the trial site as forking has been found to be the most common defect resulting from parrot damage [40]. Forking reduced harvester productivity through increased processing time as the operator had to detach and separately process each limb of the fork. In a previous study, it has been found that the presence of large branches and forks can reduce harvester productivity by up to 20% [47]. Branch thickness has been found to increase following thinning in eucalypt plantations [5,19] but Class 2 branchiness was found on very few trees in the current trial and had a negligible impact on harvester productivity. Harvester productivity for all treatments in the trial (good stem form trees only) was less than that predicted by the general harvester productivity model for Australian *E. globulus* plantations [48] using the mean tree volumes for each treatment. There are several potential reasons for these differences. The trial harvester power (123 kW) was less than the average power for the harvesters included in the general model (133 kW), which may explain why the difference between the measured and predicted harvester productivity values increased as mean tree volume increased. A link between reduced harvester productivity and time since rain was suggested by [48], based on the finding by [49] that eucalypt debarking difficulty increased with increasing soil dryness. In the current study, no rain had fallen for two weeks prior to the trial and only 7 mm had fallen in the two weeks before that. Operator performance has also been shown to have a significant impact on harvester productivity ([50]), but was not assessed in the current trial.

Harvester cycle times were significantly longer for the 400 and 500 sph treatments than for the unthinned and 700 sph treatments, which reflected the longer processing times for the larger trees in the 400 and 500 sph treatments and the longer moving times between trees in these treatments due to wider inter-tree spacing.

The unthinned treatment had the highest LEV in the study, largely because it had the shortest rotation length, the second-highest yield per hectare, and no thinning costs. For thinning to be economically justified, the additional costs incurred by thinning must be recovered by higher returns or by savings elsewhere in the rotation. As the wood from all treatments in the current study was sold as pulplogs, thinning costs could only be offset by savings. The only area where savings could be made was from decreased harvest costs resulting from the impact of increased mean tree sizes on harvester productivity in thinned stands. This resulted in the harvest cost for the 400 sph treatment being almost half that for the unthinned treatment, which substantially reduced the costs for the 400 sph treatment but was offset by an increase in annual maintenance costs resulting from a longer optimal rotation length. However, as the LEVs were found to be highly sensitive to the yield per hectare, the findings of [18] that an *Eucalyptus globulus* stand thinned to 600 sph was able to produce greater total wood volume than an unthinned stand at age 10 on a highly productive site suggested that it may be possible for the LEV of a thinned stand to equal or exceed that of an unthinned stand under some circumstances, though LEV was not calculated in that trial.

The LEVs were most sensitive to the mill gate price. In a comparison of two silvicultural regimes for *Pinus radiata* with the same MAI values, [51] found that the regime with the greater LEV was that which produced the higher proportion of sawn timber. Recent research suggested that logs from young *Eucalyptus globulus* trees have the potential to be used for higher value products such as sawn timber [52] and veneer [53]. The relatively small diameter of the trees in the current study and, more particularly, the knots resulting from the trees being unpruned would mean that most of the sawn timber or veneer products obtained from these trees would be likely to be in the poorer, low value grades [54]. However, given the sensitivity of the LEVs to mill gate price, the mean mill gate price for the 400 sph treatment has to rise by only $1.35 m^{-3} for the 400 sph LEV to exceed that of the unthinned treatment.

5. Conclusions

This paper has presented the results of an economic analysis of a *Eucalyptus globulus* stocking and harvesting trial in south-west Western Australia. The impact of modifying plantation stocking density through initial spacing and/or later thinning on individual tree size can affect final harvest costs and productivity, as numerous previous studies have shown that tree size is the major driver of harvesting costs and productivity. However, the impacts of tree size on harvest costs needs to be considered in the context of the total costs and returns for a rotation.

The early-age thinning to waste carried out on the treated plots in the study increased mean diameter up to 45% and mean height up to 19% when comparing the 400 sph treatment with the standard 1000 stocking (UTH). The thinning operation prioritised the removal of poor-form trees. The improvement in mean tree form and the increase in tree volume on the thinned plots increased harvester productivity by up to 66% for the 400 sph treatment compared with its productivity in the unthinned treatment. A preliminary analysis indicated that this could reduce direct harvesting costs by up to 40%, although this was not presented in the paper.

The mean tree volume in the 400 sph treatment was double that of the UTH treatment at time of harvest which reduced its harvest costs to approximately half of those for the UTH treatment. However, the final stand yield per hectare of the thinned stands was approximately 7–8% lower than that of the UTH treatment at time of harvest. The economic analysis showed that all of the thinning treatments resulted in a lower Land Expectation Value (LEV) and net financial loss at their theoretical optimal rotation age when compared with the control (UTH) treatment. The positive impacts on individual tree growth and form resulting from thinning and the associated reductions in harvesting costs were less than the overall value losses resulting from reduced per hectare yield. However, relatively small increases in the yield per hectare or in the mean mill gate price through the sale of a portion of the timber as higher-value products could result in the thinned treatment LEVs exceeding that of the UTH treatment.

Acknowledgments: Western Australian Plantation Resources (WAPRES) staff are thanked for their support with this study.

Author Contributions: M.A. was involved in the design and planning of the study, analysed the data, and conducted the economic analysis; he also wrote the Methodology, Results, and Conclusions sections, and parts of the Introduction section; M.S. wrote the Introduction and Discussion sections, and conducted the proofreading of the manuscript; J.W. was involved in the planning and design of the study, and data collection; he also reviewed the Methodology and Results sections of the manuscript; R.M. was involved in the review of the whole article.

Conflicts of Interest: The authors declare no conflict of interest.

References

1. Australian Bureau of Agricultural and Resource Economics and Sciences (ABARES). *Australian Plantation Statistics*; Australian Bureau of Agricultural and Resource Economics and Sciences: Canberra, Australia, 2016.

2. Sutton, W.R.J. Initial spacing and financial return of Pinus radiata on coastal sands. *N. Z. J. For.* **1968**, *13*, 203–219.

3. Peltola, H.; Gort, J.; Pulkkinen, P.; Zubizarreta Gerendiain, A.; Karppinen, J.; Ikonen, V.-P. Differences in growth and wood density traits in Scots pine (*Pinus sylvestris* L.) genetic entries grown at different spacing and sites. *Silva Fenn* **2009**, *43*, 339–354. [CrossRef]

4. McClain, K.M.; Morris, D.M.; Hills, S.C.; Buse, L.J. The effects of initial spacing on growth and crown development for planted northern conifers: 37-year results. *For. Chron.* **1994**, *70*, 174–182. [CrossRef]

5. Neilsen, W.A.; Gerrand, A.M. Growth and branching habit of Eucalyptus nitens at different spacing and the effect of final crop selection. *For. Ecol. Manag.* **1999**, *123*, 217–229. [CrossRef]

6. Schonau, A.P.G. The effect of planting espacement and pruning on growth, yield and timber density of Eucalyptus grandis. *S. Afr. For. J.* **1974**, *88*, 16–23.

7. Gascon, R.J.; Krinard, R.M. *Biological Response of Plantation Cottonwood to Spacing, Pruning and Thinning*; Symposium on Eastern Cottonwood and Related Species; Thielges, B.A., Land, S.B., Eds.; Louisiana State University: Baton Rouge, LA, USA, 1976; pp. 385–391.

8. Meade, D.J. Opportunities for improving plantation productivity. How much? How quickly? How realistic? *Biomass Bioenergy* **2005**, *28*, 249–266. [CrossRef]

9. Pollack, J.C.; Johnstone, W.C.; Coates, K.D.; LePage, P. *The Influence of Initial Spacement on the Growth of a 32-Year-Old White Spruce Plantation*; Research Note 111; BC Ministry of Forests: Fort Fraser, BC, Canada, 1992; p. 16.

10. Poynton, R.J. The Silvicultural Treatment of Eucalypt Plantations in Southern Africa. *S. Afr. For. J.* **1981**, *116*, 11–16. [CrossRef]

11. Gerrand, A.M.; Neilsen, W.A.; Medhurst, J.L. Thinning and pruning eucalypt plantations in Tasmania. *Tasforests* **1997**, *9*, 15–34.

12. Forrester, D.I.; Medhurst, J.L.; Wood, M.; Beadle, C.L.; Valencia, J.C. Growth and physiological responses to silviculture for producing solid-wood products from Eucalyptus plantations: An Australian perspective. *For. Ecol. Manag.* **2010**, *259*, 1819–1835. [CrossRef]

13. Opie, J.E.; Curtin, R.A.; Incoll, W.D. Stand Management. In *Eucalypts for Wood Production*; Hillis, W.E., Brown, A.G., Eds.; Academic: London, UK, 1984; pp. 179–197.

14. West, P.W. *Growing Plantation Forests*, 2nd ed.; Springer: Cham, Switzerland, 2014.

15. Bulman, L.S. Cyclaneusma needle-cast and Dothistroma needle blight in NZ pine plantations. *N. Z. For.* **1993**, *38*, 21–24.

16. Waring, K.M.; O'Hara, K.L. Silvicultural strategies in forest ecosystems affected by introduced pests. *For. Ecol. Manag.* **2005**, *209*, 27–41. [CrossRef]

17. Butcher, T.B.; Havel, J.J. Influence of moisture relationships on thinning practice. *N. Z. J. For. Sci.* **1976**, *6*, 158–170.

18. White, D.A.; Crombie, D.S.; Kinal, J.; Battaglia, M.; McGrath, J.F.; Mendham, D.S.; Walker, S.N. Managing productivity and drought risk in Eucalyptus globulus plantations in south-western Australia. *For. Ecol. Manag.* **2009**, *259*, 33–44. [CrossRef]

19. Medhurst, J.L.; Beadle, C.L. Crown structure and leaf area index development in thinned and unthinned Eucalyptus nitens plantations. *Tree Physiol.* **2001**, *21*, 989–999. [CrossRef] [PubMed]

20. Shepherd, K. *Plantation Silviculture*; Nijhoff Publishers: Doordrecht, The Netherlands, 1986.

21. Reed, W.; Apaloo, J. Evaluating the Effects of Risk on the Economics of Juvenile Spacing and Commercial Thinning. *Can. J. For. Res.* **1991**, *21*, 1390–1400. [CrossRef]

22. Leask, J.; Smith, R. *Guidelines for Plantation Fire Protection*; Fires and Emergency Services Authority of Western Australia, FESA: Perth, Australia, 2011; p. 36.

23. Jactel, H.; Nicoll, B.C.; Branco, M.; Gonzalez-Olabarria, J.R.; Grodzki, W.; Långström, B.; Moreira, F.; Netherer, S.; Orazio, C.; Piou, D.; et al. The influences of forest stand management on biotic and abiotic risks of damage. *Ann. For. Sci.* **2009**, *66*, 701. [CrossRef]

24. Forrester, D.I.; Elms, S.R.; Baker, T.G. Relative, but not absolute, thinning responses decline with increasing thinning age in a Eucalyptus nitens plantation. *Aust. For.* **2013**, *76*, 121–127. [CrossRef]

25. Stone, C.; Penman, T.; Turner, R. Managing drought-induced mortality in Pinus radiata plantations under climate change conditions: A local approach using digital camera data. *For. Ecol. Manag.* **2012**, *265*, 94–101. [CrossRef]

26. Medhurst, J.L.; Beadle, C.L.; Neilsen, W.A. Early-age and later-age thinning affects growth, dominance, and intraspecific competition in Eucalyptus nitens plantations. *Can. J. For. Res.* **2001**, *31*, 187–197. [CrossRef]

27. Acuna, M.A.; Kellogg, L. Evaluation of Alternative Cut-to-Length Harvesting Technology for Native Forest Thinning in Australia. *Int. J. For. Eng.* **2009**, *20*, 17–25.

28. Jiroušek, R.; Klvač, R.; Skoupy, A. Productivity and costs of the mechanised cut-to-length wood harvesting system in clear-felling operations. *J. For. Sci.* **2007**, *53*, 476–482.

29. Spinelli, R.; Owende, P.M.O.; Ward, S. Productivity and cost of CTL harvesting of Eucalyptus globulus stands using excavator-based harvesters. *For. Prod. J.* **2002**, *52*, 67–77.

30. Strandgard, M.; Walsh, D.; Acuna, M.A. Estimating harvester productivity in Pinus radiata plantations using StanForD stem files. *Scand. J. For. Res.* **2013**, *28*, 73–80. [CrossRef]

31. Plamondon, J.; Pitt, D.G. Effects of precommercial thinning on the forest value chain in northwestern New Brunswick: Part 2—Efficiency gains in cut-to-length harvesting. *For. Chron.* **2013**, *89*, 458–463. [CrossRef]

32. Strandgard, M.; Walsh, D.; Mitchell, R. Productivity and cost of whole tree harvesting without debarking in a Eucalyptus nitens plantation in Tasmania, Australia. *South For.* **2015**, *77*, 173–178.

33. Hartsough, B.; Spinelli, R.; Pottle, S.; Klepac, J. Fiber Recovery with Chain Flail Delimbing/Debarking and Chipping of Hybrid Poplar. *J. For. Eng.* **2000**, *11*, 59–68.

34. Nakagawa, M.; Hamatsu, J.; Saitou, T.; Ishida, H. Effect of tree size on productivity and time required for work elements in selective thinning by a harvester. *Int. J. For. Eng.* **2007**, *18*, 24–28.

35. Tong, Q.J.; Zhang, S.Y.; Thompson, M. Evaluation of growth response, stand value and financial return for pre-commercially thinned jack pine stands in Northwestern Ontario. *For. Ecol. Manag.* **2005**, *209*, 225–235. [CrossRef]

36. Samuelson, P. The economics of forestry in an evolving society. *Econ. Inq.* **1976**, *14*, 466–492. [CrossRef]

37. Faustmann, M. Calculation of the value which forestland and immature stands possess for forestry. *J. For. Econ.* **1995**, *1*, 7–44.

38. Brazee, R.; Mendelsohn, R. Timber Harvesting with Fluctuating Timber Prices. *For. Sci.* **1988**, *34*, 359–372.

39. Hartman, R. The Harvesting Decision when a forest stand has value. *Econ. Inq.* **1976**, *14*, 52–55. [CrossRef]

40. Shedley, E.; Adams, M. Parrot Damage in Tasmanian Bluegum Plantations in the South-West of Western Australia. In Proceedings of the Eleventh Australian Vertebrate Pest Control Conference, Bunbury, WA, Australia, 3–8 May 1998; pp. 247–253.

41. Abdi, H. Bonferroni and Šidák corrections for multiple comparisons. In *Encyclopedia of Measurement and Statistics*; Salkind, N.J., Ed.; Sage: Thousand Oaks, CA, USA, 2007.

42. Holm, S. A simple sequentially rejective multiple test procedure. *Scand. J. Stat.* **1979**, *6*, 65–70.

43. Chang, S.J. The determination of the optimal rotation age: A theoretical analysis. *For. Ecol. Manag.* **1984**, *8*, 137–147. [CrossRef]

44. Coordes, R. *Optimal Thinning within the Faustmann Approach*; Springer Vieweg: Berlin, Germany, 2014; p. 246.

45. Strandgard, M.; Mitchell, R. Impact of number of stems per stool on mechanical harvesting of a Eucalyptus globulus coppiced plantation in south-west Western Australia. *South For.* **2017**, 1–6. [CrossRef]

46. Hamilton, M.G.; Acuna, M.; Wiedemann, J.C.; Mitchell, R.; Pilbeam, D.J.; Brown, M.W.; Potts, B.M. Genetic control of Eucalyptus globulus harvest traits. *Can. J. For. Res.* **2015**, *45*, 615–624. [CrossRef]

47. Labelle, E.R.; Soucy, M.; Cyr, A.; Pelletier, G. Effect of Tree Form on the Productivity of a Cut-to-Length Harvester in a Hardwood Dominated Stand. *Croat. J. For. Eng.* **2016**, *37*, 175–183.

48. Strandgard, M.; Mitchell, R.; Acuna, M. General productivity model for single grip harvesters in Australian eucalypt plantations. *Aust. For.* **2016**, *79*, 108–113. [CrossRef]

49. Van den Berg, G.J.; Little, K.M. Effect of rainfall and under-canopy vegetation on the ability to debark Eucalyptus grandis x E. camaldulensis when felled in Zululand, South Africa. *S. Afr. For. J.* **2004**, *200*, 71–75.

50. Ovaskainen, H.; Uusitalo, J.; Väätäinen, K. Characteristics and Significance of a Harvester Operators' Working Technique in Thinnings. *Int. J. For. Eng.* **2004**, *15*, 67–76.

51. Sutton, W.R.J. Comparison of Alternative Silvicultural Regimes for Radiata Pine. *N. Z. J. For. Sci.* **1976**, *6*, 350–356.

52. Yang, J.-L.; Fife, D.; Waugh, G.; Downes, G.; Blackwell, P. The effect of growth strain and other defects on the sawn timber quality of 10-year-old Eucalyptus globulus Labill. *Aust. For.* **2002**, *65*, 31–37. [CrossRef]

53. McGavin, R.L.; Bailleres, H.; Hamilton, M.; Blackburn, D.; Vega, M.; Ozarska, B. Variation in Rotary Veneer Recovery from Australian Plantation Eucalyptus globulus and Eucalyptus nitens. *Bioresources* **2015**, *10*, 313–329. [CrossRef]

54. Nolan, G.B.; Greaves, B.L.; Washusen, R.; Parsons, M.; Jennings, S. *Eucalypt Plantations for Solid Wood Products in Australia—A Review 'If You Don't Prune It, We Can't Use It'*; PN04.3002; Forest Wood Products Research and Develoment Corporation: Melbourne, VIC, Australia, 2005; p. 130.

Article

The Effect of Logging and Strip Cutting on Forest Floor Light Condition and Following Change

Tomoya Inada [1],*, Kaoru Kitajima [1], Suryo Hardiwinoto [2] and Mamoru Kanzaki [1]

[1]   Graduate school of Agriculture, Kyoto University, Kyoto 606-8501, Japan; kaoruk@kais.kyoto-u.ac.jp (K.K.); mkanzaki@kais.kyoto-u.ac.jp (M.K.)

[2]   Faculty of Forestry, Gadjah Mada University, Bulaksumur, Yogyakarta 55281, Indonesia; suryohardiwinoto@yahoo.com

*   Correspondence: inada@kais.kyoto-u.ac.jp; Tel.: +81-757-536-361

Received: 3 September 2017; Accepted: 31 October 2017; Published: 7 November 2017

Abstract: We monitored changes in light conditions at a primary forest and two managed forest sites (one with line planting) after reduced-impact logging in Central Kalimantan, Indonesia. We also assessed the effect of the light conditions on seedlings in the planting lines. Hemispherical photographs were taken over a period of 31 months in three 50×50-m quadrats at each site and in three 100-m transects along the planting lines. The location of each photo was categorized according to the corresponding type of disturbance, including skid trails, logging gaps, and planting lines. Following logging, the level of canopy openness (CO) increased at both managed forest sites and did not differ significantly between the two. However, CO was greater in skid trails and logging gaps than in planting lines. After 31 months, the mean level of CO at each managed site had decreased significantly due to the establishment of new seedlings. Correlations between changes in CO and the growth of planted seedlings suggested that growth was inhibited by the invasion of the new species. However, the level of CO along the planting lines was greater than that at other disturbed locations. A high level of CO promoted invasion by new species that colonized the space. Line planting may influence forest dynamics and maintain a high level of CO.

Keywords: hemispherical photography; RIL; line planting; light condition; lowland dipterocarp forest

---

1. Introduction

After logging, light conditions on the forest floor are important for the growth of new seedlings. Gaps in the canopy can lead to colonization by a variety of tree seedlings, and the species composition influences the subsequent dynamics [1,2]. In an area disturbed by logging, Macaranga species often invade in response to large openings in the canopy [3–5]. Rapid colonization by particular species (e.g., Macaranga) can prevent ecologically and economically important species (e.g., dipterocarps) from becoming established. Changes in the light conditions following logging are important for predicting how the forest will recover. In Indonesia, reduced-impact logging (RIL) and line planting have been introduced to create sustainable forest management. The success of RIL in minimizing the effect of logging has been demonstrated, showing that it has less impact on the canopy than conventional logging [6–8]. In Indonesia, the line planting system, which involves selective logging using the RIL method, followed by the planting of Shorea species, has been tested for the establishment of sustainable forest management [9]. Following selective logging, 3-m wide strip cutting was performed at 20–25-m intervals, and Shorea seedlings, which belong to the family Dipterocarpaceae, were planted in the lines.

From a comparison between two sites with or without line planting after RIL, the light conditions on the forest floor changed significantly, although there was no difference in mean CO between the two managed forest sites [10]. However, strip cutting may generate differences in subsequent forest floor light conditions. In addition, different types of disturbances associated with logging activities (e.g., skid trails created by bulldozers and logging gaps) may affect the subsequent light conditions differently. Conventional selective logging can have an effect on forest light conditions even after 50 years [11]. Therefore, it is important to understand the effects of RIL and line planting on forest light conditions, especially in the period immediately after logging. The light conditions at this time will have an effect on the long-term dynamics of forest recovery. Additionally, light conditions in the planting lines are important for the planted seedlings. The initial light conditions are particularly important for Shorea seedlings. In some line planting tests, seedling mortality was very high during this initial period [12,13]. It is also necessary to understand the effects of changes in light conditions so that improvements can be made in line planting methods. Hemispherical photography is a practical method for measuring canopy openness (CO) and assessing the light conditions in a forest [11,14–16]. In this study, hemispherical photography was used to monitor the changes in light conditions at the forest floor from 2011 onward in a primary forest and at two sites with different forest management systems, as well as the effect of light conditions on line-planted seedlings.

2. Materials and Methods

2.1. Study Site

The study area was the logging concession in Central Kalimantan, Indonesia ($00°36'$–$01°10'$ S; $111°39'$–$112°25'$ E; Figure 1). The mean annual precipitation between 2001 and 2009 was 3240 mm, the altitude was 400–600 m above sea level, and the vegetation was lowland dipterocarp forest. The study sites were a primary forest (PF) and two managed forest sites. One of the latter two sites was selectively logged (S), and the other was selectively logged and line planted (SL). Selective logging was performed at both the S and SL sites during January 2011. During April 2011, 3-m-wide strip cutting at 25-m intervals was performed from North to South at the SL site. All plants were cleared except large commercially important trees. Dynamic monitoring plots were established at each site, and a tree census was performed between September and November 2011. The basal areas of trees $\geq$ 10 cm at the PF, S, and SL sites were 32, 22.9, and 18.5 m^2/ha, respectively.

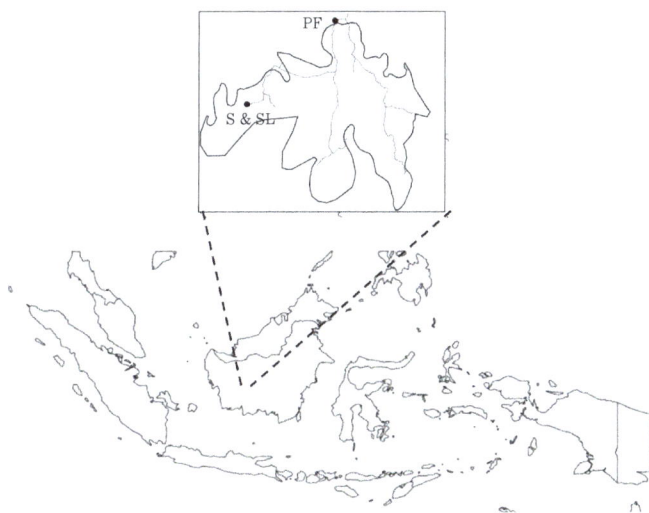

Figure 1. Study area and the location of primary forest (PF), selectively logged (S) and selectively logged and line planted (SL) sites where hemispherical photography was conducted.

2.2. Plot Setting

To monitor the light conditions, three 50 × 50-m quadrats were set up for hemispherical photography at each site (PF1–3, S1–3, and SL1–3). At the SL site, three planting lines were also marked out (Pline1, Pline2, and Pline3). To assess the impact on light conditions of different logging activities (logging gaps, skid trails, and strip cutting lines), logging trails were also mapped on each quadrat (Figure 2). The size of logging gaps in the quadrates ranged from 100 to 400 m^2, and the mean size was 200 m^2. With the RIL method, the logging gap did not exceed 500 m^2, and pioneer species could invade [17]. The skid trails were approximately 3–4 m wide. Additionally, to assess the response of seedlings to changes in light conditions, a 20 × 100-m line transect was set up along a planting line, and a tree census was performed for all trees >1 cm in diameter at breast height in 2011 and 2012.

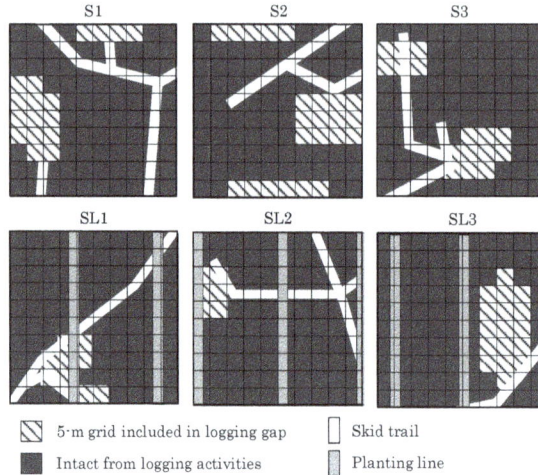

Figure 2. Logging trails in each quadrate at the S and SL sites.

2.3. Hemispherical Photography and Image Analysis

Hemispherical photography was performed to assess changes in the light conditions. The photography equipment included a Coolpix 8400 digital camera (Nikon, Tokyo, Japan) with an FC-E9 0.2× fisheye converter lens attached (Nikon, Tokyo, Japan). The camera was mounted on a tripod and oriented so that the top of each photograph would face magnetic north. The lens was positioned at a height of 1.2 m, and photographs were taken using the Open-sky Reference Method [18].

Hemispherical photographs were taken in October 2011, October 2012 (i.e., 12 months later), and May 2014 (at 31 months). The photographs were taken from a total of 49 points on a 5-m-interval grid within each 50 × 50-m quadrat. In addition, a total of 21 photo points were positioned at 5-m intervals for 100 m along each planting line transect (Figure 3).

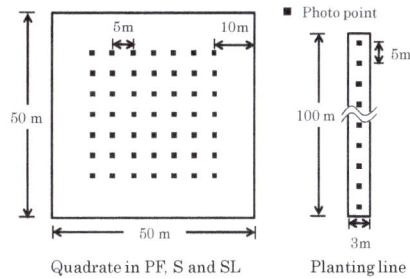

Figure 3. Hemispherical photography setting in the quadrate at the PF, S and SL sites and the planting line transect.

The hemispherical photograph data were analyzed using Gap Light Analyzer software (ver. 2.02; Frazer et al., 1999), which has been used in several previous studies [19]. CO was calculated from the hemispherical photograph taken at each point.

2.4. Monitoring Planted Seedlings in the Planting Lines

Following the first assessment of light conditions, *Shorea* seedlings were planted at 2.5-m intervals at SL sites in December 2011. Seedlings of the species *Shorea johorensis* were planted in two of the three lines (Plines 1 and 2), and *S. leprosula* seedlings were planted in the third line (Pline3). The survival and growth of planted trees at 10 months after planting were monitored beginning when the second photographs were taken. Growth was monitored by measuring tree diameters at ground level.

The hemispherical photo points in the planting lines were positioned at 5-m intervals. The seedlings were planted between two photo points. The average CO between two neighboring photo points was used to evaluate the correlation between light conditions and the growth of the planted seedlings. Multiple regression analysis was performed to assess these correlations. The objective variable was seedling growth, and the explanatory variables were the changes in CO from 2011 to 2012 and from 2011 to 2014, the initial CO in 2011, and the site conditions at the planting points recorded in 2011 (e.g., skid trails or logging gaps) as shown in equation (1). The site conditions were a categorical variable (i.e., assigned values of 0 or 1).

Equation:

$$
\begin{aligned}
\text{Seedling growth(cm)} = {}&\text{Changes in CO from 2011 to 2012}(\%)+ \\
&\text{Change in CO from 2011 to 2014}(\%) + \text{Initial CO in 2011 } (\%) + \text{Site conditions } (0,1)
\end{aligned} \tag{1}
$$

2.5. Statistical Analysis

Statistical analysis was performed using SPSS software (SPSS, Chicago, IL, USA). To compare the impact of different types of forest management, the mean CO at each study site, the PF, S, and SL sites, and the planting line, were compared using the Steel–Dwass test. To assess changes in the light conditions, the correlations between the initial CO measurements in 2011 and the reductions observed in 2012 and 2014 were analyzed using Spearman's test. A p-value < 0.05 was considered statistically significant.

3. Results

3.1. CO Change from 2011 to 2014 at Each Site

Logging and strip cutting in 2011 had a significant impact on light conditions (Figure 4). The mean CO for the PF, S, and SL sites and planting lines was $1.79 \pm 0.95\%$, $7.57 \pm 6.63\%$, $7.88 \pm 6.55\%$, and $11.17 \pm 6.6\%$, respectively. The mean CO was significantly greater at the S and SL sites and at the planting lines compared with the PF site. The difference between the two managed forest sites was not significant. The greatest mean CO was observed in the planting lines (Steel–Dwass test, $p < 0.05$).

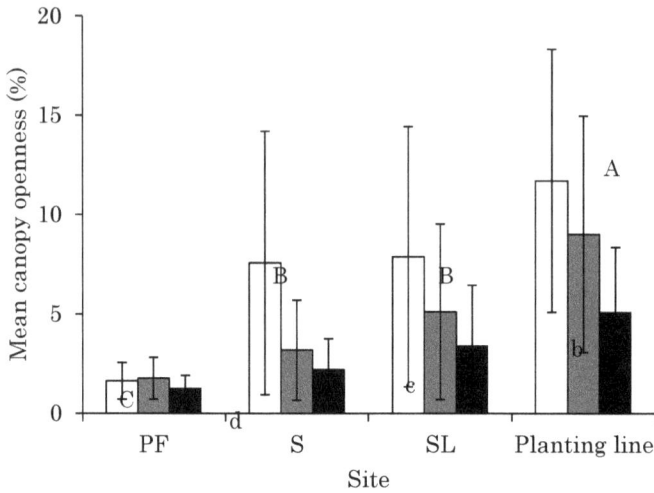

Figure 4. Mean canopy openness (CO) at each site and planting line measured in 2011 (blank bars), 2012 (shaded bars), and 2014 (filled bars). Error bars indicate the standard deviation (S.D.). Different capital and small letters indicate a statistical difference among the three sites and planting line in 2011 and 2014 (Steel–Dwass test, $p < 0.05$).

After 2011, the mean CO decreased significantly until 2012 and 2014 at the two managed sites and in the planting lines. At 31 months, the mean CO for the PF, S, and SL sites and the planting lines had changed to $1.28 \pm 0.61\%$, $2.2 \pm 1.54\%$, $3.14 \pm 1.69\%$, and $5.13 \pm 3.23\%$, respectively. With the exception of the PF site, the mean CO values had decreased significantly at all locations. The mean CO was still greater at the two managed forest sites and the planting lines than at the PF site (Steel–Dwass test, $p < 0.05$), and the difference between the S and SL sites was significant. The greatest mean CO was in the planting lines. The reductions in CO at each photo point in 2012 and 2014 were correlated with the initial CO values recorded in 2011 (Spearman's test, $p < 0.01$); these correlations were stronger in 2014 than in 2012 (Figure 5).

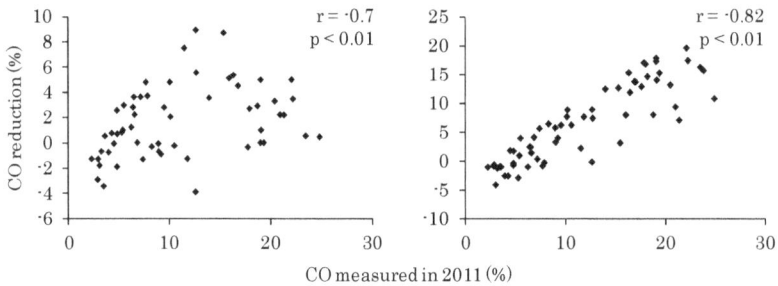

Figure 5. Correlation between the initial CO measured in 2011 and the CO reduction to 2012 (**left**) and 2014 (**right**). The correlations were tested using Spearman's method.

There was a high level of CO in 2011 associated with each type of disturbance, including skid trails, logging gaps, and planting lines (Figure 6).

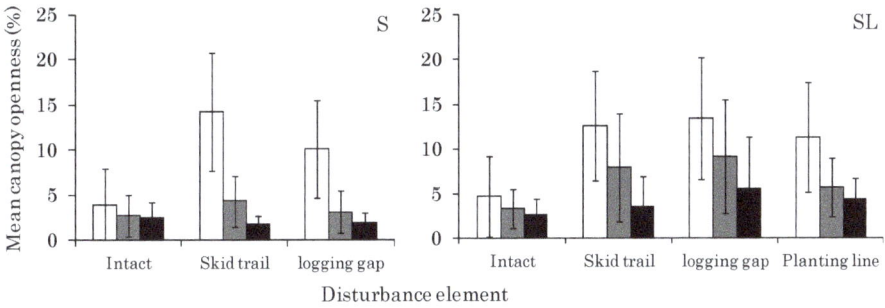

Figure 6. Mean CO under each disturbance element measured in 2011 (blank bars), 2012 (shaded bars), and 2014 (filled bars) at the S and SL sites. Error bars indicate the standard deviation.

The highest level of CO at the S site was in the skid trails. At the SL site, there was no difference in CO between the skid trails and logging gaps, although the mean CO was slightly lower in the planting lines. By May 2014, the mean CO in the skid trails and logging gaps at the S site had decreased to the same level as those in the intact forest. However, at the SL site, CO remained high in disturbed areas. At the SL3 plot, ferns had become predominant in the logging gaps. This abundance of ferns was not reflected in the photographs, which were taken at a height of 1.3 m, resulting in a high CO value at this site. When the SL3 plot was excluded from consideration, the mean CO at each type of disturbance was slightly higher at the SL than at the S site, and the mean CO was greatest in the planting lines (Figure 7).

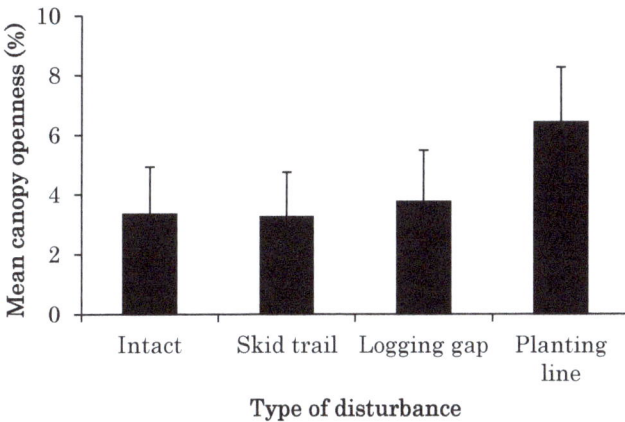

Figure 7. Mean O under each disturbance element measured at the SL site excluding plot SL3 in 2014. Error bars indicate the standard deviation.

However, when the three sites and the planting lines were compared, the mean CO was significantly higher at the SL than at the S site. This significant difference in mean CO between the S and SL sites appeared after approximately three years. The data from the tree censuses at a planting line in 2011 and 2012 demonstrated that a large number of seedlings ($n = 510$) appeared after only one year. These seedlings were abundant along the skid trails, bulldozer passes, and in logging gaps, although were less numerous along the strip cutting line (Figure 8).

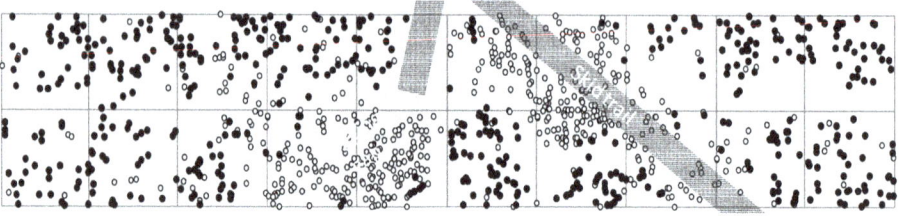

Figure 8. The positions of all trees > 1 cm diameter at breast height (DBH) in 2011 (filled dots) and seedlings that were newly established up to 2012 (blank dots) in a 20 × 100 m line transect set along a planting line.

3.2. Planted Seedling Growth and Correlation to Changes in Light Conditions

Between October 2012 and May 2014, the survival rates of the planted species *S. johorensis* (n = 82) and *S. leprosula* (n = 41) were 87% and 81.8%, respectively. The mean diameter of the *S. johorensis* and *S. leprosula* plants increased from 1.1 ± 0.27 cm and 0.86 ± 0.28 cm to 1.97 ± 0.56 cm and 1.43 ± 0.53 cm, respectively. The mean seedling height increased from 115.7 ± 38.6 cm and 254.5 ± 87.3 cm to 101.3 ± 37 cm and 230.9 ± 89.3 cm, respectively. Some seedlings were bent or damaged and became stunted. The multiple regression analysis identified no variables that were correlated with diameter and height growth. However, the changes in CO between 2011 and 2012 and changes in seedling diameters between 2012 and 2014 suggested that growth might be suppressed by a decrease in CO (Figure 9). Growth diameter decreased from 2012–2014 in association with a reduction in CO between 2011 and 2012.

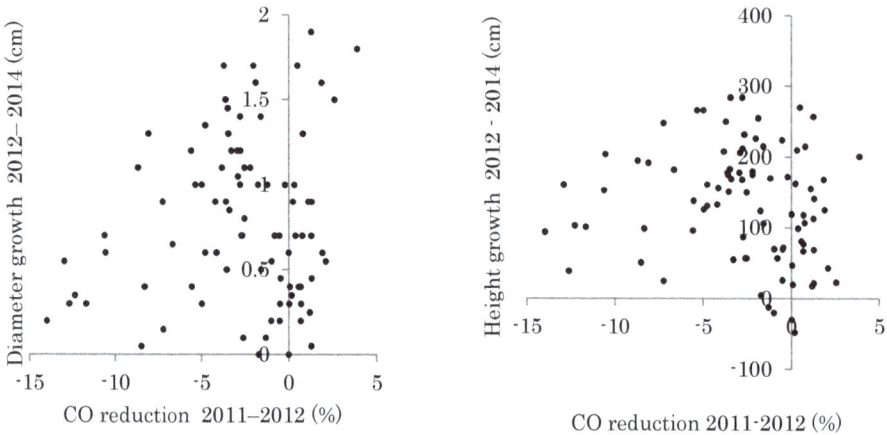

Figure 9. Correlation between the CO reduction from 2011 to 2012 at each photo point in the planting line (n = 63) and seedling growth from 2012 to 2014.

4. Discussion

4.1. Changes in Light Conditions after Logging Management

Our results were consistent with previous research that demonstrated a low level of CO in primary forest environments [3,11,20], largely due to multilayer canopies [6,16]. Logging these canopy trees significantly alters the light conditions on the forest floor. For 31 months after logging management, CO values decreased significantly in areas where the level of CO was high in 2011 (Figure 5). After only 31 months, the forest floor was no longer illuminated, even in the large clearings created by logging in 2011. Our tree census along a planting line demonstrated that the large gaps in the canopy generated by logging were closed due to the appearance of new seedlings and not because of crown expansion by neighboring trees. The higher level of CO provided opportunities for the new seedlings to take over and close the gaps in the canopy. Even in 2012, one year after logging and planting, some seedling heights exceeded 1.2 m. Sometimes, the majority of invading species were ferns, which prevented tree seedlings from becoming established [4].

By May 2014, the mean CO in the S and SL sites differed significantly, regardless of whether line planting was performed after logging. An examination of all the CO changes that occurred at each type of disturbance (i.e., skid trails, logging gaps, and planting lines) demonstrated that the highest mean levels of CO were in skid trails at the S site in 2011. When 8–15 trees/ha were removed, 15–40% of the area was typically covered by bulldozer tracks [21–23], and plant growth in these skid trails was poor [5]. Therefore, reducing the number of skid trails generated by applying RIL methods might help promote canopy recovery.

Between 2011 and May 2014, the mean level of CO returned to levels typical of intact forest, even in S-site skid trails. In October 2011, the mean level of CO in logging gaps and skid trails at the SL site was greater than that in planting lines. However, by May 2014, the mean level of CO in planting lines was greatest when a plot dominated by ferns was excluded from consideration. When strip cutting was used, the gaps in the canopy remained open. Strip cutting may be less destructive for the forest stand than logging gaps and skid trails. Large commercially important trees were not cleared from the 3-m-wide lanes, and the impact of strip cutting on changes in the light levels reaching the forest floor was probably less than that of logging gaps and skid trails. Seedlings frequently became established in and around skid trails and logging gaps. However, the relatively low impact of line planting on the light conditions produced a higher level of CO compared with other types of disturbance and also a difference between S and SL sites in 2014. Although the effect of strip cutting on forest floor light conditions was not significant, it was prolonged, and it therefore could have an effect on the growth of planted seedlings and the dynamics of canopy recovery.

4.2. Effect of Changes in Light Conditions on Planted Seedlings

Excessive light can reduce the growth rate of planted trees by photoinhibition and can also increase their mortality rates by altering the microclimate [24,25]. Survival and growth rates are also reduced by low light levels [17]. However, in this study, the observed growth characteristics were not explained by multiple regression analysis, changes in CO from 2011 onward, and the effects of different types of disturbance (e.g., logging gaps and skid trails). Over the three-year monitoring period, reductions in CO were particularly significant in areas where CO was initially very high. The growth analysis suggested that decreases in CO occur because new seedlings appear under high CO and suppress the growth of planted seedlings. Much of the decrease in CO between 2011 and 2012 was probably due to invasion by new species, whose seedlings could be competing for light with the planted seedlings. Here, large numbers of seedlings appeared and inhibited the growth of planted seedlings in 2012. In other planting tests, seedlings often died shortly after they were planted [12,13]. Growth suppression may well have increased the mortality rate among the planted seedlings.

Otherwise, some planted seedlings grew slowly with small CO reductions (Figure 9). They appeared to be planted in shaded areas, which suppressed their growth.

5. Conclusions

The light conditions of planted seedlings differed widely under systematic strip cutting because of the heterogeneity of tree distribution and logging activity traces. Shaded or opened light conditions will lead to growth suppression and high mortality in a few years.

To improve the line planting system, further examination of moderate light conditions for planted seedlings is necessary. Such an evaluation will help reduce the cost of strip cutting and planting treatment.

Acknowledgments: This study would not have been possible without the support and cooperation of the companies Sari Bumi Kusuma and Wana Subur Lestari. This work was supported by the Strategic Fund for the Promotion of Science and Technology of the Japan Science and Technology Agency (project title: Creation of a paradigm for the sustainable use of tropical rainforest with intensive forest management and advanced utilization of forest resources), JSPS KAKENHI grant number 22251004, and a Grant-in-Aid for JSPS Fellows, grant number 252464, and partly supported by Japan Science and Technology Agency (JST), Collaboration Hubs for International Research Program (CHIRP) within the framework of the Strategic International Collaborative Research Program (SICORP).

Author Contributions: Suryo Hardiwinoto conceive and arranged the experiments; Tomoya Inada, Kaoru Kitajima and Mamoru Kanzaki analyzed the data and wrote the paper.

Conflicts of Interest: The authors declare no conflict of interest.

References

1. Kuusipalo, J.; Jafarsidik, Y.; Ådjers, G.; Tuomela, K. Population dynamics of tree seedlings in a mixed dipterocarp rainforest before and after Logging and crown liberation. *For. Ecol. Manag.* **1996**, *81*, 85–94. [CrossRef]
2. Denslow, J.S. Tropical rainforest gaps and tree species diversity. *Annu. Rev. Ecol. Syst.* **1987**, *18*, 431–451. [CrossRef]
3. Nicotra, A.B.; Chazdon, R.L.; Iriarte, S.V.B. Spatial heterogeneity of light and woody seedling regeneration in tropical wet forests. *Ecology* **1999**, *80*, 1908–1926. [CrossRef]
4. Slik, J.W.F.; Verburg, R.; Keßler, P.J.A. Effects of Fire and Selective Logging on the Tree Species Composition of Lowland Dipterocarp Forest in East Kalimantan, Indonesia. *Biodivers. Conserv.* **2002**, *11*, 85–98. [CrossRef]
5. Howlett, B.E.; Davidson, D.W. Effects of seed availability, site conditions, and herbivory on pioneer recruitment after logging in Sabah, Malaysia. *For. Ecol. Manag.* **2003**, *184*, 369–383. [CrossRef]
6. Romell, E.; Hallsby, G.; Karlsson, A. Forest floor light conditions in a secondary tropical rain forest after artificial gap creation in Northern Borneo. *Agric. For. Meteorol.* **2009**, *149*, 929–937. [CrossRef]
7. Putz, F.E.; Sist, P.; Fredericksen, T.S.; Dykstra, D. Reduced-impact logging: Challenges and opportunities. *For. Ecol. Manag.* **2008**, *256*, 1427–1433. [CrossRef]
8. Putz, F.E.; Zuidema, P.A.; Synnott, T.; Peña-Claros, M.; Pinard, M.A.; Sheil, D.; Vanclay, J.K.; Sist, P.; Gourlet-Fleury, S.; Griscom, B.; et al. Sustaining conservation values in selectively logged tropical forests: The attained and the attainable. *Conserv. Lett.* **2012**, *5*, 296–303. [CrossRef]
9. Pamoengkas, P. Potentialities of line planting technique in rehabilitation of logged over area referred to species diversity, growth and soil quality. *Biodiversitas* **2010**, *11*, 34–39. [CrossRef]
10. Inada, T.; Ano, W.; Hardiwinoto, S.; Sadono, R.; Setyanto, P.E.; Kanzaki, M. Effects of logging and line planting treatment on canopy openness in logged-over forests in Bornean lowland dipterocarp forest. *Tropics* **2013**, *22*, 89–98. [CrossRef]
11. Yamada, T.; Yoshioka, A.; Hashim, M.; Liang, N.; Okuda, T. Spatial and temporal variations in the light environment in a primary and selectively logged forest long after logging in Peninsular Malaysia. *Trees* **2014**, *28*, 1355–1365. [CrossRef]
12. Ådjers, G.; Hadengganan, S.; Kuusipalo, J.; Nuryanto, K.; Vesa, L. Enrichment planting of dipterocarps in logged-over secondary forests: Effect of width, direction and maintenance method of planting line on selected *Shorea* species. *For. Ecol. Manag.* **1995**, *73*, 259–270. [CrossRef]
13. Matsune, K.; Soda, R.; Sunyoto; Tange, T.; Sasaki, S.; Suparno. Planting Techniques and Growth of Dipterocarps in an Abandoned Secondary Forest in East Kalimantan, Indonesia. In *Plantation Technology in Tropical Forest Science*; Susuki, K., Ishii, K., Sakurai, S., Sasaki, S., Eds.; Springer: Tokyo, Japan, 2006; pp. 221–229.
14. Chazdon, R.L.; Field, C.B. Photographic estimation of photosynthetically active radiation: Evaluation of a computerized technique. *Oecologia* **1987**, *73*, 525–532. [CrossRef] [PubMed]

15. Clark, D.B.; Clark, D.A.; Rich, P.M.; Weiss, S.; Oberbauer, S.F. Landscape-scale evaluation of understory light and canopy structure: Methods and application in a neotropical lowland rain forest. *Can. J. For. Res.* **1996**, *26*, 747–757. [CrossRef]

16. Silbernagel, J.; Moeur, M. Modeling canopy openness and understory gap patterns based on image analysis and mapped tree data. *For. Ecol. Manag.* **2001**, *149*, 217–233. [CrossRef]

17. Brokaw, N.V.L. Treefalls, regrowth, and community structure in tropical forests. In *The Ecology of Natural Disturbance and Patch*; Picket, A., White, P.S., Eds.; Academic Press: New York, NY, USA, 1985.

18. Tani, A.; Ito, E.; Tsujino, M.; Araki, M.; Kanzaki, M. Threshold determination by reference to open sky overcomes photographic exposure error in indirect leaf area index estimation. *Jpn. J. For. Environ.* **2011**, *53*, 41–52.

19. Jarčuška, B. Methodological overview to hemispherical photography, demonstrated on an example of the software GLA. *Folia Oecol.* **2008**, *35*, 66–69.

20. Bischoff, W.; Newbery, D.M.; Lingenfelder, M.; Schnaeckel, R.; Petol, G.H.; Madani, L.; Ridsdale, C.E. Secondary succession and dipterocarp recruitment in Bornean rain forest after logging. *For. Ecol. Manag.* **2005**, *218*, 174–192. [CrossRef]

21. Chai, D.N.P. Enrichment planting in Sabah. *Malays. For.* **1975**, *38*, 271–277.

22. Jusoff, K. A survey of soil disturbance from tractor logging in a hill forest of Peninsular Malaysia. In *Malaysian Forestry and Forest Products Research*; Appanah, S., Ng, F.S., Ismail, R., Eds.; Forest Research Institute Malaysia: Kepong, Malaysia, 1991; pp. 16–21.

23. International Timber Trade Organization (ITTO). *Pre-Project Study Report: Enrichment Planting*; International Timber Trade Organization: Yokohama, Japan, 1989.

24. Sasaki, S.; Mori, T. Growth responses of dipterocarp seedlings to light. *Malays. For.* **1981**, *44*, 319–345.

25. Van Oorschot, G.; Van Winkel, I.; Moura-Costa, P. The use of GIS to study the influence of site factors in enrichment planting with dipterocarps. In Proceedings of the 5th Round Table Conference on Dipterocarps, Chiang Mai, Thailand, 7–10 November 1994; Appanah, S., Khoo, K.C., Eds.; Forest Research Institute: Kuala Lumpur, Malaysia, 1996; pp. 267–278.

Article

Policy Recommendation from Stakeholders to Improve Forest Products Transportation: A Qualitative Study

Anil Koirala [1,2], Anil Raj Kizha [1,*] and Sandra M. De Urioste-Stone [1]

[1] School of Forest Resources, University of Maine, Orono, ME 04469, USA; anil.koirala@psu.edu (A.K.); sandra.de@maine.edu (S.M.D.U.-S.)
[2] Department of Agricultural and Biological Engineering, Pennsylvania State University, University Park, PA 16802, USA
* Correspondence: anil.kizha@maine.edu; Tel.: +1-207-581-2581

Received: 20 September 2017; Accepted: 9 November 2017; Published: 12 November 2017

Abstract: With recently announced federal funding and subsidies to redevelop vacant mills and the communities they were in, the forest products industry in Maine is poised to gain its momentum once again. One of the important components influencing the cost of delivered forest products is transportation. A recent study in the region has shown that the location and availability of markets along with lack of skilled labor force are the major challenges faced by the forest products transportation sector in Maine. This study was focused on developing a management guideline which included various field level options for improving trucking enterprises in Maine. For this, a qualitative research approach utilizing a case study research tradition was employed, with in-depth semi-structured interviews with professionals directly related to the forest products transportation sector used for data generation. Thirteen semi-structured interviews were conducted, with each being audio recorded and later transcribed verbatim. Interview transcriptions were analyzed using NVivo 11. Suggestions, like increasing benefits to drivers and providing training, were proposed for challenges related to manpower shortage, while the marketing of new forest products and adjustment in some state-level policies were proposed for challenges related to the forest products market condition of the state.

Keywords: case study; labor force; management guideline; secondary transportation

---

1. Introduction

The flow of forest products from harvesting sites to the processing facilities is a combined effort of different stakeholders. The supply chain of the forest products generally starts with foresters laying out harvest plans for forest landowners i.e., small woodlot owners, industrial land owners, and public lands. Logging operators with the direction of foresters and logging contractors take responsibilities of felling trees and piling the wood at the log landings. With guidance from a procurement manager and trucking contractor, the products are hauled from landings to facilities (usually primary forest products industries or bioenergy plants) (Figure 1). The trucking (also referred to as secondary transportation) part in this process is considered important because of its essential function of moving products from one place to another. It is also one of the expensive phases and can be crucial in fixing prices of delivered forest products [1–3]. Despite the prevalence of railroad transportation, trucking is the most common way to deliver wood products [4]. Its popularity can be associated with well-developed road networks, limited access to railway lines, and embargos in using water for timber hauling in the US and other parts of the world [5,6]. After the last log drive on the Kennebec River in 1976, the transportation of woody commodities from northern forests in Maine has predominantly been performed by trucks and tractor-trailers [6]. There are separate types of trucking fleets for specific products such as tractor trailers to haul logs, whereas chip vans to haul wood chips and comminuted biomass materials. Even with the inherent need to haul forest products, there are various challenges in this sector that needs to be addressed for its efficient operations. These challenges can be specific to the region and thereby require a local level understanding of constraints and potential mitigation strategies including policy formulation. Hence, strategic suggestions from closely related stakeholders and experts in the field are important. The recent closing of pulp and paper mills has imparted significant impacts on the entire forest products market in the state of Maine. The forest product market is highly scattered in the state; the situation has been further exacerbated by the recent closing. The increased hauling distance resulting from the closing of pulp and paper mills has increased the cost of trucking forest products compared to the situation in the past [7].

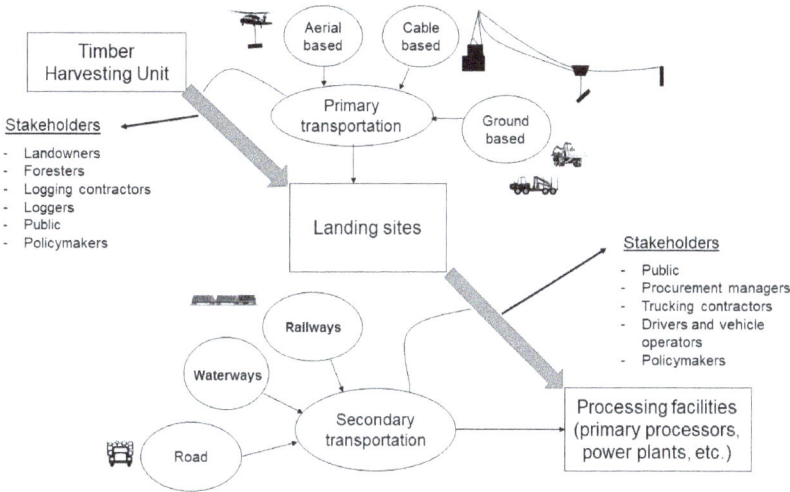

Figure 1. A sketch showing the flow of raw forest products from timber harvesting unit to the processing facilities like mills, powerplants, and industries. For this study, the focus, however, will be on the transportation from landing sites to the processing facilities.

Concerns regarding higher costs associated with forest products transportation have led to several studies in forest operations including analysis of wood products hauling costs [8–11]; increasing efficiency in transportation [12–15]; and survey analysis of logging and transportation sectors [16–20]. Similarly, there are research studies utilizing qualitative methods, such as semi-structured interviews, to comprehend views and opinions of experts [21,22]. To this end, such a qualitative research approach has not been utilized to get in-depth information on forest products transportation. The purpose of this research was to: (a) gain an in-depth understanding of stakeholders' perceptions of the problems related to trucking; and (b) identify possible measures to resolve them. Understanding related stakeholders' attitudes towards the applicability of particular solutions in the state of Maine could help the industry and policymakers implement them.

2. Materials and Methods

A qualitative research approach was selected to allow in-depth understanding of a problem within a concrete setting [23,24] and learn the interpretation of verbal experiences from stakeholders [25].

2.1. Philosophical Foundation

The methodology was based on the constructivist paradigm and used a single case study design to explain related stakeholders' perceptions and experiences. The epistemological approach of constructivism proclaims that different individuals describe the same problem in multiple ways [26]. Constructivism is based on the fact that truth is dependent on perception. Another important assumption is that problems are solved by the interaction between researcher and respondents, hence, open-ended question formats like interviews and discussions were used [27]. These questions used generally begin with how and why, rather than what and when with the intention of getting comprehensive insights on the subject [28].

2.2. Case Study Design

A case study is a research design that allows researchers to gain an in-depth understanding of a problem, process, situation, or even individual, and a group of people within a bounded system [23,24]. A common way to conduct case study research is to collect comprehensive information on the case by utilizing and triangulating across different data collection techniques such as interviews, document review, direct observations, and archival records [29,30]. The case includes bounded time, context, region, and phenomenon or topic of study [23,24]. The state of Maine was the area of study, while 2015–2016 was the timeframe for this study. The research used an instrumental case study design [23,24] to understand the phenomenon of challenges facing the forest trucking industry subsector, and local level measures adopted in different parts of Maine to mitigate forest trucking related problems. Multiple data collection methods were used, including a stakeholder questionnaire, thorough review of the literature, and in-depth semi-structured interviews with key stakeholders in Maine. Phase 1 of the study included the identification of potential solutions via a stakeholder questionnaire, unstructured interviews with key informants, and review of the scientific literature [2,7]. This phase was followed by the in-depth interviewing phase to further understand the challenges and validate the appropriateness/soundness of solutions identified in phase 1. This paper includes results from phase 2; for detailed results for phase 1 please refer to prior citations [2,7].

2.3. Participant Selection Strategy

The stakeholders were divided into four categories based on their job profile: (a) Foresters; (b) Truck owners/logging contractors (from here on referred to as contractors); (c) Representatives from forestry professional societies; and (d) Procurement managers. The categories were selected for providing appropriate and relevant responses to address the objectives from different perspectives.

The Forest Resource Association (FRA), a group of more than 500 organizations and businesses related to the forest products industry, was consulted first for participant recruitment. A public announcement for interested individuals to participate in the study was made at a FRA forum in Brewer, Maine. The process did not yield sufficient ($n = 2$) responses; hence the combination of criterion and snowball sampling techniques was used to select participants [31]. Professional contacts were utilized to recruit interviewees and enhance the gaining of entry and rapport building. First, the selection criteria were that the participants should have more than 15 years of experience in forest products handling and transportation, should have a primary workstation within the state, and be willing to participate in the study. Further, the snowball selection strategy allowed for participants and key informants from phase 1 to refer other participants to include in the study [32,33] while ensuring different regions (North, Central, South) to be included.

2.4. Ethics Statement

Approval was obtained from the Institutional Review Board (IRB), University of Maine, Orono for conducting research on human subjects prior to the interviews on 4 January 2017. A written informed consent form was given to the participants prior to the interviews which ensured confidentiality and voluntary participation.

2.5. Data Collection, Analysis, and Quality Assurance

The primary data collection method for this study phase was semi-structured face-to-face interviews with mostly open-ended questions. This method has advantages over other qualitative techniques like focus group discussions because interviews allow more privacy and a safer atmosphere to talk on dedicated issues than the later; participants have more time to express their feelings and discuss the subject matters in detail as well [34]. Interviews were helpful for the triangulation of information gained from supplementary sources from phase 1, which ensured credibility of the study results [24].

The interview protocol consisted of 13 major open-ended questions; each question included four additional probing questions (on average). The interview protocol along with the consent form were emailed to the participants three weeks prior to the interviews in order to facilitate a review of the questions and to allow time to decide a response. Interview questions were developed based on the results of the survey, unstructured interviews, and literature analysis, and were organized into four themes: (a) outlook on forest trucking sector; (b) major challenges faced; (c) potential measures; and (d) applicability of those measures in Maine. To obtain regional based information, the respondents were further categorized based on regions of their primary workplace: Northern, Central, and Southern regions.

Thirteen semi-structured interviews with open-ended questions were conducted from February to May 2017. Interviews were continued until ongoing data analysis suggested data saturation had been reached (i.e., new interviewees did not provide any additional information on the subject) [35,36]. There were no rigid rules on the number of interview participants a priori, but rather the study followed established procedures in qualitative research on data saturation [32,37]. However, the number of interviews in this study (13 interviews) was consistent with other studies that utilized a similar research approach [38].

With an average of 51 min, the total duration of the interviews ranged from 33 to 71 min. Due to the general interests on particular topics, most interviews lasted longer than the slated time frame.

The whole content of the interview was audio recorded, and the recordings were later transcribed verbatim [39] and uploaded into NVivo 11 [40]. The transcripts were meticulously read several times and important phrases/dialogues were subsequently highlighted using open coding [39] as the first coding cycle. These open codes were then abstracted into concepts identified previously through the literature review and listed in the interview protocol; this axial coding was used as the second coding cycle. The codes were generated by an iterative process that involved reviewing data multiple times [41]. This process also helped in determining the point of saturation for each question.

3. Results and Discussion

3.1. Participants' Description

Among 13 interview participants, four were based in Northern regions, three in Central regions, and three from Southern regions of the state. Based on job profiles, the majority were foresters (Table 1). The average work experience of the participants in their respective profession was 25 years, and respondents ranged from 36 to 74 years of age. The majority of the respondents worked for industrial timberland companies, while some were also small timberland owners (less than 2000 ha). All of them were Caucasian males.

Table 1. Description of the participants interviewed for the study to comprehend the challenges faced by forest trucking sector and probable resolutions to those challenges.

Stakeholder Categories	Number of Participants (by Subregions of Maine)	Experience (in Years)
Foresters (company based or consultant)	5 (Central = 3, Northern= 1, Southern = 1)	30
Truck owners/Logging contractors	2 (All subregions = 2) [1]	28
Professional society representatives	2 (Southern = 1, All regions = 1)	25
Procurement managers	4 (Northern = 3, Southern = 1)	19

[1] Participants having primary workstation in more than one subregion or working for forestry sector for the whole state.

3.2. Responsibilities and Services

The services provided by the companies or organizations that respondents were affiliated with were of diverse nature. Nearly all of them were involved in multiple forestry related tasks. One of the participants stated:

"We have forest operations of every nature. We do harvesting, trucking, chipping, loading, slashing, merchandising, building forest roads, developing forest management plans, managing our own lands, and other people's lands. We have three equipment shops; one for forest harvesting machines and two for trucks and trailers."

One of the procurement managers described his duties as overseeing entire harvesting operations, dealing with logging/trucking contractors, along with inspecting and regulating dispatch of trucks and chip vans. There was also a procurement manager just to oversee transportation-related works whose duties were, *"...to look after road maintenance for the company, and transportation of all raw materials to the mills."*

Primary duties of company foresters interviewed were managing woodlots, preparing management plans, hiring and managing temporary workers, and dealing with contractors and truck drivers. Independent consultant foresters generally worked for various landowners at any provided time.

The participants were from varying company sizes, in terms of number of employees, ranging from small (<5 employees) to large (>50 employees) companies. Basic benefits to the workers (including truck drivers) included health insurance, paid leaves, and subsidies for buying wood products. One procurement manager noted:

"We have health and many benefits like other businesses, but the additional one is the career (sic) we really enjoy and passionate about. I think there are other disciplines with higher pay (sic), but this profession provides flexibility of schedule and time. I'm not in a cubicle daily and I'm doing something different."

The above statement from the procurement manager could be applied to loggers and log truck drivers as well. There are factors other than money that drive novices to the logging and trucking industries such as the involvement of past generations of family (a family profession), the ability to work locally, and independence in the work.

Participants who hired trucking service from contractors were unaware of the exact benefits package offered to the employees and drivers working under trucking contractors. One forester specified:

"...I've not known exact details, but there should be enough to make a person sit on that giant (log trucks) and drive on rough terrain all day."

3.3. Trucking in Maine

All participants regarded trucking as an essential component, as more than 90% of the wood hauled in the state was done by road. Railway systems were also in use in northern and western parts of the state, however, trucks were still used for a certain portion of that journey. Most of the timberland owners and management companies did not own trucks but hired trucking service. The participants seemed aware of the role of trucking in determining the end price of the delivered forest products. They were also concerned about the losses incurred due to inefficient trucking. One procurement manager stated:

> "...as far as the role of trucking, it's a key to the business. When you look into harvesting and trucking of wood to our mill, it's probably one of the biggest costs both for distance and other factors like payload. It's the cost that continues to go up every year because it's something that you cannot increase productivity like in the harvesting operations. You can only put so much wood on the truck and you can only drive so far, safely and efficiently."

Several factors affected the cost of trucking, including fuel price, maintenance cost, trucking distance, and payload. Contractors were always trying to make their operations efficient enough to avoid extra expenditures: "...it's everything for us. We have more trucks than truck drivers. We have to keep an eye on every detail to make profits. All of them operates year around and are maintained timely. We are a service provider, so no compromise at all." The fewer truck drivers, in this case, was the strategy to reduce expenditures on extra drivers. It also implied that all trucks were not running at the same time.

Both above statements, although addressing different discussion points, allude to the issues faced day to day by forest trucking enterprises. As a different perspective, one participant from the professional society stated:

> "...most of the logging contractors, probably 75 percent, have a truck or two. This provides them with more stability in their services. Owning and operating trucks makes them more flexible and competitive."

There were mixed responses regarding the outlook on trucking business for the region, with participant responses articulated in terms of challenges and opportunities in the field. In general, the participants considered trucking to be a challenging business but expected that prospects will increase with a new horizon of market opportunities for products like biomass, hardwood pulp, biofuels, and others (Table 2).

Table 2. Examples of the participants' responses regarding the outlook of the forest trucking business in Maine. The blank cells indicate the absence of positive or negative attitudes of the participants.

Stakeholder Responsibility	Region in Maine	Key Ideas Showing Outlook of Trucking Business in Maine [1]	
		Positive Attitude	Negative Attitude
Forester	Central	"...the weight limit has been raised..." "...drastic drop in the oil price over recent years..."	"...having hard time finding drivers..." "...profit margins are very tight..."
Trucks owner	Entire Maine	"...trucks are in demand and will be in future."	
Professional society representative	Southern		"...more expensive than it used to be." "...much more difficult for owner operator to get started."
Procurement manager	Northern	"...the forest trucking market is growing..."	"If trucking cost doesn't work out then we're not going to be able to operate."
Forester	Northern	"...trucking business for northern climate can be profitable..." "...room for new business to enter in the market."	"...moving more volume of wood, as the payload is based on ton." "...primary concern is about the aging workforce."

[1] Participants' direct quotation.

In a different context, participants (except contractors) mentioned that for forest products companies, it might be better to contract the trucking portion of forest operations than to own and operate entire trucking fleets. One procurement manager mentioned, *"We had a fleet of trucks that we managed in the past but it's to our benefit that we hire contractors. They can run this business better than us. There are certain things that contractors are more efficient than company managed fleet (sic)."* Another forester agreed, *"This section is difficult to handle if you are dealing with many other things."*

3.4. Challenges to Trucking

All participants agreed that there were numerous challenges to efficient trucking operations. The majority (more than 80%) regarded the lack of skilled drivers as the most prominent challenge at present time (Table 3). Similar to the forest trucking industry's experiences, driver shortage was also a prominent challenge for other trucking businesses as well; a report prepared by the ATA (American Trucking Association) pointed out that the US trucking sector was short of 35,000 truck drivers in 2015 [37]. This challenge was not only getting severe due to the aging workforce, but also the difficulty to keep truck drivers in this profession for a longer period. Even with enough experience and physical abilities, many drivers did not pass the mandatory drug tests for operation, which also contributed to the cause. For international shipping, age factor was also an issue, as drivers below 21 years face restrictions in crossing the Canadian border. Because of these legal complications, participants mainly from the northern region (two from the central region) ranked the border crossing requirement as the main challenge. A forester stated:

> *"A lot of our wood goes across border into Canada. So, there are new restrictions on border crossing, their weight restrictions are different than ours (two tiers systems in US). Contract rates needs to be adjusted accordingly."*

All respondents agreed that a key challenge to the forest industry in general was due to the recent closures of five pulp mills in central and southern parts of the state within the last two years [42]. It resulted in harvesting and marketing of new forest products which were not considered valuable before. There is a need for research on the diffusion of new forest products and services to achieve global sustainability goals from the forestry sector [43]. The impact of mill closures was reflected in trucking as it increased hauling distances and transportation costs of forest products in the state. However, the effects of the closing appeared to be less severe in the northern parts compared to other regions. One respondent from northern Maine explained, *"It does have effects, but not much than (sic) the adjoining regions. It affected us in a way that wood started moving in different direction and started infringing on the markets that we always relied on."* For small landowners, the impacts were of another nature, a forester working for a small landowner stated, *"...it has affected some of our capability to market low-value products like everybody else. We have very little market cloud because we're small. So, this probably affects us more than big land owners. We don't have negotiating powers like landowners who produce 250,000 * cords a year does (sic). Last year we cut 3200 * cords."* (* 250,000 cords ~906,000 cubic meters and 3200 cords ~11,600 cubic meters).

Despite having been affected by the closing of mills, participants were optimistic about the future of forestry businesses in Maine. They agreed that certain products like hardwood pulps and biomass which were not in demand previously were in high demand at present due to mill closures. The trucking sector was directly benefited with this as trucks started delivering these new products to new markets and increased their chances to back-haul products.

Another challenge was regarding the conditions of public and forest roads. Participants from northern Maine were especially disgruntled by the condition of public roads in the region. Some of them also compared the bad conditioned roads with good Canadian roads across the border.

Other challenges mentioned were related to back-hauling, payload, high equipment owning and insurance costs, timber harvesting season, safety, and turnaround times (Table 3). Some of these were common to all forest operations, in general. Results also showed that specific challenges can be less important for one stakeholder group while representing major concerns for another. For example, all foresters were convinced that the legally allowable payload for the public highways in the state was sufficient, while the contractors and procurement managers wanted the weight limit to be increased.

Table 3. Examples of the participants' responses to the challenges faced by the forest trucking industry in Maine.

Stakeholders' Responsibility	Main Challenges	Participants' Direct Quotes
Forester	Drivers, roads, and safety	*"Finding good drivers is the main thing. We hear that all the time from our contractors. The other thing is the worse roads. I am very much concerned about the safety of drivers as well as public."*
Procurement manager	Supply chain issues and contractors' nature	*"There is a supply chain issue. Majority of logging contractors own everything, logging equipment and trucks, and they employ drivers and operators. This is somehow inefficient. They want to do that because they want their wood to reach the market first."*
Forester	Contractors' nature	*"Trucks are passing each other with same products and same origin and destination, it seems there is a competition between contractors."*
Professional society representative	Drivers and insurance cost	*"Many of these contractors could not find drivers because of the drug tests; most of them failed the test. The insurance cost goes really high if you don't have good drivers."*
Truck owner/ Logging contractor	Aging workforce and back-hauling	*"There are two major challenges. The drivers hauling wood out of the forest require a special skill, which many do not have. The ones we had are also retiring. The other challenge is too much percentage of empty drive miles that makes the transportation costs very high."*
Truck owner/ Logging contractor	Market condition and state policies	*"Closing of mills has affected our business tremendously. We lost literally a third of our business over sales in about a twelve-month period of time, which is very painful, very hard to adjust and we haven't fully adjusted yet. The other thing is the state's regulations; some of them are terrible and not business friendly."*
Procurement manager	Roads	*"Public roads in this region (northern Maine) are terrible. I mean it was really bad this time of year as the frost comes out, but they don't get a lot better anymore in the summer and fall winter. Terrible terrible!!"*
Professional society representative	Transportation distance	*"...increased hauling distance is the main issue at present"*

3.5. Possible Solutions and Applicability in Maine

The interview questions were designed to group potential solutions based on the problem type. Several options emerged when participants were asked about mitigating some specific challenges such as the shortage of skilled truck drivers and current market conditions. Participants also put forward solutions which could turn these challenges into opportunities in the future. One respondent appeared very optimistic, *"I agree there are problems now, but we will get through this. Maine is a very resilient state; we have dealt with a lot of issues in the past—take spruce-budworm outbreak."* The same individual responded to the shrinking market condition, *"It requires new investments and business models to start up. It will take time, but it will eventually. New markets are opening for new products because our market is changing."* The group representing contractors provided a suggestion for the improved market situation: *"Maine has a forest-based economy; we need to become a business-friendly state. Policies should be in favor of startup forestry businesses. There are examples that you have to wait for two to three years only to get an agreement from the state."* According to them, adjustment of policies at the state level can have a greater impact on new businesses. They were also in favor of providing subsidies to new products for struggling businesses. From the perspective of foresters, the introduction of new technologies in the business could help revamp the shrinking market. However, they were not certain that conditions would be similar to those of the last decade. One forester mentioned, *"We have been rescued by technologies in (sic) past. We started off cutting the trees with an average diameter of 12 inches * and soon we ran out of that tree diameter class. Then technology comes in where we started chipping and had mills that took smaller sized logs.*

So, our technology has changed over the years and I expect it will again." (* 12 inch = 30 cm). Like other stakeholder groups, procurement managers also believed that the present market condition could be better. Their suggestion was to utilize new products as much as possible: *"Currently it appears that there is extra fiber in the Maine wood market, I think this will help us be a good spot to build new facilities. A company in Skowhegan, Maine has decided to put almost two hundred million dollars in new products. So, something is telling those guys that this is a good place to invest."* Another procurement manager provided an opposite view towards the market condition. He believed that the market will stabilize first before getting better, and the businesses with efficient operations will stay while others might shut down. He added, *"The only solution I see is that companies should be more efficient and start harvesting and selling varieties of products. For trucking, we should figure out how to manage empty miles."*

For tackling the problem of manpower shortage, most of the participants suggested good benefits and proper training to drivers. Participants were also asked to validate solutions that had been adopted in other parts of the world. For instance, as a motivation to stay on the job, drivers were given a certain portion of truck shares to create the feeling of ownership [44]. The participants were not aware of this kind of practice in Maine. Contractors disagreed with this as a proper solution to keep drivers on the job, while all the other stakeholder groups believed it to be a novel approach. However, a large portion (actual data unknown due to an anti-trust policy of the state) of trucking fleets in Maine are run by owner-operators [45]. Truck owners mainly focused on better benefits for drivers: *"...because we offer good benefits, and have a good reputation, we have steady employment. Many logging and trucking companies don't have that."*

To attract the younger generation to the sector, some participants suggested logging and trucking companies should focus on extension activities to showcase the novelty in equipment and technologies being used, *"Trucking is becoming increasingly comfortable compared to (sic) past. Now all trucks are equipped with climate control cabs and drivers do make fairly good money based on their education."* According to a forester, the state government entities can be key players for promoting employment in trucking, *"...the Forest Service and Department of Transportation can promote trucking as a highly skilled profession like others, through different publication series and extension."*

Some foresters suggested truck owners pay their drivers using an hour-based payment system instead of the load-based system. Another forester noted the work-related pressure to drivers, pointing out the need for providing independence to drivers in regard to time scheduling and work issues. The problem can be related to another problem regarding supply chain issues and contractors' nature. In order to manage issues related to dispatching, procurement managers suggested that the trucking and the actual harvesting process should be separated (decoupled). In their view, at present, there is a trust issue between different contractors working in the same area. Separating harvesting and trucking will ensure efficiency and stability in the market. This can be helpful in minimizing competition between contractors and easing up the pressure on drivers and the supply chain.

Regarding the issues related to roads, participants from northern Maine proposed an increment on state spending on maintenance of public roads, while participants from southern Maine were more worried about public outcry and aesthetic issues created by large log hauling trucks. Some companies have started using crushed rocks on the last hundred feet of the forest roads leading to public highways to eliminate mud and clay deposition from trucks tires on the later. The problem of ruts and depreciation of roads also seemed to be associated with the legally allowable payload. When asked about their views on legally allowable payload on the pubic highways the opinions of the respondents were contradictory. Mainly trucking contractors and procurement managers were positive about increasing the payload on public highways, as it could increase trucking efficiency including fuel consumption. One of the contractors stated, *"There could be certain situations where you could have increased weight limits for certain types of trucks on certain roads and that could help the industry. I think that's something nice to keep on the table but could be hard to do politically."* As an opposing view, one forester argued, *"I agree the work they did to get interstate payload raise from 80,000 pounds * to 110,000 pounds * is important. But for safety, we must remember that my wife and daughter drive on that road. Big companies* might have different views because if they can haul more amount of woods with the same amount of fuel then it can be profitable for them. But they also want to be safe. So, I think increasing the payload is not an option." (* 80,000 pounds ~36,300 kg and 110,000 pounds ~49,900 kg).

In a different context, one participant also pointed out the benefit of having an east-west highway in the state, especially for improving the transportation of forest products and minimizing the extra costs associated with long hauls. Route I-95 is the only interstate highway (running from north to south) connecting Maine with other states. The east-west project has been a long debated topic in Maine's infrastructure development history, but the project has been rejected up to the present time due to its expected effects on wilderness and recreation [46].

Apart from the challenges mentioned above, there were other issues related to truck-turnaround times, back-hauling opportunities, and climatic adjustments. Most of the participants ($n = 8$, 62%) agreed that waiting time is a problem at mill yards. Some suggestions were to increase coordination between drivers, manage concentration yards, and use self-loading trucks. However, they also agreed that the self-loading trucks could be a bad option in terms of the extra loader (dead weight) being carried. Some of them also pointed out the $20-million investment proposed by a forest product industry in southern Maine as a potential means in which to help minimize turnaround time at their mill yard.

Although participants regarded back-hauling as a challenging job to perform, they still believed it can be carried out with some adjustments. Interestingly, the increased hauling distance incurred by the recent closing of mills seemed to be an opportunity for trucking contractors to back-haul different products. Some of the important suggestions to increase chances of back-hauling were: building more concentration yards; adopting proper networking strategies between mills within and outside the state; using self-loading trucks for short hauls, and making trucks and trailers as dynamic as possible to transport different types of products.

Similarly, participants also regarded seasonal adjustment in harvesting and trucking as an important issue for transportation. They seemed very concerned about transportation of wood in the muddy season as well as during the winter season. During the heavy winter season, log trucks used chains on their tires while driving through forest roads. Except for plowing snow and clearing public highways, no new anti-slip innovations were used in the state to mitigate this seasonal barrier.

Based on the strategies suggested and accepted by participants, a summary table was prepared, which is expected to serve as a basic guideline (managerial perspective) for trucking companies and related stakeholders (Table 4).

Table 4. Summary and highlights of potential mitigation measures. The measures are represented as views and suggestions of participants. The stakeholder groups favoring those strategies are also included.

Challenges	Views and Suggestions	Favoring Stakeholder Group(s)
Present market condition	New technologies, new investments, and marketing new products	All
	Opportunities to negotiate with new markets which were not accessible before	All
	Attracting new investors; showing the potentiality of the state in terms of forest products	All
	Favorable policies for startup businesses and subsidies in certain products	Trucking contractors and procurement managers
Manpower shortage	Good benefits, proper training, more vocational schools	All
	More extension activities; showing young generation the modern technologies currently used in forest trucking	All
	US Forest Service and Department of Transportation as lead organizations to attract youths	Forester and professional society
	Ownership sharing mechanism to drivers (giving certain percentage of truck shares)	Forester and procurement manager

Table 4. *Cont.*

Challenges	Views and Suggestions	Favoring Stakeholder Group(s)
	Flexible time schedule and independency to drivers	Forester and professional society
	Developing a well maintained and disciplined trucking fleets	Forester and professional society
	Change in payment methods to truck drivers from load based to hour based	Forester
Roads and payload	Straight forest roads as much as possible	All
	More federal and state budget for maintenance of public roads	Procurement managers and foresters from north region
	Avoiding public roads (not interstate highway) as much as possible due to aesthetic issues	Forester and professional society representative from south region
	East to west interstate highway in Maine	Professional society representative
	Different measures to clean truck tires before entering public roads	Professional society representative and foresters from south region
	Minimize repeated maintenance of private forest roads by constructing them properly at the beginning	Trucking contractors and foresters
	Increasing legal allowable payload on interstate highways for certain situations	Trucking contractors and procurement managers
	Not increasing legal allowable payload on interstate highways to insure public safety and minimize impacts on the roads	Foresters and professional society
	Light trailers to increase capacity of trucks	Trucking contractors
Turnaround time	Adding some self-loading trucks to the fleets	All
	More unloading cranes at the mill (e.g., overhead cranes used by big mills)	Trucking contractor and procurement managers
	More concentrated landing sites	Trucking contractor and forester
	Pavements in wood landing sites	Forester
	Proper coordination in dispatching between different mills in same area	Procurement manager
Backhauling of empty trucks	Long distance hauling of the forest products; an opportunity to back-haul	All
	More concentrated landing sites	All
	Proper networking between mills from different regions.	All
	Dynamic trucking configurations to accommodate various products	All
Seasonal and topographic barriers	Using trucks for other works during muddy season when timber harvesting stops	Forester
	Learning road building knowledge from other US states, mainly for steep terrain	Forester
Fuel efficiency	Use of stud tires during snow season	All
	Learning new innovations from other countries for winter transportation	Trucking contractors and procurement managers
	Using air deflectors in the trucks	Procurement manager
	Increasing payload	Trucking contractors
Contractors	Separating harvesting and trucking parts (i.e. using two different contractors for each work)	Forester and procurement manager
	Proper dispatching strategy to minimize competition between contractors	Forester and procurement manager

3.6. Limitation of the Study

This study has presented views and suggestions of people closely associated with transportation of forest products in Maine. As such, since the suggestions are explicitly based on the situation of Maine, the results of this study cannot be generalized for a broader perspective like other quantitative studies. However, these findings can have significant effects for new studies attempting to tackle these issues. The resolution identified can be validated for other regions by interviewing stakeholders operating in those regions. The fact that truck drivers were not interviewed in this study is one of the major limitations. However, the trucking contractors, who have driven their own trucks for many years before becoming contractors, can be considered as truck drivers' representation. Another concern in this study was about the representation of landowners. Some of the foresters and contractors interviewed have their own timberlands in Maine. They have experience in managing the procurement of forest products on their own lands.

4. Conclusions

This research has validated mitigation measures that can be adopted for sound forest trucking operations in the context of a forest-based economy. The study is the first of its kind for the forest products transportation sector and can serve as a basic guideline to test the technological feasibility behind the suggested resolutions. The use of a semi-structured interview method has proved to be an important approach to gain insights into the field level challenges and mitigation measures in forest products transportation.

The results of this study suggested that lack of skilled manpower and forest products market condition of the state were major challenges to this sector. However, issues like border crossing requirements and road conditions were also considered highly important. Typically, differences in regulations between the US and Canada for many issues (roads, tax, truck size, driver's age) are always a concern for businesses in northern Maine. The condition of public as well as forest roads in Maine is not satisfactory. The high movement of log trucks in public or town roads was also another concern for local residents.

Overall, due to recent changes in the forest products market condition regarding the state and shortage of skilled labor force, the trucking enterprise is a challenging business to operate. In general, the disintegration of trucking business from harvesting operations was regarded as being potentially productive for the long run. The local field-level suggestions for the mitigation of major challenges seemed crucial in the trucking sector. The region-specific suggestions can also help forest products companies and trucking enterprises to focus more on the solutions. A constant collaboration among forest products companies, contractors, and foresters is important to resolve supply chain issues like trucks dispatching, turnaround times, and backhauling. Nonetheless, coordination with the public and policymakers for issues related to public road conditions and safety is vital for the better trucking business.

Acknowledgments: This project was supported by funding from Cooperative Forestry Research Unit, University of Maine and Maine Agriculture and Forest Experiment Station: Award Number DE-EE0006297. We would like to express our gratitude to all thirteen anonymous interview participants who are the center point of this study. We would also like to thank Brian Roth, Forest Resources Association (FRA) and its Northeast division coordinator Eric Kingsley for providing us the platform to contact interview participants.

Author Contributions: A.K. and A.R.K. conducted the interview. A.K. processed and analyzed data and wrote the manuscript. A.R.K. designed the project, acquired funding, and was involved in all phases of the manuscript preparation. S.M.D.U.-S. helped in data analysis, and in formatting and editing the manuscript.

Conflicts of Interest: The authors declare no conflict of interest.

References

1. Kizha, A.R.; Han, H.-S.; Montgomery, T.; Hohl, A. Biomass power plant feedstock procurement: Modeling transportation cost zones and the potential for competition. *Calif. Agric.* **2015**, *69*, 184–190. [CrossRef]
2. Koirala, A.; Kizha, A.R.; Roth, B. Forest trucking industry in Maine: A review on challenges and resolutions. In Proceedings of the 39th Annual Meeting of the Council on Forest Engineering, Vancouver, BC, Canada, 19–21 September 2016.
3. Pan, F.; Han, H.-S.; Johnson, L.R.; Elliot, W.J. Production and cost of harvesting, processing, and transporting small-diameter ($\leq$5 inches) trees for energy. *For. Prod. J.* **2008**, *58*, 47–53.
4. Sosa, A.; Klvac, R.; Coates, E.; Kent, T.; Devlin, G. Improving Log Loading Efficiency for Improved Sustainable Transport within the Irish Forest and Biomass Sectors. *Sustainability* **2015**, *7*, 3017–3030. [CrossRef]
5. Dowling, T.N. An Analysis of Log Truck Turn Times at Harvest Sites and Mill Facilities. Master's Thesis, Virginia Polytechnic Institute and State University, Blacksburg, VA, USA, 2010.

6. Maine' Forest Products Council (MFPC). *Maine's Forest Economy*; Maine' Forest Products Council: Augusta, ME, USA, 2013; p. 28.
7. Koirala, A.; Kizha, A.R.; Roth, B.E. Perceiving Major Problems in Forest Products Transportation by Trucks and Trailers: A Cross-sectional Survey. *Eur. J. For. Eng.* **2017**, *3*, 23–34.
8. Acuna, M.; Mirowski, L.; Ghaffariyan, M.R.; Brown, M. Optimising transport efficiency and costs in Australian wood chipping operations. *Biomass Bioenergy* **2012**, *46*, 291–300. [CrossRef]
9. Grebner, D.L.; Grace, L.A.; Stuart, W.; Gilliland, D.P. A practical framework for evaluating hauling costs. *Int. J. For. Eng.* **2005**, *16*, 115–128.
10. Möller, B.; Nielsen, P.S. Analysing transport costs of Danish forest wood chip resources by means of continuous cost surfaces. *Biomass Bioenergy* **2007**, *31*, 291–298. [CrossRef]
11. Yoshioka, T.; Aruga, K.; Nitami, T.; Sakai, H.; Kobayashi, H. A case study on the costs and the fuel consumption of harvesting, transporting, and chipping chains for logging residues in Japan. *Biomass Bioenergy* **2006**, *30*, 342–348. [CrossRef]
12. Greene, W.D.; Baker, S.A.; Lowrimore, T. Analysis of Log Hauling Vehicle Accidents in the State of Georgia, USA, 1988–2004. *Int. J. For. Eng.* **2007**, *18*, 52–57.
13. Holzleitner, F.; Kanzian, C.; Stampfer, K. Analyzing time and fuel consumption in road transport of round wood with an onboard fleet manager. *Eur. J. For. Res.* **2011**, *130*, 293–301. [CrossRef]
14. Montgomery, T.D.; Han, H.-S.; Kizha, A.R. Modeling work plan logistics for centralized biomass recovery operations in mountainous terrain. *Biomass Bioenergy* **2016**, *85*, 262–270. [CrossRef]
15. Sikanen, L.; Asikainen, A.; Lehikoinen, M. Transport control of forest fuels by fleet manager, mobile terminals and GPS. *Biomass Bioenergy* **2005**, *28*, 183–191. [CrossRef]
16. Abbas, D.; Handler, R.; Hartsough, B.; Dykstra, D.; Lautala, P.; Hembroff, L. A survey analysis of forest harvesting and transportation operations in Michigan. *Croat. J. For. Eng.* **2014**, *35*, 179–192.
17. Egan, A.; Taggart, D. Public perceptions of the logging profession in Maine and implications for logger recruitment. *North. J. Appl. For.* **2009**, *26*, 93–98.
18. Egan, A.; Taggart, D. Who will log in Maine's north woods? A cross-cultural study of occupational choice and prestige. *North. J. Appl. For.* **2004**, *21*, 200–208.
19. Leon, B.H.; Benjamin, J.G. A Survey of Business Attributes, Harvest Capacity and Equipment Infrastucture of Logging Businesses in the Northern Forest. Available online: http://maineforest.org/wp-content/uploads/2013/03/Survey-on-logging-businesses.pdf (accessed on 10 November 2017).
20. Malinen, J.; Nousiainen, V.; Palojarvi, K.; Palander, T. Prospects and challenges of timber trucking in a changing operational environment in Finland. *Croat. J. For. Eng.* **2014**, *35*, 91–100.
21. Fielding, D.; Cubbage, F.; Peterson, M.N.; Hazel, D.; Gugelmann, B.; Moorman, C. Opinions of Forest Managers, Loggers, and Forest Landowners in North Carolina regarding Biomass Harvesting Guidelines. *Int. J. For. Res.* **2012**, *2012*, 1–15. [CrossRef]
22. Silver, E.J.; Leahy, J.E.; Noblet, C.L.; Weiskittel, A.R. Maine woodland owner perceptions of long rotation woody biomass harvesting and bioenergy. *Biomass Bioenergy* **2015**, *76*, 69–78. [CrossRef]
23. Creswell, J.W. *Qualitative Inquiry and Research Design: Choosing among Five Approaches*, 3rd ed.; Sage Publications Inc.: Thousand Oaks, CA, USA, 2012.
24. Stake, R.E. *The Art of Case Study Research*; Sage Publications Inc.: Thousand Oaks, CA, USA, 1995.
25. Denzin, N.K.; Lincoln, Y.S. *The Landscape of Qualitative Research: Theories and Issues*, 3rd ed.; Sage Publications Inc.: Thousand Oaks, CA, USA, 2008.
26. Crotty, M. *The Foundations of Social Research: Meaning and Perspective in the Research Process*; Sage Publications Inc.: London, UK, 1998.
27. Guba, E.G.; Lincoln, Y.S. Competing paradigms in qualitative research. In *Handbook of Qualitative Research*; Sage Publications Inc.: London, UK, 1994; pp. 105–117.
28. Seidman, I. *Interviewing as Qualitative Research: A Guide for Researchers in Education and the Social Sciences*, 3rd ed.; Teachers College Press: New York, NY, USA, 2006; ISBN 978-0-8077-4666-0.
29. Leedy, P.D.; Ormrod, J.E. *Practical Research: Planning and Design*, 11th ed.; Pearson Education: London, UK, 2014.
30. Yin, R.K. *Case Study Research: Design and Methods*, 5th ed.; Sage Publications Inc.: Thousand Oaks, CA, USA, 2013.

31. Gummesson, E. *Qualitative Methods in Management Research*, 2nd ed.; Sage Publications Inc.: London, UK, 2000.
32. Patton, M.Q. *Qualitative Research and Evaluation Methods: Integrating Theory and Practice*, 4th ed.; SAGE Publications Inc.: Thousand Oaks, CA, USA, 2015.
33. Emmel, N. *Sampling and Choosing Cases in Qualitative Research: A Realist Approach*; SAGE Publications Inc.: Thousand Oaks, CA, USA, 2013.
34. Wan, M.; D'Amato, D.; Toppinen, A.; Rekola, M. Forest Company Dependencies and Impacts on Ecosystem Services: Expert Perceptions from China. *Forests* **2017**, *8*, 134. [CrossRef]
35. Corbin, J.; Strauss, A. *Basics of Qualitative Research: Techniques and Procedures for Developing Grounded Theory*, 3rd ed.; Sage Publications Inc.: Thousand Oaks, CA, USA, 2007.
36. Miles, M.B.; Huberman, A.M. *Qualitative Data Analysis: An Expanded Sourcebook*, 2nd ed.; Sage Publications Inc.: Thousand Oaks, CA, USA, 1994.
37. Mertens, D.M. *Research and Evaluation in Education and Psychology: Integrating Diversity with Quantitative, Qualitative, and Mixed Methods*, 4th ed.; SAGE Publications Inc.: Thousand Oaks, CA, USA, 2015.
38. Guest, G. How Many Interviews Are Enough? An Experiment with Data Saturation and Variability. *Field Methods* **2006**, *18*, 59–82. [CrossRef]
39. Miles, M.B.; Huberman, A.H.; Saldana, J. *Qualitative Data Analysis: A Methods Sourcebook*, 3rd ed.; SAGE Publications Inc.: Thousand Oaks, CA, USA, 2014.
40. QSR International Pty Ltd. NVivo Qualitative Data Analysis Software. Available online: https://www.qsrinternational.com/nvivo/home (accessed on 10 November 2017).
41. Saldana, J. *The Coding Manual for Qualitative Researchers*, 3rd ed.; SAGE Publications Inc.: Thousand Oaks, CA, USA, 2016.
42. Ohm, R. Shutdown of Madison mill is state's fifth in two years. *Portland Press Herald*, 14 March 2016.
43. Hetemäki, L.; Hurmekoski, E. Forest Products Markets under Change: Review and Research Implications. *Curr. For. Rep.* **2016**, *2*, 177–188. [CrossRef]
44. Palander, T.; Vainikka, M.; Yletyinen, A. Potential Mechanisms for Co-operation between Transportation Entrepreneurs and Customers: A Case Study of Regional Entrepreneurship in Finland. *Croat. J. For. Eng.* **2012**, *33*, 89–103.
45. Irland, L.C. *Assessment of Conditions of Competition and Ratemaking in the Maine Logging and Log Trucking Industry*; The Irland Group: Wayne, ME, USA, 2011; p. 109.
46. Miller, E. Economization and beyond: (Re) composing livelihoods in Maine, USA. *Environ. Plan. A* **2014**, *46*, 2735–2751. [CrossRef]

Article

The Optimum Slash Pile Size for Grinding Operations: Grapple Excavator and Horizontal Grinder Operations Model Based on a Sierra Nevada, California Survey

Takuyuki Yoshioka [1,*], Rin Sakurai [2], Shohei Kameyama [3], Koki Inoue [1] and Bruce Hartsough [4]

[1]   College of Bioresource Sciences, Nihon University, Fujisawa 252-0880, Japan; inoue.kouki@nihon-u.ac.jp
[2]   Faculty of Agriculture, University of Miyazaki, Miyazaki 889-2192, Japan; sakurai@cc.miyazaki-u.ac.jp
[3]   Graduate School of Bioresource Sciences, Nihon University, Fujisawa 252-0880, Japan;
      kameyama.shohei.0110@gmail.com
[4]   Department of Biological and Agricultural Engineering, University of California, Davis, CA 95616, USA;
      brhartsough@ucdavis.edu
*   Correspondence: yoshioka.takuyuki@nihon-u.ac.jp; Tel.: +81-466-84-3608

Academic Editor: Raffaele Spinelli
Received: 25 September 2017; Accepted: 13 November 2017; Published: 15 November 2017

Abstract: The processing of woody biomass waste piles for use as fuel instead of burning them was investigated. At each landing of slash pile location, a 132 kW grapple excavator was used to transfer the waste piles into a 522 kW horizontal grinder. Economies of scale could be expected when grinding a larger pile, although the efficiency of the loading operation might be diminished. Here, three piles were ground and the operations were time-studied: Small (20 m long × 15 m wide × 4 m high), Medium (30 × 24 × 4 m), and Large (35 × 30 × 4 m) piles. Grinding the Medium pile was found to be the most productive at 30.65 bone dry tons per productive machine hour without delay (BDT/PMH_0), thereby suggesting that there might be an optimum size of slash pile for a grinding operation. Modeling of the excavator and grinder operations was also examined, and the constructed simulation model was observed to well-replicate the actual operations. Based on the modeling, the productivity of grinding at a landing area of 710 m^2 of slash pile location was estimated to be 31.24 BDT/PMH_0, which was the most productive rate.

Keywords: fuel reduction; slash pile; grinding operation; grapple excavator; horizontal grinder; simulation; Sierra Nevada, California; wildfire

---

1. Introduction

Increasingly fierce wildfires are currently one of the most severe problems in the western United States. California is also experiencing one of the state's worst droughts of the past century. Under natural fire conditions, a proper amount of thinning occurs and the remaining trees are thereby given a better chance to mature. In contrast, after a century of fire suppression, California's forests are denser and have fewer large trees. For example, from the 1930s to the 2000s, the number of large trees in the Sierra Nevada mountain range in California decreased by half while the density of small trees doubled [1]. Severe fires are increasing in frequency and size throughout the Sierra Nevada, and regeneration is not a given for severely burned forests where seed trees have been killed across large areas [2]. Fuel reduction operations (e.g., prescribed fire, mechanical treatment, mechanical treatment + prescribed fire) are effective to reduce the risk of high-intensity wildfires and return forests to a more fire-resilient landscape [3].

Current 'business as usual' activities for biomass disposal in much of the Sierra Nevada include pile and burn, mastication, and drop/scatter techniques. Notably, the utilization of biomass material for energy production is an appealing option for biomass disposal that can contribute to density management, forest health, and fire hazard reduction. In a previous study, the Placer County Air Pollution Control District (PCAPCD) and the Sierra Nevada Conservancy demonstrated a significant reduction in air emissions through the diversion of forest biomass that had been scheduled for open pile burning [4]. In the project entitled 'Forest Biomass Diversion in the Sierra Nevada' as a next step, the PCAPCD sponsored research that tracked the economic costs and air emissions generated from the collection, processing, and transport of forest harvest residuals generated at the Blodgett Forest Research Station, the Center for Forestry, the University of California, Berkley in 2012, with the objective of quantifying the emissions reductions gained from using the biomass for energy production compared to open pile burning (Figure 1).

The market value of forest biomass was not sufficient to cover 100% of the forecasted costs to collect, process, and transport material to the Buena Vista Biomass Power (BVBP) facility, which is the nearest biomass power generation facility located near Ione, California. The PCAPCD therefore offset the cost differential between the forest biomass market value and the actual costs of collection, processing, and transport. A forest biomass processing contractor, Brushbusters Inc., was retained to process and transport six woody biomass waste piles for use as fuel in the BVBP facility. In order to monitor the equipment operating costs and efficiencies as well as the equipment air emissions, processing the woody biomass waste piles was investigated. At each landing of slash pile location, a grapple excavator was used to transfer the piles into a horizontal grinder (Figure 2).

Figure 1. Open pile burning.

Figure 2. Grinding operation.

In contrast, in Japan, following the 'Feed-in Tariff Scheme for Renewable Energy (FIT)' that was put into practice in 2012, the building of power-generation plants that accept unused forest biomass (such as thinnings and logging residues) and the initiation of the plants' operation are progressing, since the purchase price of electricity from unused forest biomass has been set higher than that from other wood-based materials, e.g., mill residues and imported woods [5]. Thus, 1.17 million bone dry tons (BDT) of wood chips derived from thinnings and logging residues were used as energy in Japan in 2015 [6]. With respect to the FIT approval of power generation fueled by unused forest biomass, 38 plants (297 MW of total power output) were already in operation and 89 projects (436 MW) were approved as of February 2017 [7]. Because thinnings and logging residues must be comminuted before energy conversion at a power-generation plant or biomass-fired boiler, increasing numbers of the following operations are expected in Japan: the creation of large slash piles by collecting thinnings and logging residues at landings alongside forest roads or at the stockyards of power-generation plants, and the subsequent processing of the piles by chippers or grinders.

In general terms, economies of scale can be expected when grinding a larger slash pile, although the efficiency of a loading operation may be diminished. With respect to the impact of the slash pile size, Seymour and Tecle [8,9] studied the impact of burning on soil physical properties and chemical characteristics, and the impact of burning on biomass moisture change has also been tested; e.g., [10,11]. The grinding operations in the western Pacific USA were investigated and modeled; e.g., [12,13]. However, the relationship between the slash pile size and the productivity of a grinder has not been established. In the present study, three slash piles (small, medium, and large) were ground, and the operations were time-studied in the Results section by using a protocol that is similar to a protocol used by the authors of this paper previously [14–17]. In the Discussion section, based on the results of the time study, a simulation model of a grapple excavator's loading of logging residues from the varying slash piles and its unloading to the conveyor of a horizontal grinder is constructed. Thus, the optimum size of slash piles that would maximize the productivity of the grinder is discussed based on the replication of the excavator and grinder operations.

Concerning previous studies related to the modeling of forest operations by simulation, Iwaoka et al. [18] calculated the cycle time and productivity of harvesters, and Sakurai et al. [19] calculated those of tower-yarders, processors, and forwarders by determining theoretical formulae of element operations and aggregating them on the basis of a transition probability matrix of element operations. Other research groups predicted the productivity of total logging systems by determining theoretical formulae of the cycle times of forestry machines and by using the system dynamics method [20–22]. In the present study, the approach used by Iwaoka et al. and Sakurai et al. was followed in order to construct a simulation model of a grapple excavator operation by analyzing the data of element operations.

2. Materials and Methods

2.1. Study Site and Treatment

The Blodgett Forest Research Station (BFRS) is 1198 ha of Sierra Nevada forest land located east of Georgetown, California (approx. 100 km northeast of Sacramento, Figure 3). The woody biomass waste piles at the BFRS include tree tops, limbs, and small trees. The piles were generated from thinning treatments in mixed conifer plantations during the summer of 2012. The treatment objectives were to reduce the fire hazard, increase the average tree vigor, and increase the species diversity. Operations were typical of those in the Sierra Nevada, where young and dense forests have developed following wildfires or even-aged harvests. Plantations were thinned to an average of 272 trees per ha from pre-treatment stocking levels of 549 trees per ha. Four plantations were thinned, covering a total of approx. 32 ha. Because smaller trees were preferred for removal, the average stem diameter (for residual trees) at breast height (DBH) increased from 30.2 to 33.3 cm. Sawlogs with >15.2 cm dia. on the small end and ≥3.05 m long were transported to a sawmill for processing into lumber products.

Unmerchantable trees (too small to process into sawlogs) plus the tops and limbs of merchantable trees were piled at landings adjacent to the roadside for disposal by open burning; the processing residues had been piled with the intention of burning rather than grinding them, and thus no attention was paid to orienting the tops so that they could be readily fed into the grinder. The overall sizes of the piles generated were typical of thinning operations in young and mature forests, with the bulk volume averaging 1784 m^3 per pile [23].

Figure 3. Location of the study site.

At each BFRS slash pile, a grapple excavator was used to transfer the waste material into a horizontal grinder. Wood chips from the grinder were conveyed directly into chip vans operated by Brushbusters, Inc. and transported to the Buena Vista Biomass Power (BVBP) facility, typically a 105 km one-way trip. The equipment and engines used for the loading and grinding operations (Table 1) were sized for the scale of operations that a medium or large landowner might consider. Landing piles for the project contained ≥100 green tons (GT) of biomass waste (the equivalent of four chip vans each holding 25 GT). All of the biomass received at the BVBP facility had been chipped prior to transport since the BVBP facility does not have fuel-processing equipment on site. Brushbusters' operations of grinder, excavator, and chip vans were carefully observed and tracked, including the determination of the total operating hours, productive operating hours, diesel fuel use, biomass production, and distance traveled. The data of the amount and moisture content of the transported chips were derived by interviewing the BVBP staff on the day after the transport day.

Table 1. Equipment and engines for biomass processing.

Equipment	Grapple Excavator	Horizontal Grinder
Vendor, model	Link-Belt, 290 LX	Bandit, Beast 3680
Engine, horsepower	Isuzu CC-6BG1TC, 132 kW	Caterpillar C18 Tier III, 522 kW
Length	10.41 m	11.89 m
Width	3.400 m	2.845 m
Height	3.270 m	4.115 m
Weight	29,211 kg	28,122 kg
Maximum reach	10.54 m	-
Maximum feed height	-	0.890 m
Infeed conveyor	-	6.110 m × 1.520 m

2.2. Description of Slash Pile and Element Operation

For the analysis of the relationship between the slash pile size and the productivity of the grinder, the following three piles were selected from the total of six piles and studied their processing, grinding, and transport operations: Small (20 m long × 15 m wide × 4 m high; 51.41 BDT), Medium (30 × 24 × 4 m; 122.66 BDT), and Large (35 × 30 × 4 m; 173.78 BDT) piles. The following element operations of the excavator were monitored:

- **Loading** means grabbing logging residues out of a slash pile and then pivoting with load;
- **Unloading** means releasing the residues at the conveyor of a horizontal grinder and then pivoting with no load;
- **Shaking** means shaking waste material off in order to facilitate the feeding of grabbed residues;
- **Waiting** means waiting for feeding the material; the grinding operation was carried out by the interaction of excavator and grinder, so the waiting operation was essential for the excavator;
- **Pushing** means pushing the material into the grinder when it could not 'swallow' the residues because of their bulkiness;
- **Reorienting or repositioning** means reorienting or repositioning the scattered material in order to increase the amount of residue per grab when the operation proceeded and the bulk volume of pile became smaller;
- **Loading with moving** means that the loading operation shown above was done with moving;
- **Unloading with moving** means that unloading operation shown above was done with moving.

Provided that the shape of each landing slash pile location was rectangular, the amounts of logging residues per m^2 were calculated as 0.171 BDT/m^2 (=51.41 BDT/(20 m × 15 m)), 0.170 BDT/m^2 (=122.66 BDT/(30 m × 24 m)), and 0.166 BDT/m^2 (=173.78 BDT/(35 m × 30 m)) for the Small, Medium, and Large piles, respectively, and it was thus concluded that there was no significant deviation of the amount of residues among the three piles.

3. Results of the Time Study and the Monitored Productivity of a Grinder

During the period of 20 August 2013 through 4 September 2013 on eight workdays, 601 BDT (928 GT) of forest slash from the BFRS were collected, processed, and transported by Brushbusters for energy use to the BVBP facility. This comprised a total of 37 separate chip van loads, with each delivery averaging 16.3 BDT (25.1 GT). Average moisture content of the delivered chips was 55.1% on a dry basis (standard deviation = 8.01%).

The results of the time study are shown in Table 2. The times of loading and shaking would be shortened by improving the piling method, such as by orienting the tree tops and limbs so that they can most readily be fed into the grinder. Modifying the infeed conveyor of the grinder, e.g., by extending its length, would improve the times needed for waiting and pushing. With respect to the impact of the slash pile size, the average times of all element operations except for reorienting or repositioning were not influenced by the pile size. The reorienting/repositioning frequency was increased and its average time was lengthened as the size of the pile bulked up. The percentage of the time of reorienting/repositioning to the total observed time was also proportional to the pile size.

The results of the time study per BDT (Figure 4) show that grinding the Medium pile was the most productive, at 30.65 BDT/PMH_0 (=122.66 BDT/14,408 s × 3600 s/h). The productivity for the Small pile was 21.73 BDT/PMH_0 (=51.41 BDT/8519 s × 3600 s/h), and that for the Large pile was 24.49 BDT/PMH_0 (=173.78 BDT/25,545 s × 3600 s/h), thereby suggesting that there might be an optimum size of slash pile for a grinding operation. The Nordic guidelines state that the preferable size for a slash pile is 20–30 m long and a max. of 4 m high [24]; this guideline supports this paper's finding about the Medium pile, of which width was 24 m.

Table 2. Results of the time study.

Element Operation		Pile		
		Small	Medium	Large
Loading	Time (s)	3484	5312	7614
	Frequency	359	550	802
	Avg. (s)	9.70	9.66	9.49
	Std. Dev. (s)	5.55	4.29	4.56
Unloading	Time (s)	3114	4776	6848
	Frequency	383	594	863
	Avg. (s)	8.13	8.04	7.94
	Std. Dev. (s)	3.09	3.00	2.73
Shaking	Time (s)	92	95	201
	Frequency	14	15	29
	Avg. (s)	6.57	6.33	6.93
	Std. Dev. (s)	3.08	2.50	2.84
Waiting	Time (s)	479	1314	1875
	Frequency	29	71	88
	Avg. (s)	16.52	18.51	21.31
	Std. Dev. (s)	18.06	19.95	19.32
Pushing	Time (s)	1013	1190	1316
	Frequency	132	168	180
	Avg. (s)	7.67	7.08	7.31
	Std. Dev. (s)	5.02	4.83	7.06
Reorienting or repositioning	Time (s)	52	1056	6826
	Frequency	3	11	21
	Avg. (s)	17.33	96.00	325.05
	Std. Dev. (s)	2.31	126.17	732.85
Loading with moving	Time (s)	100	201	284
	Frequency	13	29	33
	Avg. (s)	7.69	6.93	8.61
	Std. Dev. (s)	3.82	2.25	3.19
Unloading with moving	Time (s)	185	464	581
	Frequency	18	47	56
	Avg. (s)	10.28	9.87	10.38
	Std. Dev. (s)	6.95	5.44	9.28
Total		8519	14,408	25,545

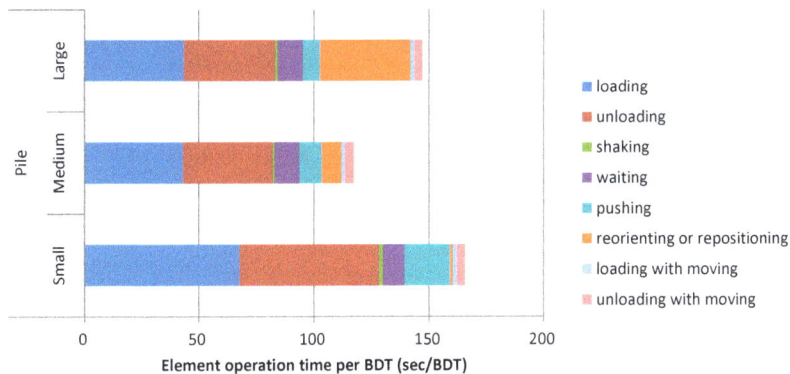

Figure 4. Element operation time per BDT.

The element operation times of reorienting/repositioning per BDT were 1.01 s/BDT (=52 s/51.41 BDT), 8.61 s/BDT (=1056 s/122.66 BDT), and 39.3 s/BDT (=6826 s/173.78 BDT) for the Small, Medium, and Large piles, respectively, thus lengthening as the size of the pile bulked up. The calculated weights of slashes per loading were 0.138 BDT/time (=51.41 BDT/359 + 13 times (this was the total frequency of element operations of loading and loading with moving)), 0.212 BDT/time (=122.66 BDT/550 + 29 times), and 0.208 BDT/time (=173.78 BDT/802 + 33 times) for the grinding of the Small, Medium, and Large piles, respectively, which suggests that reorienting or repositioning material from the pile could make the amount of slashes per loading increase and the productivity of the grinder rise. However, reorienting or repositioning from too large a pile may take too much time, resulting in a decline of the overall operational efficiency.

4. Discussion by the Simulation Model

4.1. Modeling a Grapple Excavator Operation

The respective element operations of a grapple excavator's operation were aggregated and created histograms. A theoretical formula of each element operation time was determined from the distribution of the histogram of the monitored element operation time, and each operation time was estimated by substituting random sampling numbers for the theoretical formula, such as $\exp(N(m, \sigma))$ and $N(m, \sigma)$ is an operator that generates random normal numbers of which average and standard deviation are m and σ, respectively; the theoretical formula was expected to follow a lognormal distribution according to the study by Sakurai et al. [19]. The monitored grapple excavator operation was complicated because there are so many branches in the workflow of element operations. A transition probability matrix was thus constructed based on the connectivity of element operations. In the simulation model, the next element's operation was determined by the matrix.

The average times of all of the element operations other than reorienting/repositioning were not influenced by the pile size, as mentioned above. Therefore, concerning these seven element operations, i.e., (1) loading, (2) unloading, (3) shaking, (4) waiting, (5) pushing, (6) loading with moving and (7) unloading with moving, the element operation times monitored at the three piles were put together and the transition probabilities that would indicate the probability that the next element operation would occur were calculated (Figure 5). The time distribution of each element operation was fit to a lognormal distribution, and a chi-square test of goodness of fit was conducted. The goodness of fit in none of the element operations was rejected at the significance level of 5%, so the theoretical formulae of these element operations could be determined (Table 3).

		Subsequent element operation						
		loading	unloading	shaking	waiting	pushing	loading with moving	unloading with moving
Present element operation	loading	0.002	0.714	0.033	0.039	0.163	0.037	0.012
	unloading	0.816	0.004	0.001	0.052	0.077	0.003	0.048
	shaking	0.931	0.034	0	0	0.017	0.017	0
	waiting	0.167	0.602	0	0	0.210	0	0.022
	pushing	0.004	0.923	0	0.038	0.019	0.002	0.015
	loading with moving	0.427	0.373	0	0.093	0.093	0	0.013
	unloading with moving	0.719	0.223	0	0.008	0.017	0.033	0

Figure 5. Transition probability matrix of element operations of the grapple excavator.

On the other hand, the frequency of reorienting/repositioning in the time study was low and a distinct relationship with precedent and subsequent element operations was not observed. However, reorienting/repositioning was an element operation that was definitely carried out within the workflow of the grapple excavator operation, and thus the total operation time was estimated based on a theoretical formula. From the results described in the text section above, i.e., 1.01 s/BDT for the Small pile (300 m^2), 8.61 s/BDT for the Medium pile (720 m^2), and 39.3 s/BDT for the Large pile (1050 m^2), the relationship between the landing area of slash pile location, x (m^2), and the time of reorienting or repositioning per BDT, y (s/BDT), was approximated as follows:

$$y = 0.2397 \exp(0.00489x) \ (r^2 = 0.9992) \tag{1}$$

and then the total time could be calculated by multiplying y by the amount of logging residues at the landing.

Table 3. Results of chi-square tests and theoretical formulae of element operation time.

Element Operation	Chi-Square Test			Theoretical Formula [1]
	χ^2	df	p-Value	
Loading	5.416	5	0.367	$_eN(2.140, 0.485)$
Unloading	10.985	5	0.052	$_eN(2.023, 0.370)$
Shaking	4.422	5	0.490	$_eN(1.825, 0.383)$
Waiting	6.314	5	0.277	$_eN(2.625, 0.800)$
Pushing	10.238	5	0.069	$_eN(1.819, 0.539)$
Loading with moving	8.353	5	0.138	$_eN(1.987, 0.363)$
Unloading with moving	8.009	5	0.156	$_eN(2.153, 0.530)$

[1] $N(m, \sigma)$ is an operator that generates random normal numbers of which average and standard deviation are m and σ, respectively.

In the constructed model, the mass of the landing pile is first set up, and the simulation is started at the element operation of loading. A grapple excavator grabs logging residues out of a slash pile when loading and loading with moving. If the mass of the pile falls below zero after the excavator grabs the residues, the element operation of unloading comes next; then the excavator operation is finally finished. Consequently, the total time of the excavator operation is composed of the time calculated by the simulation model and the estimated element operation time of reorienting/repositioning. Incidentally, the amount of residue per grab, z (BDT/time), was approximated as a function of the landing area of slash pile location, x, from the results described in the last section, i.e., 0.138 BDT/time for the Small pile (300 m^2), 0.212 BDT/time for the Medium pile (720 m^2), and 0.208 BDT/time for the Large pile (1050 m^2), as follows:

$$z = -2.489 \times 10^{-7}x^2 + 4.292 \times 10^{-4}x + 3.184 \times 10^{-2} \ (r^2 = 1.000) \tag{2}$$

4.2. Verification of the Replicability of the Model and an Optimum Slash Pile Size

The replicability of the constructed model was verified by comparing the monitored productivities with the values calculated by the simulation (the program was created by using Microsoft Excel VBA). The calculation was repeated 1000 times for the respective Small, Medium, and Large piles (Table 4). The maximum difference between the monitored value and the average calculated value was 1.7% for the Medium pile, and the highest ratio of the standard deviation to the average calculated productivity was only 3.2% (=0.70/21.78 × 100) for the Small pile; it was thus concluded that the constructed simulation model well-replicated the actual operations.

Table 4. Comparison between the monitored and estimated productivities.

Pile	Monitored				Estimated Productivity	
	Area of Landing (m²)	Amount of Slashes (BDT)	Productivity (BDT/PMH$_0$)	Calculation Frequency	Avg. ± Std. Dev. (BDT/PMH$_0$)	Rate of Avg. Value to Monitored (%)
Small	300	51.41	21.73	1000	21.78 ± 0.70	100.2
Medium	720	122.66	30.65	1000	31.17 ± 0.75	101.7
Large	1050	173.78	24.49	1000	24.27 ± 0.38	99.10

For the discussion of an optimum slash pile size that maximizes the productivity of a grinder, the landing area of slash pile location was focused on next. A simulation was carried out for a landing area between 300 m² (for the Small pile; 20 m long × 15 m wide) and 1050 m² (for the Large pile; 35 m long × 30 m wide) at 10 m² intervals. In the simulation of the respective landing areas, the calculation was repeated 1000 times. The productivity for each landing was determined based on the averaged total operation time. Since no significant deviation of the amount of residues among the three piles was observed in the time study, the mass of the slash pile in an initial state of simulation was calculated by multiplying 0.168 BDT/m² (=(0.171 × 300 + 0.170 × 720 + 0.166 × 1050)/(300 + 720 + 1050), which was the weighted average value of the monitored three piles) by the landing area.

Figure 6 shows the results of the simulation. The productivity of grinding at the landing area of 710 m² of slash pile location is 31.24 BDT/PMH$_0$, which is the highest productivity value obtained. However, the difference in the estimated productivities is small between the areas 690 m² (31.21 BDT/PMH$_0$) and 730 m² (31.20 BDT/PMH$_0$), and there is a range in the calculation result for each landing. It should be noted therefore that Figure 6 simply compares the average values of the 1000-times repeated calculation. Concerning the versatility of the constructed model, however, the following points should be discussed further so that the accuracy of the model can be improved:

- The shape of each landing, i.e., the ratio of its length to its width, was not considered in the simulation model;
- The theoretical formulae of (1) and (2) were both approximated from only three samples;
- The optimum size of the slash pile for a grinding operation will also depend in part on aspects of the machines used, e.g., their size, engine output, and grinding capacity.

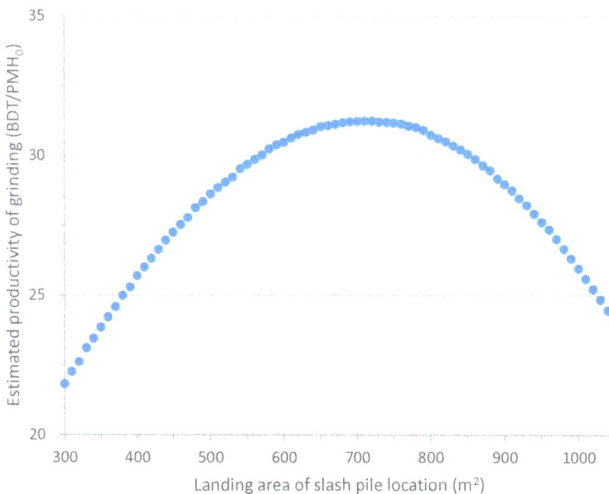

Figure 6. Relationship between the landing area of slash pile location and the estimated productivity of grinding.

5. Conclusions

The simulation model that can replicate the operations of a grapple excavator and a horizontal grinder was constructed based on a Sierra Nevada, California survey, and this paper determined the optimum size of a slash pile for a grinding operation, which is expressed in the landing area of slash pile location.

With respect to other results derived from the Blodgett Project, it was demonstrated that utilization of biomass from these large debris piles can result in energy and air quality benefits [23], as follows:

- The energy (diesel fuel) expended for processing and transport was 2.5% of the biomass fuel (energy equivalent);
- Based on measurements from a large pile burn, air emission reductions of 98–99% for PM2.5, CO, NMOC, CH_4, and BC, and 20% for NO_X and CO_2-equivalent greenhouse gases were observed;
- The delivered cost of $70/BDT exceeds the biomass plant gate price of $45/BDT. Under typical conditions, the break-even haul distance would be approx. 48 km.

Acknowledgments: The Placer County Air Pollution Control District sponsored the Blodgett Project #BF12-05RY 'Forest Biomass Diversion: Tracking the Economic Costs and Air Emissions of Forest Biomass Diversion and Allocating the Air Emissions Credits Generated' at the Blodgett Forest Research Station of the Center for Forestry, the University of California, Berkeley. This study was financially supported in part by JSPS KAKENHI Grant Numbers JP24580213 and JP15H04508, the University of California, Davis, and the Nihon University.

Author Contributions: T.Y. and B.H. conceived, designed, and performed the field experiments; T.Y., S.K., and K.I. analyzed the data; R.S. contributed the simulation model; T.Y. and B.H. wrote the paper.

Conflicts of Interest: The authors declare no conflict of interest.

References

1. McIntyre, P.J.; Thorne, J.H.; Dolanc, C.R.; Flint, A.L.; Flint, L.E.; Kelly, M.; Ackerly, D.D. Twentieth-century shifts in forest structure in California: Denser forests, smaller trees, and increased dominance of oaks. *Proc. Natl. Acad. Sci. USA* **2015**, *112*, 1458–1463. [CrossRef] [PubMed]
2. Kocher, S. Californians must learn from the past and work together to meet the forest and fire challenges of the next century. *Calif. Agric.* **2015**, *69*, 5–9. [CrossRef]
3. North, M.; Collins, B.; Stephens, S. Using fire to increase the scale, benefits, and future maintenance of fuels treatments. *J. For.* **2012**, *110*, 392–401. [CrossRef]
4. Springsteen, B.; Christofk, T.; Eubanks, S.; Mason, T.; Clavin, C.; Storey, B. Emission reductions from woody biomass waste for energy as an alternative to open burning. *J. Air Waste Manag. Assoc.* **2011**, *61*, 63–68. [CrossRef] [PubMed]
5. Present Status and Promotion Measures for the Introduction of Renewable Energy in Japan. Available online: http://www.meti.go.jp/english/policy/energy_environment/renewable/index.html (accessed on 26 October 2017).
6. Forestry Agency of Japan (Ed.) *FY 2016 White Paper in Forest and Forestry*; Zenrinkyou: Tokyo, Japan, 2017; p. 236. (In Japanese)
7. Tomari, M. (Ed.) *Biomass White Paper 2017*; Biomass Industrial Society Network (BIN), NPO: Kashiwa, Japan, 2017; p. 28. (In Japanese)
8. Seymour, G.; Tecle, A. Impact of slash pile size and burning on ponderosa pine forest soil physical properties. *J. Ariz. Nev. Acad. Sci.* **2004**, *37*, 74–82. [CrossRef]
9. Seymour, G.; Tecle, A. Impact of slash pile size and burning on soil chemical characteristics in ponderosa pine forests. *J. Ariz. Nev. Acad. Sci.* **2005**, *38*, 6–20. [CrossRef]
10. Kim, D.-W.; Murphy, G. Forecasting air-drying rates of small Douglas-fir and hybrid poplar stacked logs in Oregon, USA. *Int. J. For. Eng.* **2013**, *24*, 137–147. [CrossRef]
11. Lin, Y.; Pan, F. Effect of in-woods storage of unprocessed logging residue on biomass feedstock quality. *For. Prod. J.* **2013**, *63*, 119–124. [CrossRef]
12. Bisson, J.A.; Han, H.-S. Quality of feedstock produced from sorted forest residues. *Am. J. Biomass Bioenerg.* **2016**, *5*, 81–97. [CrossRef]

13. Zamora-Cristales, R.; Sessions, J.; Marrs, G. Economic implications of grinding, transporting, and pretreating fresh versus aged forest residues for biofuel production. *Can. J. For. Res.* **2017**, *47*, 269–276. [CrossRef]
14. Hartsough, B.; Nakamura, G. Harvesting eucalyptus for fuel chips. *Calif. Agric.* **1990**, *44*, 7–8.
15. Yoshioka, T.; Aruga, K.; Sakai, H.; Kobayashi, H.; Nitami, T. Cost, energy and carbon dioxide (CO_2) effectiveness of a harvesting and transporting system for residual forest biomass. *J. For. Res.* **2002**, *7*, 157–163. [CrossRef]
16. Yoshioka, T.; Aruga, K.; Nitami, T.; Sakai, H.; Kobayashi, H. A case study on the costs and the fuel consumption of harvesting, transporting, and chipping chains for logging residues in Japan. *Biomass Bioenerg.* **2006**, *30*, 342–348. [CrossRef]
17. Yoshioka, T.; Sakurai, R.; Aruga, K.; Nitami, T.; Sakai, H.; Kobayashi, H. Comminution of logging residues with a tub grinder: Calculation of productivity and procurement cost of wood chips. *Croat. J. For. Eng.* **2006**, *27*, 103–114.
18. Iwaoka, M.; Aruga, K.; Sakurai, R.; Cho, K.-H.; Sakai, H.; Kobayashi, H. Performance of small harvester head in a thinning operation. *J. For. Res.* **1999**, *4*, 195–200. [CrossRef]
19. Sakurai, R.; Iwaoka, M.; Sakai, H.; Kobayashi, H. Studies on yarding and hauling system of mobile-yarder, processor, and forwarder with simulation methods. *Bull. Tokyo Univ. For.* **1999**, *102*, 113–132, (In Japanese with English Summary).
20. Nitami, T. Modeling of timber harvesting operation by system dynamics and the productivity estimation function. *J. Jpn. For. Eng. Soc.* **2006**, *20*, 281–284. (In Japanese) [CrossRef]
21. Sugimoto, K.; Niinaga, S.; Hasegawa, H. Consideration on flow harvesting system utilizing system dynamics. *J. Jpn. For. Eng. Soc.* **2010**, *25*, 5–14, (In Japanese with English Summary). [CrossRef]
22. Yoshimura, T.; Hartsough, B. Conceptual evaluation of harvesting systems for fuel reduction and biomass collection on steep terrain using system dynamics. In Proceedings of the International Mountain Logging and 13th Pacific Northwest Skyline Symposium, Corvallis, OR, USA, 1–6 April 2007; Sessions, J., Havill, Y., Eds.; Department of Forest Engineering, Oregon State University: Corvallis, OR, USA, 2007; pp. 94–102. Available online: http://www.cof.orst.edu/cof/ferm/pdf/skyproceedings.pdf (accessed on 22 September 2017).
23. Springsteen, B.; Christofk, T.; York, R.; Mason, T.; Baker, S.; Lincoln, E.; Hartsough, B.; Yoshioka, T. Forest biomass diversion in the Sierra Nevada: Energy, economics and emissions. *Calif. Agric.* **2015**, *69*, 142–149. [CrossRef]
24. Nilsson, B. Costs, CO2-emissions and energy balance for applying Nordic methods of forest biomass utilization in British Columbia. M.Sc. Thesis, Department of Forest Resource Management, Swedish University of Agricultural Sciences, Umeå, Sweden, 13 April 2009. Available online: http://ex-epsilon.slu.se/id/eprint/3244 (accessed on 22 September 2017).

Article

Occupational Safety and Health Concerns in Logging: A Cross-Sectional Assessment in Virginia

Sunwook Kim [1], Maury A. Nussbaum [1],*, Ashley L. Schoenfisch [2], Scott M. Barrett [3], Michael Chad Bolding [3] and Deborah E. Dickerson [4]

[1] Department of Industrial & Systems Engineering, Virginia Tech, Blacksburg, VA 24061, USA; sunwook@vt.edu

[2] School of Nursing, Duke University, Durham, NC 27710, USA; ashley.schoenfisch@duke.edu

[3] Department of Forest Resources & Environmental Conservation, Virginia Tech, Blacksburg, VA 24061, USA; sbarrett@vt.edu (S.M.B.); bolding@vt.edu (M.C.B.)

[4] Civil and Environmental Engineering, Virginia Tech, Blacksburg, VA 24061, USA; dyoung@vt.edu

* Correspondence: nussbaum@vt.edu

Received: 29 September 2017; Accepted: 13 November 2017; Published: 15 November 2017

Abstract: Increased logging mechanization has helped improve logging safety and health, yet related safety risks and concerns are not well understood. A cross-sectional study was completed among Virginia loggers. Participants (n = 122) completed a self-administered questionnaire focusing on aspects of safety and health related to logging equipment. Respondents were at a high risk of workplace injuries, with reported career and 12-month injury prevalences of 51% and 14%, respectively. Further, nearly all (98%) respondents reported experiencing musculoskeletal symptoms. Over half (57.4%) of respondents reported symptoms related to diesel exhaust exposure in their career. Few (15.6%), however, perceived their jobs to be dangerous. Based on the opinions and suggestions of respondents, three priority areas were identified for interventions: struck-by/against hazards, situational awareness (SA) during logging operations, and visibility hazards. To address these hazards, and to have a broader and more substantial positive impact on safety and health, we discuss the need for proactive approaches such as incorporating proximity technologies in a logging machine or personal equipment, and enhancing logging machine design to enhance safety, ergonomics, and SA.

Keywords: workplace injuries; musculoskeletal disorders; diesel exhaust exposure; mechanized logging; situational awareness

---

1. Introduction

Logging is the process of harvesting trees by which workers fell, process, and transport them for further manufacture. It is an important component of the U.S. economy, in that forest products account for ~4% of the total manufacturing gross domestic product [1]. Logging is considered as one of the most dangerous occupations in the U.S., often involving heavy physical demands, nonpermanent worksites, and challenging work environments such as inclement weather, rough terrain, and being in remote or isolated locations [2]. An increase in mechanization and logger safety training have played important roles in improving logging workers' safety. These advancements have decreased adverse work-related events and injuries associated with manual tree felling and processing using a chainsaw [3–6]. However, logging machines themselves pose safety and health threats, and logging workers remain at high risk of injuries and adverse health problems. In 2015, for example, the fatality rate in the U.S. logging industry was 132.7 per 100,000 full-time equivalent (FTE) workers, the highest in any private industry [7]. Similarly, the rate of lost workday injuries was 144.1 per 100,000 FTE workers, compared to the private industry mean of 93.9 per 100,000 FTEs [8]. Common non-fatal injury mechanisms include contact with objects; slips, trips, and falls; and overexertion and bodily reaction, and which is consistent with processes related to the use of logging machinery. Specifically, machine operators are frequently injured while performing machine maintenance/repair [4,9,10] and from falls while mounting/dismounting machinery [4,6,10]. Machine operators and ground logging workers also can be exposed to contact risks with moving machinery and the risk of machine rollover [6,11]. A few studies have reported that machine operators have a high prevalence of musculoskeletal disorders (MSDs), for example, in the neck, shoulder, and lower back regions [12,13], which are attributed to working postures [13] and psychological demands [14].

The use of logging machinery may have broader safety and health implications beyond work-related injuries, in that logging workers are likely exposed to machine-related diesel exhaust. Though little information is available on such exposures in the logging industry, diesel exhaust is a pervasive airborne contaminant in workplaces where diesel-powered equipment is used [15]. Diesel engine exhaust is a highly complex and variable mixture of gases, vapors, and fine particles. The amount and composition of the exhaust vary greatly, depending on factors such as fuel type, maintenance practices, workload, and exhaust system type. Vapor constituents include hydrocarbons and oxides of carbon, sulfur, and nitrogen, while particulate components consist of liquid droplets and soot particles bearing organic compounds, sulfates, metals, and other trace contaminants. The organic fraction is mainly unburned fuel and oil and can contain thousands of compounds; most notably the polycyclic aromatic hydrocarbons, which are known to be carcinogenic and genotoxic [16].

The International Agency for Research on Cancer (IARC) classifies diesel exhaust as "carcinogenic to humans (Group 1)" [17]. Large cohort and case-control studies have yielded evidence demonstrating an association between exposure to unfiltered diesel exhaust and an increased incidence of lung neoplasm [18–20]. Diesel exhaust also has been shown to be an airway irritant, triggering release of cytokines, chemokines, immunoglobulins, and oxidants [21]. It may promote expression of the immunologic response phenotype (Th2) associated with asthma and allergic disease. This immunologic evidence is consistent with epidemiologic studies associating traffic-related air pollution, and diesel exhaust, with increased prevalence of respiratory disease.

Related to mechanized logging operations, previous work provided valuable information on common injury mechanisms as described above [4,9], workers' perception of logging risks and safe work practices [22], and concerns that safety regulations and recommendations may have little impact in practice [23]. However, workplace injuries and illnesses are still an important problem in the logging industry. We thus believe that there is a current need for more detailed characterization of the circumstances surrounding loggers' injury events, their experiences with adverse outcomes related to logging equipment (e.g., MSDs and diesel exhaust exposure), and their opinions and views regarding injury prevention approaches in practice. This will enable a better understanding of the complexities of machine-related injuries and illnesses, and support the development and implementation of effective interventions that embraces the hierarchy of hazard controls in occupational injury prevention [24]. To this end, a preliminary investigation aimed to assess the prevalence of injury, symptoms related to diesel exhaust exposures, and MSD symptoms among logging workers—both overall and across worker and work-related characteristics, and determine loggers' concerns about and recommendations to improve on-the-job safety and health.

2. Materials and Methods

A cross-sectional study was conducted during December 2014 to August 2015 through the assistance of the Virginia Sustainable Harvesting And Research Professional (SHARP) Logger program. Note that the SHARP Logger program is a market-driven program, as opposed to a legal requirement, and requires at least one logger per crew to be trained on the principles of sustainable forestry, environmental protection, and workplace safety (visit sharplogger.vt.edu for more details). All participants completed an informed consent procedure approved by the Virginia Tech Institutional Review Board.

2.1. Participants

We recruited two participant pools via convenience sampling. The first pool was from among regularly scheduled SHARP logger training classes, and the second from four different logging sites in the Mountain, Piedmont, and/or Coastal Plain regions of Virginia. For the former, we distributed a questionnaire to all attendees who indicated they worked on a logging operation during classroom-based trainings, and they were free to complete it or refuse. Logging operations differ substantially by physiographic region. However, the vast majority of all logging operations in Virginia utilize mechanized skidding and loading. In the Mountain region, operations tend to rely primarily on manual felling utilizing chainsaws while in the Piedmont and Coastal Plain, operations primarily utilize mechanized felling [25]. Questionnaires were completed at trainings in all three physiographic regions to allow for a cross-sectional assessment of operations across Virginia; however, the majority was collected in the Piedmont and Coastal Plain regions, which tend to have more mechanized operations. A total of 95 attendees completed the questionnaire. For the latter, we contacted select logging business owners for study solicitation and site visit approval. Four different logging sites were visited, at which the questionnaire was distributed. A total of 27 logging workers completed the questionnaire on-site.

2.2. Questionnaire

The self-administered questionnaire was developed based on earlier studies on mechanized logging operations [9,13], symptoms associated with diesel exhaust exposure [26,27], and existing validated instruments such as the standardized Nordic questionnaires for musculoskeletal symptoms [28]. Specifically, the questionnaire consisted of 23 yes/no, categorical, and open-ended questions covering: (1) personal and job characteristics (i.e., time in the industry, primary job, daily machine operation time); (2) perceived safety, safety training, and personal protective equipment (PPE) usage; (3) work-related injuries (e.g., circumstance surrounding injury events, post-injury care, reporting); (4) symptoms related to diesel exhaust exposure (e.g., eye irritation, wheezing); (5) MSD symptoms (i.e., pain, stiffness, spasm, aching, burning, tingling, or numbness); and (6) perceived safety concerns (machine operators and workers on the ground) and recommendations for improved safety and health. Note that the questionnaire is provided in online Supplemental Material.

2.3. Data Analysis

The distributions of close-ended questionnaire responses were summarized using means and standard deviations, or proportions, as relevant. Differences in participant characteristics between physiographic regions in Virginia, USA (i.e., Coastal Plain, Piedmont, Mountains) were examined using Kruskal-Wallis one-way analyses of variance or Pearson's Chi-squared tests as appropriate. Participants' career and 12-month prevalence of self-reported health outcomes (i.e., work-related injuries, symptoms related to diesel exhaust exposure, and MSD symptoms) were examined overall and across employment duration, primary job, daily machine operation, physiographic region, perceived safety of the job, and attitude toward PPE use. Using log-binomial regression, prevalence ratios and 95% confidence intervals (CIs) were calculated to explore relative differences in career and 12-month prevalence of the study outcomes across worker and job characteristics. Open-ended question responses were reviewed and discussed by investigators to identify naturally emerging themes, based on which the responses were sorted into content categories and sub-categories. The frequency and proportion of responses within each category are reported. All statistical analyses were performed using R statistical software [29].

3. Results

3.1. Characteristics of Questionnaire Respondents

All (n = 122) of the questionnaire respondents were male, and responses are summarized in Table 1. Participants represented each of the physiographic regions of Virginia (Coastal Plain 48.4%, Piedmont 34.4%, Mountains 17.2%), and no significant differences between physiographic regions were observed in demographic and job characteristics, perception of job safety, or PPE use, though the distribution of primary job categories did differ between regions (p = 0.008). More owners completed the questionnaire in the Mountains region (Coastal Plain 16.7%, Piedmont 31.4%, Mountains 55.6%). Overall, participants had an average of 17.3 years of experience in logging and 8 hours of daily machine operation duration, and half (50.8%) had a primary job operating a machine. Machine types reported included loader, feller-buncher, skidder, and truck. Relatively few respondents (15.6%) considered their job to be very/extremely dangerous and 45.9% considered it to be moderately dangerous, while a majority (68.9%) reported that using PPE is very/extremely important. All respondents received safety training from multiple sources, with more common sources being SHARP logger classes (48.4%) and on-the-job/safety training from co-workers, crew foremen, and/or owners (46.0%).

3.2. Prevalence of Work-Related Injury, Symptoms Related to Diesel Exhaust Exposure, and MSD Symptoms

The career prevalence of logging work-related injury among participants was 50.8%, and it increased with increasing years of experience (Table 1). The 12-month prevalence was 13.9%. Seventeen respondents experienced injuries in the prior year, and common injury causes were chainsaw use (41.2%), a slip/fall from a machine (23.5%), or being struck by a tree or machine (17.6%). Commonly affected body parts were the lower (34.5%) and upper (34.5%) extremities. A total of 10 (58.8%) injuries in the prior year resulted in missed work (range: 0.5 days to 8 months). A total of 70.1% (n = 12) of injuries in the prior year required medical care beyond first aid, and 41.2% (n = 7) were reported as a WC claim. In addition, over half (57.4%) of respondents reported symptoms related to diesel exhaust exposure in their career, and the symptom prevalence was not associated with the worker or job characteristics we considered.

Table 1. Career prevalence, prevalence ratios (PR) and 95% confidence intervals (CI) of work-related injury and symptoms related to diesel exhaust exposure among a sample of loggers in Virginia, USA.

	n	(%)	Work-Related Injury				Symptoms Related to Diesel Exhaust Exposure [1]			
			n	Prev	PR	(95% CI)	n	Prev	PR	(95% CI)
Overall	122	(100)								
Years worked in logging										
<10	41	(33.6)	12	0.29	1.00		24	0.58	1.00	
10 to <30	53	(43.4)	29	0.55	1.87	(1.14, 3.38)	33	0.62	1.06	(0.77, 1.51)
≥30	25	(20.5)	20	0.80	2.73	(1.69, 4.86)	12	0.48	0.82	(0.48, 1.30)
Missing	3	(2.5)								
Physiographic region										
Coastal Plain	59	(48.4)	32	0.54	1.00		31	0.53	1.00	
Piedmont	42	(34.4)	19	0.45	0.83	(0.54, 1.24)	25	0.60	1.13	(0.79, 1.61)
Mountain	21	(17.2)	11	0.52	0.97	(0.56, 1.48)	14	0.67	1.27	(0.82, 1.84)
Primary job										
Machine operator	62	(50.8)	30	0.48	1.00		38	0.61	1.00	
Supervisor/foreman/owner [2]	41	(33.6)	22	0.54	1.11	(0.74, 1.62)	21	0.51	0.84	(0.57, 1.18)
Deckhand [3]	5	(4.1)	3	0.60	1.24	(0.40, 2.15)	4	0.80	1.31	(0.60, 1.84)
Missing	14	(11.5)								
Daily machine operation duration (h)										
<4	12	(9.8)	7	0.58	1.11	(0.57, 1.71)	7	0.58	1.00	(0.52, 1.51)
4 to <8	20	(16.4)	7	0.35	0.67	(0.31, 1.15)	9	0.45	0.77	(0.41, 1.20)
≥8	82	(67.2)	43	0.52	1.00		48	0.59	1.00	
Missing	8	(6.6)								
Perceived safety of job										
Not at all/somewhat dangerous	43	(35.2)	22	0.51	1.00		25	0.58	1.00	
Moderately dangerous	56	(45.9)	30	0.54	1.05	(0.72, 1.56)	28	0.50	0.86	(0.59, 1.25)
Very/extremely dangerous	19	(15.6)	10	0.53	1.03	(0.57, 1.67)	15	0.79	1.36	(0.93, 1.93)
Missing	4	(3.3)								

Table 1. *Cont.*

	n	(%)	Work-Related Injury				Symptoms Related to Diesel Exhaust Exposure [1]			
			n	Prev	PR	(95% CI)	n	Prev	PR	(95% CI)
Attitude toward using PPE										
Not at all important	11	(9.0)	6	0.55	1.15	(0.54, 1.85)	5	0.45	0.75	(0.31, 1.27)
Moderately important	26	(21.3)	16	0.62	1.29	(0.85, 1.84)	14	0.54	0.89	(0.56, 1.26)
Very/extremely important	84	(68.9)	40	0.48	1		51	0.61	1	
Missing	1	(0.8)								

[1] Experience of eye/mouth irritation and/or unpleasant smell when operating a machine or while working near a machine; [2] All owners reported that they normally or at least occasionally worked in the woods and operated equipment (i.e., work alongside crew members); [3] Based on our experience and observation, the deckhand helped with moving trailers on the deck, trimming loads (using a pole saw to cut any branches or vines sticking out of the side of the load), and with maintenance in addition to operating a machine.

Nearly all (98%) respondents reported experiencing MSD symptoms in at least one body region in the past 12 months, and 93% had MSD symptoms in more than one body region. The 12-month prevalence of MSD symptoms in each of nine body regions is summarized in Table 2. The body region most commonly affected was the lower back (49.2%), followed by the knee (37.7%). About one-fifth (18.5%) of those with MSD symptoms indicated that they changed their work methods due to their MSD symptoms. Further, 10 respondents (8.4%) missed work as a result of their MSD symptoms, and the same number had considered changing jobs.

Table 2. Crude (i.e., unadjusted) 12-month prevalence of MSD symptoms by body region.

	n	Neck (%)	Shoulder (%)	Elbow (%)	Wrist (%)	Upper Back (%)	Lower Back (%)	Hip (%)	Knee (%)	Foot (%)
All	122	27.9	35.3	17.2	28.7	16.4	49.2	11.5	37.7	21.3
Experience (years)										
<10	41	31.7	39.0	22.0	31.7	24.4	51.2	17.1	46.3	31.7
10 to <30	53	26.4	28.3	18.9	28.3	86.8	52.8	7.6	34.0	13.2
≥30	25	28.0	48.0	8.0	24.0	12.0	44.0	12.0	36.0	20.0
Primary job										
Machine operator	62	22.6	30.7	14.5	30.7	16.1	40.3	6.5	41.9	16.1
Supervisor/foreman/owner	41	36.6	41.5	19.5	24.4	22.0	58.5	22.0	34.2	24.4
Deckhand	5	20.0	40.0	40.0	40.0	0.0	40.0	20.0	20.0	40.0
Daily machine operation duration (h)										
<4	12	8.3	25.0	0.0	16.7	0.0	41.7	16.7	8.3	16.7
4 to <8	20	25.0	35.0	15.0	30.0	80.0	45.0	20.0	30.0	25.0
≥8	82	32.9	37.8	20.7	31.7	81.7	51.2	9.8	46.3	23.2

3.3. Perceived Safety Concerns and Recommendations

About half (48.0%) of participants' perceived safety risks in logging operations were related to struck by/against hazards, often related to chainsaw operation and felling/delimbing/topping activities (Table 3). One fourth (26%) were related to poor situational awareness due to inattention/distraction/work speed, machine operation, and communication. When asked "when you are a machine operator, what do you see as the biggest safety risk for workers on ground?", participants indicated that the biggest safety risks involve poor situational awareness (40.2%) and visibility hazards (34.5%) (Table 4). To improve safety on their job sites or in logging operations in general, many suggested a need for improving situational awareness, more safety training and education, and use of personal protective equipment (PPE) (Table 5).

Table 3. Summary of responses (*total* = 77 responses, from 77 individuals who responded to this open-end question) regarding the biggest safety risks perceived during logging operations. Note that *n* in the table is the number of responses obtained.

Category	Sub Category	Response Examples	Freq. (%)
Struck-by/against hazards	Chainsaw operation	Cut with a chainsaw Use of chainsaw	21 (27.3)
	Felling, delimbing, or topping	Hit by tree tops Falling/flying debris Broken limbs Fell/delimb trees	13 (16.9)
	Moving/rolling logs	Logs move around/roll	3 (3.9)
		Category total = 37 (48.0)	
Poor situational awareness (SA)	Inattention/distraction/work speed	Do not pay attention Distraction Know what others do and where they are Complacency Get in a hurry Watch out for ground workers	14 (18.2)
	Machine operation	Do not run over or hit deck workers Too close to machines Work near machines (in the decking area)	4 (5.2)
	Communication	Lack of communication	2 (2.6)
		Category total = 20 (26.0)	
Machine-related hazards	Maintenance	Climb onto the machine to repair Work on the machine Slip on the machine	6 (7.8)
	Ingress and egress	In and out of machine	3 (3.9)
	Operation	Rollover Malfunction	2 (2.6)
		Category total = 11 (14.3)	
Slips, trips, and falls		Slips, trips, and falls Walk on the wood when it is wet	3 (3.9)
Others		Poor judgment Learning first time Do not wear personal protective equipment (PPE) Weather conditions Presence of non-workers	6 (7.8)

Table 4. Summary of responses (*total* = 87, responses from 86 individuals who responded to this open-end question) regarding the biggest safety risks that machine operators perceive for workers on the ground. Note one participant provided two responses, and that *n* in the table is the number of responses obtained.

Category	Sub Category	Response Examples	Freq. (%)
Poor situational awareness (SA)	Inattention/distraction	Do not pay attention Do not stay focused Do not watch out for machines Complacency	17 (19.5)
	Proximity to hazards (e.g., machine)	Too close to machines Keep safe distance from machines	15 (17.2)
	Communication	Do not make an operator see you	3 (3.4)
		Category total = 35 (40.2)	
Visibility hazards		Visible Do not stay in the blind spot of an operator Stay in the sight of an operator	30 (34.5)
Struck-by/against hazards	Felling, delimbing, or topping	Falling/flying debris/limbs Fell trees	14 (16.1)
	Machine-related (Operation)	Lose control of logs in the log grapple Knock logs (or other objects) onto ground workers	5 (5.7)
	Moving/rolling objects	Rolling objects	1 (1.1)
		Category total = 20 (23.0)	
Others		Backup alarms are too quiet Common sense	2 (2.3)

Table 5. Summary of suggestions (*total* = 92 responses, from 81 individuals who responded to this open-end question) offered for safety improvements. Note that *n* in the table is the number of responses obtained.

Category	Sub Category	Response Examples	Freq. (%)
More safety training & education		Proper training for all employees More safety meetings and classes	22 (23.9)
Improved situational awareness (SA)	Attention & comprehension	Pay more attention Slow down and think Stay alert	19 (20.7)
	Communication & Teamwork skills	Look after each other Communicate at all times Signal for warnings Better communication	12 (13.0)
	Category total = 31 (33.7)		
Machine-related enhancement	Safety & comfort	More comfortable/safe machine	5 (5.4)
	Housekeeping	Cleaner machine (e.g., no greasy surface in walk paths) Keep windows clean	2 (2.2)
	Knowledge	Know the mechanics of your machine Know how to operate a chainsaw	2 (2.2)
	Category total = 9 (9.8)		
Use of personal protective equipment (PPE)		Wear all PPE Wear hard hats all the time Use proper PPE More comfortable PPE	13 (14.1)
Improved visibility		More visible High visibility vest/shirt	11 (12.0)
Others		Teach common sense Put phone in truck Should be able to tell the rescue squad how to get to the job Fewer workers	6 (6.5)

4. Discussion

This cross-sectional study examined a sample of Virginia loggers to investigate the prevalence of workplace injuries, MSD symptoms, and symptoms related to diesel exhaust exposure, and to help understand their safety concerns and opinions with a particular focus on logging equipment.

4.1. Workplace Injuries and MSDs

Loggers included here were at a high risk of workplace injuries, with reported career and 12-month prevalences of 51% and 14%, respectively. The latter is, however, lower than earlier reports for forestry workers, including a 12-month injury prevalence of 34% for all forestry tasks/jobs (e.g., chainsaw operators, silviculturists) in New Zealand [30] and a 12-month prevalence (including work-related illnesses) of 30% for the farming, forestry, and fishing industry [31]. This discrepancy may be due to a focus on more general worker populations (i.e., not limited to loggers in predominantly mechanized logging operations) in these two earlier studies. We found that work-related injury prevalence (Table 1), but not MSD symptom prevalence (Table 2), was associated with years of experience. Nieuwenhuis et al. [9] found no significant association between years of experience and MSD prevalence for forestry workers in Ireland, and Lynch et al. [13] reported that age was not significantly associated with back pain but positively associated with neck pains for machine operations in the U.S. Southern region.

Nearly all respondents experienced MSD symptoms in at least one body region over the prior 12 months. A substantial fraction of respondents (18.5%) reported that they had changed their work practice as a result of MSD symptoms, and ~8% further indicated that they were considering changing jobs. Interestingly, knee MSD symptoms appeared to have a positive relationship with daily machine operation duration (Table 2). This may suggest operating a machine also contributes to knee MSDs, though a previous report noted that many of knee and foot injuries resulted from being struck by/against an object [6]. Additionally, the body region most commonly affected was the lower back (49.2%), followed by the knee (37.7%). These values are lower than earlier reports (12-month prevalence) of 74.3% (lower back pain) for machine operators in the US Southern region [13], and 84.6% (lower back) and 61.5% (knee) for Greek forestry workers [32]. Such a differential may have resulted from the use of a different questionnaire instrument for assessing symptoms in the former study (vs. the standardized Nordic questionnaires used here), and the fact that the study population in the latter mainly included workers who performed both manual and mechanized timber-cutting harvesting in a steep mountain forest terrain.

4.2. Symptoms Related to Diesel Exhaust Exposure

The findings of the questionnaire coupled with those of the fine particulate sampling suggest the need for additional research on the possible association of diesel exhaust exposure and adverse health effects among these workers. More than half of the study respondents reported having work-related eye/mouth irritation and/or unpleasant smells when operating a machine or when working near a machine. These irritant effects are commonly observed following exposure to diesel exhaust [26,27]. Though not statistically significant, deckhand machine operators exhibited a higher prevalence of such symptoms [PR = 1.31 (0.60, 1.84)]. Serving as a deckhand means working on the ground, which likely accounts for the observation that deckhand machine operators experienced a higher prevalence/level of diesel exhaust exposure symptoms. Working on the ground may increase the likelihood of exposure to diesel exhaust and other air pollutants.

Earlier studies demonstrated adverse health effects of diesel exhaust, including allergic reaction, asthma [19], chronic obstructive pulmonary disease [33], and lung cancer [20]. Considering this evidence, and our findings, there is a need for larger, more systematic investigations to quantify and characterize exposure to fine particulate ($PM_{2.5}$) fraction of diesel exhaust as an indicator of total exhaust exposure during logging equipment use and for different machine/job types (e.g., feller-buncher, deckhand), and with respect to work and maintenance practices (e.g., closing cabin windows, replacing air filters). For those working in a logging machine, vehicle cabin air filters can effectively reduce diesel exhaust particles and the symptoms induced by diesel exhaust, albeit depending on filter types [26]. However, additional attention is needed to address control of diesel exhaust exposure for ground workers.

4.3. Safety Concerns and Opinions of Loggers

Based on respondents' perceptions of the more common safety risks (Tables 4 and 5), as well as quantitative results regarding work-related injury prevalence, a priority of interventions may be given to address: (1) struck-by/against hazards; (2) maintaining situational awareness (SA) during logging operations (especially since such operation often require long work shifts and are quite repetitive); and (3) visibility hazards. Respondents' concerns about struck-by/against hazards are consistent with the fact that these hazards are a major source of nonfatal and fatal work-related injuries among loggers [7,34–36]. Even on these mechanized operations, participants' responses suggest chainsaw operation and felling/delimbing/topping remained frequent concerns on logging sites. Machine related struck-by events were of concern as well. Given that failures to detect and recognize hazards can be viewed as a SA problem [37], concerns regarding SA and visibility suggest a high demand on both machine operators and ground workers to process the information they perceive, and thereby to identify and prioritize hazards while maintaining good SA of the machine being operated and their surroundings (e.g., ground worker locations, physical work environments, movement of nearby machines). In addition to these three priority areas, our results support the need for efforts to understand and address loggers' musculoskeletal disorders and symptoms related to diesel exhaust exposure.

Improved SA, more training and education, and use of PPE were frequently suggested as ways to address existing logger safety risks (Table 5). With training and experience, workers can develop an efficient strategy to direct and distribute their attention to detect and recognize important stimuli [38]. However, a minimal level of SA can be determined by the attentional capacity of an individual [38], and such capacity is affected by many factors such as, for example, workload, time pressure, fatigue, and sleep deprivation [39,40]. In the case of safety training, though training can positively affect worker behaviors, large positive impacts on the incidence of adverse work-related outcomes are not generally expected from training alone [41,42]. The West Virginia Loggers' Safety Initiative (LSI) program evaluated the effectiveness of training over a 4-year period, and found no strong evidence of injury reduction [43], though loggers had increased safety knowledge [44]. Similar outcomes were reported for a multi-year video-based safety training intervention program for West Virginia loggers [45]. Interestingly, Conway et al. [22] reported that one important risk factor is human error, due to complacency, inattention, and/or underestimation of risks, and suggested that such error can be partially related to the repetitiveness of logging jobs and individual worker's motivation to work safely.

Overall, it appears that there is a need for interventions that are based not just on the behavioral changes and cognitive performance of an individual worker (e.g., training, experience). Well-designed alarms/warnings and displays can aid in efficiently allocating attention, potentially facilitating good SA [38,40]. The safety of logging machines may be intrinsically enhanced by incorporating proactive proximity warning technologies [46], and evaluating the cab design and human-machine interface of a machine for visibility and better SA during the development phases [40,47]. Further, and given the identified machine-related hazards and provided suggestions, logging machine design may benefit from user-centered and/or simulation-based design methods to enhance the safety and ergonomics, in order to achieve better operating postures [48,49], enhanced maintainability [50], and easier ingress and egress [51,52]. Similarly, proactive technologies to mitigate visibility hazards can be used among ground workers (e.g., a radio-frequency identification tag attached to a hard hat or vest as part of a proximity detection system [53]), in addition to the respondents' suggestions such as better communications (e.g., hand signal, two-way radio) and high-visibility vest use. In addition to efforts to address hazards unique to mechanized operations, approaches to prevent injuries related to chainsaw operation and felling/delimbing/topping should remain.

Limitations of the current study should be acknowledged. First, the study had a small sample size and used a convenience sampling from workers operating a variety of equipment on logging operations in different physiographic regions of Virginia. It is unknown regarding the extent to which our results will generalize to other states and countries. Our study, though, was not about specific logging practices and methods used in the test regions, but instead about logger safety and health associated with common logging operations. Second, injury and health outcome data were self-reported, which may be influenced by recall bias. To what extent such bias may have occurred is difficult to ascertain. In addition, and specific to the questions about diesel exhaust exposure, responses may have been influenced by smoking status or other preexisting health conditions.

5. Conclusions

In this cross-sectional study, using a sample of Virginia loggers, work-related injuries and MSDs were found to be quite prevalent among loggers, though many of these workers perceived their jobs as being only moderately dangerous or less so. Based on the current results, three priority areas were identified for interventions: struck-by/against hazards, situational awareness during logging operations, and visibility hazards. Though on-site training/educational materials (e.g., hand signals, high visibility vest and other PPE use) may be useful to address these hazards, we suggest a current need for proactive approaches, such as incorporating proactive proximity technologies in a logging machine or personal equipment, and enhancing logging machine design for better safety, ergonomics, and SA. We believe that proactive approaches are essential to achieve a broader and more substantial positive impact on safety and health among both machine operators and ground workers. In addition, our results are supportive of future efforts to improve awareness of the risk of musculoskeletal and diesel exhaust exposures and understand such exposures among logging workers.

Supplementary Materials: The following are available online at www.mdpi.com/1999-4907/8/11/440/s1.

Acknowledgments: The authors thank Will Saulnier and Andrew Vinson for their assistance in data collection. This work was supported through the Johns Hopkins NIOSH Education and Research Center for Occupational Safety and Health, Pilot Project Research Training Award (T42 OH0008428). The contents of this paper are solely the responsibility of the authors and do not necessarily represent the official views of the National Institute for Occupational Safety and Health.

Author Contributions: Study design and questionnaire development—All; Data collection—Michael Chad Bolding, Deborah E. Dickerson, Sunwook Kim, and Scott M. Barrett; Data analysis and interpretation—Sunwook Kim, Maury A. Nussbaum, Deborah E. Dickerson, and Ashley L. Schoenfisch; Drafting and revising the manuscript—All.

Conflicts of Interest: The authors declare no conflict of interest.

References

1. American Forest & Paper Association Economic Impact. Available online: http://www.afandpa.org/our-industry/economic-impact (accessed on 4 April 2016).

2. National Institute for Occupational Safety and Health Logging Safety. Available online: http://www.cdc.gov/niosh/topics/logging/ (accessed on 18 April 2016).

3. Albizu-Urionabarrenetxea, P.; Tolosana-Esteban, E.; Roman-Jordan, E. Safety and health in forest harvesting operations. Diagnosis and preventive actions. A review. *For. Syst.* **2013**, *22*, 392–399. [CrossRef]

4. Roberts, T.; Shaffer, R.M.; Bush, R.J. Injuries on mechanized logging operations in the southeastern United States in 2001. *For. Prod. J.* **2005**, *55*, 86–89.

5. Shaffer, B.; Roberts, T. *Logging Injuries Continue Downward Trend*; Forest Operations Review: Rockville, MD, USA, 2003; pp. 21–22.

6. Lefort, A.J., Jr.; de Hoop, C.F.; Pine, J.C. Characteristics of injuries in the logging industry of Louisiana, USA: 1986 to 1998. *Int. J. For. Eng.* **2003**, *14*, 75–89.

7. Bureau of Labor Statistics Census of Fatal Occupational Injuries (CFOI)—Current and Revised Data. Available online: https://www.bls.gov/iif/oshcfoi1.htm#2015 (accessed on 8 September 2017).

8. Bureau of Labor Statistics. *TABLE R5. Incidence Rates for Nonfatal Occupational Injuries and Illnesses Involving Days Away from Work per 10,000 Full-Time Workers by Industry and Selected Natures of Injury or Illness, Private Industry, 2015.* Available online: https://www.bls.gov/iif/oshwc/osh/case/ostb4757.pdf (accessed on 4 April 2016).

9. Nieuwenhuis, M.; Lyons, M. Health and safety issues and perceptions of forest harvesting contractors in Ireland. *Int. J. For. Eng.* **2002**, *13*, 69–76.

10. Shaffer, R.M.; Milburn, J.S. Injuries on feller-buncher/grapple skidder logging operations in the Southeastern United States. *For. Prod. J.* **1999**, *49*, 24–26.

11. Myers, J.R.; Fosbroke, D.E. Logging fatalities in the United States by region, cause of death, and other factors—1980 through 1988. *J. Saf. Res.* **1994**, *25*, 97–105. [CrossRef]

12. Axelsson, S.Å.; Pontén, B. New ergonomic problems in mechanized logging operations. *Int. J. Ind. Ergon.* **1990**, *5*, 267–273. [CrossRef]

13. Lynch, S.M.; Smidt, M.F.; Merrill, P.D.; Sesek, R.F. Incidence of MSDs and Neck and Back Pain among Logging Machine Operators in the Southern U.S. *J. Agric. Saf. Health* **2014**, *20*, 211–218. [PubMed]

14. Hagen, K.B.; Magnus, P.; Vetlesen, K. Neck/shoulder and low-back disorders in the forestry industry: Relationship to work tasks and perceived psychosocial job stress. *Ergonomics* **1998**, *41*, 1510–1518. [CrossRef] [PubMed]

15. U.S. Department of Health and Human Service. *National Institute for Occupational Safety and Health Current Intelligence Bulletin 50*; U.S. Department of Health and Human Service: Cincinnati, OH, USA, 1988.

16. Crump, K.; Van Landingham, C. Evaluation of an exposure assessment used in epidemiological studies of diesel exhaust and lung cancer in underground mines. *Crit. Rev. Toxicol.* **2012**, *42*, 599–612. [CrossRef] [PubMed]

17. Straif, K.; Benbrahim-Tallaa, L.; Baan, R.; Grosse, Y.; Secretan, B.; Ghissassi, E.F.; Bouvard, V.; Guha, N.; Freeman, C.; Galichet, L.; et al. WHO International Agency for Research on Cancer Monograph Working Group A review of human carcinogens—Part C: Metals, arsenic, dusts, and fibres. *Lancet Oncol.* **2009**, *10*, 453–454. [CrossRef]

18. Boffetta, P.; Harris, R.E.; Wynder, E.L. Case-control study on occupational exposure to diesel exhaust and lung cancer risk. *Am. J. Ind. Med.* **1990**, *17*, 577–591. [PubMed]

19. Attfield, M.D.; Schleiff, P.L.; Lubin, J.H.; Blair, A.; Stewart, P.A.; Vermeulen, R.; Coble, J.B.; Silverman, D.T. The Diesel Exhaust in Miners study: A cohort mortality study with emphasis on lung cancer. *JNCI* **2012**, *104*, 869–883. [CrossRef] [PubMed]

20. Bhatia, R.; Lopipero, P.; Smith, A.H. Diesel Exhaust Exposure and Lung Cancer. *Epidemiology* **1998**, *9*, 84–91. [CrossRef] [PubMed]

21. Pandya, R.J.; Solomon, G.; Kinner, A.; Balmes, J.R. Diesel exhaust and asthma: Hypotheses and molecular mechanisms of action. *Environ. Health Perspect.* **2002**, *110*, 103–112. [CrossRef] [PubMed]

22. Conway, S.H.; Pompeii, L.A.; Casanova, V.; Douphrate, D.I. A qualitative assessment of safe work practices in logging in the southern United States. *Am. J. Ind. Med.* **2016**, *60*, 58–68. [CrossRef] [PubMed]

23. Bordas, R.M.; Davis, G.A.; Hopkins, B.L.; Thomas, R.E.; Rummer, R.B. Documentation of hazards and safety perceptions for mechanized logging operations in East Central Alabama. *J. Agric. Saf. Health* **2001**, *7*, 113–123. [CrossRef] [PubMed]

24. Lipscomb, H.J.; Schoenfisch, A.L. Reflections on Occupational Injury Control. *Saf. Sci. Monit.* **2014**, *18*, 1–8.

25. Bolding, M.C.; Barrett, S.M.; Munsell, J.F.; Groover, M.C. Characteristics of Virginia's logging businesses in a changing timber market. *For. Prod. J.* **2010**, *60*, 86–93. [CrossRef]

26. Rudell, B.; Wass, U.; Hörstedt, P.; Levin, J.O.; Lindahl, R.; Rannug, U.; Sunesson, A.L.; Ostberg, Y.; Sandström, T. Efficiency of automotive cabin air filters to reduce acute health effects of diesel exhaust in human subjects. *Occup. Environ. Med.* **1999**, *56*, 222–231. [CrossRef] [PubMed]

27. Laumbach, R.J.; Kipen, H.M.; Kelly-McNeil, K.; Zhang, J.; Zhang, L.; Lioy, P.J.; Ohman-Strickland, P.; Gong, J.; Kusnecov, A.; Fiedler, N. Sickness Response Symptoms among Healthy Volunteers after Controlled Exposures to Diesel Exhaust and Psychological Stress. *Environ. Health Perspect.* **2011**, *119*, 945–950. [CrossRef] [PubMed]

28. Kuorinka, I.; Jonsson, B.; Kilbom, A.; Vinterberg, H.; Biering-Sørensen, F.; Andersson, G.; Jørgensen, K. Standardised Nordic questionnaires for the analysis of musculoskeletal symptoms. *Appl. Ergon.* **1987**, *18*, 233–237. [CrossRef]

29. Core Team R. *A Language and Environment for Statistical Computing*; R Foundation for Statistical Computing: Vienna, Austria, 2015.

30. Lilley, R.; Feyer, A.-M.; Kirk, P.; Gander, P. A survey of forest workers in New Zealand. Do hours of work, rest, and recovery play a role in accidents and injury? *J. Saf. Res.* **2002**, *33*, 53–71. [CrossRef]

31. Fan, Z.J.; Bonauto, D.K.; Foley, M.P.; Silverstein, B.A. Underreporting of Work-Related Injury or Illness to Workers' Compensation: Individual and Industry Factors. *J. Occup. Environ. Med.* **2006**, *48*, 914–922. [CrossRef] [PubMed]

32. Gallis, C. Work-related prevalence of musculoskeletal symptoms among Greek forest workers. *J. Ind. Ergon.* **2006**, *36*, 731–736. [CrossRef]

33. Sehra, G.; Barnes, P.; Rogers, D.; Donnelly, L. Effect of diesel exhaust particles (DEP) on monocyte-derived macrophage (MDM) mediator release and phagocytosis in chronic obstructive pulmonary disease (COPD). *Eur. Respir. J.* **2013**, *42*, 618.

34. Bureau of Labor Statistics Injuries, Illnesses, and Fatalities. Available online: http://www.bls.gov/iif/osh_nwrl.htm (accessed on 23 August 2017).

35. Roberts, E.T.; Bolding, M.C. *A Ten Year Analysis of Timber Harvesting Injuries in the Southeastern United States*; Master of Forestry Paper; Virginia Tech: Blacksburg, VA, USA, 2012; pp. 1–28.

36. Lagerstrom, E.; Magzamen, S.; Rosecrance, J. A mixed-methods analysis of logging injuries in Montana and Idaho. *Am. J. Ind. Med.* **2017**, *25*, 1–11. [CrossRef] [PubMed]

37. Kaber, D.B.; Endsley, M.R. The effects of level of automation and adaptive automation on human performance, situation awareness and workload in a dynamic control task. *Theor. Issues Ergon. Sci.* **2007**, *5*, 113–153. [CrossRef]

38. Durso, F.T.; Rawson, K.A.; Girotto, S. Comprehension and Situation Awareness. In *Handbook of Applied Cognition*; Durso, F.T., Nickerson, R.S., Dumais, S.T., Lewandowsky, S., Perfect, T.J., Eds.; John Wiley & Sons Inc.: New York, NY, USA, 2007; p. 163.

39. Wickens, C.D.; Hollands, J.G.; Banbury, S.; Parasuraman, R. *Engineering Psychology & Human Performance*; Psychology Press: London, UK, 2015.

40. Endsley, M.R.; Bolte, B.; Jones, D.G. *Designing for Situation Awareness*; CRC Press: Boca Raton, FL, USA, 2003.

41. Robson, L.S.; Stephenson, C.M.; Schulte, P.A.; Amick, B.C.I.; Irvin, E.L.; Eggerth, D.E.; Chan, S.; Bielecky, A.R.; Wang, A.M.; Heidotting, T.L.; et al. A systematic review of the effectiveness of occupational health and safety training. *Scand. J. Work Environ. Health* **2011**, *38*, 193–208. [CrossRef] [PubMed]

42. Nelson, A.; Baptiste, A.S. Evidence-Based Practices for Safe Patient Handling and Movement. *Clin. Rev. Bone Miner. Metab.* **2006**, *4*, 55–69. [CrossRef]

43. Bell, J.L.; Grushecky, S.T. Evaluating the effectiveness of a logger safety training program. *J. Saf. Res.* **2006**, *37*, 53–61. [CrossRef] [PubMed]

44. Helmkamp, J.C.; Bell, J.L.; Lundstrom, W.J.; Ramprasad, J.; Haque, A. Assessing safety awareness and knowledge and behavioral change among West Virginia loggers. *Inj. Prev.* **2004**, *10*, 233–238. [CrossRef] [PubMed]

45. Mujuru, P.; Helmkamp, J.C.; Mutambudzi, M.; Hu, W.; Bell, J.L. Evaluating the impact of an intervention to reduce injuries among loggers in West Virginia, 1999–2007. *J. Agric. Saf. Health* **2009**, *15*, 75–88. [CrossRef] [PubMed]

46. Teizer, J.; Allread, B.S.; Fullerton, C.E.; Hinze, J. Autonomous pro-active real-time construction worker and equipment operator proximity safety alert system. *Autom. Constr.* **2010**, *19*, 630–640. [CrossRef]

47. Ahn, S.Y.; Jeon, Y.W.; Yun, J.M.; Kang, P.; Ko, J.H.; Park, P. Development of a forward visibility assessment tool based on visibility angle. *Int. J. Autom. Technol.* **2015**, *16*, 1051–1055. [CrossRef]

48. Reed, M.P.; Manary, M.A.; Flannagan, C.A.; Schneider, L.W. Effects of vehicle interior geometry and anthropometric variables on automobile driving posture. *Hum. Factors* **2000**, *42*, 541–552. [CrossRef] [PubMed]

49. Vogt, C.; Mergl, C.; Bubb, H. Interior layout design of passenger vehicles with RAMSIS. *Hum. Factors Ergon. Manuf. Serv. Ind.* **2005**, *15*, 197–212. [CrossRef]

50. Di Gironimo, G.; Monacelli, G.; Patalano, S. A Design Methodology for Maintainability of Automotive Components in Virtual Environment. In Proceedings of the 8th International Design Conference (DESIGN 2004), Dubrovnik, Croatia, 16–19 May 2004; pp. 1–12.

51. Reed, M.P.; Ebert, S.M.; Hoffman, S.G. *Modeling Foot Trajectories for Heavy Truck Ingress Simulation*; CRC Press: Miami, FL, USA, 2010.

52. Choi, N.-C.; Lee, S. Discomfort Evaluation of Truck Ingress/Egress Motions Based on Biomechanical Analysis. *Sensors* **2015**, *15*, 13568–13590. [CrossRef] [PubMed]

53. Teizer, J. Wearable, wireless identification sensing platform: Self-Monitoring Alert and Reporting Technology for Hazard Avoidance and Training (SmartHat). *J. Inf. Technol. Constr.* **2015**, *20*, 295–312.

Article

Incidence of Trailer Frame Structure on Driver's Safety during Log Transportation

Marco Manzone * and Angela Calvo

Department of Agricultural, Forest and Food Sciences, University of Torino, Largo Paolo Braccini, 2, 10095 Grugliasco, Italy; angela.calvo@unito.it
* Correspondence: marco.manzone@unito.it; Tel.: +39-011-670-8638

Received: 29 September 2017; Accepted: 11 November 2017; Published: 18 November 2017

Abstract: The frame structure of the trailer may influence both the traction and the tractor-trailer stability, especially along sloped paths. The aim of this research was to analyze a trailer overturning and the strains on the connected tractors (wheeled, or crawled) during log transportation (loose or tied) along a hillside. Two two-axle trailers were used: tandem and turntable steering. Three types of measurements were carried out during the field tests: (i) the detachment from the ground of the rear upstream wheels (or crawler); (ii) the transversal and longitudinal strains occurring when the trailer overturned (and released the hooking system of the tractor); (iii) the lateral deviation of the rear wheels (or crawler) of the tractor. The study highlighted that the two-axle trailer with turntable steering combined with the crawl tractor gave better results in terms of safety during trailer overturning. In addition, independent of the type of trailer, a tied load was found to be more dangerous than a load restrained only by steel struts, because when overturning, the load forms a single unit with the trailer mass which increases the strains.

Keywords: frame structure; two-axle trailer; crawl tractor; wheel tractor; safety

---

1. Introduction

In the NACE (Nomenclature of the Economic Activities) statistics in Europe, the number of fatal accidents in agriculture and forestry lies in third place (14.8%), after the activities of construction (21.5%) followed by transportation and storage (16.7%) [1]. Nevertheless, many accidents are not officially recorded [2].

Forestry and logging are among the most hazardous activities in agriculture: compared to other agricultural activities, the percentage of fatal accidents at work in this sector is never lower than 16% (Figure 1).

Tractors are among the highest causes of fatal accidents [1] and the most severe accidents are caused while the machine is operating in the field [3,4].

The main cause of deaths by tractors is machine overturning [5], especially when it is operating along sloped terrains for logging operations [6]. Machine overturning is due to the displacement of the center of gravity of the vehicle outside the stability baseline of the machine when it is moving forward. There are many reasons that affect the dynamic stability of the tractor [7,8]. They may depend on driver behavior (driving style, forward speed), on the environment (slope, rough terrain, stones, stems, potholes) or on the presence of additional weights connected to the tractor (trailers, towed implements, ballasts).

In a study conducted in 1991 it was demonstrated that the above factors were the cause of more than 50% of the tractor rollover accidents [9].

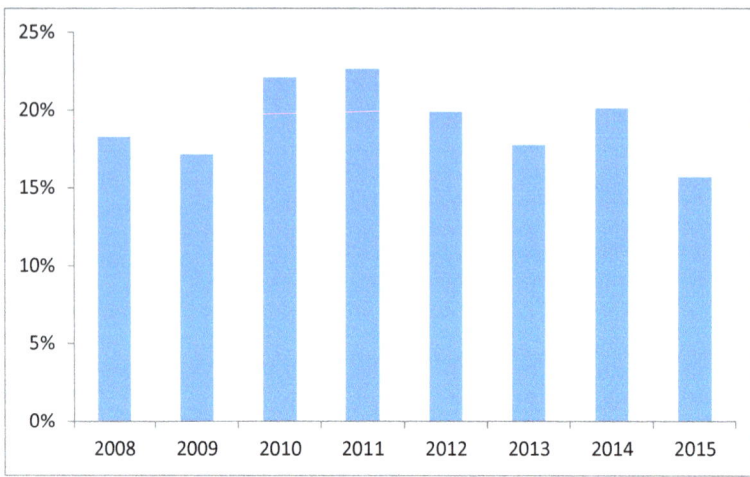

Figure 1. Fatal accidents during forestry and logging activities recorded in Europe, compared to other agricultural tasks (Source: Authors' elaboration of Eurostat data [1]).

Concerning trailers, the higher the center of gravity and the loading capacity, the higher is the risk of the convoy (tractor-trailer) overturning [10]. Studies on the stability of the tractor-trailer system have been carried out by several researchers [11–14]. Pereira et al. in 2011 [15] considered the critical conditions during the transportation of logs, while Lindroos and Wasterlund [16] analyzed the risks caused by the gross weight of a heavy trailer connected to a tractor with a low traction capacity. In the Italian alpine North-West regions (where sloped and rough terrains are spread) small old tractors and equipment are used [17], particularly in order to guarantee environmental resiliency and wood regeneration.

For many years in forestry and logging activities, trailers equipped with motor axles have been used, in order to improve the traction of the whole convoy. Moreover, the use of a crawler tractor, instead of a wheeled tractor, may well improve the convoy stability.

Also the frame structure of the trailer influences the traction and the stability, especially along sloped paths.

The aim of this research was to analyze trailer overturning and the strains on the connected tractor (one wheeled, one crawled) during log transportation (loose and tied) along a hillside, using two two-axles trailers with different characteristics: the first tandem, the second with turntable steering. The detachment from the ground of the rear upstream wheels and the side-slipping of the rear wheels (or crawler) of the two tractors as well as the transversal and longitudinal strains occurring when the trailer overturned were analyzed.

2. Material and Methods

2.1. Tractors

The technical characteristics of the two tractors used in the field tests (a 2WD and a crawler each with a mass of about 4200 kg) are displayed in Table 1. Both the tractors satisfied the minimum safety requirements required by the European Directive 2006/42 and they were equipped with roll over protection systems and seat belts. These tractors are very common for carrying out logging operations in the South of Europe. Moreover, two tractors with different characteristics were considered in evaluating the different behavior of the trailers during overturning.

Table 1. Technical characteristics of the tractors.

	NH T5.115 (A)	NH TK 4040M (B)
Power (kW)	84	65
Mass (kg)	4250 *	4250 *
Propulsion system	wheels	tracks
Driving wheels	4	-
Rear wheel type	540/65-34	-
Front wheel type	440/65-24	-
Overall width (m)	2.20	1.65

Note: (*) The mass of the driver (90 kg) is not included in these values.

2.2. Trailers

A two-axle trailer in tandem and a two-axle trailer with turntable steering were used to perform the tests (Table 2).

Table 2. Technical characteristics of the used trailers.

	Two-Axle in Tandem (1)	Two-Axle with Turntable Steering (2)
Mass (kg)	3150	3000
Flatbed width (m)	2.20	2.20
Flatbed height (m)	1.20	1.20
Flatbed length (m)	4.50	4.50
Wheel dimension	385/65 R195	385/65 R195
Hooking height (m)	0.55	1.10

Note: (*) Calculated as the distance from the towing eye.

The height of the platform was 1200 mm, the floor was 4.5 m long and 2.2 m width. The tracks and the wheel sizes were equal for both the trailers.

During the tests the hooking height of the two-axle trailer in tandem was 0.55 m, while it was at 1.10 m for the two-axle trailer with turntable steering (Table 2).

2.3. The Trailer's Load

Cylinder coded logs of 1.5 m length (with a diameter ranging between 150 and 250 mm) were piled on the trailers, transversely positioned in the forward direction. All the tests were performed with the same coded logs, positioned in the same former places on the trailer, to guarantee the same load distribution and to avoid different balances of the convoys. In addition, the tests were carried out both with the logs freely movable on the load floor (held by supports of steel fixed to the front and rear extremities of the floor) and with the logs tied (using two ropes, diagonally placed to the longitudinal axis of the trailer).

The gross mass (trailer and logs) did never exceed fifty percent of the tractor mass (about 6300 kg), to ensure a safe convoy movement in very sloped conditions.

2.4. The Crossed Path

A 21 m sloped country dirt road with a flat area at its base was chosen to perform the trailer rollovers. The flat area was used to right the trailers after overturning and to collect the spread wood on the ground after the rollover. The crossed trail was transversal to the hillside and the path was 18% transversally and 32% longitudinally sloped.

An artificial wooden wedge (320 mm length, 110 mm height, and 220 mm width) was used as an obstacle to cause the trailer rollover. The obstacle was positioned in a place where the tractor was already in the flat area, to guarantee the safety of the tractor driver when the trailer overturned.

The trailer rollover, in fact, was determined by the impact of the front wheels of the trailer when it alighted along the path.

In Figure 2 there is a sketch of an additional safety measure to avoid possible unpredictable accidents to the driver during the trailer rollover. For this purpose, an operator held the extremity of a rope twisted to the base of a stump and anchored at the other extremity to the top of the tractor's rollover protection structure (ROP) with a node.

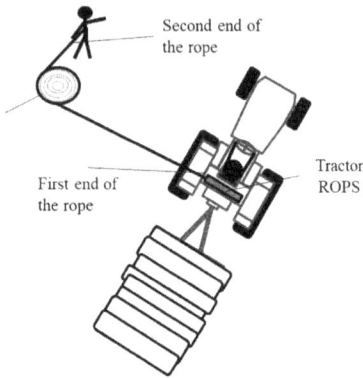

Figure 2. The used safety anchorage.

The rope was free to slide around the stump while the tractor moved forward. In the case of the risk of the tractor overturning, the operator could stop the rope sliding, thereby preventing the possible overturning of the tractor.

The average forward speed was about 3.2 km h^{-1}. For each type of tractor, trailer, and load configuration (untied and tied logs), three overturning repetitions were carried out, with a total of 24 field tests.

2.5. Measures

Three types of measurement were carried out during the field tests:

-   the detachment from the ground of the rear upstream wheels (or crawler)
-   the transversal and longitudinal strains occurring when the trailer overturned (and released the hooking system of the tractor)
-   the side-slipping (lateral deviation) of the rear wheels (or crawler) of the tractor

2.6. Instruments

A measurement tool, built ad hoc, was used to measure the height of the detachment of the wheels (or crawler) from the ground. The tool was a graduated plastic strip (scale: 1 mm), rolled onto a reel without a return spring. A steel support (mass: 1 kg) was used to connect the strip and was free to run on the ground by using a small rope linked to another support fixed to the tractor frame. The strip was maintained fully stretched before starting the test until the overturning of the trailer. When the wheels (or crawler) started to detach from the ground, the strip began to extend and the length difference before and after the overturning was a measure of the detachment height (Figure 3).

A mechanical tool was used to measure transversal and longitudinal strains. It was composed of a mechanical pendulum (length: 100 mm) linked to a goniometer (diameter: 120 mm) by a center hinge, where two metallic pointers were positioned and free to move (Figure 4). The zero point referred to the static position before starting the test, when the pointers were aligned to the pendulum. At the unrestrained extremity of the pendulum was fixed a mass of 20 g.

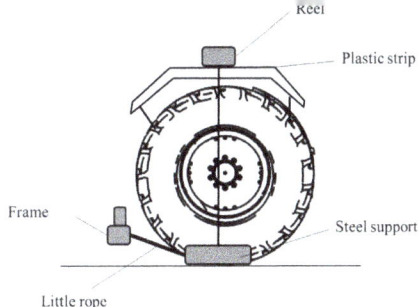

Figure 3. Scheme of the tool used to measure the detachment height of the wheel/crawler from the ground during the overturning of the trailer.

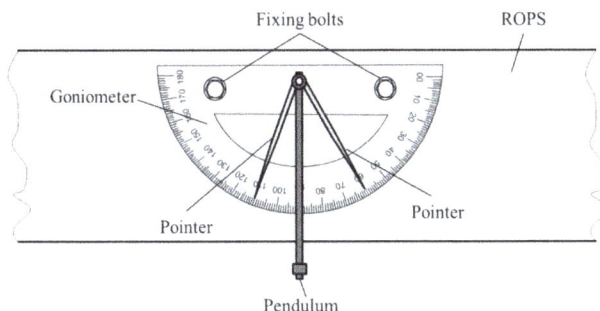

Figure 4. The tool used to measure the transversal and longitudinal strains.

Two mechanical tools were used, fixed by two bolts to the middle of the top of the ROP. The first pendulum was positioned orthogonally to the tractor's forward speed for measuring the transversal strains. To measure the longitudinal train the second pendulum was positioned parallel to the tractor's forward speed.

Positive measurements were clockwise. Transversal strains were positive at the left, negative at the right, while longitudinal measured strains were positive in the front, negative at the rear.

In the data elaboration only linear measurements were considered, because the goniometer measured the small displacements in linear units (mm).

A graduated steel ruler (precision: 1 mm) was used to measure the side-slipping (lateral deviation) of the wheels (or crawler) of the tractors.

2.7. Data Elaboration

The elaboration of the data was performed using both Microsoft Excel and IBM SPSS V. 24.0 Statistic package [18]. In particular, the GLM (General Linear Model) was adopted in order to highlight differences between the considered treatments with a significance coefficient of 0.05.

3. Results

3.1. Lateral Deviation and Ground Detachment Height of the Tractor Wheels/Crawler

In all the tests the lowest lateral deviation of the tractor wheels was produced by the two-axle trailer with turntable steering. The lateral deviations were limited to several centimeters and they were observed always on the rear axle of the tractor (Table 3).

Table 3. Lateral deviation and ground detachment of the rear wheels (crawler) of the tractor in the different test configurations.

Tractor	Trailer	Load	Lateral Deviation (cm)				Ground Detachment (cm)			
			Mean	SD	Min	Max	Mean	SD	Min	Max
Wheel	Tandem	loose	3.5	0.6	2.8	3.9	3.6	0.7	2.8	4.1
	Turntable steering		2.2	0.4	1.8	2.6	0.0	0.0	0.0	0.0
Crawl	Tandem		1.7	0.2	1.6	1.9	1.7	0.2	1.6	1.9
	Turntable steering		1.1	0.2	0.9	1.2	0.0	0.0	0.0	0.0
Wheel	Tandem	tied	6.4	0.6	5.7	6.8	4.8	0.3	4.6	5.1
	Turntable steering		4.0	0.3	3.8	4.3	0.0	0.0	0.0	0.0
Crawl	Tandem		3.8	0.2	3.6	3.9	3.5	0.7	2.9	4.2
	Turntable steering		2.7	0.2	2.5	2.9	0.0	0.0	0.0	0.0

Lateral deviations ranged between 1.1 and 6.4 cm: the trailer with turntable steering highlighted the lowest lateral deviations in the crawled tractor, whereas the two-axle in tandem trailer showed the highest lateral deviation values when it was coupled with the wheeled tractor. The lateral deviation data doubled in the presence of the tied load, independent of the type of trailers.

The trailer with turntable steering did not cause rear wheel (or crawlers) detachment of the tractor from the ground, independent of the load type (tied or loose). On the contrary, the two-axle in tandem trailer gave a value of 3.6 cm with the wheel tractor and 1.7 cm with the crawler tractor when they were travelling with the loose load. These values increased up to two times in the presence of the tied load in the crawler tractor (Table 3).

3.2. Longitudinal Strain

Also measuring longitudinal strains, the highest values were found using the two-axle in tandem trailer when the load was tied. In detail, the maximum total longitudinal strain (45 mm) was recorded when the two-axle in tandem trailer was coupled with the wheeled tractor and the tied load. The minimum total longitudinal strain (14 mm) was observed using the two-axle with turntable steering trailer coupled to the crawled tractor and with the loose load. The highest longitudinal strains were observed with the tied load, independent of the trailer type. The crawler tractor better absorbed the strains caused by the trailer overturning (Table 4).

Table 4. Descriptive statistics of the longitudinal strains measured at the right and at the left sides of the wheel (crawler) of the tractor.

Tractor	Trailer	Load	Long. Strain Right (mm)				Long. Strain Left (mm)				Total
			Mean	SD	Min	Max	Mean	SD	Min	Max	
Wheel	Tandem	loose	14.3	2.1	12.0	16.0	−8.0	1.0	−9.0	−7.0	25.0
	Turntable steering		6.0	2.6	3.0	8.0	−8.3	0.6	−9.0	−8.0	17.0
Crawl	Tandem		9.0	1.0	8.0	10.0	−8.3	0.6	−9.0	−8.0	19.0
	Turntable steering		7.0	1.0	6.0	8.0	−5.3	1.2	−6.0	−4.0	14.0
Wheel	Tandem	tied	19.0	1.0	18.0	20.0	−21.7	3.1	−25.0	−19.0	45.0
	Turntable steering		14.3	1.5	13.0	16.0	−15.0	3.0	−18.0	−12.0	34.0
Crawl	Tandem		16.3	1.5	15.0	18.0	−15.3	2.1	−17.0	−13.0	35.0
	Turntable steering		12.0	1.0	11.0	13.0	−11.0	1.0	−12.0	−10.0	25.0

3.3. Transversal Strain

In general, transversal strains were higher than longitudinal ones (Table 5) and the two-axle in tandem trailer caused the highest boosts on the tractor. The trailer with turntable steering generated transversal strains, always about 30% lower than the two-axle in tandem trailer. Also in this case, the highest values (between 28 and 48 mm) were observed with the tied load.

Table 5. Descriptive statistics of the transversal strains measured at the right and at the left sides of the wheel (crawler) of the tractor.

Tractor	Trailer	Load	Trans. Strain Right (mm)				Trans. Strain Left (mm)				Total
			Mean	SD	Min	Max	Mean	SD	Min	Max	
Wheel	Tandem	loose	14.7	1.5	13.0	16.0	−17.3	2.1	−19.0	−15.0	31.0
	Turntable steering		8.0	1.0	7.0	9.0	−6.0	1.0	−7.0	−5.0	14.0
Crawl	Tandem		15.0	1.0	14.0	16.0	−13.7	2.1	−16.0	−12.0	28.0
	Turntable steering		8.0	1.0	7.0	9.0	−8.0	1.0	−9.0	−7.0	16.0
Wheel	Tandem	tied	18.3	2.5	16.0	21.0	−28.0	1.0	−29.0	−27.0	48.0
	Turntable steering		16.3	1.5	15.0	18.0	−16.7	2.1	−19.0	−15.0	33.0
Crawl	Tandem		17.0	2.0	15.0	19.0	−20.0	2.0	−22.0	−18.0	37.0
	Turntable steering		14.0	1.7	12.0	15.0	−14.3	1.5	−16.0	−13.0	28.0

3.4. Influence of the Tractor and Trailer Structure on Measured Parameters

Data processing highlighted that the tractor and the trailer structure play an important role in the safety of the operator during trailer overturning. Concerning the lateral deviations, similarities were revealed by the General Linear Model (GLM) procedure with the same type of tractor and load system (Table 6).

Using the same statistical procedure, instead, all the parameters (tractor, trailer, and load type) produced different results for the measured ground rear wheel/crawler detachment (Table 7).

On the other hand, longitudinal strains were statistically similar with the tractor-trailer, trailer-load, and tractor-load combinations of parameters (Table 8). On the contrary, transversal strains were different considering all the combination of the parameters (Table 9).

Table 6. Statistical analysis (General Linear Model (GLM)) of the lateral deviation.

	SS	DF	SM	F Value	p Value
intercept	239.402	1	239.402	1859.430	<0.0001
tractor	17.340	1	17.340	134.680	<0.0001
trailer	10.935	1	10.935	84.932	<0.0001
load	26.042	1	26.042	202.265	<0.0001
tractor * trailer	1.307	1	1.307	10.149	0.006
tractor * load	0.427	1	0.427	3.314	0.087
trailer * load	0.882	1	0.882	6.848	0.019
tractor * trailer * load	0.167	1	0.167	1.294	0.272

Notes: SS = Sum of Squares; DF = Default Freedom; MS = Squared Mean.

Table 7. Statistical analysis (GLM) of the ground detachment.

	SS	DF	SM	F Value	p Value
intercept	69.700	1	69.700	550.266	<0.0001
tractor	3.920	1	3.920	30.951	<0.0001
trailer	69.700	1	69.700	550.266	<0.0001
load	3.450	1	3.450	27.240	<0.0001
tractor * trailer	3.920	1	3.920	30.951	<0.0001
tractor * load	0.120	1	0.120	0.951	0.344
trailer * load	3.450	1	3.450	27.240	<0.0001
tractor * trailer * load	0.120	1	0.120	0.951	0.344

Notes: SS = Sum of Squares; DF = Default Freedom; MS = Squared Mean.

Table 8. Statistical analysis (GLM) of the longitudinal strain.

	SS	DF	SM	F Value	p Value
intercept	13,680.375	1	13,680.375	1977.886	<0.0001
tractor	187.042	1	187.042	27.042	<0.0001
trailer	408.375	1	408.375	59.042	<0.0001
load	1276.042	1	1276.042	184.488	<0.0001
tractor * trailer	12.042	1	12.042	1.741	0.206
tractor * load	26.042	1	26.042	3.765	0.070
trailer * load	18.375	1	18.375	2.657	0.123
tractor * trailer * load	0.042	1	0.042	0.006	0.939

Notes: SS = Sum of Squares; DF = Default Freedom; MS = Squared Mean.

Table 9. Statistical analysis (GLM) of the transversal strain.

	SS	DF	SM	F Value	p Value
intercept	20,768.167	1	20,768.167	3134.818	<0.0001
tractor	88.167	1	88.167	13.308	0.002
trailer	1040.167	1	1040.167	157.006	<0.0001
load	1093.500	1	1093.500	165.057	<0.0001
tractor * trailer	37.500	1	37.500	5.660	0.030
tractor * load	60.167	1	60.167	9.082	0.008
trailer * load	28.167	1	28.167	4.252	0.036
tractor * trailer * load	0.167	1	0.167	0.025	0.876

Notes: SS = Sum of Squares; DF = Default Freedom; MS = Squared Mean.

4. Discussion

The study showed that the trailer with a fixed drawbar (two-axle in tandem) caused higher strains (longitudinal and transversal) to the tractor during overturning of the trailer, compared to the trailer equipped with an "articulate" drawbar (presence of a turntable steering). In addition, adopting a trailer with turntable steering also permits the reduction of the risk of the tractor overturning, because in the event of the trailer overturning this type of trailer does not cause wheel or crawl detachment from the ground.

Nevertheless, these results are in contrast to what usually occurs in forestry yards, where trailers with fixed drawbars are preferred to trailers with an articulate drawbar [13,17]. In fact, the use of the first type of trailer guarantees a higher traction force because part of the load of the trailer weighs on the tractor; moreover, this type of trailer shows greater simplicity in field maneuvers due to the lower articulation points of the convoy structure.

The best performance of the trailer with turntable steering could be attributable to the higher number of junction joints (four swivel joints) in comparison to the trailers with fixed drawbar (two swivel joints) (Figure 5). In fact, the higher number of joint points guarantees greater quote compensation of the critical points during the trailer overturning (support points, coupling point, etc.)

(Figure 6a). In contrast, a drawbar fixed to the trailer frame can cause higher instability of the tractor due to the overturning force on the coupling point of the tractor during trailer overturning (Figure 6b).

Articulate drawbar   **Fixed drawbar**

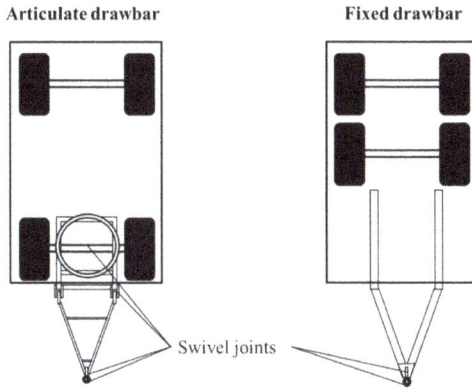

Figure 5. Swivel joints in the different drawbar types.

Figure 6. Scheme of the trailer overturning in flat soil in different configurations (two-axle trailer with turntable steering (**a**) and two-axle tandem trailer (**b**)).

Furthermore, the study highlighted that the loose load guaranteed better operator safety during trailer overturning; conversely, the tied load is dangerous because during trailer overturning its mass is added to the trailer mass and causes higher strains on the tractor.

Concerning the tractor propulsion system, the crawls guaranteed a better absorption of the strains caused by the trailer overturning: in fact, the crawler tractor, having a higher contact surface with the ground, had a better performance than the wheel tractor.

On the basis of these considerations, it is desirable that future studies on log transportation should also be focused on trailers with turntable steering. With this regard, tests were carried out on a double steering trailer prototype composed of two single-axle trailers [19]. This solution is very suitable for the transport of logs of high length (more than 8–10 m long) on narrow and steep forest roads to reduce the curve radius and to improve the convoy maneuverability. Nevertheless, considering the results obtained in this experiment, this prototype may produce the same strains for the trailer with a fixed drawbar during trailer overturning, because the first module of the prototype has a fixed single-axle drawbar.

5. Conclusions

In the Italian alpine territory two-axle tandem trailers are usually used for their technical characteristics (ease of maneuverability and high traction force due to the effect of part of their mass on the tractor), this study highlighted that the two-axle trailer with turntable steering is safer for the tractor driver during passages along transversal sloped terrains, where the risk of trailer overturning is high. This result is enhanced if a crawler tractor is used instead a wheeled one.

In addition, considering the effects due to the overturning of trailers tested in this study, it was found that, independent of the trailer type, a tied load is more dangerous than a load where the logs are freely movable on the load floor. In the first case, in fact, the logs form a single unit with the trailer mass which increases the transversal and longitudinal strains during the trailer overturning. The authors suggest that this study might be considered a valuable contribution if adopted for improving operators' safety in forestry yards.

Acknowledgments: We want to thank all the Italian forestry farmers that were available to conduct the field tests.

Author Contributions: The authors equally contributed to all experimental design, elaboration, and paper editing.

Conflicts of Interest: The authors declare no conflicts of interest.

References

1. Eurostat. *Agriculture, Forestry and Fishery Statistics*; Statistical Books; Publications Office of the European Union: Luxembourg, 2015.
2. Pessina, D.; Facchinetti, D. A survey on fatal accidents for overturning of agricultural tractors in Italy. *Chem. Eng. Trans.* **2017**, *58*, 79–84.
3. Cole, H.P.; Myers, M.L.; Westneat, S.C. Frequency and severity of injuries to operators during overturns of farm tractors. *J. Agric. Saf. Health* **2006**, *12*, 127–138. [CrossRef] [PubMed]
4. Myers, M.L.; Cole, H.P.; Westneat, S.C. Injury severity related to overturn characteristics of tractors. *J. Saf. Res.* **2009**, *40*, 165–170. [CrossRef] [PubMed]
5. Bernik, R.; Jerončič, R. The Research of the Number of Accidents with the Agriculture and Forestry Tractors in the Europe and the Main Reasons for those Accidents. *Stroj. Vestn. J. Mech. Eng.* **2008**, *54*, 557–564.
6. Blombäck, P.; Poschen, P.; Lövgren, M. *Employment Trends and Prospects in the European Forest Sector*; Geneva Timber and Forest Discussion Papers; United Nations: Geneva, Switzerland, 2003.
7. Spencer, H.B.; Gilfillan, G. An approach to the assessment of tractor stability on rough sloping ground. *J. Agric. Eng. Res.* **1976**, *21*, 169–176. [CrossRef]
8. Yisa, M.G.; Terao, H.; Noguchi, N.; Kubota, M. Stability criteria for tractor-implement operation on slopes. *J. Terramech.* **1998**, *35*, 1–19. [CrossRef]
9. Hunter, A.G.M. Stability of agricultural machinery on slopes. In *Progress in Agricultural Physics and Engineering*; Matthews, J., Ed.; CAB International: Wallingford, UK, 1991.
10. Chou, T.; Chu, T.W. An improvement in rollover detection of articulated vehicles using the grey system theory. *Veh. Syst. Dyn.* **2014**, *52*, 679–703. [CrossRef]
11. Melemez, K.; Di Gironimo, G.; Esposito, G.; Lanzotti, A. Concept design in virtual reality of a forestry trailer using a QFD-TRIZ based approach. *Turk. J. Agric. For.* **2013**, *37*, 789–801. [CrossRef]
12. Manzone, M.; Balsari, P. Electronic control of the motor axles of the forestry trailers. *Croat. J. For. Eng.* **2015**, *36*, 131–136.
13. Manzone, M. Performance of an electronic control system for hydraulically driven forestry tandem trailers. *Biosyst. Eng.* **2015**, *130*, 106–110. [CrossRef]
14. Bietresato, M.; Carabin, G.; Vidoni, R.; Mazzetto, F.; Gasparetto, A. A Parametric Approach for Evaluating the Stability of Agricultural Tractors Using Implements during Side-Slope Activities. *Contemp. Eng. Sci.* **2015**, *8*, 1289–1309. [CrossRef]
15. Pereira, D.; Fiedler, N.C.; de Souza Lima, J.S.; De Oliveira Bauer, M.; Rezende, A.V.; Missiaggia, A.A.; Pavesi Simão, J.B. Lateral stability limits of farm tractors for forest plantations in steep areas. *Sci. For.* **2011**, *39*, 433–439.
16. Lindroos, O.; Wasterlund, I. Theoretical potentials of forwarder trailers with and without axle load restrictions. *Croat. J. For. Eng.* **2014**, *35*, 211–219.
17. Spinelli, R.; Magagnotti, N.; Facchinetti, D. A survey of logging enterprises in the Italian Alps: Firm size and type, annual production, total workforce and machine fleet. *Int. J. For. Eng.* **2013**, *24*, 109–120.

18. Keppel, G.; Wickens, T.D. *Design and Analysis: A Researchers Handbook*, 4th ed.; Pearson: Upper Saddle River, NJ, USA, 2004.
19. Marinello, F.; Grigolato, S.; Sartori, L.; Cavalli, R. Analysis of a double steering forest trailer for long wood transportation. *J. Agric. Eng.* **2013**, *44*, 10–15. [CrossRef]

Article

Influence of Chain Filing, Tree Species and Chain Type on Cross Cutting Efficiency and Health Risk

Jurij Marenče, Matevž Mihelič and Anton Poje *

Department of Forestry and Renewable Resources, Biotechnical Faculty, University of Ljubljana, Večna pot 83, 1000 Ljubljana, Slovenia; jurij.marence@bf.uni-lj.si (J.M.); matevz.mihelic@bf.uni-lj.si (M.M.)
* Correspondence: anton.poje@bf.uni-lj.si; Tel.: +386-1-320-3516

Received: 11 October 2017; Accepted: 21 November 2017; Published: 24 November 2017

Abstract: As one of the major parts of the chainsaw, the cutting chain has an important impact on productivity and health risk in motor-manual harvesting. The efficiency of cross cutting and quantity of sawdust produced in relation to different cutting chain settings, chain producers and wood species has been measured. The trial was set up to include two tree species (fir and beech) and saw chains from two different producers. The chains were filed at three different top plate filing angles and depth height gauges. All factors were significant in terms of cutting efficiency and wood dust production. The top plate angle recommended by producers proved to be the most efficient, with the smallest quantity of inhalable wood dust. Cutting chain settings can be adapted to the specific requirements of the user; however, safe working practices should be followed. Significant differences between chain producers mean that users should conduct rational decision making when choosing a saw chain.

Keywords: forestry; chainsaw; filing; sawdust; safety

1. Introduction

The introduction of mechanized harvesting has been hindered by several factors, including steep terrain, ownership fragmentation and close-to-nature management [1]. Motor-manual tree felling and processing using a chainsaw therefore remains the predominant harvesting method in many regions of the world (South East Europe, Asia, Africa) [2–4]. Additionally, the low investment cost and the chainsaw's versatility play a significant role when harvesting technology is chosen in private forests [5–7]. The chainsaw's simplicity and inherent reliability make it ideally suited for the harsh working conditions commonly encountered in forest operations.

Petrol chainsaws have recently witnessed a steady, but slow increase in cutting efficiency related to increases in engine power and speed, and improvements in the control of engine performance. Higher efficiency reduces the amount of time and energy required for the production of one unit of goods. Therefore, increased productivity reduces greenhouse gas emissions, which are considered to be the main cause of climate change [8].

Unfortunately, motor-manual forest work is inherently dangerous [9,10], physically demanding [11,12] and involves physical environmental factors which have an extremely detrimental effect on worker health. Workers engaged in motor-manual cutting are exposed to excessive noise and vibration [13–15] and to the effects of exhaust gases [16], floating particles of mineral oil and airborne wood dust [17–19].

Although the air in logging operations can be very polluted [20], very few motor-manual workers are aware of the negative consequences of exposure to dust. The risk, however, is significant, and the International Agency for Research on Cancer has classified hardwood dust as a human carcinogen [21]. It is estimated that in 25 member states of the European Union, 2% of the work force is occupationally exposed to inhalable wood dust [22]. In the European Union the Directive 99/38/EC [18], is setting the legal limit for the exposure to inhalable wood dust at 5 mg/m³, as an average of an 8-h working day. In regard to the size of particles inhalable, thoracic and respirable particles are to be distinguished [23]. In our study however, as a health risk indicator, the share of the inhalable wood dust in regard to the total amount of saw dust was used [24].

Potential health effects from exposure to wood dust are numerous and well documented [25] and include changes in pulmonary function and allergic respiratory response (asthma). The most serious problem arising from exposure to wood dust is the risk of developing cancer, particularly nose and sinus adenocarcinoma [26].

The saw chain, as the main sawing part, influences the efficiency, safety and ergonomic suitability of the chainsaw. The main problem for the user is that only the technical data that are needed for compliance of the chain with the saw bar are available for any single chain. There is no, or insufficient, qualitative data needed for rational decision making. Producer information about quality, i.e., "chain with lower vibration," is more a function of faith in a trademark than of rational thinking. The problem is exacerbated because of the lack of knowledge available about basic principles of how efficiency or health risk change depending on chain maintenance practices.

Practice and science agree that proper filing and maintenance of the saw chain is crucial for safe and efficient work [27]. The word "proper" usually means that users have followed the producer's directions. In practice, however, modification of filing direction and depth is common, with implications for chain durability and work efficiency.

Since the cutting chain is one of the most important parts of the chainsaw, its preparation and maintenance have a direct impact on work efficiency and health risk. The aim of the study was therefore to determine the factors affecting efficiency and the amount of inhalable wood dust produced during cross cutting with a chain saw. To achieve these goals, the experiment has been set up including two filing factors, two tree species and two saw chains from different producers. The effect of wood moisture was controlled, as it has been established that it affects particle size distribution [28].

According to our experience and previous studies, it was expected that the chain setting recommended by the producer would yield the greatest efficiency and the lowest health impact, as the indicated values are presumably the result of in-depth analysis performed by the chain producer. The chains were chosen from two established producers, and tree species were chosen according to their predominance in Central Europe, although no specific hypotheses were proposed based on these factors.

2. Materials and Methods

The experiment was designed as a full factorial design (Table 1) with two factors at two levels and two factors at three levels ($2^2 3^2$). To eliminate the possible effect of wood structure and moisture, each combination was repeated five times (Scheme 1). The positive effect of repetition of cross cutting on experimental design was confirmed with an insignificant difference in wood moisture between all factors involved in the experiment with the exception of tree species. The average wood moisture was 32.9% for fir (*Abies alba* Mill.) and 50.6% for beech (*Fagus sylvatica* L.). The total number of crosscuts with repetitions was 180 or 10 by each saw chain setup. For the purpose of wood dust analyses, additional crosscuts (11th and 12th crosscut, Scheme 1) were done at all factorial levels but without repetitions (n = 36).

Table 1. Description of factors, their respective levels, coding and specification of variables for saw chain comparison test.

Factors	Variable Description	Levels	Coding	Unit	N [1]	Mean	Standard Deviation
Cross cutting time	Continuous variable indicating duration of cross cutting.	-	t	s	180/0	26.39	10.44
Sawdust structure	Continuous variable indicating weight structure of chips distributed in three classes.	>3 mm	wtB	%	0/36	52.87	10.34
		0.125 mm <>3 mm	wtM	%	0/36	45.66	10.14
		<0.125 mm	wtS	%	0/36	1.49	0.87

215

Table 1. *Cont.*

Factors	Variable Description	Levels	Coding	Unit	N [1]	Mean	Standard Deviation
Top plate filing angle	Dummy variable indicating angle of the cutting tooth. The filing angle was between 15–40°, depending on the type of chain and producer. The producer proposed filing angle is marked as 0°.	−10°	A − 10°	-	60/12	-	-
		0°	A0°	-	60/12	-	-
		+10°	A + 10°	-	60/12	-	-
Depth gauge height	Dummy variable indicating height difference between the depth gauge and the tip of the tooth cutting tooth. The height difference was 0.45, 0.65 and 0.85 mm. The proposed producer's difference is 0.65 mm.	0.45 mm	G0.45	-	60/12	-	-
		0.65 mm	G0.65	-	60/12	-	-
		0.85 mm	G0.85	-	60/12	-	-
Tree species	Dummy variable indicating two tree species (beech and fir) indicative of Central European conditions.	Beech	SB	-	90/18	-	-
		Fir	SF	-	90/18	-	-
Chain producer	Dummy variable indicating two producers of saw chains.	Producer 1	P1	-	90/18	-	-
		Producer 2	P2	-	90/18	-	-

[1] Sample size: cutting efficiency analyses/wood dust analyses.

In the trial, air dried, two first class logs of fir and beech were used. The logs were standardized into a beam with length of 6 meters and sides of 32 × 32 cm, in compliance with ISO standards [29]. The beams were cut into 6 shorter (about 2 m long) sections just before the trial (Scheme 1).

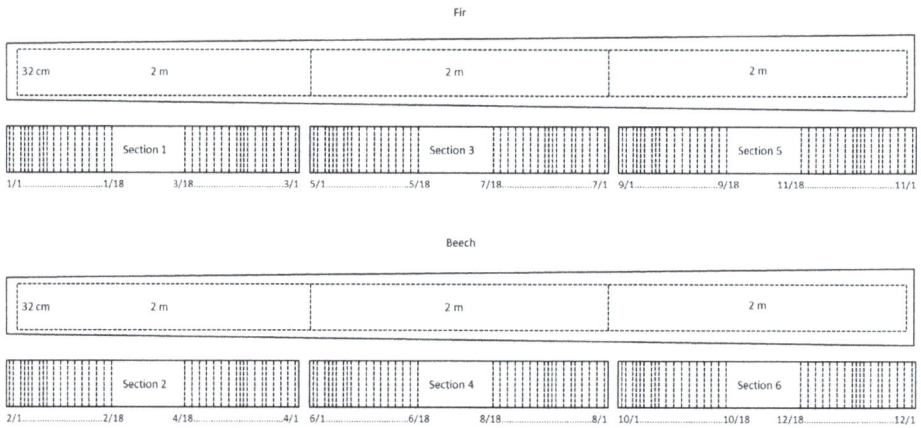

Scheme 1. Preparation and division of wood beams from two trees species into sections and cross cutting sequence. Note: 1/1-cross-cut/chain setup sequence.

A new Husqvarna 372 XP professional chain saw with a 45-cm-long Stihl Rollomatic E cutting bar was used in the experiment, and a skilled worker was employed as the chain saw operator. Eighteen 3/8″ chisel chains with 72 drive links from two different producers were used and filed in accordance with the trial scheme (Table 1). The top filing angles recommended by the chain producers differ for 5°. The top plate filing angle was done using a Stihl USG filing tool, while correct depth gauge height was achieved using hand filing and controlled with calipers.

Cross cutting took place in turns according to tree species and always in the same sequence (Scheme 1). The sequence of used chain setups was randomly selected. To further guarantee comparable conditions, cutting chain tension was controlled and the weight of the saw was kept at the same level by assuring the same level of lubricant and gasoline in the saw. The saw was controlled only through the rear handle, assuring that the pressure applied to all sample logs was constant. Visible defects in the wood were avoided when cross cutting.

The whole trial was recorded using a camera (Sony HDR-XR200; Sony, Tokyo, Japan), while the time studies were performed in the office to ensure the highest precision of measurement. A total amount of saw dust was directly captured in a plastic bag using an adapter attached to the chain saw. Moisture content was determined with the gravimetric method, according to European standards [30]. Fresh sample weight was determined immediately after sample collection with a portable scale to avoid the bias caused by moisture loss during storage and transport to the laboratory. Particle size distribution was determined with the oscillating screen method according to European standards [31]. Entire samples were sieved, and from the mass of fractions the shares of particle size distributions were calculated. The share of fine particles from the particle size distribution analysis was used as an indicator value.

Parametric multivariate analysis of variance (MANOVA), univariate analysis of variance (ANOVA) and ordinary least squares regression were used for statistical data processing. Before analysis, dependent variables were checked for normality (Kolmogorov–Smirnov and Shapiro–Wilks tests) and homogeneity of variance (Levene's test). In case of violation of data considerations, variables were transformed.

For the analysis of cross cutting duration, all samples were used (n = 180). To meet considerations for use of parametric tests (MANOVA, ANOVA), the dependent variable (time) was transformed with the inverse square root transformation.

Because of the interdependent nature of particle size distributions, basic approaches of compositional data analysis were used [28]. Compositional data were transformed using Isometric Log Ratio transformation in the CoDaPac software (2.02.04, 2017, University of Girona, Girona, Spain) [32]. The data were analyzed with IBM SPSS Statistics software (21.0, 2012, IBM, Armonk, NY, USA).

3. Results

3.1. Factors Influencing Cross Cutting Time

Statistical analysis of cross cutting time shows that cross cutting time is significantly dependent on all four factors: top plate filing angle, depth gauge height, tree species and cutting chain producer (Table 2).

Table 2. ANOVA results between cross cutting time (square root transformed) and independent factors.

Source	SS [1]	df [2]	MS [3]	F value	p-level
Corrected Model	0.100	6	0.017	29.646	0.000
Intercept	7.483	1	7.483	13,339.983	0.000
Depth gauge height	0.016	2	0.008	14.108	0.000
Chain producer	0.004	1	0.004	7.380	0.007
Tree species	0.046	1	0.046	81.318	0.000
Top plate filing angle	0.034	2	0.017	30.480	0.000
Error	0.097	173	0.001		
Total	7.679	180			
Corrected Total	0.197	179			

[1] Type III sum of squares, [2] Degrees of freedom, [3] Mean square.

In a four-factorial model (Equation (1)) with a constant level of other factors, the cross cutting time with the smallest gauge depth height is significantly shorter than the cross cutting time at the largest gauge depth height, while shorter ($p = 0.10$) than with the factory-recommended gauge depth. When comparing the top plate filing angle, the cross cutting time is the shortest when compared with the other two settings. It takes significantly longer to cut through beech logs than fir logs. Cross cutting time is significantly longer when using the saw chain of producer 2.

$$t(\text{s}) = \cfrac{1}{(0.206 - 0.023^{***} \times G0.45 - 0.007^{\circ} \times G0.65 + 0.010^{**} \times P1 - 0.032^{***} \times SB + 0.023^{***} \times A - 10^{\circ} + 0.033^{***} \times A0^{\circ})^2} \tag{1}$$

Note: * p-level of significance of the effect—$^{\circ} \leq 0.1$, * ≤ 0.05, ** ≤ 0.01, *** ≤ 0.001; variables defined in Table 1.

With simultaneous changing of all factors in the model it can be determined which factor combinations yield the longest and the shortest cross cutting time (Figure 1). Cross cutting time reaches a maximum when the filing angle is the largest and the depth gauge height is the smallest, when using the saw chain of producer 2 and when cutting beech wood. The minimum cross cutting time is achieved using the factory setting filing angle and maximal depth gauge height when using the saw chain of producer 2 and when cutting fir wood. The maximum time (43.29 s) is 2.7 times longer than the minimum time (16.15 s).

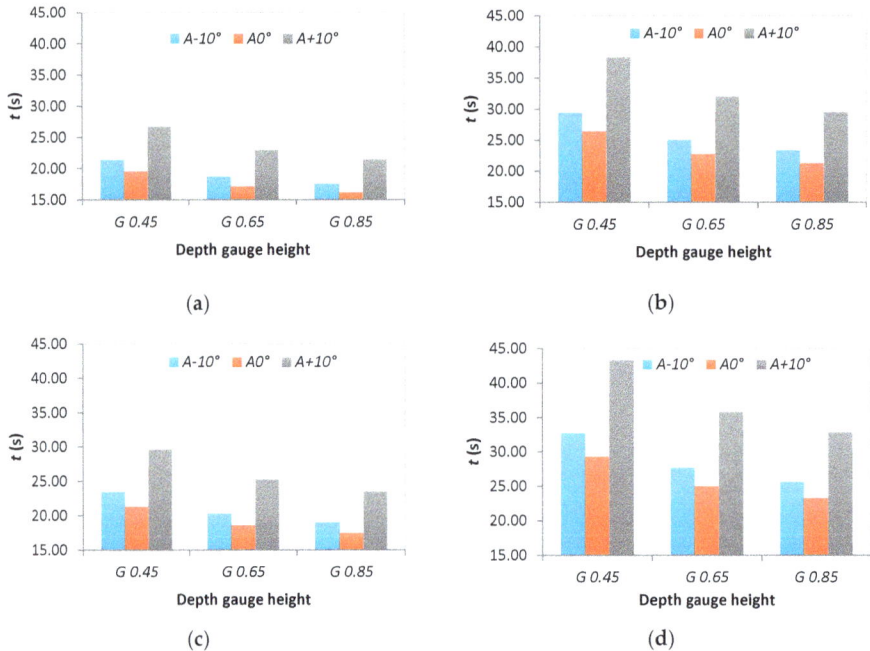

(a)

(b)

(c)

(d)

Figure 1. Dependency of cross cutting time (t) on depth gauge height (G) and top plate filing angle (A): (**a**) chain producer 1 and fir wood (minimum cross cutting time); (**b**) chain producer 1 and beech wood; (**c**) chain producer 2 and fir wood and (**d**) chain producer 2 and beech wood (maximum cross cutting time).

3.2. Factors Influencing Sawdust Structure

Using multivariate data analysis (Table 3), it has been determined that three factors, i.e., top plate filing angle, tree species and chain producer, influence the particle size distribution of sawdust.

Table 3. MANOVA results between particle size distribution and independent factors.

Effect		Value	F Value	Hypothesis df [1]	Error df [1]	p-Level
Intercept		0.997	5476.936	2	28	0.000
Top plate filing angle		0.613	6.404	4	58	0.000
Depth gauge height	Pillai's Trace	0.170	1.345	4	58	0.264
Chain producer		0.323	6.668	2	28	0.004
Tree species		0.912	145.325	2	28	0.000

[1] Degrees of freedom.

Additional univariate analysis (Table 4) shows that the relationship between the largest and middle-sized fractions of particle size distributions (ILR1) is influenced by the top plate filing angle and producer, while the relationship between the largest and middle fraction versus the smallest fractions (ILR2) is influenced by the top plate filing angle and tree species.

Table 4. ANOVA results between particle size distribution (ILR1, ILR2) and independent factors.

Source		SS [1]	df [2]	MS [3]	F Value	p-Level
Corrected Model	ILR1	1.739	6	0.290	5.551	0.001
	ILR2	9.200	6	1.533	52.511	0.000
Intercept	ILR1	0.386	1	0.386	7.398	0.011
	ILR2	322.640	1	322.640	11,049.729	0.000
Top plate filing angle	ILR1	1.092	2	0.546	10.456	0.000
	ILR2	0.450	2	0.225	7.699	0.002
Depth gauge height	ILR1	0.108	2	0.054	1.035	0.368
	ILR2	0.140	2	0.070	2.399	0.109
Chain producer	ILR1	0.529	1	0.529	10.129	0.003
	ILR2	0.049	1	0.049	1.669	0.207
Tree species	ILR1	0.010	1	0.010	0.198	0.660
	ILR2	8.561	1	8.561	293.204	0.000
Error	ILR1	1.514	29	0.052		
	ILR2	0.847	29	0.029		
Total	ILR1	3.640	36			
	ILR2	332.687	36			
Corrected Total	ILR1	3.254	35			
	ILR2	10.046	35			

[1] Type III sum of squares, [2] Degrees of freedom, [3] Mean square.

Further analysis shows that the percentage of the largest fraction is maximal when the producer-recommended top plate filing angle is used, and that it is larger when using the saw chain of producer 1 (Figure 2). The percentage of inhalable dust is lowest when using the chain of producer 1, with the filing angle recommended by the producer and when cutting fir wood (0.55%). The largest proportion of inhalable fraction (2.62%) was formed when the filing angle was the lowest, when cross cutting beech wood and using the saw chain of producer 2. The largest percentage of inhalable fraction is therefore 4.7 times higher than the sample with the lowest dust content.

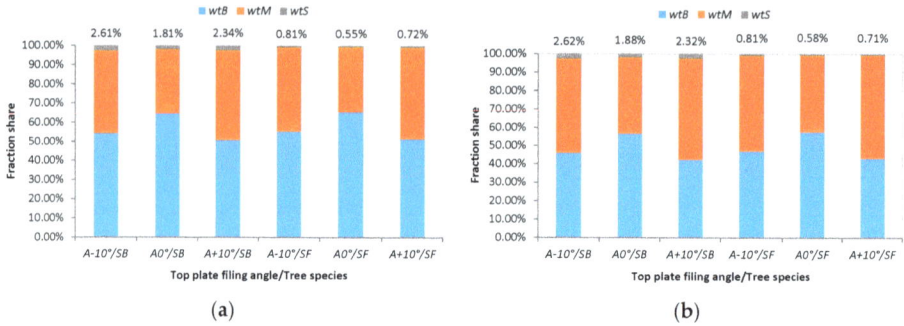

Figure 2. Dependency of sawdust weight structure on top plate filing angle (*A*) and tree species (*S*): (**a**) chain producer 1 and (**b**) chain producer 2. Note: *wtB*—big particles (more than 3 mm), *wtM*—medium-sized particles (3–0.125 mm) and *wtS*—small (inhalable) particles (less than 0.125 mm).

4. Discussion

The results show that the work efficiency and health risk due to inhalable dust is influenced by all of the researched factors: chain filing, tree species and chain producer. Ignorance of the above factors can result in significant differences in work efficiency and increased health risk.

From the viewpoint of efficiency and dust particles, the best top plate filing angle was the angle recommended by the chain producer. With this angle the cross cutting time and the amount of inhalable dust particles reached a minimum. These results are in direct contradiction with research that found that the top plate filing angle recommended by the producer corresponded to the greatest exposure to hand-arm vibration [33]. However, it was established that from the viewpoint of efficiency, it is better that the top plate filing angle is smaller than the angle recommended by producer rather than larger. The real-world implications of these findings are that mistakes made by less-experienced chain saw operators depend on the deviation from the recommended angle, and it is not as critical when the angle is smaller than that prescribed.

The depth gauge height has been found to deviate strongly in real-world operations, as only 15% of chains were found to have the proper gauge depths [34]. The consequence of large depth gauge height is higher cutting efficiency, while no statistically significant effect has been found on the quantity of hazardous dust particles. Excessive depth gauge heights are not recommended, as they lead to higher levels of kickback forces [35] and larger vibration loads [36]. Higher depth gauge setting increases the chain saw power requirements, and the consequences are higher fuel consumption, higher weight, higher vibration exposure and potentially greater negative impact on the health of the environment and worker. The open question remains the optimal relation between depth gauge setting and all of the above factors.

The effect of different wood species is important because of different wood density, as it was established that wood density has a significant effect on hand-arm vibration exposure [15]. The influence of the user on this factor is relatively small, as this factor depends on the worksite. However, a detailed analysis of interactions between factors has shown that the species influences cutting efficiency through the factors of top plate filing angle and saw chain producer. The dependency was confirmed by the experience of cutters, who often alter chain angles depending on whether the work is in hardwood or softwood forests. Another possible cause for differences in work efficiency and health risk is the wood moisture content, which is significant between tree species.

Few studies have examined exposure of forest operators to wood dust, but it was determined that exposure to dust was 1.5 mg/m³ for all operations except coppicing, where loads were greater [18]. Most loads, however, were lower than the EU occupational exposure limit (OEL) of 5 mg/m³. This level has been disputed, as some countries have lower allowed OEL levels, and the figures reported in the above study are above the allowed levels in some EU countries. In the article it has been shown that the amount of inhalable dust particles can be effectively lowered with the use of different filing angles. Also, salvage cuts of trees that have remained in the forest long enough for the wood to become dry should be avoided, as such trees are associated with a higher risk of accidents [10] and the risk of high concentrations of inhalable dust particles [37]. Additionally, the cutting chain preparation in the studies is usually not controlled, but it has been demonstrated that it is indeed a very important factor when conducting sawdust analysis, as it influences sawdust composition and therefore the amount of inhalable dust produced. The measured quantities of inhalable wood dust are to be considered as potential risk for health, since it is assumed that only part of inhalable saw dust enters the air and a part of that dust enters the worker's body.

One of the main conclusions is that producers do not produce equivalent chains. The chain type therefore influences the cutting efficiency and the health risk for the chain saw operator. Furthermore, after a number of cuts the performance and risk for health [15] may change because of different durability of chains. The results clearly demonstrate that a rational decision when purchasing a saw chain is crucial, and force producers into continued technological development.

It has to be acknowledged that in the study only two factors were taken into consideration from the pool of factors that influence the suitability of a saw chain for professional use. For instance, the viewpoint of safety at work in the sense of vibration exposure has not been included, and protection against kickback. It is assumed that the producer carefully weighs all of the above considerations when designing the product, and chooses the optimum accordingly. The results, however, show that it would be reasonable to provide not only one filing angle, but several, or to develop chains specifically intended for conifers or broadleaves.

From the lessons learned, the extension of scope in similar future studies is suggested, especially the inclusion of additional parameters, such as fuel and lubricant consumption, hand-arm vibration, different wood moisture levels, different lengths of cutters, direct measurements of airborne wood dust emissions and durability and life span of saw chain. Also, there are other, smaller chain saws, i.e., non-professional and electric chainsaws that use smaller chains and that could significantly alter the results. It has been found that the wood species also influences kickback forces [35], and an important safety consideration is how the filing angle influences the kickback of the saw in relation to wood species. Certainly, it would have the highest value for practice by testing different chains, their lifespan and their properties in real-life conditions during forest operations.

Unfortunately, the safety improvements offered by high mechanization are unattainable for motor-manual work [38,39]. It is therefore necessary to continually improve machines to make work as friendly to the worker as possible. Although all of the saw chain producers advertise safety features, and it is clear that it is necessary for producers to make complex decisions related to optimization in order to achieve these goals, more work should be done on the topic in order to provide more accessible and transparent information about saw chains.

5. Conclusions

The study demonstrated that cutting chain selection and proper chain preparation are crucial for achieving high productivity and reducing health risk. There was a more than two-fold difference in efficiency and quantity of inhalable wood dust between the best and worst cutting chain set up, making the chain one of the most important parts of the chainsaw which critically influences work with this machine. Additionally, the results proved that the chain settings recommended by the chain producers yield the greatest efficiency and the least health impact. It is evident that the recommended settings are the result of the optimization of the factors examined in the study. The results also suggest that the filing angle should change according to tree species, but safe working practices must be followed.

Acknowledgments: The authors wish to thank the Pahernik Foundation for supporting the scientific work and publishing the study results.

Author Contributions: A.P. and J.M. conceived, designed and performed the experiments; A.P. analyzed the data; A.P., M.M., and J.M. contributed materials and analysis tools, and wrote the paper.

Conflicts of Interest: The authors declare no conflict of interest and also declare that the funding sponsors had no role in the design of the study, either in the collection, analyses, or interpretation of the data or in the writing of the manuscript, or in the decision to publish the results.

References

1. Spinelli, R.; Magagnotti, N.; Nati, C. Options for the mechanized processing of hardwood trees in Mediterranean forests. *Int. J. For. Eng.* **2009**, *20*, 39–44.
2. Montorselli, N.B.; Lombardin, C.; Magagnotti, N.; Marchi, E.; Neri, F.; Picchi, G.; Spinelli, R. Relating safety, productivity and company type for motor-manual logging operations in the Italian Alps. *Accid. Anal. Prev.* **2010**, *42*, 2013–2017. [CrossRef] [PubMed]
3. Vusić, D.; Šušnjar, M.; Marchi, E.; Spina, R.; Zečić, Ž.; Picchio, R. Skidding operations in thinning and shelterwood cut of mixed stands—Work productivity, energy inputs and emissions. *Ecol. Eng.* **2013**, *61*, 216–223. [CrossRef]
4. Karjalainen, T.; Zimmer, B.; Berg, S.; Welling, J.; Schwaiger, H.; Finér, L.; Cortijo, P. *Energy, Carbon and other Material Flows in the Life Cycle Assessment in Forestry and Forest Products*; European Forest Institute: Joensuu, Finland, 2001; p. 68.
5. Jourgholami, M.; Majnounian, B.; Zargham, N. Performance, capability and costs of motor-manual tree felling in Hyrcanian hardwood forest. *Croat. J. For. Eng.* **2013**, *34*, 283–293.
6. Liepiņš, K.; Lazdiņš, A.; Liepiņš, J.; Prindulis, U. Productivity and cost-effectiveness of mechanized and motor-manual harvesting of grey alder (*Alnus incana* (L.) Moench): A case study in Latvia. *Small-Scale For.* **2015**, *14*, 493–506. [CrossRef]
7. Koutsianitis, D.; Tsioras, P.A. Time consumption and production costs of two small-scale wood harvesting systems in northern Greece. *Small-Scale For.* **2017**, *16*, 19–35. [CrossRef]
8. Wallington, T.J.; Srinivasan, J.; Nielsen, O.J.; Highwood, E.J. Greenhouse gases and global warming. In *Environmental and Ecological Chemistry—Volume 1*; Sabljic, A., Ed.; Eolss Publishers: Oxford, UK, 2009; pp. 36–63.
9. Lindroos, O.; Burström, L. Accident rates and types among self-employed private forest owners. *Accid. Anal. Prev.* **2010**, *42*, 1729–1735. [CrossRef] [PubMed]
10. Poje, A.; Potočnik, I.; Košir, B.; Krč, J. Cutting patterns as a predictor of the odds of accident among professional fellers. *Saf. Sci.* **2016**, *89*, 158–166. [CrossRef]
11. Melemez, K.; Tunay, M. Determining physical workload of chainsaw operators working in forest harvesting. *Technology* **2010**, *13*, 237–243.
12. Parker, R.J.; Bentley, T.A.; Ashby, L.J. Human factors testing in forest industry. In *Handbook of Human Factors Testing and Evaluation—Second Edition*; Charlton, S.G., O'Brien, T.G., Eds.; Lawrence Erlbaum Associates: Mahwah, NJ, USA, 2008; pp. 319–340.
13. Minetti, L.J.; de Souza, A.P.; Machado, C.C.; Fiedler, N.C.; Baêta, F.d.C. Evaluation of noise and vibration effects of forest cutting on chainsaw operators. *Rev. Árvore* **1998**, *22*, 325–330.
14. Fonseca, A.; Aghazadeh, F.; de Hoop, C.; Ikuma, L.; Al-Qaisi, S. Effect of noise emitted by forestry equipment on workers' hearing capacity. *Int. J. Ind. Ergon.* **2015**, *46*, 105–112. [CrossRef]
15. Rottensteiner, C.; Tsioras, P.; Stampfer, K. Wood density impact on hand-arm vibration. *Croat. J. For. Eng.* **2012**, *33*, 303–312.
16. Hooper, B.; Parker, R.; Todoroki, C. Exploring chainsaw operator occupational exposure to carbon monoxide in forestry. *J. Occup. Environ. Hyg.* **2017**, *14*, D1–D12. [CrossRef] [PubMed]
17. Horvat, D.; Kos, A.; Zečič, Z.; Jazbec, A.; Šušnjar, M.; Očkajová, A. Tree cutters' exposure to oakwood dust—A case study from Croatia. *Die Bodenkult.* **2007**, *58*, 59–65.
18. Marchi, E.; Neri, F.; Cambi, M.; Laschi, A.; Foderi, C.; Sciarra, G.; Fabiano, F. Analysis of dust exposure during chainsaw forest operations. *iForest-Biogeosci. For.* **2017**, *10*, 341–347. [CrossRef]
19. Neri, F.; Foderi, C.; Laschi, A.; Fabiano, F.; Cambi, M.; Sciarra, G.; Aprea, M.C.; Cenni, A.; Marchi, E. Determining exhaust fumes exposure in chainsaw operations. *Environ. Pollut.* **2016**, *218*, 1162–1169. [CrossRef] [PubMed]

20. Mitchell, D. Air quality on biomass harvesting operations. In Proceedings of the 34th Council on Forest Engineering Annual Meeting, Quebec City, QC, Canada, 12–15 June 2011; p. 9.

21. IARC. *IARC Monographs on the Evaluation of the Carcinogenic Risks to Humans: Wood Dust and Formaldehyde;* WHO: Lyon, France, 1995; Volume 62.

22. Kauppinen, T.; Vincent, R.; Liukkonen, T.; Grzebyk, M.; Kauppinen, A.; Welling, I.; Arezes, P.; Black, N.; Bochmann, F.; Campelo, F.; et al. Occupational exposure to inhalable wood dust in the member states of the European Union. *Ann. Occup. Hyg.* **2006**, *50*, 549–561. [PubMed]

23. Sánchez, J.A.; van Martie, T.; Cherrie, J.W. *A Review of Monitoring Methods for Inhalable Hardwood Dust;* IOM (Institute of Occupational Medicine): Edinburgh, UK, 2011; p. 21.

24. Harper, M.; Muller, B.S. An evaluation of total and inhalable samplers for the collection of wood dust in three wood products industries. *J. Environ. Monit.* **2002**, *4*, 648–656. [CrossRef] [PubMed]

25. WHO. *Hazard Prevention and Control in the Work Environment: Airborne Dust;* WHO: Geneva, Switzerland, 1999; p. 96.

26. Charbotel, B.; Fervers, B.; Droz, J.P. Occupational exposures in rare cancers: A critical review of the literature. *Crit. Rev. Oncol. Hematol.* **2014**, *90*, 99–134. [CrossRef] [PubMed]

27. Maciak, A. Wpływ geometrii ostrza żłobikowego na jego obciążenie podczas skrawania drewna sosnowego [Effects of the cutting link geometry on its load during the cutting of pine wood]. *Przeg. Tech. Roln. Leśn.* **1998**, *5*, 18–21.

28. Mihelič, M.; Spinelli, R.; Magagnotti, N.; Poje, A. Performance of a new industrial chipper for rural contractors. *Biomass Bioenergy* **2015**, *83*, 152–158. [CrossRef]

29. ISO. *Acoustics—Measurement at the Operator's Position of Airborne Noise Emitted by Chain Saws;* ISO: Geneva, Switzerland, 1984; Volume 7182.

30. EN. *Solid Biofuels—Methods for the Determination of Moisture Content—Oven Dry Method—Part 2: Total Moisture—Simplified Method;* European Committee for Standardization: Brussels, Belgium, 2010; Volume 14774-2.

31. EN. *Solid Biofuels—Determination of Particle Size Distribution—Part 1: Oscillating Screen Method Using Sieve Apertures of 1 mm and above;* European Committee for Standardization: Brussels, Belgium, 2011; Volume 15149-1.

32. Comas-Cufí, M.; Thio-Henestrosa, S. Codapack 2.0: A stand-alone, multi-platform compositional software. In Proceedings of the CoDaWork'11: 4th International Workshop on Compositional Data Analysis, Sant Feliu de Guíxols, Spain, 10–13 May 2011; Egozcue, J.J., Tolosana-Delgado, R., Ortego, M.I., Eds.; Universitat de Girona: Sant Feliu de Guíxols, Spain, 2011; pp. 1–10.

33. Stempski, W.; Jabłon'ski, K.; Wegner, J. Relations between top-plate filing angle values of cutting chains and chain saw vibration levels. *Acta Sci. Pol. Silv. Colendar. Rat. Ind. Lignar.* **2010**, *9*, 31–39.

34. Trzciński, G. Assessment of the technical condition of chain saws owned by forestry workers. *Przeg. Tech. Roln. Leśn.* **1995**, *1*, 21–23.

35. Dabrowski, A. Reducing kickback of portable combustion chain saws and related injury risks: Laboratory tests and deductions. *Int. J. Occup. Saf. Ergon.* **2012**, *18*, 399–417. [CrossRef] [PubMed]

36. Dessureault, P.C.; Laperrière, A.; Vincent, J.Y. The control of chain saw vibration. *Sound Vib.* **1988**, *22*, 32–34.

37. Magagnotti, N.; Nannicini, C.; Sciarra, G.; Spinelli, R.; Volpi, D. Determining the exposure of chipper operators to inhalable wood dust. *Ann. Occup. Hyg.* **2013**, *57*, 784–792. [PubMed]

38. Bell, J.L. Changes in logging injury rates associated with use of feller-bunchers in West Virginia. *J. Saf. Res.* **2002**, *33*, 463–471. [CrossRef]

39. Axelsson, S.-Å. The mechanization of logging operations in Sweden and its effect on occupational safety and health. *Int. J. For. Eng.* **1998**, *9*, 25–31.

Article
Characteristics of Logging Businesses across Virginia's Diverse Physiographic Regions

Scott M. Barrett *, M. Chad Bolding and John F. Munsell

Department of Forest Resources and Environmental Conservation, Virginia Polytechnic Institute and State University, 310 W. Campus Drive, Blacksburg, VA 24061, USA; bolding@vt.edu (M.C.B.); jfmunsel@vt.edu (J.F.M.)
* Correspondence: sbarrett@vt.edu; Tel.: +1-540-231-6702; Fax: +1-540-231-3330

Received: 29 September 2017; Accepted: 21 November 2017; Published: 28 November 2017

Abstract: Logging businesses play an important role in implementing forest management plans and delivering the raw material needed by forest products mills. Understanding the characteristics of the logging workforce can help forest managers make better decisions related to harvesting operations. We surveyed logging business owners across Virginia's three physiographic regions (Mountains, Piedmont, and Coastal Plain). Overall, logging businesses reported an average production rate of 761.37 t/business/week, but this varied substantially by region, with the highest production rates in the Coastal Plain (1403.55 t/business/week), followed by the Piedmont (824.69 t/business/week) and the Mountains (245.42 t/business/week). Many operations in the Mountains rely primarily on manual felling (66.6% of respondents) and these operations often have lower production rates. Across all regions, 81.7% of reported production came from operations that primarily utilized rubber-tired feller-bunchers for felling. Logging businesses were sorted based on reported production capacity and then divided into three groups (high, medium, and low production) based on total reported production. Across all regions, the majority of reported production was produced by the high production logging businesses. This was highest in the Piedmont, where the high production businesses accounted for 74.8% of total reported production.

Keywords: logging business; woody biomass; forest harvesting

1. Introduction

The forest industry is an important component of Virginia's economy and contributes over $17 billion in total economic value annually [1]. More than 60% of Virginia is forested, with 16 million acres of forestland predominantly owned by private forest landowners [2]. Virginia's logging businesses primarily consist of independent contractors who work with forest landowners and others in the wood supply chain to provide the raw material needed to supply mills. Logging businesses provide critical harvesting services that help forest landowners carry out their management plans and deliver wood to Virginia's diverse wood-consuming industries.

In 2016, there were over 5000 tracts harvested in Virginia [2]. Logging businesses that harvest these tracts serve as a link between forest landowners and the mills that utilize the products coming from their land. As a result, the characteristics of logging businesses are important for forest managers and those involved in the wood supply chain. Forest management and harvesting decisions can benefit from a better understanding of the capabilities and characteristics of the local logging workforce.

Surveys of logging businesses are important for establishing information on their characteristics and can be useful for documenting and examining changes over time. Logger surveys have been used throughout the USA for determining characteristics of logging businesses. For example, Milauskas and Wang surveyed West Virginia loggers on their operational characteristics as well as their opinions on training needs [3]. Dirkswager et al. used a phone survey of loggers to determine practices and perspectives related to utilizing logging residues for energy in Minnesota [4]. Egan et al. surveyed New England loggers regarding their perceptions of increasing population on logging opportunities [5] and Munsell et al. studied relationship between attitudes and biomass harvesting in Virginia and North Carolina [6]. Additional logger surveys have been conducted related to specific topics such as harvesting on parcelized forestland in the Midwest [7] as well as in South Carolina [8]. Logger surveys that are repeated over time can show changes in logging practices. For example, repeated logger surveys have been conducted in Minnesota [9] and Georgia [10].

A survey was administered to Virginia logging businesses in 2009 with the intention of repeating the survey at five-year intervals [11]. The original survey established a baseline of Virginia logging business characteristics and highlighted some of the operational differences between physiographic regions in the state. This survey was performed approximately five years after the initial survey. Some additional questions were added but many relating to logging business characteristics remained the same, so comparisons can be made over time.

This is the second comprehensive survey of the characteristics of Virginia logging businesses. While some comparisons can be made to the results of the previous survey, this manuscript focuses primarily on the current survey and much of the value of repeated surveys will come in the future as multiple surveys will enable analysis of long-term trends.

2. Methods

A questionnaire was designed for delivery to participants in the Virginia Sustainable Harvesting and Resource Professional (SHARP) Logger Program [12]. The objective was to collect information on characteristics of Virginia logging businesses as well as to evaluate the SHARP logger program. In addition to logging business owners, the SHARP logger program includes participants who are logging business employees as well as foresters and others that choose to participate in the program. Therefore, the questionnaire included two sections. The first section was a program evaluation section that included 12 questions related to the participants' demographics and to program evaluation and training needs. All respondents were able to categorize their primary occupation in the program evaluation section, while logging business owners completed a second section in order to provide additional characteristics of their harvesting operations. In the case of logging businesses with multiple owners, participants were asked to choose one owner to complete the business owner section and complete one questionnaire per business. In Virginia, logger training is not required by law, so the exact number of logging businesses is not known. However, loggers are required to notify the Virginia Department of Forestry (VDOF) of timber harvests so harvested tracts can be inspected for compliance with the silvicultural water quality law. Based on a study of reported data from VDOF harvest notifications during 2015–2016, there were approximately 1000 individuals who notified the VDOF of at least one timber harvest [13]. We would therefore assume there are a maximum of 1000 active logging businesses.

The logging business owner section included a total of 36 questions related to their harvesting operations such as harvesting equipment, the size and characteristics of their operation, as well as the markets and business environment they work in and tracts they harvest. The questionnaire included a combination of multiple choice, fill-in-the-blank, and open-ended questions. The questionnaire was mailed to current participants in the Virginia SHARP Logger Program. The survey was conducted based on the Dilman method [14] using a series of mailings including a pre-notice letter, followed by the questionnaire, then a reminder letter, and a second questionnaire mailed to those who had not returned the initial questionnaire. A total of 1607 questionnaires were mailed in the third and fourth quarters of 2014. Seven were returned as undeliverable or invalid, resulting in a total of 1600 questionnaires mailed to participants in the SHARP Logger Program.

Written responses from questionnaires were coded and entered into a spreadsheet then analyzed using JMP Pro version 13 [15]. Means testing utilized analysis of variance and the Tukey-Kramer HSD (honest significant difference) test. Categorical responses were compared using the Chi Square test.

Respondents were grouped into physiographic regions (Figure 1) based on the county the respondent indicated as their center of operations. Counties were grouped into physiographic regions based on regions used by the U.S. Forest Service [16], with the northern and southern Piedmont grouped into a single Piedmont region and northern and southern mountains grouped into a single mountain region. These are the same regional boundaries used in the 2009 survey of Virginia loggers [11]. Respondents were asked to report production levels in the units they commonly use, which was either tons or truck loads produced per week. Results from the previous survey [11] were reported as truck loads produced, which is the most common metric reported by loggers. However, because of differences in average payloads depending on trucking configurations, results for this survey were converted to a common unit on the basis of tonnes (metric tons, t) produced. Production levels reported in truckloads were converted to tonnes based on the reported trucking configuration the business most commonly uses. Average payloads for different trucking configurations were determined based on discussion with local industry professionals. A payload conversion factor of 22.7 t (25 tons) per load was used for tractor-trailers, as well as tandem or tri-axle trucks with a pup trailer. A payload conversion factor of 8.2 t (9 tons) per load was used for single-axle trucks, 12.7 t (14 tons) for tandem-axle trucks, and 17.2 t (19 tons) for tri-axle trucks. Based on standard practices for this region, all weights reported by logging businesses were assumed to be in green tonnes and typical truck payloads were based on green tonnes.

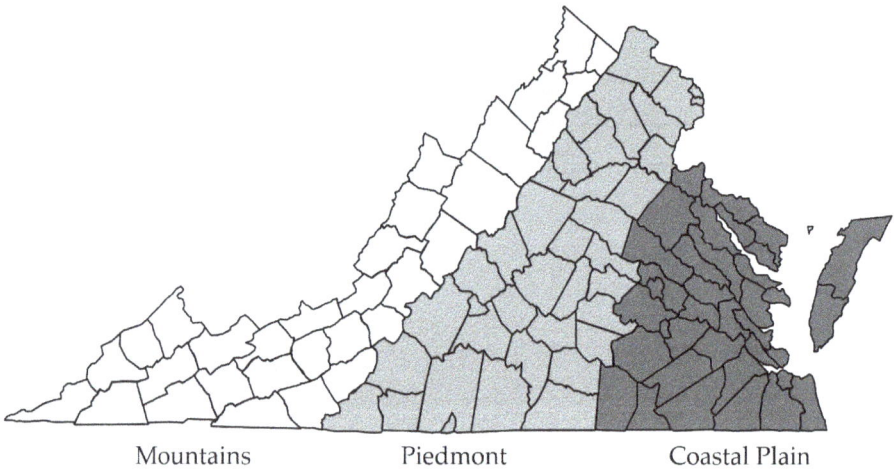

Figure 1. Virginia county boundaries used to group logging businesses by physiographic region (map not to scale).

3. Results and Discussion

3.1. Response Rate and Owner Demographics

A total of 847 questionnaires were completed and returned for a 53% response rate. Of the 847 responses, 400 (47.2%) indicated they were logging business owners. Data reported here are from the 400 logging business owner responses. Based on VDOF harvest notification data [13], if we assume there are a maximum of approximately 1000 active logging businesses, then survey responses would represent about 40% of all logging businesses in Virginia. Logging businesses included both full-time (87%) as well as part-time (13%) logging operations. Virginia logging business owners were predominantly men (99.5%), with an average age of 51.2 years (standard deviation (SD) = 12.3). Respondents indicated they had been operating their own logging business for an average of 22.6 years (SD = 13.9). The majority (91.2%) indicated they were White/Caucasian, with 8.3% Black/African

American and 0.5% other. Owner demographics are similar to those from the 2009 survey [11]; however, the average age shows an increase from 48.6 years in 2009 to 51.2 in 2014.

Respondents represented logging businesses operating across all three physiographic regions of Virginia including the Mountains (105), Piedmont (192), and Coastal Plain (70), as well as adjoining states (24); in some cases they did not specify a region (9). These three physiographic regions of Virginia would be similar to physiographic regions occurring throughout much of the Southeastern USA. Owners were asked to indicate their level of formal education. Across all regions, high school graduates were the most common (Table 1). Business owners in the Coastal Plain tended to have the highest level of formal education, with 26.1% indicating they were college graduates.

Table 1. Logging business owner education level by physiographic region.

	Percent of Responses for Level of Formal Education			
	Some High School	High School Graduate	Some College	College Graduate
Mountains	29.5	55.3	7.6	7.6
Piedmont	24.2	54.2	13.7	7.9
Coastal Plain	13.0	47.9	13.0	26.1
Overall	24.3	53.5	11.1	11.1

3.2. Harvest Types and Tract Characteristics

Respondents were asked to report the type of harvests they performed over the past year (Figure 2). Similar to the 2009 survey of Virginia loggers [11], the highest percentage of harvests reported in the Mountains were hardwood selection cuts. However this survey showed that the most common harvest type in the Coastal Plain was a pine clearcut, whereas in 2009 it was a pine thinning [11].

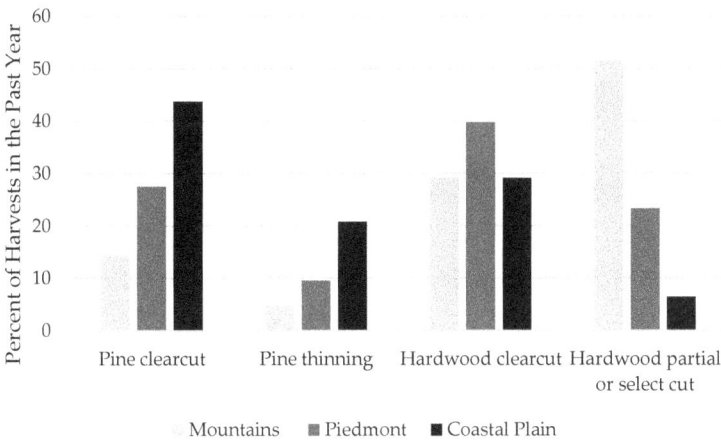

Figure 2. Reported type of harvests performed in the past year by physiographic region.

As expected, based on the most prevalent types of harvests reported (Figure 2), the primary product harvested in the Mountains was hardwood sawtimber while in the Coastal Plain, the most commonly reported product harvested was pine pulpwood (Figure 3). The most commonly reported average hauling distance for their primary product was between 20 and 60 miles and the most commonly reported distance moved between tracts was 20–40 miles.

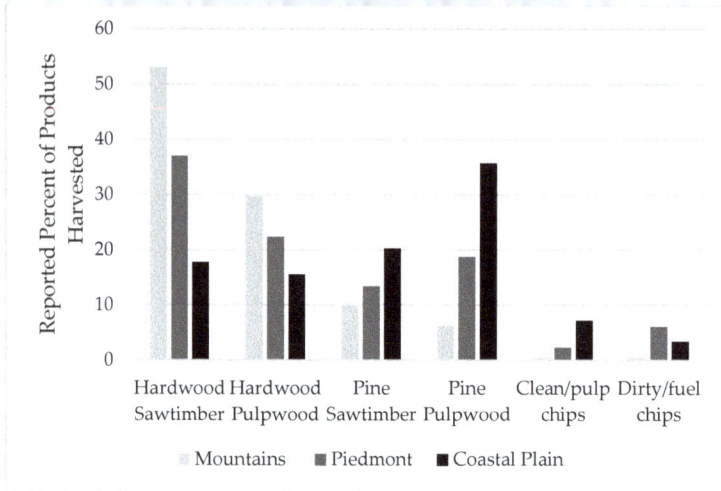

Figure 3. Reported percentage of products harvested in the past year by physiographic region.

Across all regions, the most commonly reported average harvest size over the past year was 40–80 acres. These harvests occur predominantly on forestland owned by private individuals. Harvests on forest industry owned land represented a relatively small portion of all harvests, with 5.2% or less of harvests in any region (Table 2).

Table 2. Type of land ownership where business owners harvested timber in the past year.

	Mean Response Percentage			
	Mountains	Piedmont	Coastal Plain	Overall
Private individuals	92.0	91.3	83.2	89.6
Forest industry	2.9	5.2	4.9	4.9
US Forest Service	4.4	0	0	1.3
State Forests	0	0.1	0.3	0.1
TIMO/REIT[1]	0	3.3	9.6	3.3
Other	0.7	0.1	2.0	0.8

TIMO = Timberland Investment Management Organization, REIT = Real Estate Investment Trust.

The way loggers acquired the timber they harvested varied by region as well. Business owners were asked to indicate the percent of their harvests in the past year that derived from timber where they bought or negotiated the sale directly, and the percent that was contract logging of timber bought by someone else. The majority of harvests across all regions were bought or negotiated directly by the logger. The coastal plain had the highest reported percentage of sales where the logger was contract logging timber that was bought by someone else (Figure 4).

For those sales where the logger indicated that they bought the timber, they were asked specifically how the timber was purchased. Options for the type of sale included lump sum, cut on "shares", which was defined as the "landowner receives a set percentage of delivered price received as the tract is harvested", or per unit/pay as cut. Per unit sales were defined as "landowner receives a set price per ton/MBF (thousand board feet) for each ton/MBF as tract is harvested". In the Mountains, the majority of sales were cut on "shares", whereas in the Coastal Plain the majority of sales were per unit/pay as cut (Figure 5).

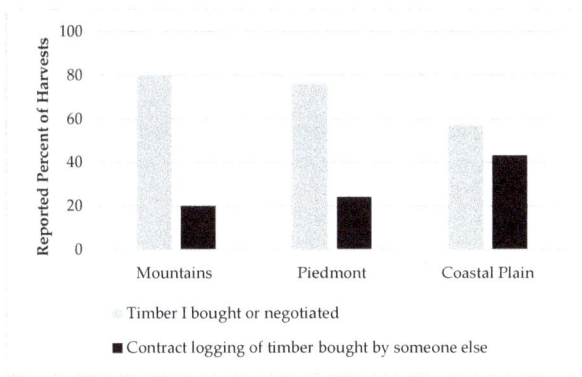

Figure 4. Reported source of timber harvested in past year.

Legend (Figure 4):
- Timber I bought or negotiated
- Contract logging of timber bought by someone else

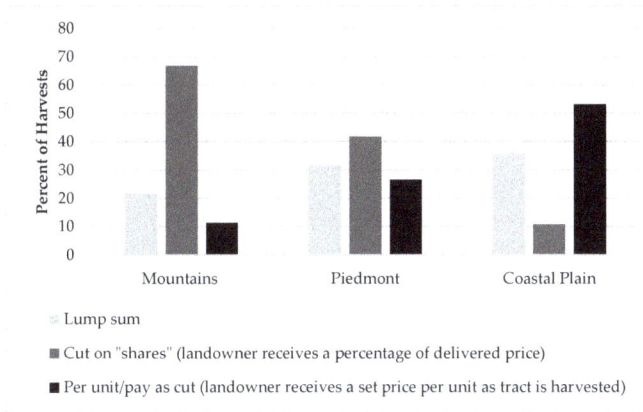

Figure 5. Reported sale type for timber bought by loggers.

Legend (Figure 5):
- Lump sum
- Cut on "shares" (landowner receives a percentage of delivered price)
- Per unit/pay as cut (landowner receives a set price per unit as tract is harvested)

3.3. Harvesting System and Production Characteristics

The majority of harvests reported were on forestland owned by private individuals (Table 2). Forest landowners often rely on harvests to meet their forest management objectives. Without a viable logging workforce capable of implementing multiple types of harvests, many landowners would not be able to meet their forest management objectives. The production level and type of equipment utilized on a logging operation are two important factors for determining the capabilities of a logging operation. Awareness of the characteristics and types of logging businesses within a region and the equipment they utilize can help forest managers select the most appropriate operation to meet their forest harvesting and management objectives.

Logging business owners were asked to report production in terms of loads or tons produced per crew per week, and number of crews owned. Values reported in loads per week were converted to tonnes per week based on average payloads for the trucking configuration they most commonly used. Tonnes per business per week was based on production per crew and the number of crews owned per business. Total reported production in the three regions was 240,727 tonnes per week with the majority of the total production reported in the Piedmont region (57.9%) and only 8.9% from the Mountains (Table 3). There were considerable differences in average production rates by region. Production in terms of t/crew/week and t/business/week both showed significant differences between regions ($P < 0.0001$). The Coastal Plain had the highest average tonnes per business per week (1403.55), tonnes per crew per week (864.01), and also had more crews per business (1.48). Compared to reported production values from the 2009 survey [11], the overall average number of crews per business increased from 1.16 to 1.20. The number of crews per business stayed the same in the Mountains, but increased in the Piedmont (from 1.12 to 1.2) and Coastal Plain (from 1.43 to 1.48). The number of workers per crew was nearly identical to 2009 levels in the Mountains and Piedmont and decreased from an average of 4.21 to 4.16 in the Coastal Plain. Production levels in 2009 were reported in loads per week with a statewide average of 20.42 loads per crew per week. If we assumed this production was entirely from tractor-trailers (highest average payloads) with a payload of 22.7 t/load (25 tons) then that would be 463.1 t/crew/week. Assuming tractor-trailer payloads for all production would result in the highest possible production levels from the 2009 survey results, which is still lower than the 2014 calculated overall average production level of 530.9 t/crew/week. Increases in production per crew since the 2009 survey could be from increased mechanization and overall efficiency, but are also likely a result of improved markets and increased demand following the recession, which likely impacted operations during the previous survey timeframe.

Business owners were asked to estimate the total current value for all logging equipment used for producing wood (excluding trucks and trailers). We used a method similar to Baker and Greene's [10] to evaluate production in terms of annual tonnes per $1000 investment and found an overall average of approximately 112.6 tonnes per $1000 investment. This was highest in the Coastal Plain at over 125 tonnes annually per $1000 investment (Table 3). Baker and Greene [10] looked only at feller-buncher/grapple skidder crews in Georgia between 1987 and 2007 and reported values for tonnes per $1000 investment that ranged between 112 and 165. They calculated a value for logging equipment based on reported equipment type and age, whereas we asked respondents to estimate the value of their equipment. However, the results are similar to their findings, especially in the Coastal Plain, which predominantly uses feller-buncher/grapple skidder operations.

Table 3. Logging business production characteristics by region.

	n	Mean	MIN	MAX	SD
Mountains					
Crews per business	99	1.05	1	2	0.22
Workers per crew	98	2.80	1	12	1.50
Tonnes per crew per week	89	224.62	16	1247	201.35
Tonnes per business per week	87	245.42	16	1247	251.27
Total current investment [1] (thousand USD)	87	257.68	2	3000	514.39
Annual tonnes per $1000 investment [2]	73	99.18	10	499	92.02
Piedmont					
Crews per business	180	1.20	1	4	0.51
Workers per crew	178	3.31	1	15	1.85
Tonnes per crew per week	169	581.84	11	2948	509.77
Tonnes per business per week	169	824.69	11	6804	1125.81
Total current investment [1] (thousand USD)	158	491.81	5	6000	712.42
Annual tonnes per $1000 investment [2]	149	113.01	4	653	97.50
Coastal Plain					
Crews per business	64	1.48	1	5	0.91
Workers per crew	64	4.16	1	11	1.85
Tonnes per crew per week	57	864.01	23	2268	509.45
Tonnes per business per week	57	1403.55	23	6350	1396.08
Total current investment [1] (thousand USD)	56	794.73	10	5000	919.48
Annual tonnes per $1000 investment [2]	49	125.17	9	680	115.51
Overall [3]					
Crews per business	373	1.20	1	5	0.55
Workers per crew	369	3.29	1	15	1.77
Tonnes per crew per week	339	530.94	11	3062	505.56
Tonnes per business per week	336	761.37	11	6804	1105.24
Total current investment [1] (thousand USD)	328	462.07	2	6000	704.14
Annual tonnes per $1000 investment [2]	291	112.56	4	680	99.48

[1] Indicates current value of harvesting equipment excluding trucking. [2] Annual production (tonnes) calculated using reported weekly production ×50 weeks/year. [3] Includes responses from all Virginia regions and adjoining states or regions not specifically identified.

Reported production levels (Table 3) indicate that there is considerable variation in weekly production between businesses. To further compare production levels, the businesses were sorted by region based on the total reported tonnes of production per business. Businesses were then categorized into groups based on reported production levels. The categories were high (upper 1/3 of businesses), medium, and low (lower 1/3 of businesses) (Figure 6). Results indicate that production is not evenly distributed across logging businesses, and a relatively small proportion of the businesses account for the majority of production. Across all regions, the high-production businesses accounted for over two-thirds of all production and the low-production businesses accounted for less than 10% of all reported production. This was most pronounced in the Piedmont, where the high-production group accounted for nearly 75% of production.

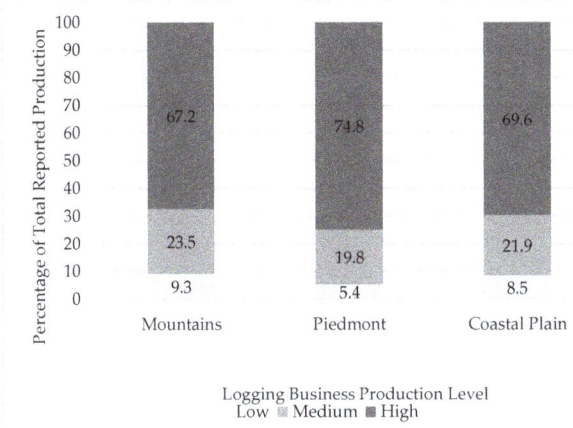

Figure 6. Percentage of total reported production by region associated with low, medium, and high production logging businesses when categorized based on the upper, middle, and lower thirds of logging business production levels.

Business owners were asked to indicate the equipment utilized for different harvesting functions on their operations. They were provided a list of common equipment options for each harvesting function and were asked to circle the one that they most commonly used. They were also provided an option for listing "other" equipment used for that function. While almost all types of equipment can be found on operations in each region, the most common types of equipment utilized varied by region (Table 4). Chainsaws were the most common felling method reported by businesses in the Mountains (88.4%), while rubber-tired feller-bunchers were most common in the Coastal Plain (84.1%). The Mountains showed the most diversity in terms of trucking configurations, while operations in the Piedmont and Coastal Plain predominantly utilized tractortrailers.

The percent of all responses by logging businesses shows how there is diversity in terms of the types of equipment utilized on harvesting operations (Table 4). However, there are also substantial differences in the overall production rates between different types of operations (Figure 6). Therefore, we also determined the percent of overall total reported production that was produced by businesses commonly utilizing that type of equipment for a specific harvesting function. This shows the relative proportion of all timber harvested using that method. Some harvesting methods tend to be less productive in terms of tonnes produced. For example, overall, 43.7% of respondents indicated that they primarily used a chainsaw for felling; however, only 10.1% of all reported production was from operations primarily felling with chainsaws (Table 4). While there are multiple equipment types used in each region, the overall majority of production is felled with rubber-tired feller-bunchers (81.7%), skidded with grapple skidders (93.2%), delimbed using pull-through delimbers (64.6%), bucked with a buck/slasher saw (88.9%), loaded with a trailer mounted knuckleboom loader (90.0%), and trucked using a tractor-trailer (90.9%).

Table 4. Equipment used for each harvesting function. Percent of reported production represents the proportion of overall total reported production (tonnes) associated with businesses indicating they most commonly utilized that equipment.

	% Responses (% of Overall Total Reported Production in Tonnes)							
	Mountains		Piedmont		Coastal Plain		Overall	
Felling								
Chainsaw	88.4	(66.6)	29.6	(6.7)	11.1	(0.6)	43.7	(10.1)
Rubber-tired feller-buncher	4.9	(13.8)	60.7	(85.6)	84.1	(93.7)	48.2	(81.7)
Tracked feller-buncher	3.9	(15.4)	1.1	(0.8)	1.6	(1.3)	1.8	(2.1)
Cut-to-length harvester	0	(0)	0.5	(0.3)	0	(0)	0.3	(0.2)
Other/Multiple [1]	2.9	(4.1)	8.1	(6.6)	3.2	(4.4)	6.0	(5.9)
Skidding								
Cable skidder	48.0	(35.4)	11.4	(1.7)	6.3	(0.4)	22.1	(4.4)
Grapple skidder	38.2	(56.1)	78.8	(96.0)	92.1	(99.6)	68.2	(93.2)
Forwarder	1.0	(0.5)	0.5	(0)	0	(0)	0.5	(<0.1)
Bulldozer	2.0	(0.3)	2.2	(<0.1)	0	(0)	1.6	(<0.1)
Other/Multiple [1]	10.8	(7.7)	7.1	(2.3)	1.6	(<0.1)	7.6	(2.4)
Delimbing								
Chainsaw	84.4	(61.9)	36.9	(12.5)	14.1	(0.9)	46.6	(13.0)
Delimbing gate	1.0	(2.1)	5.6	(5.5)	7.8	(13.1)	4.5	(7.5)
Pull-through delimber	8.7	(18.0)	46.9	(63.1)	70.3	(76.3)	40.2	(64.6)
Chain-flail delimber	0	(0)	1.1	(0.9)	4.7	(5.4)	1.3	(2.2)
Stroke delimber	1.0	(4.2)	2.8	(10.1)	0	(0)	1.6	(5.6)
Other/Multiple [1]	4.9	(13.8)	6.7	(7.9)	3.1	(4.3)	5.8	(7.1)
Bucking								
Chainsaw	40.2	(21.0)	18.1	(3.2)	13.6	(5.2)	23.7	(5.5)
Buck/Slasher saw	52.6	(76.0)	79.1	(89.6)	77.9	(90.0)	71.3	(88.9)
Swing-boom processor	0	(0)	0	(0)	1.7	(<0.1)	0.3	(<0.1)
Tree length only (no bucking)	0	(0)	1.7	(4.1)	3.4	(2.5)	1.4	(3.0)
Other/Multiple [1]	7.2	(3.0)	1.1	(3.1)	3.4	(2.3)	3.3	(2.6)
Loading								
Trailer mounted knuckleboom	50.5	(56.6)	78.4	(90.3)	88.7	(97.1)	73.0	(90.0)
Mobile knuckleboom	30.1	(22.4)	10.8	(8.2)	3.2	(0.4)	14.4	(6.5)
Self-loading trucks	6.8	(4.9)	0.5	(0.4)	0	(0)	2.1	(0.6)
Front-end loader	5.8	(1.4)	4.9	(0.3)	3.2	(0.2)	4.7	(0.4)
Other/Multiple [1]	6.8	(14.7)	5.4	(0.8)	4.9	(2.3)	5.8	(2.5)
Trucking								
Tractor-trailer	23.0	(35.5)	70.6	(95.1)	89.5	(99.2)	58.5	(90.9)
Single axle	7.0	(1.8)	8.3	(0.7)	3.5	(0.1)	7.4	(0.7)
Tandem axle	16.0	(5.7)	9.4	(1.3)	1.8	(0.7)	10.9	(1.7)
Tandem with pup trailer	8.0	(11.3)	2.8	(1.1)	0	(0)	4.4	(1.9)
Tri-axle	14	(8.8)	5.0	(1.6)	0	(0)	6.3	(1.6)
Tri-axle with pup trailer	15	(20.6)	1.1	(.2)	0	(0)	4.6	(1.8)
Other/Multiple [1]	17.0	(16.3)	2.8	(0)	5.2	(<0.1)	7.9	(1.4)
Chipping (or biomass production)								
Whole tree (dirty chips)	7.6	(24.1)	29.2	(51.5)	18.6	(29.7)	20	(39.8)
Whole tree w/flail (clean chips)	1.0	(3.2)	2.6	(1.4)	1.4	(4.3)	1.7	(2.3)
Whole tree w/flail + grinder	0	(0)	0.5	(1.6)	2.9	(10.2)	0.8	(4.1)
Other/Multiple [1]	1.0	(0)	1.6	(3.9)	4.3	(5.7)	2.0	(4.2)
No chipper reported	90.4	(72.7)	66.1	(41.6)	72.8	(50.1)	75.5	(49.6)

[1] Respondents were asked to select the most common one, but sometimes selected multiple options.

3.4. Biomass Harvesting

Respondents were asked to report if they harvested biomass or "fuel chips" for energy. Overall, 25% of businesses reported they harvested biomass. Previous studies have noted that biomass harvests in Virginia were typically performed on integrated harvesting operations, where biomass is produced from residues concurrently with roundwood production [17]. With integrated harvesting operations, biomass harvesting is often more feasible for businesses with higher production levels because there is a larger quantity of residues available and it is more likely that the operation could justify adding a chipper to utilize residues for energy. Biomass harvesting was compared by region across production categories (high, medium, and low production). Production of biomass was more common among the higher production levels, with the majority of the high-production businesses producing biomass in the Piedmont and Coastal Plain (Table 5). Significant differences across business production categories were observed in the Mountains ($X^2 = 0.0054$), Piedmont ($X^2 < 0.0001$), and Coastal Plain ($X^2 = 0.0421$). This question related to harvesting fuel chips was independent of the question related to type of equipment used for chipping (Table 4), so while the percentages are similar to those reporting chippers they do not match exactly because respondents who indicated they produced fuel chips may not have all reported the primary type of equipment utilized.

Table 5. Percent of businesses indicating harvest of biomass/fuel chips by region across production category when sorted by production level.

	Percent Harvesting Biomass		
Production Category	**Mountains**	**Piedmont**	**Coastal Plain**
All respondents	8.7	35.2	29.2
High Production (Upper third)	20.7	60.7	52.6
Medium Production (Middle third)	3.4	39.3	26.3
Low Production (Lower third)	0	11.1	15.8

3.5. Harvest Planning and Implementation of BMPs

Harvest planning is an important tool to ensure efficient and effective logging operations. Pre-harvest planning of items such as roads, skid trails, landings, and stream crossings can impact the productivity of the operation, and can also have an impact in terms of compliance with implementation of Best Management Practices (BMPs) for water quality. Implementation of BMPs for water quality is important throughout the USA [18] and responses indicate that Virginia logging businesses invest a substantial amount of time in pre-harvest planning and the implementation of BMPs to protect water quality.

Loggers in the Mountains and Piedmont tended to spend more time on pre-harvest planning than those in the Coastal Plain. The most common response indicated that they typically spend a half to a full day per tract on pre-harvest planning, while those in the Coastal Plain tended to spend a half day or less (Figure 7). This is likely due to the more challenging terrain in the Mountains and Piedmont. Similarly, logging businesses in the Mountains tended to spend more time per tract closing out the site and implementing BMPs. The most common response in the Mountains was 1–2 days spent closing out the tract, whereas the most common time reported in the Piedmont and Coastal Plain was a half to a full day (Figure 8).

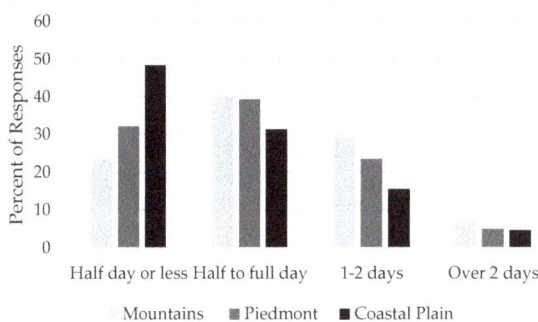

Figure 7. Typical time spent per tract on pre-harvest planning to locate roads, skid trails, landings, and stream crossings.

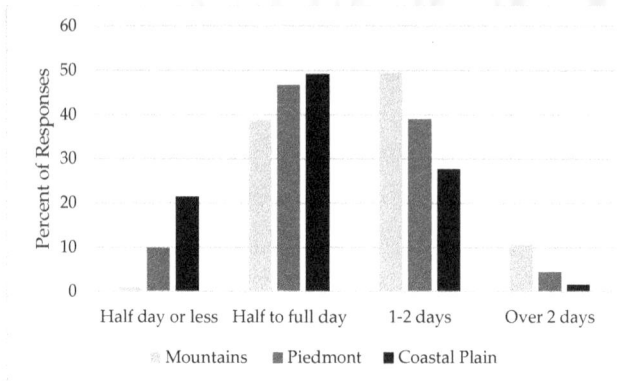

Figure 8. Typical time spent per tract on closing out the site and implementing Best Management Practices (BMPs) for water quality.

3.6. Use of Technology and Common Operational Challenges

Owners were asked about the use of various types of technology in their business (Table 6). Similar to other operational characteristics, this also varied by region. Use of all technologies across all regions showed increases from the 2009 survey. It is interesting to note that the reported percentage of e-mail use is higher than use of the Internet for businesses in the Piedmont and Coastal Plain. This could potentially be because the owners were not necessarily thinking of e-mail as using the Internet and thought of Internet usage as more related to things such as company websites, social media, online purchasing, or other web-based activities.

Table 6. Percentage by physiographic region of logging business owners indicating use of different technologies in their business.

Technology	Percent Who Indicated Use of This Technology [1]		
	Mountains	Piedmont	Coastal Plain
GPS	18.8	29.8	44.6
Computer mapping	13.7	27.0	27.7
Internet	43.1	51.1	75.4
E-mail	39.6	53.9	80.0

[1] Respondents were asked to report all technologies they used, so percentages may be greater than 100%.

At the end of the questionnaire, respondents were given an open-ended question related to the biggest problem they faced in their business. There was considerable variation in these responses and many were hard to categorize; however, there were some common responses. Common issues cited by business owners often related to challenges with labor and dealing with employees, e.g., "getting reliable help", or "lack of qualified personnel at a reasonable rate", or "finding people who are willing to show up for work". Other common problems identified included complying with regulations, especially regulations related to transportation and trucking. With most logging businesses, fuel for trucks and in-woods equipment can be a major expense, so it is not surprising that one of the common problems identified was the cost of fuel. Other common expenses in logging operations were identified as a problem, such as the cost of insurance, parts, and new equipment. Prices received for their products at the mill were also a common problem identified by business owners. While some problems such as finding qualified employees or dealing with regulations may be common in any business, some problems identified by logging businesses are somewhat unique to the industry. For example, some owners reported challenges related to procuring the future tracts they will harvest, e.g., "running out of timber to harvest", "lack of timber", or "trying to find timber close to home". Weather in general was listed as a problem, as well as finding "winter tracts" or tracts with suitable ground conditions that could be harvested during the typically wetter winter months.

4. Conclusions

Logging businesses vary substantially by physiographic region in Virginia. Overall, the majority of harvests occurred on forest lands owned by private individuals (89.6%), and private landowners often have multiple objectives for their forest lands. A diverse logging workforce in terms of size, equipment, and type of products typically harvested may better enable logging businesses to effectively work with a diverse group of private landowners who often have multiple forest management objectives.

Logging business production rates were one of the major differences by region. Average production rates for businesses in the Coastal Plain were more than five times those in the Mountains. This could be the result of a number of operational differences, including smaller crews, fewer crews per business, and a greater reliance on manual felling in the Mountains. While a wide variety of logging operations can be found in any region, the vast majority of reported production across Virginia was felled with a rubber-tired feller-buncher (81.7%), skidded with a grapple skidder (93.2%), loaded with a trailer mounted knuckle boom loader (90.0%), and delivered to the mill using a tractor-trailer (90.9%). Regardless of the region, the high-production (upper third) businesses accounted for the majority of production. Over two-thirds of all production is harvested by the high-production (upper third) logging businesses. As markets for biomass increased, many loggers responded by adding a chipper to produce biomass; this was especially true among the high-production operations, where the majority in the Piedmont (60.7%) and Coastal Plain (52.6%) reported producing biomass.

Logging is a challenging business and many owners report a number of significant challenges to their business. However, this survey also indicates that the average production per crew and number of crews owned per business increased since the previous survey (2009). The average size of businesses appears to be increasing and much of the production capacity is concentrated in the high-production (upper third) businesses. The future health of the industry and the ability of the logging workforce to provide the wood needed by mills may depend on the success and viability of this group of higher production loggers. Despite many previous challenges, the logging workforce has continued to adapt with changing markets and other factors and will likely continue to do so in the future.

Acknowledgments: This project received financial and logistical support from the Virginia SHARP Logger Program, the Department of Forest Resources and Environmental Conservation (FREC) at Virginia Polytechnic Institute and State University, and the McIntire-Stennis Program of the National Institute of Food and Agriculture, U.S. Department of Agriculture. The Virginia Tech Open Access Subvention Fund (OASF) provided funding to publish in open access.

Author Contributions: Scott M. Barrett was involved in the original survey design, administered this survey, analyzed the data and wrote the manuscript. M. Chad Bolding and John F. Munsell were involved in the design of the original survey and reviewed and edited the manuscript.

Conflicts of Interest: The authors declare no conflict of interest.

References

1. Rephann, T.J. *The Economic Impacts of Agriculture and Forest Industries in Virginia*; Weldon Cooper Center for Public Service, University of Virginia: Charlottesville, VA, USA, 2013; pp. 31–37.
2. Virginia Department of Forestry. 2016 State of the Forrest. Available online: http://dof.virginia.gov/infopubs/sof/SOF-2016pub.pdf (accessed on 12 September 2017).
3. Milauskas, S.J.; Wang, J. West Virginia logger characteristics. *For. Prod. J.* **2006**, *56*, 19.
4. Dirkswager, A.L.; Kilgore, M.A.; Becker, D.R.; Blinn, C.; Ek, A. Logging business practices and perspectives on harvesting forest residues for energy: A Minnesota case study. *North. J. Appl. For.* **2011**, *28*, 41–46.
5. Egan, A.; Taggart, D.; Annis, I. Effects of population pressures on wood procurement and logging opportunities in northern New England. *North. J. Appl. For.* **2007**, *24*, 85–90.

6. Munsell, J.F.; Barrett, S.M.; Bolding, M.C. An exploratory study of biomass harvesting among logging firms in Virginia and North Carolina. *For. Sci.* **2011**, *57*, 427–434.

7. Allred, S.; Michler, C.; Mycroft, C. Midwest logging firm perspectives: Harvesting on increasingly parcelized forestlands. *Int. J. For. Res.* **2011**, *2011*. [CrossRef]

8. Moldenhauer, M.C.; Bolding, M.C. Parcelization of South Carolina's private forestland: Loggers' reactions to a growing threat. *For. Prod. J.* **2009**, *59*, 37.

9. Blinn, C.R.; O'Hara, T.J.; Chura, D.T.; Russell, M.B. *Status of the Minnesota Logging Sector in 2011*; Retrieved from the University of Minnesota Digital Conservancy; University of Minnesota: Saint Paul, MN, USA, 2014.

10. Baker, S.A.; Greene, W.D. Changes in Georgia's logging workforce, 1987–2007. *South. J. Appl. For.* **2008**, *32*, 60–68.

11. Bolding, M.C.; Barrett, S.M.; Munsell, J.F.; Groover, M.C. Characteristics of Virginia's logging businesses in a changing timber market. *For. Prod. J.* **2010**, *60*, 86–93. [CrossRef]

12. Virginia SHARP Logger Program. 2017. Available online: http://sharplogger.vt.edu/ (accessed on 26 October 2017).

13. Dangle, C.; Vinson, J.A.; Barrett, S.M. Regional Forest Harvest Characteristics across Virginia. Virginia Cooperative Extension Publication Number ANR-264NP. 2017. Available online: http://www.pubs.ext.vt. edu/ANR/ANR-264/ANR-264.html (accessed on 11 July 2017).

14. Dillman, D.A. *Mail and Internet Surveys: The Tailored Design Method*; John Wiley & Sons: New York, NY, USA, 2000; p. 480.

15. JMP, Version 13.0.0. Available online: https://www.jmp.com/en_us/home.html (accessed on 27 November 2017).

16. Cooper, J.; Becker, C. *Virginia's Timber Industry—An Assessment of Timber Product Output and Use, 2007*; Resour. Bull. SRS—155; Department of Agriculture Forest Service, Southern Research Station: Asheville, NC, USA, 2009; 33p.

17. Barrett, S.M.; Bolding, M.C.; Aust, W.M.; Munsell, J.F. Characteristics of logging businesses that harvest biomass for energy production. *For. Prod. J.* **2014**, *64*, 265–272. [CrossRef]

18. Cristan, R.; Aust, W.M.; Bolding, M.C.; Barrett, S.M.; Munsell, J.F.; Schilling, E. Effectiveness of forestry best management practices in the United States: Literature review. *For. Ecol. Manag.* **2016**, *360*, 133–151. [CrossRef]

Article

Aboveground Biomass Equations for Small Trees of Brutian Pine in Turkey to Facilitate Harvesting and Management

Mehmet Eker [1,*], Krishna P. Poudel [2] and Ramazan Özçelik [1]

[1] Forest Engineering Department, Faculty of Forestry, Süleyman Demirel University, 32260 Isparta, Turkey; ramazanozcelik@sdu.edu.tr

[2] Department of Forest Engineering, Resources, and Management, Oregon State University, Corvallis, OR 97333, USA; Krishna.Poudel@oregonstate.edu

* Correspondence: mehmeteker@sdu.edu.tr; Tel.: +90-505-923-8203

Received: 12 October 2017; Accepted: 30 November 2017; Published: 3 December 2017

Abstract: Brutian pine (*Pinus brutia* Ten.) is the most widespread conifer species in the Eastern Mediterranean. Aboveground biomass equations for small diameter brutian pine trees are needed for accurate fuel inventory and to assess carbon sequestration potential. In this study, we developed tree biomass models based on 143 brutian pine saplings measured in 11 research plots. Aboveground biomass (AGB) was modeled with a nonlinear mixed effects model which accounted for the variability among plots. The predicted total AGB was then distributed into foliage, branch and stem components. The Beta, Dirichlet, and multinomial logistic regressions were unbiased in their estimates of biomass component proportions. The Dirichlet regression has the advantage of an additive property and does not require non-standard data.

Keywords: biomass equations; beta regression; multinomial logistic regression; Dirichlet regression; small trees

1. Introduction

Brutian pine is the most important tree species in Turkey, both ecologically and economically. Brutian pine forests cover about 25% of Turkey's total forest area which is about 5.6 million hectares with a current standing volume of approximately 270 million m^3 [1]. Because of its valuable wood properties, it is one of the most important pine species for the forest products industry in Turkey [2]. Furthermore, brutian pine plays a key role in providing important benefits and environmental services such as protection of soil and water resources, conservation of biological diversity, and climate change mitigation and adaptation in Turkey [3]. Therefore, detailed information about stand structure, total biomass, or biomass of different tree components is needed for sustainable forest management and harvesting of utilizable potential of the brutian pine forest.

Accurate estimation of tree or forest biomass is a key requirement for calculating biomass energy, carbon sequestration, as well as for studying climate change, forest health, site productivity, and nutrient cycling [4]. Furthermore, the increasing use of weight or biomass as a measure of forest productivity with ever changing market conditions has heightened the need for accurate estimates of total and component biomass of trees.

The approaches in biomass estimation depend on the scale of analysis, need for detail, user group interest, and purpose of estimation [5]. Generally, there are three approaches used to estimate total and component biomass. In the first group of methods, total tree and component biomass (e.g., stem, crown, branches, leaves, and bark) are regressed against easily measurable tree attributes such as diameter at breast height (DBH) or DBH and height using linear and nonlinear regression.

Such methods, however, do not ensure that the sum of biomass predictions from component models is equal to the biomass prediction from the total aboveground biomass (AGB) model. This issue of additivity can be resolved by fitting component and total biomass equations as a simultaneous system. Such methods, if fitted with the ordinary least squares approach, ignore the inherent correlations among the component models [6]. Therefore, the second group of methods is a regression-based approach that uses a system of equations to deal with this issue of non-additivity or incompatibility. Different estimation methods have been suggested to ensure the additivity in a system of biomass equations for both linear and nonlinear models. In this framework, seemingly unrelated regression (SUR) and non-linear seemingly unrelated regression (NSUR) have become more popular in recent years [4,7]. The third group of methods, fairly new in biomass estimation, predicts the proportions of biomass in each component using generalized linear models such as beta, Dirichlet, and multinomial logistic regression. Predicted proportions are then applied to the observed total AGB [8] or the total AGB obtained from fitting a separate equation [9].

Information regarding estimations of total and component tree biomass is currently lacking in Turkey. Four published sources for brutian pine biomass include studies based on a sample of 30 trees from Eastern Mediterranean Region [10], a sample of 24 trees in north and south Aegean Islands of Greece [11], a sample of 201 trees from Syria and Lebanon [12], and a sample of 164 trees in southern of Mediterranean Region of Turkey [13]. However, two of these studies were conducted outside of Turkey. Allometric equations used in these studies are simple expressions relating tree level biomass to expressions of tree size except for Özçelik et al. [13]. Common independent variables of biomass models are DBH and tree height, although some studies have used crown dimensions as well. Poudel and Temesgen [8] indicated that factors that affect growth in tree diameter and height (e.g., genetics, site quality, environmental factors, stand density, tree and stand age) also affect AGB and component biomass. Therefore, there is a need to develop tree-level biomass models using data from areas within the natural distribution of this species in the Mediterranean Region.

Recently, Chaturvedi and Raghubanshi [14] indicated that woody individuals of small diameter classes have a significant role in the estimation of total AGB, since this component of the forest comprises a significant proportion, by number, of the tree population. These trees are not of significance for volumetric production but can contribute substantially towards biomass and bioenergy as they have a faster growth than the trees in larger diameter classes. The information about small diameter trees can be used in inventories of fuel or wood energy, to assess the potential of young stands as fiber sources and the carbon sequestration potential of natural stands, and as indicators of net primary production [15]. Additionally, accurate assessment of wildfire behavior requires quantitative estimates of available fuel load by size class and condition in terms of forest management [16].

Many of the studies concerning biomass estimation have focused solely on the estimation of individual trees having DBH greater than 8 cm, ignoring the AGB of small diameter trees at the sapling stage. As a result, available woody biomass and the carbon stored at early stages are often neglected. Only a few studies address the estimation of small diameter tree biomass in tropical dry forests [14,17], in temperate deciduous forest [15,18], and in temperate pine forest [19]. Ideally, the biomass equations should be developed covering all size classes without discontinuity at any tree size.

Small diameter tree biomass estimates in Turkey are limited to a few studies and their predictions are mainly for crown biomass components and are based on a small sample size [20,21]. Trees with less than 8 cm DBH are considered small diameter trees in Turkey and are not measured in regular forest inventory applications such as industrial roundwood production. Therefore, there is no reliable information about tree volume and total tree biomass or biomass components, such as stem and branches, for such trees. Reliable small tree biomass models are especially important in fire-prone forest ecosystems such as brutian pine forests in the Mediterranean Region of Turkey, where nearly 15% of the forested area is dominated by sapling sized (0.1 cm–8 cm DBH) stands. The lack of aboveground small diameter tree biomass equations has also affected the accuracy of assessing the amount of utilizable woody biomass, forest fuel inventories, and carbon sequestration potential.

AGB is commonly divided into three major components: stem, stem bark and crown (branch and leaves/foliage). The component biomass models are useful to account for the variability within the tree. In addition, tree component biomass is used for different purposes that require separate estimates. The stem wood is used for industrial wood, chip-board wood, and fuel/energy wood production; crown biomass can provide information on fuel load and wildfire assessment, and woody sections of branches are also useful in bioenergy production. Therefore, component biomass estimates are necessary to determine available forest products within the concept of sustainable management and harvesting of small trees.

In this study, destructive sampling was used to measure the biomass of foliage, branch, and bole (stem) of sapling stage brutian pine in the Mediterranean Region of Turkey. The objective of the study was to develop estimation models for total AGB and component biomass of small diameter brutian pine. Thus, the aim was to estimate the type and amount of biomass that emerged after silvicultural interventions were applied to young and small diameter trees. A nonlinear mixed effects model was fitted to predict total tree biomass as a function of DBH and total tree height accounting for the variation among plots. Predicted total tree biomass was then apportioned into different components according to the predicted proportions from beta, Dirichlet, and multinomial logistic regressions.

2. Materials and Methods

2.1. Data

This study was carried out in the natural and pure brutian pine forest at the Bucak Model Forest Enterprises (37°38' N–30°50' W) located in Isparta Forest District, part of the Mediterranean Forest Region of Turkey. Tree species, brutian pine, was selected based on its relative abundance and because it is a fast growing and industrially most valuable species for this region. The study site/plots were randomly selected from fully operational stands where regular (early) thinning operations had been carried out. A heavy thinning was conducted in the 11–26 year-old young stands. This treatment is considered part of common management practices adopted in commercial forests at the sapling stages. The younger stands experience a similar thinning treatment once they reach the respective stand development stage.

The studied sites are found on cracked limestone, marnly and flysch deposits with alternating sandy, silt and limey layers [22]. The mean annual temperature varies from 15 °C to 20 °C, while the annual precipitation ranges from 400 mm to 900 mm, with climates ranging from semi-arid to humid zone. The mean tree density, for trees with DBH smaller than 8 cm, was between 1600 and 3100 per ha. The thinning intensity was between 58–72% per ha, i.e., removal was between 1200 and 2200 trees per ha. Mean slope gradient varied from 30 to 65 percent. Ground condition was mountainous, uneven, rough, and undulating.

Detailed biomass data were collected by destructively sampling 143 brutian pine trees from 11 research plots installed in natural stands. All plots were located within a radius of 30 km and with similar environmental conditions. The fieldwork was carried out between the summers of 2014 and 2015. The plots were subjectively selected to represent the existing range of ages, yield class, stand densities, and sites throughout the area of distribution of brutian pine in the Western Mediterranean Region. The plot size ranged from 225 to 400 m^2, depending on the number of trees per hectare. All trees in each sample plot were labeled with a number. Descriptive variables such as status (alive or dead) of each tree in the plots were also recorded. In each research plot, seven to fifteen trees with DBH < 8 cm were selected and flagged as sample trees with the aim to cover the range of DBH in each plot. Before felling, DBHs were measured. Two perpendicular diameters outside-bark (1.3 m above ground level) were measured with a digital caliper to the nearest 0.1 cm and were arithmetically averaged to obtain DBH (cm). The trees were later felled, leaving stump of average height 0.10 m, and total bole height was measured to nearest 0.01 m to calculate the total tree height (in meters). During the felling of a sample tree, it was held by one or more workers to avoid deformation and material loss. Trees deformed during the felling were discarded. As soon as trees were cut, weights of total AGB were recorded using a precise portable scale (30 kg capacity and ±1 g accuracy) in the field. AGB was then separated into stem and crown components. The crown was further divided into branch and foliage. Subsequently, stems were stripped of their branches and foliage was also separated from them. The fresh weight of each component (stem, branches, and foliage) was also recorded in the field.

Representative sub-samples were taken from each biomass component to determine fresh to oven-dried biomass ratios. For each tree, after all the branches were separated, the whole stem was measured and cut with a chainsaw into 1 m long sections. Lost sawdust mass during tree and stem cutting or other sections was neglected because it was generally less than 0.03 kg per tree. To avoid the losses, newly sharpened chainsaws and axes were used in partitioning stem and branches. Subsequently, discs were (5 cm thick) cut and taken from the stem section at 1 m regular intervals along the full length of stem. Additional stem discs were also taken from the previously marked gravity center of the boles in order to guarantee reliable results. For branch and foliage, four 10–15 cm branch sub-samples were immediately taken from the representative branches at the center of gravity of crown and from different parts of the tree, distributed evenly within the crown based on a randomized branch sampling procedure. Half of the branch sub-samples were further stripped of all needles in order to determine foliage to branch biomass ratio. Discs and representative sub-samples of branch and foliage were then weighed and taken to the laboratory. Discs taken from the stem were oven dried at 104 °C until constant weight for a minimum of two days depending on the amount of sample. Foliage and branches were dried at 65 °C until constant weight for at least two days. After drying, sub-samples of the biomass components (discs, branches, and foliage) were reweighed to determine the average fresh to oven-dried biomass ratio which was later applied to convert fresh biomass into dry biomass for each component as well as for the whole tree. The summary statistics of DBH, height, age, total and component dry weights (biomass) are given in Table 1. In Figure 1, the values of each of the biomass components and total AGB are plotted against the tree level variables of DBH and tree height. This figure indicates that, for a given tree height, there is a large variation in the component and total biomass values.

Table 1. Summary statistics of sample trees used in estimating total aboveground biomass and its components.

Variables	n	Minimum	Maximum	Mean	Standard Deviation
DBH	143	1.50	7.8	4.40	1.53
HT	143	1.30	7.70	3.86	1.42
AGE	143	11.00	26.00	14.87	4.68
AGB	143	0.42	15.48	4.45	3.03
STM	143	0.18	10.29	2.32	1.85
BCH	143	0.15	3.86	1.42	0.86
FOL	143	0.05	2.82	0.71	0.52

DBH, diameter at breast height (cm); HT, total tree height (m); AGE, age in years; AGB, aboveground biomass (kg); STM, stem biomass (kg); BCH, branch biomass (kg); FOL, foliage biomass (kg).

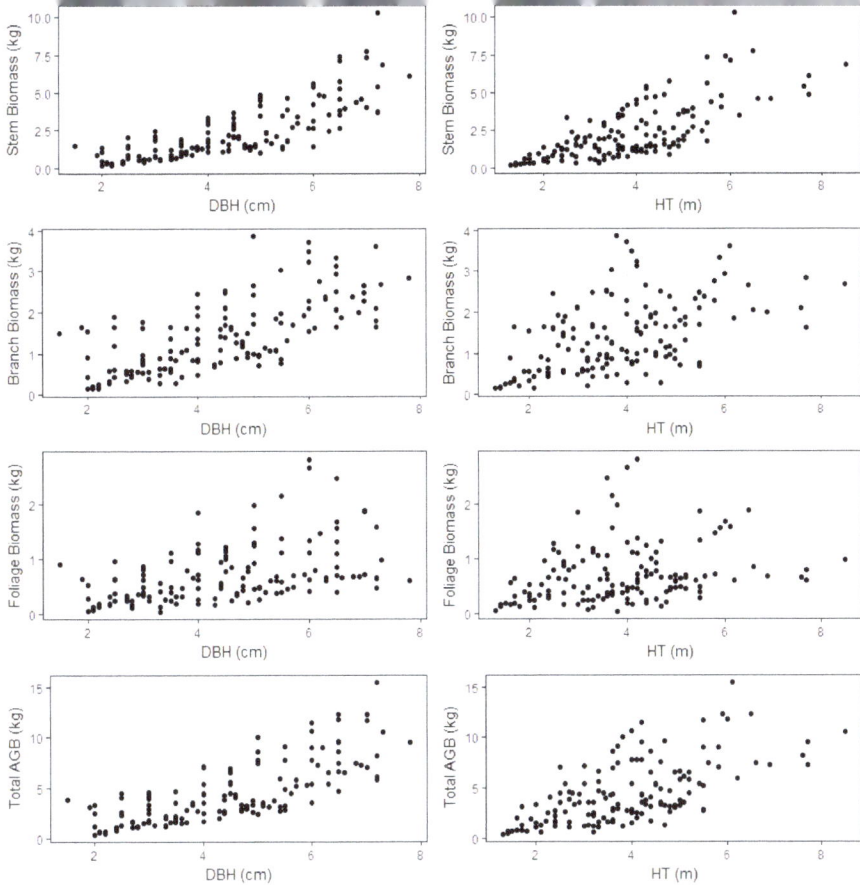

Figure 1. Sample trees of each of biomass component and total aboveground biomass (AGB) versus tree level variables for analyzed tree species. DBH and HT are diameter at breast height (cm) and total tree height (m).

2.2. Data Analysis

A two-step process was used to estimate AGB and its components. In the first step, a model to predict total AGB was fitted and in the second step, models to predict the proportions of biomass in different components were fitted. The methods used in each step are discussed below.

2.2.1. Estimating AGB

DBH is the most commonly used predictor variable of AGB. However, including height in addition to DBH improves the accuracy of such models [8,23]. A variety of model forms such as power function (e.g., Picard et al. [23]), simple logarithmic model resulting from logarithmic transformation of power function (e.g., Poudel and Temesgen [9]), and exponential model (e.g., Ritchie et al. [24]; Poudel and Temesgen [8]) have been used to relate such dendrometric variables to AGB. A simple linear model in the form of Equation (1) was first tested.

$$AGB_i = \beta_0 + \beta_1 \times DBH_i + \beta_2 \times HT_i + \epsilon_i \tag{1}$$

where AGB_i is the total aboveground biomass of ith tree; β_0, β_1, and β_2 are regression parameters to be estimated from the data; DBH_i is the diameter at breast height of the ith tree; HT_i is the height of ith tree, and ϵ_i is the random error.

The regression coefficient for height (estimate of β_2 in Equation (1)) was not statistically significant (p-value = 0.514). A logarithmic model in the form of Equation (2) was then tested.

$$ln(AGB_i) = \beta_0 + \beta_1 \times ln(DBH_i) + \beta_2 \times ln(HT_i) + \epsilon_i \tag{2}$$

where $ln(\cdot)$ is the natural log and all other variables are the same as defined previously.

Parameters β_0, β_1, and β_2 in Equations (1) and (2) are not necessarily the same. Coefficient of height was still not statistically significant (p-value = 0.237). Scatterplots of DBH and height against AGB, however, did not show severe departure from the linear relationship (Figure 1). In addition, power and exponential models were also deemed insufficient.

Data for this study originated from 11 research plots installed in natural stands. Thus, our dataset has a hierarchical nature and trees within a plot have more similar allometry than trees from different plots. Analysis of such a dataset is best done by separating variance due to the plots using mixed effects models. Therefore, a nonlinear mixed model in the form of Equation (3) was selected as the final model for predicting AGB using DBH and tree height. All the regression coefficients of this model were statistically significant at the 0.05 level of significance. In addition, the likelihood ratio test indicated that the random effect was warranted ($\chi^2_{(1)} = 181.67$, p–$value < 0.001$).

$$AGB_{ij} = exp\left(\beta_0 + b_i + \beta_1 ln(DBH_{ij}) + \beta_2 ln(HT_{ij})\right) + \epsilon_{ij} \tag{3}$$

where AGB_{ij} is the total aboveground biomass of ith tree in jth plot; b_i is the random plot effect and $b_i \sim N(0, \sigma_b^2)$; DBH_{ij} is the diameter at breast height of the ith tree in jth plot; HT_{ij} is the height of ith tree in jth plot. In addition, b_i is independent of $\epsilon_{ij} \sim N(0, \sigma_e^2)$.

2.2.2. Estimating Component Biomass

Traditionally, the amount of biomass present in different tree components has been modeled using similar equations as the models used to predict total tree biomass. To ensure the additivity of such equations, the constrained seemingly unrelated regression has been popular. Recently, component biomass has been estimated as the product of predicted proportion obtained from different generalized linear models [5,8,9,25] and the predicted total biomass obtained from the method described in the previous section. In this study, the proportion of biomass in different tree components was estimated using three generalized linear models: beta regression, Dirichlet regression, and multinomial logistic regression. Biomass in different components was obtained as the product of predicted proportions and the total AGB obtained from Equation (3).

Beta Regression

Selection of a regression method depends on the type of the dependent variable. Proportions of component biomass, the dependent variables in our study, are continuous and restricted to the (0, 1) interval. In linear regression, the dependent variable is assumed to have a distribution following a normal distribution making it unsuitable for modeling proportions. Beta regression, introduced by Ferrari and Cribari-Neto [26], assumes that the dependent variables are beta distributed. The beta distribution is a continuous distribution defined on the unit interval. It has been used in forestry to model percent canopy cover [27], riparian percent shrub cover [28], component biomass proportions [5,8,9], and to model basal area mortality due to fire [29].

With mean and precision parameters defined as $\mu = \frac{\alpha}{(\alpha+\beta)}$ and $\phi = (\alpha + \beta)$ respectively, Ferrari and Cribari-Neto [26] defined beta density function as:

$$f(y; \mu, \phi) = \frac{\Gamma(\phi)}{\Gamma(\mu\phi)\Gamma((1-\mu)\phi)} y^{\mu\phi-1}(1-y)^{(1-\mu)\phi-1} \text{ for } 0 < y < 1 \tag{4}$$

where $0 < \mu < 1$ and $\phi > 0$.

The beta regression model can then be written as:

$$g(\mu_i) = x_i^T \beta = \eta_i \tag{5}$$

where $g(\cdot)$ is a strictly increasing and double differentiable link function that maps $(0, 1)$ in to the real line $\mathbb{R}$, $x_i = (x_{i1}, \ldots\ldots, x_{ik})^T$ is a vector of k explanatory variables, $\beta = (\beta_1, \ldots\ldots, \beta_k)^T$ is a vector of unknown k unknown regression parameters $(k < n)$, and η_i is a linear predictor (i.e., $\eta_i = \beta_1 x_{i1} + \ldots + \beta_k x_{ik}$, usually $x_{i1} = 1$ for all i so that the model has an intercept) [26].

Beta regression is a generalized linear model and different link functions are available to link the dependent variable with the linear predictor. A logit link function $g(\mu) = \log\left(\frac{\mu}{1-\mu}\right)$, was used, thus the predicted proportions are obtained as $\mu_i = \frac{exp(\eta_i)}{1+exp(\eta_i)}$. The beta regression was performed in R 3.4.1 [30] with function betareg in the library betareg [26].

Dirichlet Regression

In Dirichlet regression, the dependent variable is assumed to follow a Dirichlet distribution which is a multivariate generalization of the beta regression. Therefore, it is useful when the dependent variable is a vector of proportions that represent the components as percentage of the total, thus the component proportions sum to 1. In forestry, it has been used to model component biomass [8,9,25,31] and to assess the potential of using photogrammetric data for species-specific forest inventories [32].

Maier [33] used similar parameterization of Ferrari and Cribari-Neto [26] to represent the Dirichlet distribution as follows:

$$f(y; \mu, \phi) = \frac{1}{B(\mu\phi)} \prod_{c=1}^{C} y_c^{(\mu_c\phi-1)} \tag{6}$$

where $0 < \mu_c < 1$ and $\phi > 0$ ($\mu_c = \frac{\alpha_c}{\phi}$ and $\phi = \alpha_0 = \sum_{c=1}^{C} \alpha_c$ are mean and precision parameters respectively); $\alpha_c > 0$, $\forall c$ are the shape parameters for each components, $y_c \in (0, 1)$, $\sum_{c=1}^{C} y_c = 1$, and $B(\cdot)$ is the multinomial beta function. The regression model for mean is formulated as follows:

$$g_\mu(\mu_c) = \eta_{\mu c} = X\beta_c \tag{7}$$

where $g_\mu(\cdot)$ is the link-function and again with the logit link function, the predicted values are calculated as $\mu_r = \frac{1}{1+\sum_{a=1}^{C} exp(X\beta_a)}$ for reference component and $\mu_c = \frac{exp(X\beta_a)}{1+\sum_{a=1}^{C} exp(X\beta_a)}$ for other components. For the details on parameterization of the Dirichlet distribution and the method of parameter estimation in Dirichlet regression, refer to Maier [33]. The Dirichlet regression was performed in R 3.4.1 [30] with function DirichReg in the library DirichletReg [33] using component stem biomass as a reference group.

Multinomial Logistic Regression

The multinomial logistic regression provides the conditional probabilities of observing different components [34] and can be considered as the proportion of biomass in each component and estimated by model parameters [35]. The conditional probabilities of each component assuming component stem biomass as reference category are given by Equations (8)–(10). The multinomial logistic regression has been used by Poudel and Temesgen [8] to estimate the proportions of component biomass in Douglas-fir and lodgepole pine, by Poudel and Temesgen [9] to estimate biomass proportions in red alder and western hemlock, and by Huff et al. [36] to estimate proportion of biomass in different fuel class categories.

$$p_{STM} = \frac{1}{1 + e^{(a1 + a2 \times X_1 + a3 \times X_2)} + e^{(b1 + b2 \times X_1 + b3 \times X_2)}} \tag{8}$$

$$p_{BCH} = \frac{e^{(a1 + a2 \times X_1 + a3 \times X_2)}}{1 + e^{(a1 + a2 \times X_1 + a3 \times X_2)} + e^{(b1 + b2 \times X_1 + b3 \times X_2)}} \tag{9}$$

$$p_{FOL} = \frac{e^{(b1 + b2 \times X_1 + b3 \times X_2)}}{1 + e^{(a1 + a2 \times X_1 + a3 \times X_2)} + e^{(b1 + b2 \times X_1 + b3 \times X_2)}} \tag{10}$$

where p_{STM}, p_{BCH} and p_{FOL} are proportions of aboveground biomass in stem, branch, and foliage respectively; X_1 = DBH; X_2 = total tree height; and ai, bi ($i = 1, 2, 3$) are model parameters. The multinomial logistic regression was performed in R 3.4.1 [30] with function multinomial in the library net with biomass present in each component was used as frequency weight.

2.3. Evaluation

Performances of all the methods were evaluated based on the bias (mean difference in observed and predicted values), bias percent, root mean squared error (RMSE), and RMSE percent produced by each method.

$$Bias = \frac{\sum_{i=1}^{n}(y_i - \hat{y}_i)}{n} \tag{11}$$

$$Bias\ \% = 100 \times \frac{Bias}{\overline{Y}}\% \tag{12}$$

$$RMSE = \sqrt{\frac{\sum_{i=1}^{n}(y_i - \hat{y}_i)^2}{n}} \tag{13}$$

$$RMSE\ \% = 100 \times \frac{RMSE}{\overline{Y}}\% \tag{14}$$

where n is the number of trees, y_i and $\hat{y}_i$ are observed and predicted values of AGB or its component, and $\overline{Y}$ is the mean AGB or component biomass.

3. Results and Discussion

Parameter estimates and their standard errors for the nonlinear mixed effects model used to predict total AGB are given in Table 2. The boxplot of AGB in each plot shows the variability in total AGB among 11 plots (Figure 2).

The relationship between AGB and dendrometric variables such as DBH and height varies by stands or plots. Therefore, the mixed effects model was appropriate in our study because it addressed the hierarchical nature of the data by incorporating plot level variation in the model. The bias and RMSE of this model for the modeling data were −0.01 kg and 0.84 kg (−0.17% and 18.84% of mean AGB, respectively). The residual analysis did not show any problems with the model fit (Figure 3).

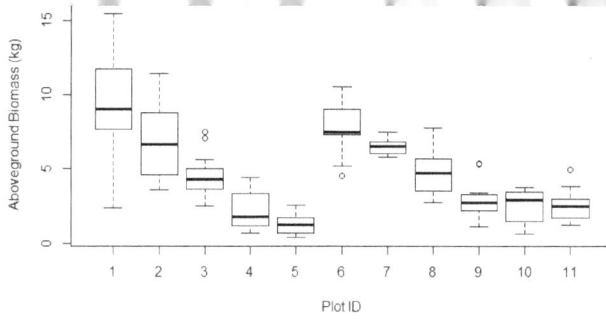

Figure 2. Boxplot of aboveground biomass in different plots in which the samples were destructively collected.

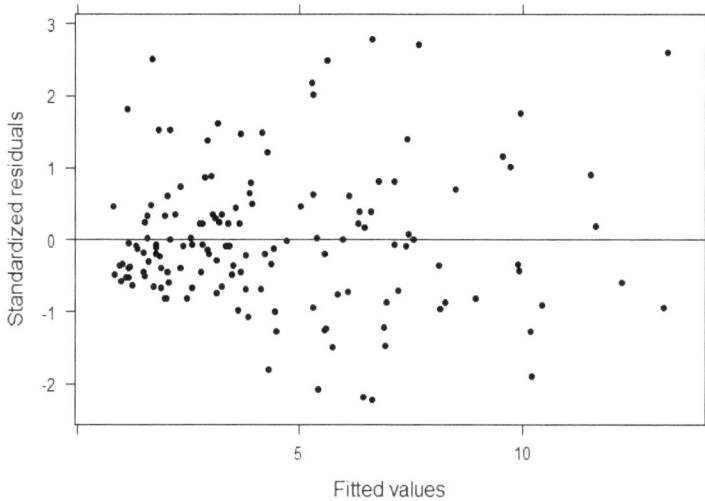

Figure 3. Scatter plots of fitted values against the standardized residual of the model for predicting total aboveground biomass.

Table 2. Parameter estimates and their standard errors for the model used to predict total AGB (Equation (3)) of small DBH trees. $\beta_i's$ $(i = 0, 1, 2)$ are regression coefficients and σ_e and σ_b are variance components of mixed models.

Coefficient	Estimate	Standard Error
β_0	−0.99599	0.17875
β_1	1.19234	0.10157
β_2	0.44547	0.11328
σ_e	0.87035	
σ_b	0.35789	

Allocation of component biomass is influenced by various site factors such as stand density, site productivity, competition at the tree level, soil characteristics such as texture and moisture content, and tree characteristics such as species and age [8]. In mature stands and trees, most of the biomass is contained in the main stem. For our sample trees, stem, branch, and foliage biomass, on average, accounted for 50%, 33%, and 17% of total AGB. Stem biomass ranged from 27% to 68%, branch biomass ranged from 18% to 53%, and foliage biomass ranged from 4% to 29% of total AGB. The proportion of stem biomass increased with increasing DBH and height (Figure 4). Similar findings for brutian pine were reported by de-Miguel et al. [12]. They found that the proportion of stem biomass is lower in small or young trees whereas the proportion of crown biomass diminishes as the tree grows. Note that the proportions in compositional data are inversely related, i.e., if the proportion of one component increases, the proportion of the other components decreases—also seen in Figure 4. Foliage biomass decreased with increasing diameter and height. However, the rate of decline in foliage biomass was higher with increasing height (in meter) than with increasing DBH (in centimeter). This could be because the vertical competition has more effect on crown biomass than the horizontal competition. Branch biomass proportion showed a similar trend as the foliage proportion. After 4 cm DBH and 4 m height, both branch and foliage biomass declined monotonically while the stem biomass increased monotonically. This can have both ecological significance as well as management implications. One such implication is assessing the potential for supplying branch and foliage biomass (the logging residues) for bioenergy production. Kuuluvainen [37] found that the proportion of stem biomass from total AGB increased from smaller to larger trees and then stabilized. This reflects, in line with the pipe-model theory [38], the increasing need for biomass allocation into stem at early stages of tree development until a balance between stem and crown biomass accumulation is achieved. The results are consistent with current biological knowledge on stand dynamics of light-demanding species. Brutian pine is managed under even-aged schedules and trees growing in such stands have longer stems and smaller crowns because of the competition for light.

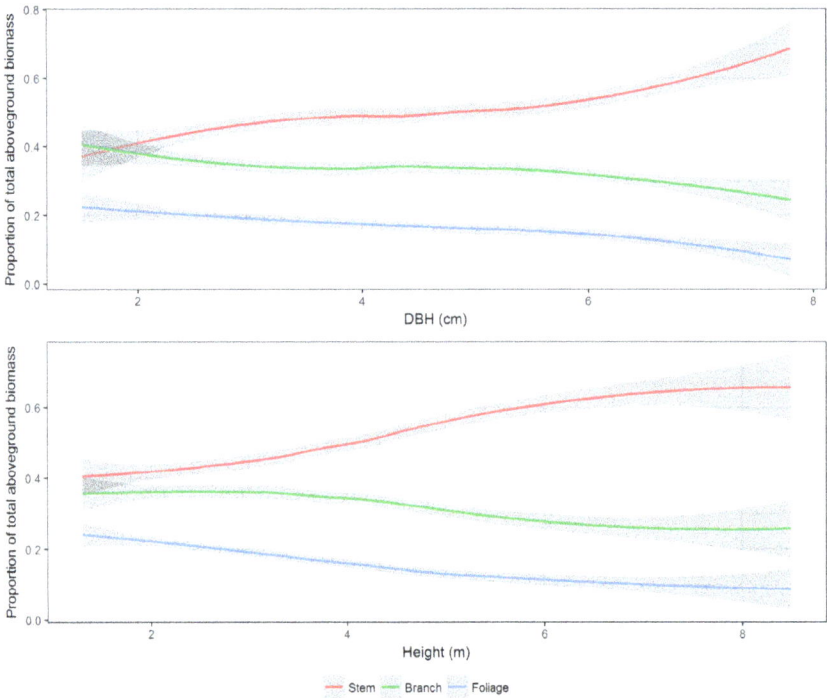

Figure 4. Trend in component biomass proportion with respect to DBH and total height. Trend lines are obtained through loess fit.

In this study beta, Dirichlet, and multinomial logistic regressions were used to predict the proportion of biomass in stem, branch, and foliage. These are all generalized linear models and were unbiased in predicting biomass proportions. However, the prediction of proportion and error in proportion itself is not as relevant and requires that both estimates and error in biomass units are obtained. This can be obtained by applying predicted proportions to the total biomass obtained from AGB equation. This underscores the importance of developing the best possible model to obtain total AGB because partitioning an inaccurate AGB would provide inaccurate estimates of the component masses as well. On the other hand, accurate models to predict component biomass are essential to meet other purposes such as to assess availability of biomass feedstock. Therefore, biomass in different components was obtained by applying predicted proportions to AGB predicted from the nonlinear mixed effects model.

The beta regression models predicted proportion of biomass in stem, branch, and foliage independently of each other. Parameter estimates and their standard errors of beta regression models are given in Table 3.

Table 3. Parameter estimates, their standard error (in parenthesis), and *pseudo-R²* (an R^2-like measure calculated based on estimated likelihood) values for the small DBH trees using beta regression.

Model	Parameter Estimates			*pseudo-R²*
	Intercept	DBH	HT	
Foliage	−0.91797 (0.07839)	0.06866 (0.03182)	−0.26256 (0.03567)	0.4245
Branch	−0.35913 (0.07030)	0.01450 (0.02795)	−0.10239 (0.03045)	0.1827
Stem	−0.68726 (0.07219)	−0.04483 (0.02826)	0.22690 (0.03089)	0.4780

Component models are generally not as good as the model to fit AGB. The model to predict branch biomass proportion had the smaller *pseudo-R²* (0.1827) than the models to predict stem and foliage biomass. This is justified by the flatter smooth line observed in Figure 4. The stem and foliage proportion models had *pseudo-R²* 0.4245 and 0.4780, respectively. The evaluation statistics produced by the beta regression models are given in Table 4. Branch and stem biomass were over predicted by the beta regression models by 1.62% and 0.25% whereas the foliage biomass was under predicted by 2.93%. RMSEs for the beta models were 39.11%, 31.50%, and 18.77% for foliage, branch, and stem biomass estimation, respectively. Note that, even though the foliage model had a higher *pseudo-R²* value than the branch model, it had a higher RMSE percent than the branch model.

Table 4. Bias and root mean squared error (RMSE) obtained from beta regression.

Component	Bias (kg)	Bias Percent	RMSE (kg)	RMSE Percent
Foliage	0.0207	2.9291	0.2764	39.1106
Branch	−0.0230	−1.6228	0.4464	31.4956
Stem	−0.0059	−0.2541	0.4357	18.7655

Dirichlet regression assumes that the dependent variable is a vector of proportions with unit sum and follows a Dirichlet distribution which is a multivariate generalization of the beta distribution. Unlike in beta regression, component models are fitted simultaneously, thus ensuring the unit sum of the predicted proportions. Parameter estimates and their standard errors along with fit statistics are presented in Table 5. The biomass in stem was used as the reference group, hence there were no parameter estimates for stem biomass. One can change the reference group to obtain model coefficients for stem biomass. However, the component proportions estimated in such a manner would not necessarily have unit sum.

Table 5. Parameter estimates for the component biomass models for small DBH trees using Dirichlet regression. Component stem biomass was used as the reference group.

Model	Parameter Estimates			R^2
	Intercept	DBH	HT	
Foliage	0.16631 (0.07383)	0.02972 (0.02884)	−0.17838 (0.03135)	0.4602
Branch	−0.16228 (0.09182)	0.07570 (0.03647)	−0.32911 (0.04184)	0.1880

Similar to beta regression, the Dirichlet regression over predicted branch and stem biomass by 1.52% and 0.12% whereas the foliage biomass was under predicted by 2.36%. RMSEs produced by the Dirichlet regression (Table 6) were practically identical to that produced by the beta regression but the Dirichlet regression should be preferred to the beta regression due to the assurance of the desired additive property.

Table 6. Bias and root mean squared error obtained from Dirichlet regression.

Component	Bias (kg)	Bias Percent	RMSE (kg)	RMSE Percent
Foliage	0.0167	2.3631	0.2763	39.0965
Branch	-0.0216	−1.5240	0.4468	31.5238
Stem	−0.0028	−0.1206	0.4351	18.7396

Multinomial logistic regression is similar to the Dirichlet regression in the sense that both fit the components simultaneously and ensures that the sum of predicted proportions is equal to one. Parameter estimates and their standard errors of the multinomial logistic regression are given in Table 7.

Table 7. Parameter estimates for the component biomass models for small DBH trees using multinomial logistic regression model. Component stem was used as the reference group.

Model	Parameter Estimates			R^2
	Intercept	DBH	HT	
Foliage	−0.10942 (0.41677)	0.05607 (0.13695)	−0.31204 (0.14648)	0.4516
Branch	0.28112 (0.33235)	−0.00041 (0.10719)	−0.17256 (0.10946)	0.1766

Multinomial logistic regression over predicted the biomass proportions for all components by no more than 0.2% (Table 8). It produced the smallest RMSEs, compared to both beta and Dirichlet regressions. However, these RMSEs were within 0.5% of each other.

Table 8. Bias and root mean squared error obtained from multinomial logistic regression.

Component	Bias (kg)	Bias Percent	RMSE (kg)	RMSE Percent
Foliage	−0.0004	−0.0566	0.2744	38.8276
Branch	−0.0017	−0.1199	0.4395	31.0087
Stem	−0.0055	−0.2369	0.4309	18.5587

4. Conclusions

Using the data collected by destructively sampling 143 trees in 11 research plots installed in natural stands that were 11 to 26 years old, total and component biomass of brutian pine trees in Turkey were modeled. Total AGB was modeled using a nonlinear mixed effects model which accounted for the variability among plots. The predicted total AGB was then distributed into different tree components using the predicted proportions obtained from generalized linear models.

Stem, branch, and foliage biomass, on average, accounted for 50% (range 27–68%), 33% (range 18–53%), and 17% (range 4–29%) of total AGB. Biomass of different components did not follow a consistent trend with respect to tree age (Figure 5). Proportion of stem wood biomass increased until age 16 years, then declined until age 22 years, and increased again. Proportions of branch and foliage biomass declined until age 16, then increased until age 21, and declined again thereafter. The foliage proportion had a similar trend to the branch proportion but the second decline began after around age 25 years (Figure 5).

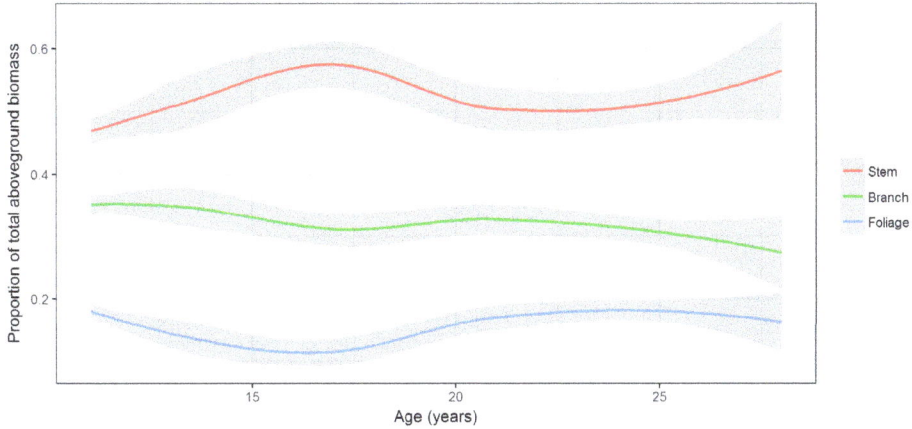

Figure 5. Distribution of biomass in different aboveground components by age in small diameter brutian pine trees sampled in this study.

The beta, Dirichlet, and multinomial logistic regressions produced unbiased estimates of biomass proportions. These methods produced similar bias and root mean squared error values. The beta regression fits the component models independently. Therefore, it does not guarantee that the sum of the predicted proportions is equal to one. However, the Dirichlet and multinomial logistic regressions fit component models simultaneously and ensure the additivity of component masses. Note that the use of multinomial logistic regression requires arbitrary categorization and provides the predicted probabilities of those categories. In component biomass modeling, such predicted probabilities are considered the predicted proportions. The Dirichlet regression has the additive property, does not require such non-standard data categorization, and has similar performance as the multinomial logistic regression and may be preferred over the multinomial logistic regression. Models developed in this study can also be used in feasibility analysis of theoretical potential of establishing bioenergy plants, assessment of wildfire fuel load, and potential of brutian pine for carbon sequestration. Since the prediction accuracy of component biomass is dependent on the accuracy of the model to predict total AGB, future work on testing model forms and modeling approaches for AGB prediction with larger datasets is also critical.

Acknowledgments: This research was conducted within the scope of the project "Investigation on supply possibilities of logging residues" (Project No. 110O435), funded by Scientific and Technological Research Council of Turkey (TUBITAK). We are grateful to TUBITAK. A special thanks to all project members. We thank Bryce Frank of the Forest Measurement and Biometrics Lab at Oregon State University for providing diligent proofreading of the manuscript. Finally, we thank the anonymous reviewers of this article for their constructive comments, which have contributed to improving this manuscript.

Author Contributions: M.E. collected data and coordinated the research project; R.Ö. and K.P.P. analyzed data and co-wrote the results. M.E., R.Ö., and K.P.P. reviewed all results and wrote a draft. All authors discussed, revised, and designed the last version of the manuscript.

Conflicts of Interest: The authors declare no conflict of interest. The funding sponsors had no role in the design of the study; in the collection, analyses, or interpretation of data; in the writing of the manuscript, and in the decision to publish the results.

References

1. GDF. Forestry Statistics–Wood Based Forest Products in Year of 2015. General Directorate of Forestry: Ankara. Available online: http://www.ogm.gov.tr/ekutuphane/Istatistikler/ (accessed on 20 February 2016).
2. Bozkurt, A.Y.; Göker, Y. *Forest Products Utilization*; Istanbul University Press: Istanbul, Turkey, 1996.
3. Fischer, R.; Lorenz, M.; Köhl, M.; Becher, G.; Granke, O.; Christou, A. *The Condition of Forests in Europe: 2008 Executive Report*; United Nations Economic Commission for Europe, Convention on Long-Range Transboundary Air Pollution, International Co-Operative Programme on Assessment and Monitoring of Air Pollution Effects on Forests (ICP Forests), Institute for World Forestry: Hamburg, Germany, 2008; p. 23.
4. Dong, L.; Zhang, L.; Li, F. A compatible system of biomass equations for three conifer species in Northeast China. *For. Ecol. Manag.* **2014**, *329*, 306–317. [CrossRef]
5. Zhou, X.; Hemstrom, M.A. *Estimating Above-Ground Tree Biomass on Forest Land in the Pacific Northwest: a Comparison of Approaches*; Res. Pap. PNW-RP-584; Department of Agriculture, Forest Service, Pacific Northwest Research Station: Portland, OR, USA, 2009; p. 18.
6. Parresol, B.R. Assessing tree and stand biomass: A review with examples and critical comparisons. *For. Sci.* **1999**, *45*, 573–593.
7. Parresol, B.R. Additivity of nonlinear biomass equations. *Can. J. For. Res.* **2001**, *31*, 865–878. [CrossRef]
8. Poudel, K.P.; Temesgen, H. Methods for estimating aboveground biomass and its components for Douglas-fir and lodgepole pine trees. *Can. J. For. Res.* **2016**, *46*, 77–87. [CrossRef]
9. Poudel, K.P.; Temesgen, H. Developing biomass equations for western hemlock and red alder trees in western Oregon forests. *Forests* **2016**, *7*, 88. [CrossRef]
10. Durkaya, A.; Durkaya, B.; Atmaca, S. Predicting the above-ground biomass of Scots pine (*Pinus sylvestris* L.) stands in Turkey. *Energy Source Part A Recovery Utili. Environ. Eff.* **2009**, *32*, 485–493. [CrossRef]
11. Zianis, D.; Xanthopoulos, G.; Kalabokidis, K.; Kazakis, G.; Ghosn, D.; Roussou, O. Allometric equations for aboveground biomass estimation by size class for Pinus brutia Ten. trees growing in North and South Aegean Islands, Greece. *Eur. J. For. Res.* **2011**, *130*, 145–160. [CrossRef]
12. De-Miguel, S.; Pukkala, T.; Assaf, N.; Shater, Z. Intra-specific differences in allometric equations for aboveground biomass of eastern Mediterranean Pinus brutia. *Ann. For. Sci.* **2014**, *71*, 101–112. [CrossRef]
13. Özçelik, R.; Diamantopoulou, M.J.; Eker, M.; Gürlevik, N. Artificial Neural Network Models: An Alternative Approach for Reliable Aboveground Pine Tree Biomass Prediction. *For. Sci.* **2017**, *63*, 291–302.
14. Chaturvedi, R.K.; Raghubanshi, A.S. Aboveground biomass estimations of small diameter woody species of tropical dry forest. *New For.* **2013**, *44*, 509–519. [CrossRef]
15. Daryaei, A.; Sohrabi, H. Additive biomass equations for small diameter trees of temperate mixed deciduous forests. *Scand. J. For. Res.* **2017**, *31*, 394–398. [CrossRef]
16. Murray, R.B.; Jacobson, M.Q. An evaluation of dimension analysis for predicting shrub biomass. *J. Range Manag.* **1982**, *35*, 451–454. [CrossRef]
17. Singh, V.; Tewari, A.; Kushwaha, S.P.S.; Dadhwal, V.K. Formulating allometric equations for estimating biomass and carbon stock in small diameter trees. *For. Ecol. Manag.* **2011**, *261*, 1945–1949. [CrossRef]
18. Nelson, A.S.; Weiskittel, A.R.; Wagner, R.G.; Saunders, M.R. Development and evaluation of aboveground small tree biomass models for naturally regenerated and planted species in eastern Maine, U.S.A. *Biomass Bioenergy* **2014**, *68*, 215–227. [CrossRef]
19. Schuler, J.; Bragg, D.C.; McElligott, K. Biomass estimates of small diameter planted and natural–origin Loblolly pines show major departures from the National biomass estimator equations. *For. Sci.* **2017**, *63*, 319–330. [CrossRef]
20. Küçük, Ö.; Bilgili, E. Crown fuel load for young Calabrian pine (Pinus brutia Ten.) trees. *Kastamonu Univ. J. Fac. For.* **2007**, *7*, 180–189.
21. Bilgili, E.; Küçük, O. Estimating above-ground fuel biomass in young Calabrian pine (*Pinus brutia* Ten.). *Energy Fuels* **2009**, *23*, 1797–1800. [CrossRef]

22. Boydak, M. Silvicultural characteristics and natural regeneration of Pinus brutia Ten.—A review. *Plant Ecol.* **2004**, *171*, 153–163. [CrossRef]

23. Picard, N.; Rutishauser, E.; Ploton, P.; Ngomanda, A.; Henry, M. Should tree biomass allometry be restricted to power models? *For. Ecol. Manag.* **2015**, *353*, 156–163. [CrossRef]

24. Ritchie, M.W.; Zhang, J.; Hamilton, T.A. Aboveground tree biomass for Pinus ponderosa in northeastern California. *Forests* **2013**, *4*, 179–196. [CrossRef]

25. Dahlhausen, J.; Uhl, E.; Heym, M.; Biber, P.; Ventura, M.; Panzacchi, P.; Tonon, G.; Horváth, T.; Pretzsch, H. Stand density sensitive biomass functions for young oak trees at four different European sites. *Trees* **2017**, *31*, 1–16. [CrossRef]

26. Ferrari, S.; Cribari-Neto, F. Beta regression for modelling rates and proportions. *J. Appl. Stat.* **2004**, *31*, 799–815. [CrossRef]

27. Korhonen, L.; Korhonen, K.T.; Stenberg, P.; Maltamo, M.; Rautiainen, M. Local models for forest canopy cover with beta regression. *Silva Fenn.* **2007**, *41*, 671. [CrossRef]

28. Eskelson, B.N.; Madsen, L.; Hagar, J.C.; Temesgen, H. Estimating Riparian understory vegetation cover with Beta regression and copula models. *For. Sci.* **2011**, *57*, 212–221.

29. Hoe, M.S. Multi-Temporal Lidar Analysis of Landscape Fire Effects in Southwestern Oregon. Master's Thesis, Oregon State University, Corvallis, OR, USA, 2016; p. 162.

30. R Core Team. A Language and Environment for Statistical Computing. Available online: http://www.R-project.org/ (accessed on 4 August 2017).

31. Zhao, D.; Kane, M.; Teskey, R.; Markewitz, D. Modeling aboveground biomass components and volume-to-weight conversion ratios for loblolly pine trees. *For. Sci.* **2016**, *62*, 463–473. [CrossRef]

32. Puliti, S.; Gobakken, T.; Ørka, H.O.; Næsset, E. Assessing 3D point clouds from aerial photographs for species-specific forest inventoires. *Scand. J. For. Res.* **2017**, *32*, 68–79. [CrossRef]

33. Maier, M.J. DirichletReg: Dirichlet Regression for Compositional Data in R. Available online: http://epub.wu.ac.at/4077/ (accessed on 29 January 2014).

34. Hosmer, D.W.; Lemeshow, S.; Sturdivant, R.X. *Applied Logistic Regression*, 3rd ed.; John Wiley & Sons Inc.: New York, NY, USA, 2013; p. 529.

35. Boudewyn, P.; Song, X.; Magnussen, S.; Gillis, M. *Model-Based, Volume to Biomass Conversion for Forested and Vegetated Land in Canada*; Information Report BC-X-411; Natural Resources Canada, Canadian Forest Service, Pacific Forestry Centre: Victoria, BC, Canada, 2007.

36. Huff, S.; Ritchie, M.; Temesgen, H. Allometric equations for estimating abovegorund biomass for common shrubs in northeastern California. *For. Ecol. Manag.* **2017**, *398*, 48–63. [CrossRef]

37. Kuuluvainen, T. Relationships between crown projected area and components of above-ground biomass in Norway spruce trees in even-aged stands: Empirical results and their interpretation. *For. Ecol. Manag.* **1991**, *40*, 243–260. [CrossRef]

38. Makela, A. Implications of the pipe model theory on dry matter partitioning and height growth in trees. *J. Theor. Biol.* **1986**, *123*, 103–120. [CrossRef]

Article

Suitability of Soil Erosion Models for the Evaluation of Bladed Skid Trail BMPs in the Southern Appalachians

J. Andrew Vinson *, Scott M. Barrett, W. Michael Aust and M. Chad Bolding

Department of Forest Resources and Environmental Conservation, Virginia Polytechnic Institute and State University, 310 West Campus Drive, Blacksburg, VA 24061, USA; sbarrett@vt.edu (S.M.B.); waust@vt.edu (W.M.A.); bolding@vt.edu (M.C.B.)
* Correspondence: josephav@vt.edu; Tel.:+540-231-6494

Received: 29 September 2017; Accepted: 30 November 2017; Published: 6 December 2017

Abstract: This project measured soil erosion rates from bladed skid trails in the mountains of Virginia following a timber harvest, and compared measured erosion to four erosion model predictions produced by Universal Soil Loss Equation—Forest (USLE-Forest), Revised Universal Soil Loss Equation, v.2 (RUSLE2), Water Erosion Prediction Project—Road (WEPP-Road) using default files, and WEPP-Road using modified files in order to assess the utility of the models for these conditions. Skid trails were segregated into six blocks where each block had similar trail slopes and soils. Each block contained four skid trail closure treatments: (1) bare soil (Control); (2) residual limbs and tops (Slash); (3) grass seed (Seed); and (4) fertilizer, seed, and straw mulch (Mulch). All treatments had waterbars, the minimum trail closure best management practice (BMP), to provide upslope and downslope borders of experimental units. Site cover characteristics on each experimental unit were collected quarterly as input parameters for erosion models. The suitability of soil erosion models were evaluated based upon statistical summaries, linear relationships with measured erosion rates, Nash-Sutcliffe Model Efficiency, and a nonparametric analysis. Treatments were measured to have erosion rates of 15.2 tonnes ha^{-1} year^{-1} (Control), 5.9 tonnes ha^{-1} year^{-1} (Seed), 1.1 tonnes ha^{-1} year^{-1} (Mulch), and 0.8 tonnes ha^{-1} year^{-1} (Slash). It was determined that WEPP-Road: Modified (p-value = 0.643) and USLE-Forest (p-value = 0.307) were the most suitable models given their accuracy; however USLE-Forest may be better for making management decisions given its practicality.

Keywords: bladed skid trails; forest operations; forest harvesting; soil erosion modeling; best management practices

---

1. Introduction

The United States Environmental Protection Agency (USEPA) has identified sediment as the most damaging nonpoint-source pollutant in the U.S. [1]. Forest operations have the potential to produce substantial amounts of soil erosion that may be delivered as sediment in streams [2], thus a variety of forestry best management practices have been developed to either reduce soil erosion or interrupt delivery of eroded material to streams. In the southern Appalachians of the U.S., primary sources of soil erosion associated with forest operations are forest roads [3], overland [4] and bladed skid trails [5], and stream crossings [6]. Roads and skid trails are potentially highly erosive due to exposure of bare soil, terrain slope steepness, low road drainage standards [7,8], and traffic during poor weather conditions. The combination of these factors are known to increase erosion; therefore increasing the possibility of stream sedimentation and degradation [9,10]. Skid trails are potentially of more concern than haul roads because skid trails typically have lower standards than roads and skid trails may comprise a larger percentage of the harvest area [11]. Bladed skid trails are often used in the steep terrain of the region to facilitate skidder operator safety and operational efficiency. They differ from overland skid trails in that a bulldozer is used to construct the road, as opposed to having equipment simply drive on the surface of the soil [3,4]. Kochenderfer [12] estimated that up to 84% of exposed mineral soil in a harvest area was due to skid trails. More recently Worrell et al. [13] reported that bladed skid trails comprised approximately 8% of harvest area in the Appalachian Mountains. Wade et al. [5] measured erosion produced by bladed skid trails in the Piedmont region and determined that sediment production was strongly influenced by the application of forestry best management practices (BMPs). Trails with only waterbars produced 1.1 tonnes ha^{-1} year^{-1} of erosion whereas trails using slash or mulch cover produced <4 tonnes ha^{-1} year^{-1}.

Best management practices for skid trails have been developed to reduce the impacts of forest operations on water quality [14]. Skid trail BMPs include pre-harvest planning (e.g., layout of bladed skid trails), water control structures (e.g., water bars), and the use of ground cover on skid trails [15]. Commonly suggested methods of ground cover for bladed skid trails include grass seed, straw mulch, and residual limbs and tops from the forest harvest (slash) [16–19]. These methods of ground cover have been found to be both effective and economical in the past [4,5,20].

Soil erosion has the potential to reduce site productivity [21,22] and negatively impact water quality [2], thus quantification of the effects of forest best management practices on soil erosion are clearly important. However, on-site measurement of erosion is often costly and time consuming, thus models are commonly used to estimate erosion potentials [23,24]. Several models were developed to allow agricultural land managers to estimate and prioritize erosion issues and have been adapted to forest use over time [25]. Erosion models can be used by forest managers to make silvicultural, management, or even forest engineering decisions [26]. They are frequently modified to maintain and increase their accuracy and dependability [23]. One of the oldest and most widely applied soil erosion models is the Universal Soil Loss Equation (USLE) that was originally developed by the USDA in 1954 to estimate potential sheet and rill erosion from agricultural lands. The USLE is an empirical model that has been adapted to predict erosion from rangelands, minelands, watershed, and forest lands [27]. Dissmeyer and Foster [28] modified the USLE for use on forestlands (USLE-Forest). The USLE-Forest is relatively simple to use and has been widely used successfully on skid trails in the Piedmont physiographic region [4,5,20]. The USLE-Forest equation components are:

$$A = RKLSCP \tag{1}$$

where A is the annual soil loss per unit area, R is the rainfall and runoff factor, K is the soil erodibility factor, L represents the slope-length factor, S is the slope-steepness factor, C is the cover and management factor, and P represents the support practices factor [28]. R is determined based upon the average weather conditions at the location of interest. K is a function of multiple soil characteristics: soil texture, organic matter content, structure, and permeability. K values can be found in soil surveys or soil descriptions [29]. More accurate K-value estimates can be obtained by completing a nomograph included in the USLE-Forest manual. The L value is "the ratio of soil loss from the field slope length to that from a 72.6-foot (22.13 m) length under identical conditions" [28]. Likewise, slope-steepness factor (S) is defined as "the ratio of soil loss from the field slope gradient to that from a 9-percent slope under otherwise identical conditions" [28]. These two variables can be determined from a table found in A Guide to Predicting Sheet and Rill Erosion on Forest Land, written by Dissmeyer and Foster [28]. Cover and management (CP) factors are based upon the amount of bare soil, presence of canopy, soil reconsolidation, high organic matter content, fine roots, residual binding effects, onsite storage, and natural sediment trapping resulting in steps, and can be derived from tables published by Dissmeyer and Foster [28].

The USLE was later revised and converted to a computerized format, labeled the RUSLE or Revised Universal Soil Loss Equation. This model was first produced in the early 1990's, and RUSLE1.06 and RUSLE2 were both released in 2003. Although the original empirical algorithm from the USLE was kept, it was modified for improved accuracy by deriving soil loss factors in new ways. This revision included changes to make the model more suited for use with forest lands. Other improvements included updated rainfall coefficients, after changing some of the R factors in the eastern US based on weather data collected from more than 1200 weather stations. Soil erodibility (K) is varied seasonally for increased accuracy. The LS factor is improved in that it takes into account the "susceptibility of the soil to rill erosion relative to interrill erosion" and the cover factor uses a new algorithm for determining cover based on prior land use, canopy cover, soil cover, and soil surface roughness [30]. RUSLE2 has no specific data files for forest roads, however there are "highly disturbed land" files that can be modified to suit different forest road treatments [24].

The Water Erosion Prediction Project (WEPP) is a physically-based model produced by the U.S. Department of Agriculture Natural Resource Conservation Service (NRCS) and U.S. Forest Service (USFS) to replace the USLE formula. WEPP "models soil erosion as a process of rill and interrill detachment and transport" [31] as opposed to empirically modeling the ground conditions [32,33]. The WEPP model had additional potential utility because it estimates daily conditions that affect erosion, over the course of a year. In this, senescence, plant growth, residue accumulation and decomposition, as well as daily temperatures and soil water availability are taken into account to provide a very detailed estimate of soil loss over time. An additional benefit is the ability to model complex slopes and forest road profiles, with features such as cutslopes and fillslopes, ditches, and road surfaces [23]. Four types of data files are required to run WEPP: (1) a climate file, to include data on daily precipitation and temperature; (2) a hillslope file, which can contain multiple points to describe a slope's shape; (3) a soils file, which can include multiple soil types across the hillslope; and (4) a management file containing information on soil disturbances and vegetative conditions present [26]. Weather data are obtained through Cligen, the USDA's weather resource. This weather file models weather data on a daily basis for more than 1000 climates [34]. Using the hillslope file, WEPP determines the erosion or deposition rates for at least 100 points of the hillslope if there is any runoff predicted that day [35]. Because WEPP, like other models, was originally intended for cropland or rangelands, there have been many efforts to adapt it for forest uses [36–41]. One of these interfaces is the WEPP-Road model interface. This program allows the user to determine the amount of sediment delivered to the stream through the forest buffer and amount of sediment eroded from each portion of the road, as well as determine the presence of a sediment plume in the forest [42]. At this time, the selections for cover and land use scenarios appear to limit WEPPs utility for estimation of erosion for many eastern forest management regimes [43].

There have been several attempts to assess the performance of these three models. Wade et al. [24], compared sediment trap data to predictions by all three models. Erosion rates were estimated from different sections of bladed skid trail in the Piedmont of Virginia using sediment traps, and were then compared to erosion rates predicted by USLE, RUSLE2, and WEPP models. It was found that overall, all three models performed well enough for identifying erosion hazards and making management decisions. When comparing the modeled data, it was determined that USLE-Forest ranged from $0.9\times$ to $2.2\times$ the actual erosion rates from data collected from the sediment traps. RUSLE2 ranged from $0.4\times$ to $2\times$ the actual erosion, and WEPP-Road ranged from $2.3\times$ to $7.5\times$ [24]. These data indicated that the USLE-Forest and RUSLE2 can be useful at approximating erosion rates, but WEPP-Road values should only be used for ranking purposes on bladed skid trails. Foster, Toy, and Renard [44] found similar results when comparing USLE, RUSLE1.06, and RUSLE2. WEPP modeling efforts can be improved with laborious programming, but is time consuming and requires many measurements to modify the working files [45]. Lang et al. [45] found that soil erosion models worked best when estimating erosion rates less than 11.2 Mg ha^{-1} year^{-1}; however when erosion rates surpassed this amount model estimates varied widely. Croke and Netherly [25] compared the USLE and WEPP on skid trails in Australia and concluded that the USLE was more user friendly while the WEPP model was a better predictor of erosion on skid trails. However, their investigation indicated that neither method was wholly satisfactory for estimation of erosion. One important distinction to note is the difference between empirical models, which are simpler to use but are based on observations and measurements; and physically-based models which replicate erosion processes as equations [32,33]. Both types of models have their own advantages and applications [32].

Overall, the literature clearly indicates that erosion from skid trails can be a significant source of nonpoint source pollution from forestry operations [2,46–48] and that rates of erosion for different types of skid trail BMPs are warranted in order to evaluate BMP efficacy. This aspect of the problem is addressed by a companion paper [49]. Furthermore, the literature indicates that erosion models have been used with varying success to estimate erosion from skid trails, but modeled erosion rates from bladed skid trails in mountainous terrain have not been compared to direct erosion measures.

The primary objective of this study was to compare measured erosion rates from four bladed skid trail closure methods in mountainous terrain with those produced by the Universal Soil Loss Equation (USLE-Forest), the Revised Universal Soil Loss Equation (RUSLE2), the Water Erosion Prediction Project (WEPP-Road: Default), and a more modified version of WEPP (WEPP-Road: Modified).

2. Materials and Methods

2.1. Research Site

The study site is located in the Ridge and Valley physiographic province, on Virginia Tech's Fishburn Forest located in Montgomery County, Virginia (Figure 1). This physiographic province is characterized by long mountain ridges with constant linear valleys in between them. The average yearly precipitation is 103.86 cm [50]. The average high and low temperatures for this location in January are 5.3 °C and −5.9 °C. The average high and low temperatures in July are 27.9 °C and 15.6 °C [50]. Rainfall data were collected from a nearby weather station for the duration of the study period (Figure 2) [51] and were used to compare the effects of rainfall on erosion rates [49].

The soils are very shallow, well drained silt loams, being derived mostly from shale, siltstone, and sandstone residuum. Berks, Weikert, Berks-Weikert and Clymer soil series (*Lithic Dystrudepts*) dominate the site [29]. The harvested stands were primarily mixed upland hardwood-pine stands, composed of white pine (*Pinus strobus* L.), chestnut oak (*Quercus montana* Willd.), and white oak (*Quercus alba* L.). Slopes in this region range from 0% to 100%.

The site was harvested in late 2014-early 2015 using a shelterwood overstory removal of upland hardwoods and pine. Skid trails were laid out in a "logger's choice" arrangement. Skid trail slopes ranged from 0–35%, with an average slope of 16%. Skid trail sideslopes ranged from 5–45%. The skid trails were divided into 6 blocks based on slope class. Two blocks were arranged in each slope class: Gentle (0–10%), Moderate (11–20%), and Steep (>20%). Each block of treatments contained the four closure methods that were compared in this experiment.

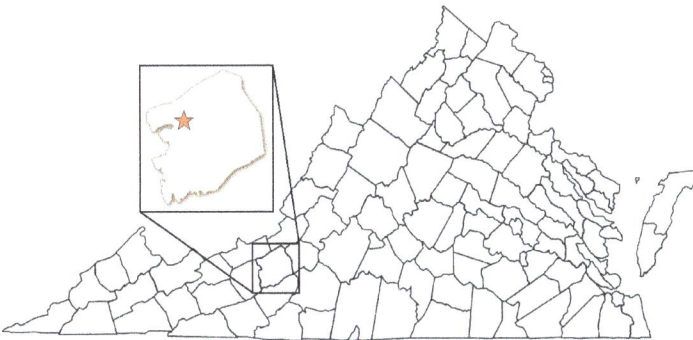

Figure 1. Timber harvest was located in Montgomery County of southwestern Virginia, located on the southeastern coast of the United States. Map is not to scale.

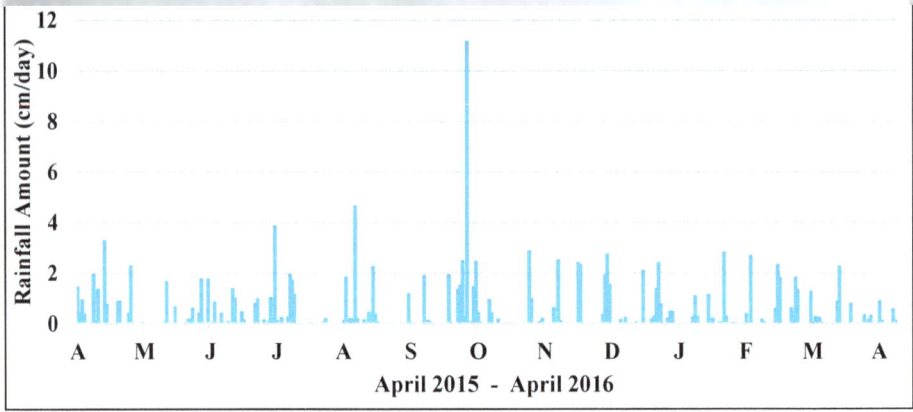

Figure 2. Daily rainfall amounts over the course of 1 year of data collection.

Treatments consisted of 15.2 m segments of skid trail, approximately 2 m wide. On steep slopes (>20%) the treatments were shortened to 12.2 m in order to comply with state BMP guidelines [14]. Closure methods were randomly assigned to each of the 24 experimental units. Units were separated by waterbars. Skid trail treatments were closed to vehicle traffic over the course of the study period in order to avoid any effects of heavy trafficking on soil erosion rates [52].

2.2. BMP Treatments

Four types of treatments were used in this study: (1) waterbars only (Control), (2) waterbars with grass seed (Seed), (3) waterbars with grass seed, fertilizer, and straw mulch (Mulch), and (4) waterbars with slash (Slash) (Figure 3). The Control treatment consists of waterbars with no ground cover treatments and represents the minimum acceptable BMPs as a control reference to which the other treatments were compared. For the Seed treatment, grass seed was applied at the time of skid trail closeout (April 2015) using a mix of 50% perennial ryegrass (*Lolium perenne* L.) seed and 50% K-31 fescue (*Festuca arundinacea* Schreb.), based on suggestions from the VDOF BMP manual [14]. Seeds were spread with a hand operated seeder to ensure adequate coverage, at a rate of approximately 168 kg/ha. For the Mulch treatment, the same grass seed mixture was applied, followed by fertilizer and straw mulch. Mulch was spread by hand to ensure near total coverage, at a depth of 3–6 cm across each experimental unit [14]. Fertilizer [5-10-10 (N, P_2O_5, K_2O)] was added at a rate of 336 kg/ha to provide sufficient nutrient availability for the grass. Slash treatments utilized residual slash from on-site logging operations, and was primarily composed of yellow-poplar (*Liriodendron tulipifera* L.), hickory (*Carya spp.*), scarlet oak (*Quercus coccinea* Münch.), chestnut oak (*Quercus montana* Willd.), white oak (*Quercus alba* L.), white pine (*Pinus strobus* L.), and Virginia pine (*Pinus virginiana* Mill.). Slash was hand applied onto skid trails to ensure similar coverage and then compacted with a bulldozer to make contact with the ground. After being tracked in by the bulldozer, slash was at a depth of 0.6–0.9 m.

Figure 3. A comparison of the four skid trail closeout best management practices (BMPs) used in the study.

2.3. Sediment Trap Installation and Measurement

A full description of the collection of field data and the effectiveness of skid trail closure methods is available from Vinson et al. [49]. Sediment traps were used to measure erosion rates in the field over the course of the year. These sediment traps consisted of silt fences that were joined to the downslope waterbars so that they collected all runoff from the skid trail treatment (Figure 4). Berms were constructed on either side of the skid trail to limit overland flow and to ensure runoff from the treatment made it into the sediment trap. Within each sediment trap, metal pins were driven into the ground at regular intervals in a grid pattern. The depth of the sediment was measured at each sediment pin on a monthly basis, as was the area of the sediment collected. From this a volumetric accumulation of sediment was determined over time. Bulk density samples were taken from the accumulated sediment, and this was used to convert the volume of collected sediment to a gravimetric amount.

Figure 4. An example of the silt fence sediment traps used in the study. Sediment pins were driven into trap area on a grid pattern and measured at regular intervals.

2.4. Erosion Model Parameters

For modeling, each experimental unit was divided into three sections. The first section being the downhill side of the upslope waterbar, the second section being the actual skid trail surface, and the third being the uphill side of the downslope waterbar (Figure 5a,b). Sections 1 and 2 were modeled together. Since the two waterbars have sides that are contributing to the area, they needed to be accounted for in the modeling as well. The slope and length of every section was measured using a total station. The USLE was used to estimate erosion from each section of each treatment, and estimates were combined in a weighted average total erosion estimate for each treatment. Grass treatments had model estimates determined both before and after seed germination for a comparison, as ground cover values were measured in the field every 3 months to account for variations in seasons, the establishment of grass, and decomposition of slash and mulch. Slope, climate data, soil characteristics, and cover practices were determined for each experimental unit and input into all three models to estimate soil loss. Actual erosion rates were converted to tonnes ha^{-1} year^{-1} in order to compare estimates provided by all three models. For each treatment area the following data were collected: ground cover, slope gradient and slope length, percent of soil in clay, sand, and silt, soil rock content, and rainfall data.

2.5. USLE-Forest Parameters

A rainfall runoff factor of 150 was used as it was derived from a rainfall contour map provided by the USLE-Forest manual [28]. A soil erodibility factor of 0.43 was obtained from the Montgomery County, VA Soil Survey [29]. A total station was used to measure the slope length and gradient for the upper and lower waterbars, and the section of bladed skid trail located between the two. Slope lengths were often too small to be found in the USLE-Forest manual, and therefore were obtained using the equation:

$$LS = (\lambda/72.6)^m (65.41\sin\theta^2 + 4.65\sin\theta + 0.065) \tag{2}$$

where λ is the slope length in feet, θ is the slope angle in degrees, and m is 0.2 for <1% slopes, 0.3 for 1% to 3% slopes, 0.4 for 3.5% to 4.5% slopes, and 0.5 for $\geq$5% slopes [28]. The bladed skid trails were considered to be tilled soil, therefore having CP factors to include bare soil, residual binding, and soil reconsolidation; canopy effect; steps; onsite storage; invading vegetation; and contour tillage. Bare soil percentages were calculated by creating transects across the treatment, with evenly spaced points. At each point, ground cover was determined to be either bare or covered, and ground cover percentage was calculated. Ground cover included vegetation, straw mulch, woody residues, rock fragments, and leaf litter. These measurements were collected quarterly over the course of a year to cover the span of four seasons. A weighted average of the four periods was used to determine a final annual erosion rate for each treatment.

2.6. RUSLE Parameters

Erosion estimates were also predicted using RUSLE2. Montgomery county weather and soil files were imported into the program to more accurately estimate soil loss. Climatic data were accessed from the NRCS database [29] for Montgomery County, Virginia. Daily and monthly average rainfall rates were included in these data. Montgomery county soil survey indicated the Berks-Weikert complex as the soil series for the site [29], the soil file for which was then imported into the program. The soil file contains information on the erodibility of the soil, the soil texture, and acceptable loss rates. For every treatment, a slope profile was created based on the measured slope and length of each section of the treatment area. Management files had to be created for each BMP treatment, as there were no pre-made files to represent forest roads or skid trails. All operations were set to occur in late April to coincide with the initial site installation. The "highly disturbed land/blade cut" option was selected to represent the Control treatments. Seed treatments used this file, but with the modification of "broadcast seed operation" also used. "Fescue" and "Ryegrass" were used as the species of seed applied, and the "live surface cover" was modified to represent the percentage of ground cover contributed by the germination of the grass seed as time increased. Mulch treatments used this file; however it was modified to include the "add mulch" operation in the form of "bale straw or residue." The type of mulch chosen in this instance was "wheat straw." The option "specify cover directly" was chosen and modified for each treatment to correspond with cover percentages measured in the field. Slash treatments were best represented by the "highly disturbed/blade cut" option, followed by the "add mulch" operation, with "prunings, orchard and vineyard, flail shredded" chosen as the material. The cover was again manipulated by modifying the "specify cover directly" parameter, and by modifying the decomposition half-life of the material to 1800 days, as based on rates used by Wade et al. [24] to represent the decomposition rate of woody debris from southern Appalachian hardwood forests.

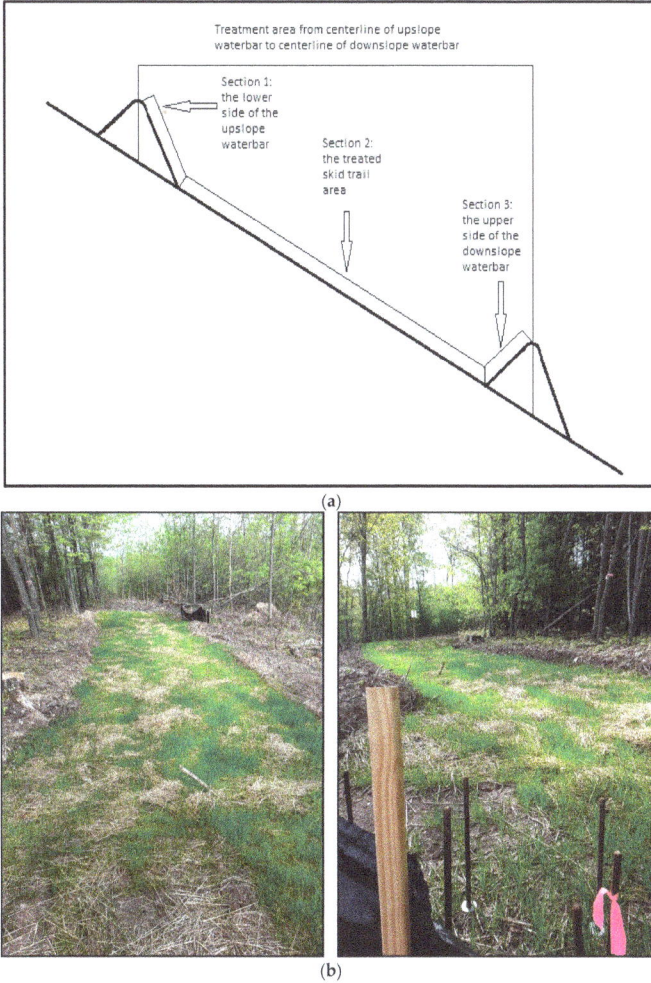

(a)

(b)

Figure 5. (a) Profile of skid trail treatment. Sections 1 and 2 (lower side of upslope waterbar and skid trail surface) were modeled together; Section 3 (upper side of downslope waterbar) was modeled separately; (b) Photographs of a Mulch treatment from the upslope waterbar looking down toward treatment, downslope waterbar, and sediment trap (first photograph), and from the sediment trap looking upslope to waterbar at top of treatment (second photograph).

2.7. WEPP-Road: Default Parameters

WEPP-Road is dependent upon four different types of files to predict soil erosion rates. The software features a database that contains basic files for each of these that can be easily modified to best represent the site. The four types of files are: (1) climate; (2) soil characteristics; (3) slope length and gradient; and (4) land management operations. A climate file for Blacksburg, Virginia is embedded into the software and was therefore chosen as the best representative of the site conditions, as the weather station is less than 8 km (5 miles) away from the study site. Within the WEPP-Road soils database, the file most similar to a Berks-Weikert complex was the "Disturbed Skid Clay Loam," which was chosen for modeling on this site. Soil rock content for each treatment varied from 10–36%, and was directly correlated with slope steepness. Therefore, it was determined that rock content of the soil would be a parameter which needed modification for each treatment, as well as factored into the ground cover in the initial conditions and management files. Slope length and gradient values were modified for each treatment as they were measured with the total station. The "Forest Bladed Road" management file was used for the Control treatments, and was modified for the others. Initial conditions were modified by their initial rill and interrill ground cover percentage, as measured in the field. Seed treatments used this base of "Forest Bladed Road" as the initial conditions and then were modified with the "fescue" and "annual ryegrass at a low fertilization rate." Mulch treatment management files used this file as a base, however "fescue residue" was added as a mulch at a rate of 0.788 kg m^{-2}. Similar to RUSLE2, there are no management files in WEPP-Road that represent Slash treatments. Since there are no woody residue mulch treatments, the same "fescue residue" mulch was chosen. The actual application rate (by weight) that was used to apply slash in the field was used to model this treatment, similar to methods used by Wade et al. [24]. All treatments were modeled for one year.

2.8. WEPP-Road: Modified Parameters

WEPP-Road was then used to model these treatments a second time, using files that were modified to more accurately represent the soil and treatment conditions throughout the year. The primary reason for this being that WEPP-Road has a large number of parameters that can be manipulated when using the model. We wished to determine if collecting more data and making use of more of the model parameters would provide a significantly more accurate soil erosion estimate, and how much more labor would be needed to accomplish this. A soils file was created for each experimental unit based on the "Disturbed Skid Clay-Loam" file used earlier, but modified with the soil rock content and soil particle size present in each of the experimental units. The model was used to calculate interrill erodibility, rill erodibility, critical shear, and effective hydrologic conductivity instead of using the default preset values in place for that particular file. The weather file remained the same, as the Blacksburg, Virginia climate file was determined to be accurately representative of the study site based on its geographic proximity. Slope gradient and length were created once again based on measurements taken in the field with a total station. For the Control treatment, the same "Forest Bladed Road" management file was used for the control treatments, and was modified for the others. The Forest Bladed Road file was modified with initial rill and interrill ground cover percentage, as well as bulk density of the experimental units. This time, the "Initial Plant" field in the "Initial Conditions" file was changed to "Skid Trail-Disturbed," and the "Days Since Last Tillage" field was modified to reflect that the disturbance had just occurred (0 days). Seed treatments used this base of "Forest Bladed Road" as the initial conditions and then were modified with the "fescue" and "annual ryegrass at a low fertilization rate." Mulch treatment management files used this file as a base, however "annual ryegrass at a high fertilization rate" was used instead of "annual ryegrass at a low fertilization rate;" and "fescue residue" was added as a mulch at a rate of 0.788 kg m^{-2}. Once again, there are no files in WEPP-Road that represent woody material for Slash treatments. In this instance, the "Rock" file was chosen in the "Residue Added" field, and was modified to represent the decomposition rate of woody material. This file was chosen because it is the closest available file that could be modified to represent a slash treatment. The actual application rate (by weight) that was used to apply slash in the field was used to model this treatment. All treatments were modeled for one year.

2.9. Data Analysis

Treatment effects for each erosion model were analyzed using JMP statistical software [53]. A variety of methods were used to compare the trapped and modeled estimates including: (1) summary statistics; (2) linear relationships; (3) Nash-Sutcliffe Model Efficiency (NSE) [54]; and (4) a nonparametric analysis. Summary statistics were analyzed to examine means and standard deviations for each treatment using each erosion model. Linear relationships, and NSE were evaluated to determine the accuracy of the models when compared to the actual trapped erosion rates, and a nonparametric comparison for each pair using the Wilcoxon method was conducted to compare these models to each other. Similar comparisons have been conducted by Wade et al. [24] and Croke and Netherly [25].

3. Results

3.1. BMP Treatment Effectiveness

The sediment collected in traps clearly indicates an overall effectiveness of the BMPs compared. Control treatments with waterbars only were measured to have an erosion rate of 15.2 tonnes ha^{-1} year^{-1}. Seed treatments were measured to have an erosion rate of 5.9 tonnes ha^{-1} year^{-1}, and Mulch treatments eroded at a rate of 1.1 tonnes ha^{-1} year^{-1}. Slash treatments eroded at a rate of 0.8 tonnes ha^{-1} year^{-1}. Each model ranked the BMP treatments as having the Control as the most erosive, and the Mulch treatment as least erosive. All models tended to over-estimate the erosion rates of Slash treatments. The Control treatments represent the minimum level of BMPs that are acceptable for skid trail closeout, and was measured to have eroded at rates 2.8× to 8× that of seeded treatments, the next most erodible treatment [49]. Mulch and Slash treatments both reduced average sediment rates to minimal amounts. Adding mulch and fertilizer provided the trail with immediate ground cover, which was not attained by the Seed treatments due to the time necessary for germination. Mulch also aided in the retention of soil nutrients and moisture, as well as reduced predation of the grass seeds from wildlife. Slash provided immediate ground cover, and offers the additional benefits of reducing traffic on the trail, in the form of four-wheelers and pedestrians. After cost analysis, Slash was also shown to provide the greatest benefit in soil erosion reduction per dollar spent in installation [49]. This is due to the fact that no materials are needed to be purchased to install a slash treatment, since slash is already present following the harvest. For all treatments, as ground cover increased, soil erosion decreased. Slope and length did have effects upon the erosion rates, as did rock content of the soil. Steeper slopes in this soil series tended to feature higher rock fragment contents, which acted to increase soil cover over time as the soil around them was eroded away. More information on BMP effectiveness and erosion rates over time may be found in Vinson et al. [49].

3.2. Model Suitability

Models were evaluated using the four different techniques outlined earlier. Each of the model predictions was compared to the trapped sediment data after one year (Table 1). WEPP-Road: Modified had the closest overall mean erosion rate estimate for the Control treatments, while RUSLE2 had the closest overall mean erosion rate for the Seed and Mulch treatments and USLE-Forest provided the closest overall mean erosion rate for the Slash treatments. It is to be noted that for the Control and Slash treatments, the estimates provided by RUSLE2 were more than double the next closest estimate, indicating that its results may be very inconsistent based on the conditions being modeled.

Linear relationships were also used to determine model accuracy. Each of the sets of modeled data were compared to the data collected by sediment traps. Accurate models are expected to have a linear relationship to the collected data [24]. In this study, RUSLE2 was shown to have the highest

R^2 value amongst the three, at 0.6069 (Figure 6). This indicates that RUSLE2 has the best estimated linear relationship with the trapped data. The linear relationship of WEPP-Road: Default to measured data has the second highest R^2 value of 0.5855 (Figure 7), followed by USLE-Forest with an R^2 value of 0.4652 (Figure 8). When compared to the 1:1 line, the inclination of the trend lines of these models indicate that they both tend to overestimate erosion rates. Lastly, the relationship of WEPP-Road: Modified to the measured erosion data has the lowest R^2 value (0.0977) which is indicative of a poor model accuracy (Figure 9). This could have occurred due to inadequacies in modeling just one specific treatment.

Table 1. Summary statistics for each of the models analyzed. For each treatment, there is an asterisk (*) next to the model prediction that is closest to the measured amount.

Treatment	Method	Erosion Rate (tonnes ha^{-1} year^{-1})	Std Dev [a]	Std Error Mean [b]	Lower 95%	Upper 95%
	Measured	15.2	12.1	5.0	2.4	27.9
	USLE-Forest [c]	24.1	11.3	4.6	12.3	36.0
Control	RUSLE2 [d]	66.4	29.2	11.9	35.8	97.1
	WEPP-Road [e]: Default	27.1	6.9	2.8	19.8	34.3
	WEPP-Road [e]: Modified	* 10.8	9.5	3.9	0.8	20.7
	Measured	5.9	5.4	2.2	0.2	11.6
	USLE-Forest [c]	16.5	12.5	5.1	3.4	29.7
Seed	RUSLE2 [d]	* 6.4	3.6	1.5	2.6	10.2
	WEPP-Road [e]: Default	12.7	6.2	2.5	6.2	19.2
	WEPP-Road [e]: Modified	2.8	2.3	0.9	0.4	5.2
	Measured	1.1	1.0	0.4	0.1	2.1
	USLE-Forest [c]	0.3	0.3	0.1	0.0	0.7
Mulch	RUSLE2 [d]	* 0.6	0.4	0.2	0.2	1.1
	WEPP-Road [e]: Default	1.6	0.7	0.3	0.9	2.4
	WEPP-Road [e]: Modified	0.5	0.2	0.1	0.2	0.7
	Measured	0.8	0.6	0.2	0.2	1.4
	USLE-Forest [c]	* 2.3	1.9	0.8	0.3	4.3
Slash	RUSLE2 [d]	21.8	11.0	4.5	10.3	33.3
	WEPP-Road [e]: Default	7.3	3.6	1.5	3.5	11.0
	WEPP-Road [e]: Modified	2.4	1.8	0.7	0.5	4.2

[a] Standard Deviation; [b] Standard Error of the Mean; [c] Univsersal Soil Loss Equation—Forest; [d] Revised Universal Soil Loss Equation, v.2; [e] Water Erosion Prediction Project—Road.

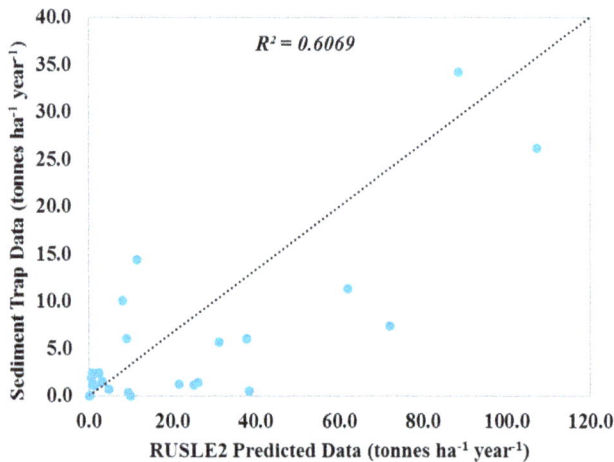

Figure 6. Linear relationship of RUSLE2 modeled data and actual measured erosion data. Line on graph represents a 1:1 relationship.

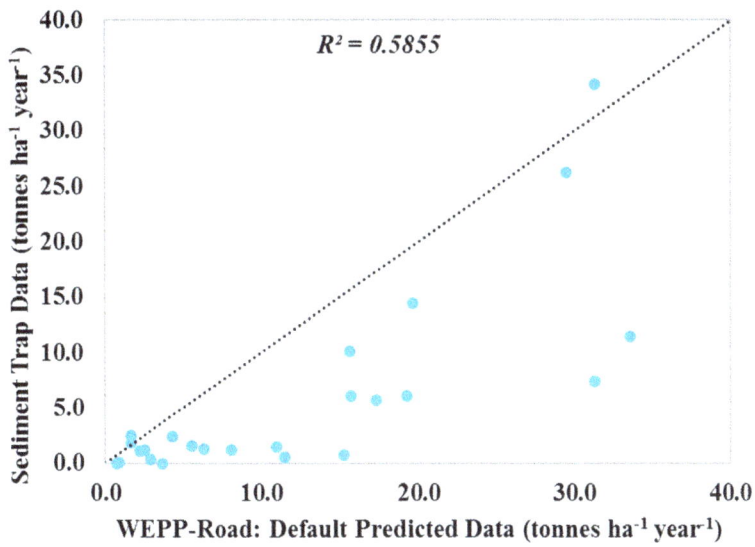

Figure 7. Linear relationship of WEPP-Road modeled data and actual measured erosion data. Line on graph represents a 1:1 relationship.

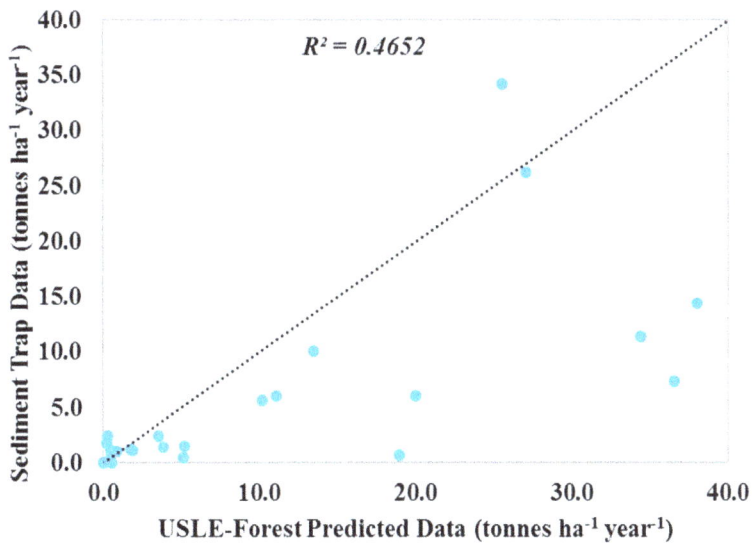

Figure 8. Linear relationship of USLE-Forest modeled data and actual measured erosion data. Line on graph represents a 1:1 relationship.

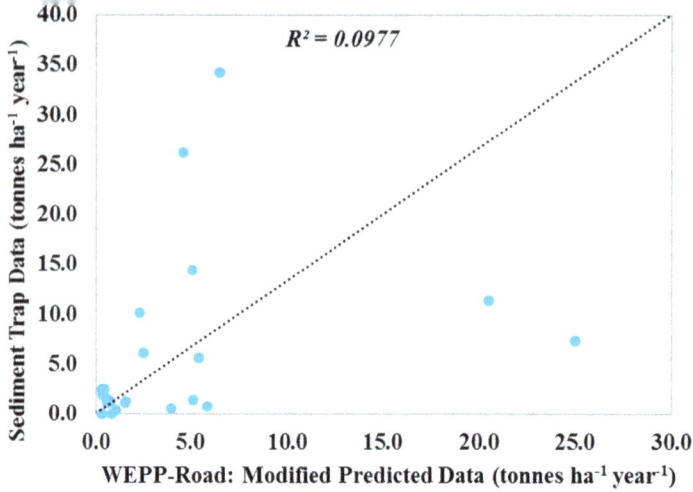

Figure 9. Linear relationship of WEPP-Road: Modified modeled data and actual erosion data. Line on graph represents a 1:1 relationship.

The Nash-Sutcliffe Model Efficiency (NSE) is commonly used to evaluate hydrologic models. The range of efficiency is from $-\infty$ to 1, with values from 0–1 indicating that the model is a good predictor of the measured values. As values approach 1, the model is a more accurate representation of true values. Negative values indicate that the mean of the measured values are a better predictor than the model itself [54], with lower values representing less suitable models. NSE was calculated for each of the treatments and each of the models as a whole (Table 2). All values were negative with the exception of the WEPP-Road: Default (0.15) and RUSLE2 (0.23) models at predicting Mulch treatment erosion rates. The NSE values for the other two models are negative for this treatment (-0.29, and -0.24), however they are substantially greater in value than most other treatment categories. This is evidence that the models did reasonably well at predicting erosion from Mulch treatments. Control treatments were found to have the lowest values for each model, indicating that all models were insufficient at predicting soil loss from bare soil treatments. When evaluating the entire model over all types of treatments, RUSLE2 has a much lower NSE value (-1174.15) than USLE-Forest (-146.35), WEPP-Road: Default (-115.01), and WEPP-Road: Modified (-102.72) indicating that it is the least suitable of the three models. Using this "whole model evaluation," WEPP-Road: Modified has the highest NSE score and would be ranked the most accurate of the four.

Table 2. A comparison of predicted erosion rates and their Nash-Sutcliffe Model Efficiency (NSE) values for the whole model and for each treatment type.

		Control	Seed	Mulch	Slash	Whole Model NSE [a]
Trapped Sediment	(tonnes ha⁻¹ year⁻¹)	15.15	5.87	1.10	0.78	–
USLE-Forest [b]	(tonnes ha⁻¹ year⁻¹)	24.14	16.54	0.33	2.30	–
	NSE [a]	-476.81	-276.44	-0.29	-4.87	-146.35
RUSLE2 [c]	(tonnes ha⁻¹ year⁻¹)	66.44	6.36	0.62	21.77	–
	NSE [a]	-5681.25	-8.90	0.23	-651.77	-1174.15
WEPP-Road [d]: Default	(tonnes ha⁻¹ year⁻¹)	27.06	12.70	1.62	7.25	–
	NSE [a]	-442.32	-94.39	0.15	-60.95	-115.01
WEPP-Road [d]: Modified	(tonnes ha⁻¹ year⁻¹)	10.75	2.05	0.45	2.37	–
	NSE [a]	-520.42	-46.55	-0.24	-4.49	-102.72

[a] Nash-Sutcliffe Model Efficiency; [b] Univsersal Soil Loss Equation—Forest; [c] Revised Universal Soil Loss Equation, v.2; [d] Water Erosion Prediction Project—Road.

Lastly, the models were analyzed using a nonparametric comparison for each pair using the Wilcoxon method (Table 3). For this method, each model was individually compared to the measured data to find significance. In this instance, WEPP-Road: Default and RUSLE2 were considered to be significantly different to the measured data (p-value = 0.0046, p-value = 0.0154).

Table 3. Nonparametric analysis comparing each model to the measured data using Wilcoxon method. $\alpha = 0.05$.

Model	Score Mean Difference	Standard Error Difference	p-Value
USLE-Forest [a]	4.13	4.04	0.3074
RUSLE2 [b]	9.79	4.04	* 0.0154
WEPP-Road [c]: Default	11.45	4.04	* 0.0046
WEPP-Road [c]: Modified	−1.88	4.04	0.6427

[a] Univsersal Soil Loss Equation—Forest; [b] Revised Universal Soil Loss Equation, v.2; [c] Water Erosion Prediction Project—Road.

4. Discussion

Results indicate that the BMPs effectively provided ground cover necessary to reduce erosion. Generally, as ground cover increased, erosion rates decreased. It was seen in the field that rock fragments had a major impact on ground cover and therefore erosion rates [49], which may have been difficult for the models to assess. Slash, seed, and straw mulch have been shown to reduce erosion from skid trails and temporary roads. Slash and straw mulch both provide immediate cover, especially during the initial months at which one can expect erosion rates to be the highest. Slash is readily available, lasts longer than straw mulch, and is more effective at reducing trail traffic. Both slash and straw mulch may also improve the chemical and physical properties of the soil through decomposition. This study indicates that additional road closure BMPs can be used to enhance the minimal effects of waterbars only. Erosion models were shown to have varying degrees of accuracy and suitability based upon their use. Similar conclusions were also reached by Wade et al. [24], Brown et al. [43], and Lang et al. [45].

USLE-Forest was slightly different than the other models in terms of management and cover practices. Whereas RUSLE2 and WEPP-Road model a specific operation and make assumptions based upon its effects, USLE-Forest models these effects directly. This has a noticeable impact on the accuracy of the model. However, many field measurements are required to produce a feasible value from this model. Soils, ground cover, and canopy cover must all be measured in the field. However, this does allow for a more "field available" prediction, whereas RUSLE2 and WEPP-Road both require the use of computer software. The USLE was shown to be the most user-friendly of the three models, in that it can easily be performed with a manual in the field with relatively minimal training and still provide an acceptable estimate of soil loss. Of all the models compared, USLE-Forest provided a consistently, reasonably reliable prediction with minimal difficulty.

RUSLE2 was determined to be the least suitable of the four models assessed, in that its NSE values and nonparametric p-values are all the least favorable of the models. One of the factors affecting the model accuracy is the aforementioned soil rock content. While soils files were accurate enough for the model, it did not take into account the increased soil ground cover from the high soil rock content over time. Other factors include the fact that operations are modeled as such instead of the effects that those operations had upon the ground surface [44]. The primary reason for poor performance of this model would be the fact that there are no management files available for bladed roads or slash treatments. However, RUSLE2 was able to model Seed and Mulch treatments exceptionally well. This shows that RUSLE2 can sufficiently model soil loss for certain ground conditions, but overall may be too inconsistent to be trusted.

WEPP-Road (both Default and Modified) was shown to be the most accurate of the four models based on NSE. This can be attributed to a number of factors. This is the only model that takes into account soil rock content in its analysis, which could have helped to make predictions more accurately. In addition to this, there are forest road and skid trail treatment files available, which gives WEPP-Road an advantage over RUSLE2. One major disadvantage to WEPP is that it does not feature any wood or wood-fiber based mulches to represent slash treatments. Both WEPP-Road and RUSLE2 are at a disadvantage, in that when compared to USLE-Forest, they are difficult to learn initially. They also require significant computer use, which is not always practical for field management decisions.

WEPP-Road: Modified outperformed WEPP-Road: Default. Of the predictions that the modified WEPP model produced, 71% were closer to the measured value than the default WEPP predictions. Lang et al. [45] found similar results when comparing soil erosion models to trapped data from forest haul roads. However, there were some treatments that WEPP-Road: Default modeled better than WEPP-Road: Modified. Inaccuracies in the modified version may have arisen from issues with certain parameters, resulting in the low correlation of modeled points in a linear relationship. It was noted that when using the WEPP model, as the rock content of the experimental unit was increased, the predicted erosion rate dramatically increased. This is not reflective of what was measured or observed in the field measurements, and is also contrary to what other studies have found regarding the effects of soil rock content on the erosion of soils [55,56]. For this reason, we perceive WEPP to be limited in its uses of producing accurate soil erosion predictions on steep, rocky slopes.

5. Conclusions

The primary objective of this study was to evaluate models based on the similarity of their predictions to erosion data collected in the field. After having modeled 24 experimental units over the course of a year using all four models, they were analyzed to determine accuracy. Four BMP treatments were compared to show that adding grass seed; fertilizer, grass seed, mulch; or slash were able to significantly reduce the amount of soil erosion from a bladed skid trail. Mulch and Slash treatments were both the most effective at reducing soil erosion, as they provide immediate ground cover. Based on the Nash-Sutcliffe Model Evaluation and a nonparametric analysis, USLE-Forest and WEPP-Road (both Default and Modified) were significantly better than the other models applied to this site. RUSLE2 was shown to be insufficient for use in modeling bladed skid trails, having over-predicted almost every value. However, of all the soil erosion models, RUSLE2 featured the best linear relationship with the measured erosion data. Each model was able to rank the BMP treatments as having the Control as the most erosive, and the Mulch treatment as least erosive. All models overestimated the erosion rates for Slash treatments, with RUSLE2 placing it at the second-highest erosion rate.

USLE-Forest and WEPP-Road (both Default and Modified) were shown to have been the best suited for this site. With improvements in management and soil files for RUSLE2 and the Default WEPP-Road models, we can expect model accuracy to significantly increase, therefore broadening their applicability to more varied sites. However, as can be seen with results from our Modified WEPP-Road model, as more files are modified and accuracy is further increased the labor involved and time required to complete the modeling drastically increases. This challenge could lead to other models like USLE-Forest being better suited for making forestland management decisions due to their ease of use and ability to provide an acceptable erosion estimate with fewer field measurements and less time required. There are additional research opportunities for comparing these models under different conditions globally.

Acknowledgments: This work was sponsored by the Virginia Polytechnic Institute and State University Department of Forest Resources and Environmental Conservation along with the USDA National Institute of Food and Agriculture McIntire-Stennis program. The Virginia Tech Open Access Subvention Fund (OASF) provided funding to publish in open access.

Author Contributions: J. Andrew Vinson, Scott M. Barrett, W. Michael Aust, and Chad M. Bolding conceived and designed the experiments; J. Andrew Vinson performed the experiments; J. Andrew Vinson and Scott M. Barrett analyzed the data; J. Andrew Vinson wrote the manuscript, and all co-authors edited numerous drafts.

Conflicts of Interest: The authors declare no conflict of interest.

References

1. U.S. Environmental Protection Agency (USEPA). *Nonpoint-Source Pollution: The Nation's Largest Water Quality Problem*; US Environmental Protection Agency, Office of Water, Nonpoint Source Control Branch: Washington, DC, USA, 2003. Available online: http://www.epa.gov/owow/nps/facts/point1.htm (accessed on 12 November 2014).
2. Yoho, N.S. Forest management and sediment production in the south—A review. *South. J. Appl. For.* **1981**, *4*, 27–36.
3. Brown, K.R.; Aust, W.M.; McGuire, K.J. Sediment delivery from bare and graveled forest road stream crossings approaches in the Virginia Piedmont. *For. Ecol. Manag.* **2013**, *310*, 836–846. [CrossRef]
4. Sawyers, B.C.; Bolding, M.C.; Aust, W.M.; Lakel, W.A. Effectiveness and implementation costs of overland skid trail closure techniques in the Virginia Piedmont. *J. Soil Water Conserv.* **2012**, *67*, 300–310. [CrossRef]
5. Wade, C.R.; Bolding, M.C.; Aust, W.M.; Lakel, W.A. Comparison of five erosion control techniques for bladed skid trails in Virginia. *South. J. Appl. For.* **2012**, *36*, 191–197. [CrossRef]
6. Aust, W.M.; Carroll, M.B.; Bolding, M.C.; Dolloff, C.A. Operational forest stream crossings effects on water quality in the Virginia Piedmont. *South. J. Appl. For.* **2011**, *35*, 123–130.
7. Anderson, C.J.; Lockaby, B.G. Effectiveness of forestry best management practices for sediment control in the Southeastern United States: A literature review. *South. J. Appl. For.* **2011**, *35*, 170–177.
8. Grace, J.M. Effectiveness of vegetation in erosion control from forest road sideslopes. *Trans. ASAE* **2002**, *45*, 681–685.
9. Grace, J.M. Forest operations and water quality in the south. *Trans. ASAE* **2005**, *48*, 871–880. [CrossRef]
10. Swift, L.W. Forest road design to minimize erosion in the Southern Appalachians. In Proceedings of the Forestry and Water Quality: A Mid-South Symposium, Little Rock, AR, USA, 8–9 May 1985; Blackmon, B.G., Ed.; University of Arkansas: Monticello, AR, USA, 1985; pp. 141–151.
11. Nolan, L.; Aust, W.M.; Barrett, S.B.; Bolding, M.C.; Brown, K.; McGuire, K. Estimating costs and effectiveness of upgrades in Forestry Best Management Practices for stream crossings. *Water* **2015**, *7*, 6946–6966. [CrossRef]
12. Kochenderfer, J.N. Area in skidroads, truck roads, and landings in the Central Appalachians. *J. For.* **1977**, *75*, 507–508.
13. Worrell, W.C.; Bolding, M.C.; Aust, W.M. Potential soil erosion following skyline yarding versus tracked skidding on bladed skid trails in the Appalachian region of Virginia. *South. J. Appl. For.* **2011**, *35*, 131–135.
14. Virginia Department of Forestry. *Virginia's Best Management Practices for Water Quality*, 5th ed.; Virginia Department of Forestry: Charlottesville, VA, USA, 2011; 165p.
15. Cristan, R.; Aust, W.M.; Bolding, M.C.; Barrett, S.M.; Munsell, J.F.; Schilling, E. Effectiveness of forestry best management practices in the United States: Literature review. *For. Ecol. Manag.* **2015**, *360*, 133–151. [CrossRef]
16. Foltz, R.B. A comparison of three erosion control mulches on decommissioned forest road corridors in the northern Rocky Mountains, United States. *J. Soil Water Conserv.* **2012**, *67*, 536–544. [CrossRef]
17. Grushecky, S.T.; Spong, B.D.; McGill, D.W.; Edwards, J.W. Reducing sediments from skid roads in West Virginia using fiber mats. *North. J. Appl. For.* **2009**, *26*, 118–121.
18. Lyons, K.; Day, K. Temporary logging roads surfaced with mulched wood. *West. J. Appl. For.* **2009**, *24*, 124–127.
19. McGreer, D.J. *Skid Trail Erosion Tests—First-Year Results*; Potlatch Company: Spokane, WA, USA, 1981.
20. Wear, L.R.; Aust, W.M.; Bolding, M.C.; Strahm, B.D.; Dolloff, C.A. Effectiveness of Best Management Practices for Sediment Reduction at Operational Forest Stream Crossings. *For. Ecol. Manag.* **2013**, *289*, 551–561. [CrossRef]
21. Crosson, P.R. *Productivity Effects of Cropland Erosion in the United States*; Routledge: London, UK, 2016.
22. Weil, R.R.; Brady, N.C. *The Nature and Property of Soils*; Pearson: London, UK, 2016; pp. 516–517.
23. Fu, B.; Newham, L.T.; Ramos-Scharrón, C.E. A review of surface erosion and sediment delivery models for unsealed roads. *Environ. Model. Softw.* **2010**, *25*, 1–14. [CrossRef]

24. Wade, C.R.; Bolding, M.C.; Aust, W.M.; Lakel, W.A.; Schilling, E.B. Comparing sediment trap data with the USLE-Forest, RUSLE2, and WEPP-Road erosion models for evaluation of bladed skid trail BMPs. *Trans. ASABE* **2012**, *55*, 403–414. [CrossRef]

25. Croke, J.C.; Nethery, M. Modelling runoff and erosion in logged forests: Scope and application of some existing models. *Catena* **2006**, *67*, 35–49. [CrossRef]

26. Elliot, W.J. WEPP internet interfaces for forest erosion prediction. *J. Am. Water Resour. Assoc.* **2004**, *40*, 299–309. [CrossRef]

27. Toy, T.J.; Osterkamp, W.R. The application of RUSLE to geomorphic studies. *J. Soil Water Conserv.* **1995**, *50*, 498–503.

28. Dissmeyer, G.E.; Foster, G.R. *A Guide for Predicting Sheet and Rill Erosion on Forestland*; General Technical Report R8-TP-6; U.S. Department of Agriculture, Forest Service: Washington, DC, USA, 1980; 40p.

29. United States Department of Agriculture Natural Resource Conservation Service. Web Soil Survey. Natural Resources Conservation Service, 2013. Available online: http://websoilsurvey.nrcs.usda.gov/app/ (accessed on 9 October 2014).

30. Renard, K.G.; Foster, G.R.; Weesies, G.A.; Porter, J.P. RUSLE revised universal soil loss equation. *J. Soil Water Conserv.* **1991**, *46*, 30–33.

31. Laflen, J.M.; Elliot, W.J.; Simanton, J.R.; Holzhey, C.S.; Kohl, K.D. WEPP soil erodibility experiments for rangeland and cropland soils. *J. Soil Water Conserv.* **1991**, *46*, 39–44.

32. Morgan, R.P.C. *Soil Erosion and Conservation*; John Wiley and Sons: Hoboken, NJ, USA, 2009.

33. Amore, E.; Modica, C.; Nearing, M.A.; Santoro, V.C. Scale effect in USLE and WEPP application for soil erosion computation from three Sicilian basins. *J. Hydrol.* **2004**, *293*, 100–114. [CrossRef]

34. Cligen Overview. 2015. Available online: https://www.ars.usda.gov/midwest-area/west-lafayette-in/national-soil-erosion-research/docs/wepp/cligen/ (accessed on 12 June 2015).

35. Elliot, W.J.; Hall, D.E.; Graves, S.R. Predicting sedimentation from forest roads. *J. For.* **1999**, *8*, 23–29.

36. Dun, S.; Wu, J.Q.; Elliot, W.J.; Robichaud, P.R.; Flanagan, D.C.; Frankenberger, J.R.; Brown, R.E.; Xu, A.C. Adapting the Water Erosion Prediction Project (WEPP) model for forest applications. *J. Hydrol.* **2009**, *366*, 46–54. [CrossRef]

37. Elliot, W.J.; Foltz, R.B.; Luce, C.H. Applying the WEPP erosion model to timber harvest areas. In Proceedings of the ASCE Watershed Management Symposium, San Antonio, TX, USA, 14–16 August 1995; ASCE: New York, NY, USA, 1995; pp. 83–93.

38. Elliot, W.J.; Hall, D.E. *Water Erosion Prediction Project (WEPP) Forest Applications*; General Technical Report INT-GTR-365; Moscow, ID; U.S. Department of Agriculture: Washington, DC, USA; U.S. Forest Service: Washington, DC, USA; Intermountain Research Station: Ogden, UT, USA, 1997; 11p.

39. Elliot, W.J.; Foltz, M. Validation of the FS WEPP Interfaces for forest roads and disturbances. In Proceedings of the Transactions of the ASAE Annual International Meeting, Sacramento, CA, USA, 30 July–1 August 2001.

40. Morfin, S.; Elliot, B.; Foltz, R.; Miller, S. Predicting effects of climate, soil, and topography on road erosion with WEPP. In Proceedings of the Transactions of the ASAE Annual International Meeting, Phoenix, AZ, USA, 14–18 July 1996.

41. Tysdal, L.M.; Elliot, W.J.; Luce, C.H.; Black, T. Modeling insloped road erosion processes with the WEPP Watershed Model. In Proceedings of the Transactions of the ASAE Annual International Meeting, Minneapolis, MN, USA, 10–14 August 1997.

42. Elliot, W.J.; Scheele, D.L.; Hall, D.E. The Forest Service WEPP Interfaces. In Proceedings of the Transactions of the ASAE Annual International Meeting, Milwaukee, WI, USA, 9–12 July 2000.

43. Brown, K.R.; McGuire, K.J.; Hession, W.C.; Aust, W.M. Can the Water Erosion Prediction Project model be used to estimate best management practice effectiveness from forest roads? *J. For.* **2016**, *114*, 17–26. [CrossRef]

44. Foster, G.R.; Toy, T.J.; Renard, K.G. Comparison of the USLE, RUSLE 1.06c, and RUSLE2 for Application to Highly Disturbed Lands. In Proceedings of the First Interagency Conference on Research in the Watersheds, Benson, AZ, USA, 27–30 October 2003; pp. 154–160.

45. Lang, A.J.; Aust, W.M.; Bolding, M.C.; McGuire, K.J.; Schilling, E.B. Comparing sediment trap data with erosion models for evaluation of forest haul road stream crossing approaches. *Trans. ASABE* **2017**, *60*, 393–408.

46. Fransen, P.J.; Phillips, C.J.; Fahey, B.D. Forest road erosion in New Zealand, overview. *Earth Surf. Process. Landf.* **2001**, *26*, 165–174. [CrossRef]
47. Croke, J.; Mockler, S. Gully initiation and road-to-stream linkage in a forested catchment, southeastern Australia. *Earth Surf. Process. Landf.* **2001**, *26*, 205–217. [CrossRef]
48. Luce, C.H.; Black, T.A. Sediment production from forest roads in western Oregon. *Water Resour. Res.* **1999**, *35*, 2561–2570. [CrossRef]
49. Vinson, J.A.; Barrett, S.M.; Aust, W.M.; Bolding, M.C. Evaluation of bladed skid trail closure methods in the ridge and valley region. *For. Sci.* **2017**, *63*, 432–440. [CrossRef]
50. National Oceanographic and Atmospheric Administration. *Summary of Monthly Normals 1981–2010*; Blacksburg National Weather Station Office: Blacksburg, VA, USA, 2010. Available online: http://www.weather.gov/rnk/MonthlyClimateNormals (accessed on 13 May 2015).
51. Weather Underground Service. *Weather History for April 2015 through April 2016*; the Weather Channel Interactive, Inc.: Atlanta, GA, USA, 2016; Available online: www.wunderground.com/history (accessed on 9 April 2016).
52. Cambi, M.; Certini, G.; Neri, F.; Marchi, E. The impact of heavy traffic on forest soils: A review. *For. Ecol. Manag.* **2015**, *338*, 124–138. [CrossRef]
53. *JMP®*, version 11.0; SAS Institute Inc.: Cary, NC, USA, 1989–2015.
54. Nash, J.E.; Sutcliffe, J.E. River flow forecasting through conceptual models: Part 1. A discussion of principles. *J. Hydrol.* **1970**, *10*, 282–290. [CrossRef]
55. Poesen, J.W.; Torri, D.; Bunte, K. Effects of rock fragments on soil erosion by water at different spatial scales: A review. *Catena* **1994**, *23*, 141–166. [CrossRef]
56. Cerdá, A. Effects of rock fragment cover on soil infiltration, interrill runoff, and erosion. *Eur. J. Soil Sci.* **2001**, *52*, 59–68. [CrossRef]

Article

Use, Utilization, Productivity and Fuel Consumption of Purpose-Built and Excavator-Based Harvesters and Processors in Italy

Natascia Magagnotti [1], Luigi Pari [2] and Raffaele Spinelli [1,3,]*

1   CNR Ivalsa, Via Madonna del Piano 10, I-50019 Sesto Fiorentino (FI), Italy; magagnotti@ivalsa.cnr.it
2   CRA-ING Via della Pascolare 16, I-00015 Monterotondo Scalo (Roma), Italy; luigi.pari@crea.giv.it
3   AFORA, University of the Sunshine Coast, Locked Bag 4, Maroochydore, QLD 4558, Australia
*   Correspondence: spinelli@ivalsa.cnr.it; Tel.: +39-335-542-9798

Received: 4 November 2017; Accepted: 4 December 2017; Published: 6 December 2017

Abstract: Annual use, utilization, productivity and fuel consumption of three purpose-built and three excavator-based harvesters and processors were monitored for one work year. All machines were owned and operated by private contractors and were representative of the Italian machine fleet. Despite challenging mountain terrain, annual use ranged from 675 to 1525 h per year, and production from 3200 to 27,400 m^3 per year. Productivity was lower for excavator-based units, and for machines working under a yarder, due to limited yarder capacity. Purpose-built machines offered higher utilization, productivity and fuel efficiency compared with excavator-based machines. Fuel consumption per m^3 was 2.4 times greater for excavator-based units, compared with purpose-built machines. Excavator-based units offered financial and technical advantages, but their long-term market success will likely depend on future improvements in fuel efficiency, in the face of increasing fuel prices.

Keywords: efficiency; contractors; benchmark; logging

---

1. Introduction

Mechanized harvesting is performed by specialized machines that fell and process the trees (harvesters) or process already felled trees (processors) into commercial assortments, and by other units (forwarders) that extract these assortments to a landing [1]. Compared with motor-manual operations, mechanized harvesting allows dramatic progress in terms of value recovery [2] labour productivity [3] and operator safety and comfort [4]. Even where motor-manual harvesting techniques are still competitive due to cheap labour, mechanization may increase production capacity and anticipate future labour shortages [5].

For these reasons, harvesters and processors are now common in many countries [6], and not only in Northern Europe where they were first developed and adopted [7,8]. While originally designed for low-land conifer forests, today these machines operate in steep terrain [9,10], hardwood stands [11] and fast-growing plantations [12].

In Italy, mechanized harvesting was introduced at the beginning of the new century [13], and by 2013 the Italian fleet counted over 200 units, including harvesters, processors and forwarders [14]. These machines are used under different and peculiar conditions compared with those encountered further North, and namely steep terrain alpine forests, industrial poplar plantations, close-to-nature forests, coppice stands, and non-industrial private forestry. Despite a relatively difficult work environment, mechanized harvesting technology has enjoyed much success and many logging firms have already purchased their second or third machine [15].

On the other hand, some logging contractors still have some doubts about acquiring a harvester or a processor, mainly due to the high purchase cost of the equipment and the limited investment capacity of their firms [16]. Before they buy, operators would like to obtain reliable information about productivity, actual use potential and fuel consumption, for costing purposes. Productivity references are mostly available for Nordic and Central European conditions, and these figures may not correctly represent the conditions of Southern Europe, where operator training, work environment and technology type can be quite different.

In fact, productivity figures can be derived from published short-term studies lasting a few days; however, these figures can be representative of actual work time only, because short-term studies offer a poor representation of delay incidence, and of long-term productivity in general. This is best gauged through long-term follow-up studies [17]. Besides, long-term follow-up studies are generally immune from the so-called "Hawthorne effect", i.e., the tendency of observed workers to change their behavior as a result of being observed. Even when operators know their performance is being observed, work pace tends to stabilize to normal levels if the observation period is very long and the knowledge of being observed fades into the general background.

Existing literature offers few examples of long-term follow-up studies of harvesters and processors. Most of these studies tap into State company records, due to the general practice of State companies to keep accurate records of their own activities. However, machine use conditions are likely different between large State or private companies and individual small-scale contractors, who still represent the backbone of the forestry contracting sector in many countries, inside and outside Europe. Therefore, data obtained from large companies may not provide an accurate representation of the use pattern, productivity and fuel consumption normally experienced by individual private contractors.

Furthermore, machine use and performance may differ between machine types, and especially between purpose-built machines and excavator-based units. In particular, excavator-based units seem to be poorly represented in the Northern and Central European machine fleets, while they are quite popular in Southern Europe due to their lower investment cost and their higher operational flexibility. The former facilitates penetration of new markets such as the Southern European market, while the latter favors introduction to radically different work chains, such as those encountered in mountain operations or in short-rotation tree farms established on ex-arable land. This is also the case for Italy, where three-quarters of the harvester and processor heads are mounted on adapted excavators, often pre-owned, which indicates the strong interest in minimizing investment cost and the associated financial risk [15]. Very little long-term information is available about these machines, nor can one find any comparison between machine types that could help derive estimates from the figures already available for purpose-built harvesters.

Therefore, the goals of this study were (1) to provide reliable information about machine use (hours per year), utilization (percent of productive time over scheduled time), productivity and fuel efficiency, as typical of individual small-scale private contractors operating in Southern Europe; (2) to detect any significant differences between purpose-built and excavator-based units. While the study was centered on small-scale Italian logging contractors, some of these contractors conducted cross-border business and also operated in the neighboring Austrian and French regions, which may give some more general value to the figures obtained in the process.

2. Materials and Methods

The study involved five logging contractors, of which one was based in Central Italy and four in Northern Italy, where about two-thirds of the Italian harvester and processor fleet is stationed [15]. The company based in Central Italy was one of the early adopters, and had accumulated over 15 years of experience with mechanized harvesting technology, which they had acquired already in the early 2000s.

Six machines were selected for the study, of which three were purpose-built machines and three were excavator-based machines (Figure 1). Two of the purpose-built machines were owned by the

Figure 1. Example of the purpose-built (**top**) and excavator-based (**bottom**) units in the study. These are respectively unit # 2 and # 6.

The two machine types represented in the study (i.e., purpose-built and excavator-based) reflect different technical choices available to contractors working with mechanized harvesting technology. In general, contractors opt for an excavator-based machine because it is cheaper to acquire and easier to re-sell if the business does not grow as expected, compared with a purpose-built machine. Deployment conditions were similar across the two groups: within each group, two machines out of three were deployed to assist a yarder and worked as processors, whereas one was working independently as a harvester (i.e., machines 3 and 4). The six machines in the study were considered representative of the entire population of harvesters and processors used in Italy, and in much of Southern Europe [15]. Their technical characteristics are described in Table 1, while Figure 2 shows their work areas (e.g., home range).

Table 1. Technical characteristics of the machines in the study.

Unit	#	1	2	3	4	5	6
Machine	Type		Purpose-Built			Excavator-Based	
Base	Make	Skogsjan	Ecolog	John Deere	Liebherr	Daewoo	JCB
Base	Model	495	580	1470	912	225 NLCV	180
Base	kW	165	205	180	80	110	81
Base	Kg	15,500	18,500	19,700	16,000	21,500	20,200
Head	Make	Woody	Woody	Waratah	Valmet	Zoeggeler	Woody
Head	Model	60H	60H	290	965 II	ZBH 70	60H
Age	Years	9	8	6	6	1	3

Each machine was considered as a study unit and was operated by an individual driver, often the main company owner. All drivers were male Italian nationals, aged between 30 and 45 years. They all had at least three years of experience running a harvester or a processor, often the same machine included in the study. The drivers on machines # 3 and 6 were employees working for the machine owner, while all the others were also the main company and machine owners. Each driver agreed to keep a detailed logbook, where he recorded the main work data on a daily basis, for one year. Each daily record included date, location, travel hours, work hours, delay hours, processed volume and fuel used. Travel distance between sites was calculated on maps, and so was the home range of each unit—i.e., the work area explored during one year. Units # 1, 2, 5 and 6 worked in the Alps and negotiated the maturity cut of typical steep-terrain alpine forests, generally consisting of an uneven-aged mix of spruce (*Picea abies* Karst.), fir (*Abies alba* L.) and beech (*Fagus sylvatica* L.), and treated with selection cutting. These machines were associated with a yarder and operated as processors on pre-felled trees. In contrast, units # 3 and 4 negotiated the clear-cutting of mature low-land pine (*Pinus pinaster* L.) and poplar (*Populus x euroamericana* sp.) plantations, and worked as harvesters, felling and processing the trees at the stump site.

Figure 2. Work area (i.e., home range) explored by the machines during this study (notes: polygons represent the area explored by each study unit, as reported in the legend near each individual polygon).

Daily records were consolidated and analyzed as monthly averages, to smooth extreme differences. This was especially important for fuel consumption, because machines were not refueled every day, which made it difficult to produce exact daily consumption estimates. Furthermore, monthly records were summed into annual totals, offering an immediate view of annual work load.

Data were analyzed with the Statview 5.01 advanced statistics software, in order to check the statistical significance of the eventual differences between unit types—purpose-built vs. excavator-based. The significance of any differences between annual totals was tested with non-parametric techniques, because the data points were few and their distribution violated the normality assumption. In contrast, the significance of any differences resulting from the comparison of monthly averages was tested with a conventional analysis of variance, because the data was normally distributed. The elected significance level was $\alpha < 0.05$.

3. Results

Annual use ranged from 675 to 1525 h (Table 2). Purpose-built machines worked more hours, had a higher percent utilization and produced larger log volumes than excavator-based machines, but these differences had no statistical significance when taken as annual totals. In that case, the only statistically significant difference concerned fuel consumption per cubic meter, which was almost twice as high for excavator-based units compared with purpose-built units (Table 3).

Table 2. Annual totals by test unit.

Unit	#	1	2	3	4	5	6
Machine	Type	Purpose-Built			Excavator-Based		
		Use					
Work	h year^{-1}	792	813	1260	604	366	701
Delays	h year^{-1}	229	371	265	648	309	314
Total time	h year^{-1}	1021	1184	1525	1253	675	1015
Utilization	%	78	69	83	48	54	69
Work days	n° year^{-1}	123	123	162	152	88	140
Work days	h day^{-1}	8.3	9.6	9.4	8.3	7.7	7.3
Idling	month year^{-1}	1	0	0	0	4	2
		Productivity and consumption					
Fuel	L year^{-1}	7807	6851	19,279	10,803	5465	9170
Volume	m^3 year^{-1}	10,533	10,337	27,432	10,487	3191	4146
Productivity	m^3 pmh^{-1}	13.3	12.7	21.8	17.4	8.7	5.9
Productivity	m^3 smh^{-1}	10.3	8.7	18.0	8.4	4.7	4.1
Fuel use	L pmh^{-1}	9.9	8.4	15.3	17.9	14.9	13.1
Fuel use	L m^{-3}	0.74	0.66	0.70	1.03	1.71	2.21

Notes: Utilization = productive time/scheduled time; pmh = productive machine hours, excluding delays; smh = scheduled machine hours, including delays (same here as total time).

The area explored by the different units showed a large variation, ranging between 52 and 1571 km^2, while the number of relocation trips per year varied from 3 to 10. Purpose-built units travelled for longer distances compared with excavator-based units. Most machines experienced some seasonal stop, which was generally very short except for unit 5. In fact, the two machines that performed both felling and processing in low-land forests experienced no prolonged seasonal stops, which were recorded only for machines working alongside a yarder, possibly because these machines operated on mountain forests where snow could impose prolonged seasonal interruptions of harvesting activities.

Table 3. Medians of the annual records by machine type.

Machine	Type	Purpose-Built	Excavator-Based	p-Value	DF
Total hours	h year^{-1}	1184	1015	0.2752	6
Utilization	%	78	54	0.1266	6
Work day	d year^{-1}	123	140	0.8273	6
Fuel	L year^{-1}	7806	9170	0.8273	6
Volume	m^3 year^{-1}	10,533	4146	0.1266	6
Productivity	m^3 h^{-1}	13.3	8.7	0.2752	6
Fuel use	L pmh^{-1}	9.8	14.9	0.2752	6
Fuel use	L m^{-3}	0.70	1.71	0.0495	6

Notes: Utilization = productive time/scheduled time; pmh = productive machine hours, excluding delays; DF = degrees of freedom.

The analysis of monthly averages confirmed the main findings that already emerged from the analysis of the annual means, offering further insights into working hours, utilization, productivity and fuel use (Table 4).

Table 4. Means of the monthly totals by machine type.

Machine	Type	Purpose-Built	Excavator-Based	p-Value	DF	Eta^2
Work	h month^{-1}	86	56	0.0003	59	0.27
Delays	h month^{-1}	26	42	0.0156	59	MW
Total time	h month^{-1}	112	98	0.1897	59	0.03
Work days	n$^\circ$ month^{-1}	12	13	0.7042	59	0.00
Work days	h day^{-1}	9.2	7.7	<0.0001	59	0.59
Utilization	%	77.1	58.9	<0.0001	59	MW
Product	m^3	1490	584	<0.0001	59	MW
Productivity	m^3 h^{-1}	17	11	0.0004	59	MW
Fuel use	L month^{-1}	1043	846	0.1716	59	0.03
Fuel use	L h^{-1}	12	15	0.0002	59	0.29
Fuel use	L m^{-3}	0.72	1.70	<0.0001	59	MW
Relocation	km month^{-1}	104	28	0.9814	47	MW
Relocation	trips month^{-1}	0.50	0.72	0.4432	47	MW
Max distance	km	126.6	30.2	0.9243	41	MW

Notes: DF = degrees of freedom; Eta^2 = Effect size for parametric tests, or type of non-parametric test used (MW = Mann–Whitney); Utilization = productive time/scheduled time.

Monthly records offered higher resolution, and the higher number of data points allowed significant differences to be disclosed between machine types with regard to use, utilization, production, productivity and fuel use. Operators used purpose-built machines more intensely, which generally turned out to be more efficient than excavator-based units in terms of productivity, fuel use and utilization. As an average, the utilization of purpose-built machines was 18 percentage points higher than that of excavator-based machines (i.e., 30% more in relative terms). Furthermore, the mean productivity of purpose-built machines was 50% higher than that of excavator-based machines, while fuel consumption per hour was 20% lower: as a result, mean fuel consumption per cubic meter was half as high as that recorded for excavator-based machines. Productivity was also dependent on work organization: machines that worked independently (i.e., units 3 and 4) were 60% to 100% more productive than machines that worked under a yarder and were limited by yarder output.

Furthermore, analysis of monthly records disclosed seasonal trends, which were different for different machine types. In general, all machines experienced a lull in activities at the end of winter (February) and at the peak of summer (August), which can be justified by the combination of climatic factors and calendar holidays, especially for August. However, peaks and lulls were very different, with excavator-based units experiencing deeper and longer drops than purpose-built units (Figure 3).

As an average, excavator-based machines were deployed on much smaller jobs than purpose-built machines, which may have contributed to their lower utilization. In fact, the two machine types seem to underline different use patterns, with excavator-based machines apparently targeting smaller jobs within a smaller distance from each other, and purpose-built machines accepting longer relocation distances (three times longer) in order to acquire significantly larger jobs (three times larger). This is confirmed by the area covered by the contractors, which averaged 3400 km^2 for purpose-built machines and 1300 km^2 for excavator-based machines.

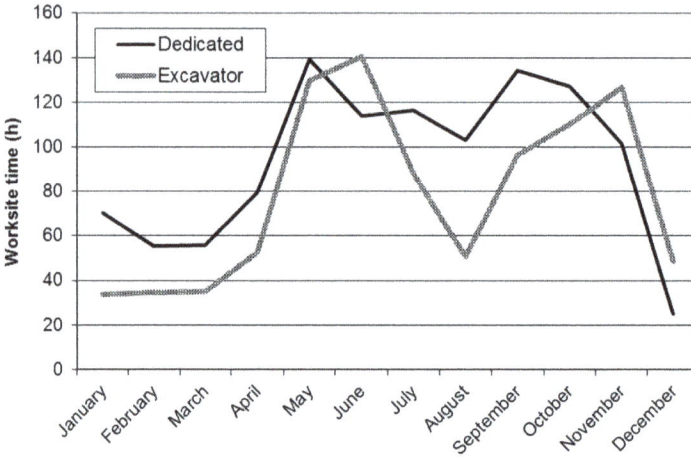

Figure 3. Total working hours per month and per machine type (mean of three samples for each type).

4. Discussion

First of all, it is important to state upfront the limitations of the study: (a) the use of a small and widely variable sample and (b) the reliance on company records, with the inherent variability derived from the different recording practices adopted by different companies. Obtaining detailed, long-term records from small-scale private contractors is rather difficult, because many contractors keep partial records only, and they are often unwilling to share data that may be considered sensitive. Therefore, the small sample size is a justified limitation, and its negative effect has been contained by selecting representative cases based on a detailed knowledge of the sector, which was obtained from three previous surveys of Italian contractors [14–16]. These surveys produced a detailed picture of the contracting sector in Italy, as well as a list of contractors that was used for extracting representative samples. Concerning the variability of individual records, this limitation may have affected the resolution of the data more than its reliability, because this study extracted only basic figures that were accounted for in a common way by all contractors, such as place names, worked volumes, fuel inputs and time inputs. If at all, variability could have affected the estimates for work hours, because different operators may have labelled different tasks under different headings, potentially attributing some delay time to work time or vice versa. However, all operators were carefully instructed about what pertained to work time and what pertained to delay time, and it is unlikely that they made major errors.

In any case, this is one of the very few studies representing small-scale private contractors. Most other available studies tap into the records of State companies, which often benefit from a formal support infrastructure and long-term operational planning. This may explain the much larger annual use figures recorded specifically for State companies in Austria and Germany (Table 5). If the difference was just a national one, then the European surveys recently conducted by Malinen et al. [18] and by Spinelli et al. [19] would not report use figures that are so close to those reported in this study. At the same time, similarity with the figures already available for the general pool of Italian harvesters and processors indicates the selection of a representative sample, despite its small size [15]. Further corroboration comes from the analysis of the utilization figures presented in this study: these are generally comparable with those reported in the existing relevant literature, which fall in the bracket between 50% and 80% [20,21].

Productivity is affected by many factors besides machine characteristics, such as tree size and form, operator experience, assortment length and number, branch diameter and management objectives [22–25]. Given the limited detail included in a long-term study, it is difficult to thoroughly discuss the similarities and the differences between the figures obtained from this study and those reported in the general literature on the subject. However, the figures in this study are generally compatible with those indicated in the literature, and they match quite well the Italian productivity standards for harvesters and processors [26]. This said, readers must be aware that direct comparison can be deceiving, because most of the existing productivity figures are obtained from short-term time study sessions, not from long-term company records. The fact that the latter match quite well a productivity standard derived from a compilation of short-term time studies bodes well for the capacity of time studies to reflect actual long-term performance, at least when more individual study sessions are consolidated into a cluster and analyzed as a group.

Concerning productivity, this study shows that purpose-built machines regularly outperform excavator-based units. The reason for the superiority of purpose-built machines is likely in the adoption of a specialized design and a much larger engine. While it may be difficult to separate the effects of design and power, it is worth recalling that the purpose-built machines in this study are twice as powerful as excavator-based machines, and that must have a strong impact on performance. In any case, readers must be aware that the figures in this report represent one element in a more complex supply chain and are likely affected by deployment conditions, which are not described in much detail in the records. Therefore, it may be difficult to produce a detailed productivity analysis from this study, but such analysis was never one of the goals of the project. However, this study clearly discriminates between units used independently from other machines and units routinely used in association with a yarder. The inevitably lower productivity and utilization of the latter raises the question about what processor one would best detach to serve a yarder. Obviously, high productivity is not a primary requirement, because this machine will not be able to express its full potential. On the other hand, a smaller, cheaper and less productive machine may be unable to handle the large trees associated with mountain operations—the only trees that can justify the higher cost of yarding [27]. One solution is to task the processor with additional duties, such as fleeting, stacking, loading and general landing management. That is already implemented by most operators. Readers will notice that all units deployed at yarder operations in this survey (i.e., 1, 2, 5 and 6) carry Konrad or Zoeggeler heads, which can effectively double as log grapples. This is not casual: all processors that worked under a yarder did perform additional duties, such as fleeting and stacking. However, this measure does not seem to fully offset the higher productive potential of the processor, and additional solutions must be considered.

Like productivity, fuel consumption is affected by machine and work type, as well as by operator skills and technique [28]. Again, the range of fuel consumption figures reported in the existing literature is wide enough that the results of this study are fully corroborated. Investigations conducted on purpose-built harvesters operating under similar conditions in neighboring Austria indicate a mean fuel consumption of 15.6 L h^{-1}, with a range between 10 and 24 L h^{-1} [20]. More importantly, the present study highlights a sharp difference between purpose-built and excavator-based machines and points at the potential fuel savings that could be accrued if excavator-based machines could be improved, or replaced with purpose-built units. The lower fuel consumption incurred by purpose-built harvesters is the result of a sophisticated machine control system designed to adjust power output to power demand in real time. Theoretically, excavators are equipped with similar systems but the connection between a processor head and an excavator that was not originally designed to receive it does not allow the same optimization level as achieved by integrated machines built specifically for this task. Extending such benefits to excavator-based machines would require that some excavator manufacturer finally decided to invest time and money into developing specific harvester capability options for one or more of their excavator models, or that a separate manufacturer—possibly the same building the harvester head—developed an effective adaptation kit comprising both hardware and software. The latter solution has already been attempted in the past, and some harvester head manufacturers have developed such kits, but that is often a one-sided effort that has received little support from excavator manufacturers and has produced limited benefits. Manufacturers should keep in mind that contractors are very much concerned with fuel consumption because they cannot control fuel price, and surveys have already shown that fuel-efficiency is a main driver when purchasing new machines [29]. Therefore, the development of fuel-efficient excavator-based processors should represent a strategic goal for excavator manufacturers, and for those companies that build processor heads designed for fitting to an excavator. Otherwise, the increase of fuel price may erode their market shares, in favor of those companies that sell purpose-built units.

Another important question is the extent to which these results can be generalized, and whether they could be used to represent Italy alone, or could be extended to other countries as well. It is true that Italian forestry presents peculiar conditions in terms of extreme ownership fragmentation, conservative silviculture, low product value, and poor integration between forest management and wood industry—all of which affect the progress of mechanized harvesting, when they do not limit it [16]. That is the main reason why the Italian harvester and processor fleet is still much smaller than in the neighboring alpine countries, such as Austria [30] France [31] or Germany [32]. On the other hand, mechanized harvesting has made rapid and significant inroads in Italian forestry over the last few years [14] as the old generations of foresters and loggers are being replaced by new young professionals who are not ready to accept the same taxing work conditions that their elders had to cope with [33]. Therefore, Italian forestry may soon become intensely mechanized, which would justify the attempt to establish benchmark figures for the Italian case alone, even if the work conditions encountered in Italy could not be assimilated to those of any other countries. This said, the case of Italy seems to be a classic example of modernization, where logging is transitioning from a traditional small-scale business to an industrial activity. In that case, results obtained in Italy could be extended to other countries where forestry is experiencing the same transition.

Table 5. Annual utilization of purpose-built harvesters in some European Countries.

Hours Year^{-1}	Country	Population	Source
1184	Italy	Private contractors	This study
1328	Italy	Private contractors	Spinelli et al., 2010 [15]
1439	Europe	General	Spinelli et al., 2011 [19]
1323	Western Europe	General	Malinen et al., 2016 [18]
2042	Austria	State forests	Holzleitner et al., 2011 [20]
1560	Austria	General	Pröll 2005 [30]
1750	Germany	State forests	Forbrig 2000 [34]
1900	Germany	State forests	Denninger 2002 [35]
2036	Germany	General	Findeisen 2002 [36]
1865	Germany	General	Nicks and Forbrig 2002 [37]
1300	Germany	General	Drewes and Jacke 2005 [38]

While producing a benchmark, one should also define its characteristics. In particular, the question is whether the reference figures produced with this study represent best practice or ordinary practice—which are conceptually different. If one has obtained these figures from a representative sample of contractors, then the resulting benchmark must refer to ordinary practice. Otherwise, it should have been obtained from the top operators in the general contractor pool. On the other hand, mechanization is still adopted by a minority of contractors, who generally represent an elite: therefore, the benchmark figures estimated in this study approach best practice, at least for Italian forestry.

The same "natural selection" principle may apply to the differences found between excavator-based and purpose-built machines. The latter are generally adopted by the largest and most skillful contractors, who may have a higher level of professionalism compared with the contractors resorting to cheaper excavator-based units. Therefore, part of the efficiency difference between machine types might result from a combination of machine and operator characteristics, both of which have a strong effect on performance. Even if that were the case, there would still be no reason to try to separate the two effect types, because they would be inherently associated and an eventual separation would not achieve any practical purpose.

5. Conclusions

This study offers reference figures that can be used for benchmarking the performance of harvesters and processors used in Italy, or under work conditions that can be assimilated with those encountered there. These figures are quite reliable, due to the selection of a representative sample and to the long duration of the study itself. Furthermore, these figures are consistent with the results of previous studies, conducted on the same subject with different methods. This corroborates the estimates obtained from the study, and supports confidence in its results. The study also describes the substantial difference between purpose-built harvesters and excavator-based units, in terms of use intensity, productivity and fuel efficiency. Shifting from excavator-based units to purpose-built machines would allow a dramatic reduction of fuel consumption, and the long-term market success of excavator-based technology will likely depend on the ability to increase their fuel efficiency. Since excavator-based units may offer specific financial and technical benefits, there is scope for new research aiming to reduce their fuel consumption. The findings of this research will be useful to contractors who are considering shifting from manual to mechanized harvesting, for increased productivity, safety and comfort.

Acknowledgments: This project has been partly supported by the Bio Based Industries Joint Undertaking under the European Union's Horizon 2020 research and innovation program under grant agreement No. 720757 Tech4Effect. The Authors thank Dott. M. De Stefano (Conaibo-Italian Association of Logging Contractors) for her support with data collection and verification.

Author Contributions: Natascia Magagnotti, Raffaele Spinelli and Luigi Pari conceived and designed the experiments; Natascia Magagnotti and Raffaele Spinelli performed the experiments; Natascia Magagnotti, Raffaele Spinelli and Luigi Pari analyzed the data; Natascia Magagnotti and Raffaele Spinelli wrote the paper.

Conflicts of Interest: The authors declare no conflict of interest.

References

1. Wang, J.; Ledoux, C. Simulating cut-to-length harvesting operations in Appalachian hardwoods. *Int. J. For. Eng.* **2005**, *16*, 11–27.
2. Spinelli, R.; Magagnotti, N.; Nati, C. Work quality and veneer value recovery of mechanised and manual log-making in Italian poplar plantations. *Eur. J. For. Res.* **2011**, *130*, 737–744. [CrossRef]
3. Chiorescu, S.; Grönlund, A. Assessing the role of the harvester within the forestry-wood chain. *For. Prod. J.* **2001**, *51*, 77–84.
4. Bell, J. Changes in logging injury rates associated with use of feller-bunchers in West Virginia. *J. Saf. Res.* **2002**, *33*, 436–471. [CrossRef]
5. Spinelli, R.; Owende, P.; Ward, S. Productivity and cost of CTL harvesting of Eucalyptus globulus stands using excavator-based harvesters. *For. Prod. J.* **2002**, *52*, 67–77.
6. Gellerstedt, S.; Dahlin, B. Cut-to-length: The next decade. *J. For. Eng.* **1999**, *10*, 17–25.
7. Brunberg, T. *Basic Data for Productivity Norms for Single-Grip Harvesters in Thinning*; Skogsforsk Redogörelse; Skogsforsk: Uppsala, Sweden, 1997; Volume 8, p. 18.
8. Nurminen, T.; Korpunen, H.; Uusitalo, J. Time consumption analysis of the mechanized cut-to-length harvesting system. *Silva Fenn.* **2006**, *40*, 335–363. [CrossRef]
9. Visser, R.; Stampfer, K. Expanding ground-based harvesting onto steep terrain: A review. *Croat. J. For. Eng.* **2015**, *36*, 321–331.

10. Frutig, F.; Fahmi, F.; Settler, A.; Egger, A. Mechanisierte Holzernte in Hanglagen. *Wald Holz* **2007**, *5*, 47–52.

11. Cuchet, E.; Morel, P.J. Evolution du bucheronnage mechanisé, cas de la Bourgogne et perspectives nationales (Evolution of mechanized processing, the case of Burgundy). *Inf. Foret* **2001**, *2*, 6. (In French)

12. McEwan, A.; Magagnotti, N.; Spinelli, R. The effects of number of stems per stool on cutting productivity in coppiced Eucalyptus plantations. *Silva Fenn.* **2016**, *50*, 1448. [CrossRef]

13. Spinelli, R.; Magagnotti, N. The effects of introducing modern technology on the financial, labour and energy performance of forest operations in the Italian Alps. *For. Policy Econ.* **2011**, *13*, 520–524. [CrossRef]

14. Spinelli, R.; Magagnotti, N.; Facchinetti, D. Logging companies in the European mountain: An example from the Italian Alps. *Int. J. For. Eng.* **2013**, *24*, 109–120. [CrossRef]

15. Spinelli, R.; Magagnotti, N.; Picchi, G. Deploying mechanized cut-to-length technology in Italy: Fleet size, annual usage and costs. *Int. J. For. Eng.* **2010**, *21*, 23–31.

16. Spinelli, R.; Magagnotti, N.; Jessup, E.; Soucy, M. Perspectives and challenges of logging enterprises in the Italian Alps. *For. Policy Econ.* **2017**, *80*, 44–51. [CrossRef]

17. Eriksson, M.; Lindroos, O. Productivity of harvesters and forwarders in CTL operations in northern Sweden based on large follow-up datasets. *Int. J. For. Eng.* **2014**, *25*, 179–200. [CrossRef]

18. Malinen, J.; Laitila, J.; Väätäinen, K.; Viitamäki, K. Variation in age, annual ussage and resale price of cut-to-lenght machinary in different regions of Europe. *Int. J. For. Eng.* **2016**, *27*, 97–102.

19. Spinelli, R.; Magagnotti, N.; Picchi, G. Annual use, economic life and residual value of cut-to-length harvesting machines. *J. For. Econ.* **2011**, *17*, 378–387. [CrossRef]

20. Holzleitner, F.; Stampfer, K.; Visser, R. Utilization rates and cost factors in timber harvesting based on long-term machine data. *Croat. J. For. Eng.* **2011**, *32*, 501–508.

21. Spinelli, R.; Visser, R. Analyzing and estimating delays in harvester operations. *Int. J. For. Eng.* **2008**, *19*, 35–40.

22. Hiesl, P.; Benjamin, J.G. Applicability of international harvesting equipment productivity studies in Maine, USA: A literature review. *Forests* **2013**, *4*, 898–921. [CrossRef]

23. Ovaskainen, H.; Uusitalo, J.; Väätäinen, K. Characteristics and significance of a harvester operator's working techniques in thinnings. *Int. J. For. Eng.* **2004**, *15*, 67–77.

24. Evanson, T.; McConchie, M. Productivity measurements of two Waratah 234 hydraulic tree harvesters in radiata pine in New Zealand. *J. For. Eng.* **1996**, *7*, 41–52. [CrossRef]

25. Glöde, D. Single- and double-grip harvesters: Productive measurements in final cutting of shelterwood. *Int. J. For. Eng.* **1999**, *10*, 63–74.

26. Spinelli, R.; Hartsough, B.; Magagnotti, N. Productivity standards for harvesters and processors in Italy. *For. Prod. J.* **2010**, *60*, 226–235. [CrossRef]

27. Spinelli, R.; Visser, R.; Riond, C.; Magagnotti, N. A survey of logging contract rates in the southern European Alps. *Small-Scale For.* **2016**, *16*, 179–193. [CrossRef]

28. Miyata, E.M. *Determining Fixed Costs and Operating Costs of Logging Equipment*; General Technical Report NCFES-NC-55; USDA Forest Service: St. Paul, MN, USA; p. 18.

29. Moldenhauer, M.C.; Bolding, M.C. Identifying loggers' reactions and priorities in an increasingly fragmented landscape. In Proceedings of the 32nd Annual Council on Forest Engineering Meeting, King Beach, CA, USA, 15–18 June 2009; p. 5.

30. Pröll, W. Harvestereinsatz steigt. *Forstzeitung, Arbeit im Wald* **2005**, *116*, 4–6.

31. Nguyen, T.N.; Perinot, C.; Duprat, M.; Villette, A. *Managing Waste Generated by Logging Operations: An Environmental, Legal and Economic Necessity*; FIF 712GB; Afocel: Paris, France, 2005; p. 6. ISSN 0336-0261.

32. Borchert, H.; Kremer, J. Maschinenausstattung der forstunternehemen in Bayern (Mechanical equipment of forest contractors in Bavaria). *Forst Tech.* **2007**, *8*, 6–10.

33. Ferrari, E.; Spinelli, R.; Cavallo, E.; Magagnotti, N. Attitudes towards mechanized Cut-to-Length technology among logging contractors in Northern Italy. *Scand. J. For. Res.* **2012**, *27*, 800–806. [CrossRef]

34. Forbrig, A. Konzeption und anwendung eines informationssystems über forstmaschinen auf der grundlage von maschinenbuchführung, leistungsnachweisen und technischen daten. *KWF-Ber.* **2000**, *29*, 119–125.

35. Denninger, W. Stand der hochmechanisierte holzernte in Niedersachsen. *Forst Tech.* **2002**, *7*, 14–17.

36. Findeisen, E. ThüringenForst: Erfahrungen zur teilautonomen gruppenarbeit in der hochmechanisiert holzernte. *KWF Forsttech. Inf.* **2002**, *4*, 37–44.

37. Nick, L.; Forbrig, A. Forsttechnikerhebung—Stand, bewertung, bedarf, entwicklung; zwischenergebnis. *KWF Forsttech. Inf.* **2002**, *9*, 93–99.

38. Drewes, D.; Jacke, H. Einsatzzeiten—Die nutzung von selbstfahrenden forstmaschinen (auch) in Deutschland. *KWF Forsttech. Inf.* **2005**, *4*, 47–49.

Article

Value Retention, Service Life, Use Intensity and Long-Term Productivity of Wood Chippers as Obtained from Contractor Records

Raffaele Spinelli [1,2,*], Lars Eliasson [3] and Natascia Magagnotti [1,2]

[1] CNR Ivalsa, Via Madonna del Piano 10, 50019 Sesto Fiorentino (FI), Italy; magagnotti@ivalsa.cnr.it
[2] Australian Forest Operations Research Alliance (AFORA), University of the Sunshine Coast, Locked Bag 4, Maroochydore DC, QLD 4558, Australia
[3] Skogforsk, Uppsala Science Park, 75183 Uppsala, Sweden; Lars.Eliasson@skogforsk.se
* Correspondence: spinelli@ivalsa.cnr.it; Tel.: +39-335-5429798

Received: 30 November 2017; Accepted: 18 December 2017; Published: 20 December 2017

Abstract: Acknowledging the absence of up-to-date empirical data on the value retention, service life and annual use of chipping machinery, in 2017 the authors surveyed the records kept by 50 contractors offering biomass chipping services. The machine fleet and operations in this survey could be taken as representative for most of Europe, where the biomass sector is well established and is facing further expansion. Data collection included the whole chipping unit, comprised of chipper, carrier and loader. Manually-fed units were excluded from the survey. The data pointed at a service life up to and exceeding 10,000 h and 10 years, which relieved any concerns about poor durability. Value retention was good, and may exceed that of other mainstream forestry equipment. Engine power was the main explanatory variable in any models to predict purchase price and productivity. The effect of this variable could explain most of the variability (>80%) in the purchase price and productivity data. Results also pointed at the essential equivalence in price and productivity between PTO-driven (i.e., tractor powered) and independent-engine chippers, once differences in engine power are accounted for. However, the distribution of purchase price between different components of the chipping unit was different between the two unit types, with the chipper accounting for a larger proportion of the total investment in independent-engine units. Machine power was also different, with most PTO-driven units being significantly smaller than independent-engine units, due to the limitations of existing tractors. Furthermore, half of the carriers assigned to a PTO-driven unit were subject to flexible use, i.e., they were not solely used for chipping work.

Keywords: work efficiency; biomass; benchmarking; costing

---

1. Introduction

Driven by a growing demand for renewable energy feedstock, chip production has expanded rapidly all over Europe providing work to logging, chipping and hauling companies that operate near the new plants. For this reason, many contractors have invested in mobile chippers, and more are prospecting the purchase of such equipment.

The acquisition of a mobile chipper involves a significant capital investment, which makes the formulation of a correct machine rate a crucial task. Machine cost is estimated using standard economic methods that divide total cost into capital and operating cost components [1]. Capital cost is dependent on the size of the investment, the expected economic life of the machine and the interest rate charged on borrowed capital, and it is incurred whether or not the machine is working. Conversely, the operational cost depends on all the expenses incurred when the machine is actually working. To calculate a cost per unit of product (m^3 or ton), a productivity estimate is also needed. Contractors need reliable data for pricing their machines, so that they can make informed decisions when planning new investments, and offer competitive contract rates with a minimised risk of going bankrupt. Standard cost calculation methods are often assembled into dedicated spreadsheet calculators capable of returning reliable estimates of chipping productivity and cost, based on user-defined input data [2].

Chipper productivity [3–5] and fuel consumption [6–8] have been documented in many studies, performed on a variety of wood chipping equipment under a wide range of work conditions. Information is also available about utilization [9] and annual use [6]. However, the information about annual use and fuel consumption available so far has been obtained from relatively small samples, which makes it difficult to model these figures as a function of machine type, age or use intensity. Even worse, any estimates for service life and resale value are based on older regional studies, when they do not derive from guesswork or anecdotal information. Due to the large capital investment, the assumptions made for economical service life, annual use and resale value also have a large impact on the calculation of interest charges [10].

The combination of accurate productivity estimates coupled with largely hypothetical cost assumptions makes chipping cost predictions quite unreliable. There is a fundamental contradiction between the large amount of work invested in developing accurate chipper productivity benchmarks that can reflect machine, job and feedstock characteristics, and the limited attention paid to determining equally accurate figures for chipper cost.

Chipping emerged as an important business sector in Italy already in the late 1990s [11]. Many contractors have now worked with chipping for almost two decades, changing several machines and gaining significant experience with this technique. Much knowledge is now available among the contractors themselves, and their records can provide a wealth of important information, because they often cover the whole life time of their machines. As this study is based on these records, it has a strong Italian bias. However, in many respects it can be generalized and extended to other countries, because the Italian chipper fleet represents a whole range of machine makes and models that are popular outside Italy, and the local business structures are similar to those found elsewhere in Europe.

Therefore, the goals of this study were to: (1) produce reliable benchmark figures for the service life and the resale value of wood chippers, which may reflect machine type, age and intensity of use, and (2) produce reliable estimates of productivity and annual use, based on long-term contractor records. This information is considered essential to estimating a correct chipper management cost, which may reflect the specific conditions encountered by each individual user. This work will enable chipping contractors to calculate fair chipping rates, which will allow them to stay competitive while accruing reasonable profits.

2. Materials and Methods

The study covered a whole range of chipping units, from light tractor-powered machines (PTO-driven, or driven through the tractor power take-off) originally designed for part-time use, to powerful independent-engine industrial units best suited to the specialized chipping contractor (Figure 1). For the purpose of the study, the chipping unit represented a complete self-supported operation designed for mechanical feeding with a hydraulic loader. Therefore, each unit comprised of a chipper, a loader and a carrier. The loader could be integral or separate, i.e., a self-propelled loader, an adapted excavator or a farm tractor equipped with its own loader. The carrier could be a tractor, a truck or a forwarder. However, if the chipper was equipped with its own independent engine and was moved to the work site using a low-bed trailer, then no carrier was included with the chipping unit.

All machines in the survey were owned and operated by Italian entrepreneurs, and were deployed in Italy, with few exceptions of occasional cross-border activity. Despite its strong Italian character, the study covered a whole range of makes and models, both Italian and foreign. Overall, the survey included 50 units, where the chipper was powered by an engine ranging from 59 to 460 kW.

For the purpose of the study, machine owners were requested to supply data about: machine price at purchase and at the moment of the interview in 2017, or at the time of resale if the machine had already been sold; date of purchase and of the eventual resale; total machine hours (motor hours) worked from purchase to present, or to resale; total fresh tons of chips produced in the same period. Price and use data were to be provided separately for the chipper, the carrier and the loader—if one was used.

When the carrier and the loader were not permanently attached to the chipper, and thus could be used to perform other jobs, then the entrepreneurs were asked to indicate the percent of chipping work performed with this machinery, so that their total separate costs could be pro-rated in terms of chipping work only, avoiding redundant attribution. Additional cost figures were also collected, such as: insurance, fuel, lube, repair and maintenance, and labor. That was done in order to determine the proportion of fixed cost over total cost, while keeping the latter confidential not to weaken the contractors' negotiating power.

In total, 116 operators were contacted by phone before sending the survey form. All were private entrepreneurs who performed chipping as a full-time specialized occupation, or as a part-time job complementary with a main occupation in the logging sector. Once the survey form was received, follow-up phone calls allowed clarifying any doubts and/or integrating missing data. Understandably, not all entrepreneurs had kept records of the required quality, or were willing to search their files in order to track all relevant machine use records for the whole life of their machines. For this reason, the study was based on 50 surveys only, which represented those respondents who could guarantee sufficient data quality.

(a)   (b)

Figure 1. Example of a tractor-powered—or PTO-driven (a) and an independent-engine chipper (b).

All surveys were consolidated into a single data file for statistical analysis, which was conducted using SAS Statview (Version 5.1) [12]. As a first step, descriptive statistics were calculated, separately for the main machine classes. The statistical significance of any differences between machine classes was checked with parametric and non-parametric tests, depending on the distribution of data. Then, regression analysis was used for establishing the significance of any relationships between purchase price, value retention, annual use, productivity and all independent variables that were expected to have some effect on these parameters. The elected significance level was $\alpha < 0.05$.

3. Results

3.1. Characteristics of the Machine Pool

The 50 units included in the study represented a whole range of operations, and comprised 29 PTO-driven and 21 independent-engine machines (Figure 2). The former were powered most often by a wheeled farm tractor, of variable size and power. Only two PTO-driven machines were powered by a truck, and could be defined as chipper trucks [13].

Out of 50 sample units, 26 (52%) operated in Northern Italy, 20 (40%) in Central Italy and only 4 (8%) in Southern Italy. Austrian, German and Italian makes accounted for the largest proportion of the sample, offering a good representation of the current Italian fleet, which is quite international (Figure 3). The Italian *Pezzolato* was the most common brand (19 machines), immediately followed by the German *Jenz* (11 machines). Other common brands were: Farmi (Nordic, 3 units), Gandini (Italian, 3 units), Heizo-hack (German, 3 units) and Mus-Max (Austrian, 3 units).

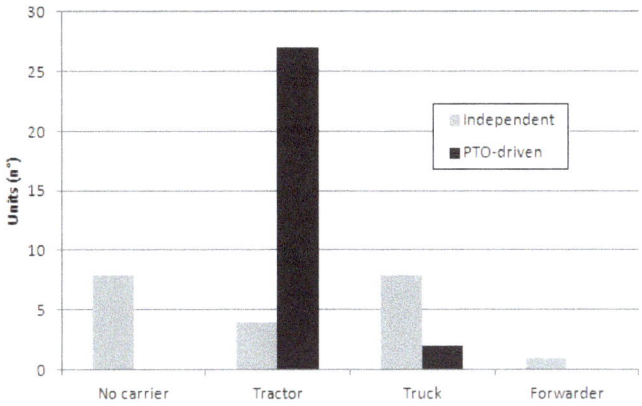

Figure 2. Number of chipping units by carrier and chipper type.

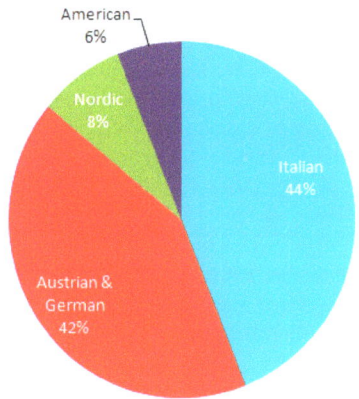

Figure 3. Distribution of the sample between different machine makes.

The sample showed a clear distinction between PTO-driven and independent-engine machines. Compared with PTO-driven units, independent-engine chippers were significantly more powerful, expensive and productive (Table 1). In contrast, no significant differences between these two machine types were found concerning value retention, duration in service and intensity of use. Furthermore, use of an integral loader was twice as common among independent-engine units (60% of the cases) as among PTO-driven units (30% of the cases). Together with the possibility of de-coupling the prime mover from the chipper, that enabled about half of the PTO-driven operations to clock additional work hours on the loader and the carrier, while only 10% of the independent-engine units followed a similar strategy—largely because disconnecting the loader and/or the carrier was technically difficult and would not pay off, even when chipper use was not very intense.

None of the PTO-driven machines represented small, part-time operations: on the contrary seven among them were coupled with tractors delivering between 170 kW and 280 kW, and two were powered by large trucks with the same 350 kW engine installed on some of the most common and powerful independent-engine units in the sample.

Table 1. Main characteristics of the chipping units on test, separated by chipper type (n = 50).

		Independent Engine		PTO-Driven		MW-Test
		Mean	Median	Mean	Median	p-Value
Power	kW	316	328	147	118	<0.0001
Purchase price	k€	326	304	170	120	0.0008
Resale value	k€	146	113	80	56	0.0059
Value retention	%	45	43	48	45	0.6301
Service	Years	7	6	7	6	0.9448
Service	h	4231	2000	3123	2000	0.4315
Production	Kt	81.8	42.0	31.6	16.0	0.0046
Annual use	h year^{-1}	669	625	524	363	0.1878
Annual production	t year^{-1}	14,067	12,000	5118	1920	0.0021
Productivity	t h^{-1}	18.8	18.0	7.7	6.7	0.0001
Use Carrier	%	94	100.0	70	100	0.0268
Use Loader	%	90	100.0	64	50	0.0054

Notes: Purchase price ex. VAT; MW test = Mann-Whitney test (non-parametric); h = meter hours; t = green tons; Use Carrier = % of total carrier time invested with chipping work; Use Loader = % of total loader time invested with chipping work.

3.2. Purchase Price and Value Retention

Purchase price varied from 35,000 to 850,000 €, or most commonly from 110,000 to 360,000 € (lower and upper quartile, respectively). This price was the 2017 equivalent, obtained by revaluating actual figures to 2017 currency values with the dedicated calculator provided by the Italian Statistical Agency [14]. The purchase price included the chipper, the loader and the carrier, when one was used (all cases except for eight). If a carrier or a loader were also used in other tasks than chipping, then their revaluated purchase price was allocated to the chipping unit as a proportion of actual use with the chipper.

The distribution on the total purchase price among these main components varied significantly between chipping unit types (Figure 4). While the price of the chipper was the dominant component for both PTO-driven and independent-engine machines, it was significantly more important for the latter (p = 0.0003). In contrast, the proportion of total price invested in the carrier was significantly higher for PTO-driven machines, compared with independent-engine machines (p = 0.0007). The loader accounted for ca. 20% of the total price, with no significant differences between machine types (p = 0.6870).

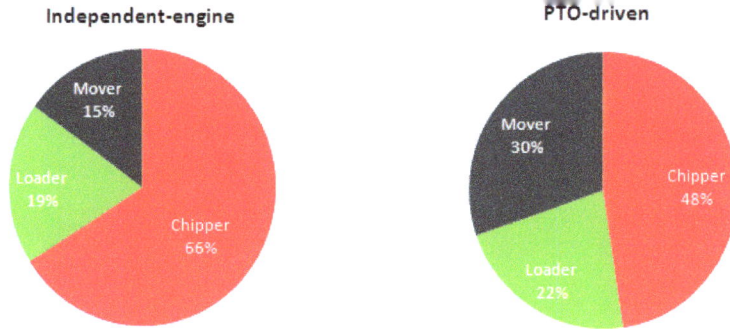

Figure 4. Distribution of purchase price between different chipping unit types.

Purchase price was strongly associated with engine power, which explained almost 90% of the variability in the dataset (Table 2). The remaining variability was likely explained by differences in other machine characteristics, including optional equipment features. Furthermore, different negotiating power and capacity at the time of buying must have had their impact on price formulation [15]. Regression analysis showed that chipper type (independent-engine or PTO-driven) had no significant effect on purchase price, indicating that on a kW by kW basis a PTO-driven chipper was as expensive to purchase as an independent-engine chipper. Of course, that accounted for the whole unit, comprised of chipper, loader and carrier—not for the chipper component only.

Table 2. Regression equations for estimating purchase price and value retention.

	Coeff	SE	T	p-value
Purchase price = a kW				
R^2 adjusted = 0.887; n = 50; F = 395.382; $p < 0.0001$				
a	1124.395	56.547	19.884	<0.0001
Purchase price = a kW + b kW^2				
R^2 adjusted = 0.892; n = 50; F = 207.667; $p < 0.0001$				
a	794.169	195.86	4.055	0.0002
b	1.002	0.570	1.758	0.0852
Retention = a + b LOG Age + c Use Intensity				
R^2 adjusted = 0.618; n = 50; F = 40.588; $p < 0.0001$				
a	87.183	4.802	18.157	<0.0001
b	−47.124	5.337	−8.830	<0.0001
c	−0.010	0.003	−3.173	0.0027

Where: Purchase price (complete chipping unit) = €; n = number of valid observations; SE = Standard error; kW = engine power in kW; Retention = Value retention, i.e., Resale value/Purchase price in %; Age = years: Use intensity = h year^{-1}.

Regression analysis yielded two different equations for estimating purchase price as a function of engine power: a linear equation and a quadratic equation (Table 2). The linear equation was probably the safest one, because one of the terms in the quadratic equation was not highly significant and the coefficient of determination R^2 for the quadratic equation was only marginally higher than that of the linear equation. This was most likely an effect of the increased number of variables rather than an indicator of a better capacity to represent dataset variability. However, the quadratic equation was also reported in the paper because visual analysis of the data hinted at an upward turning curve, which was best represented by a quadratic equation rather than a linear one (Figure 5). In any case, readers are warned against extrapolating the results of these models beyond the range of variation of the independent variable: that is generally wrong and may produce incorrect estimates, especially when the curve represents a quadratic function.

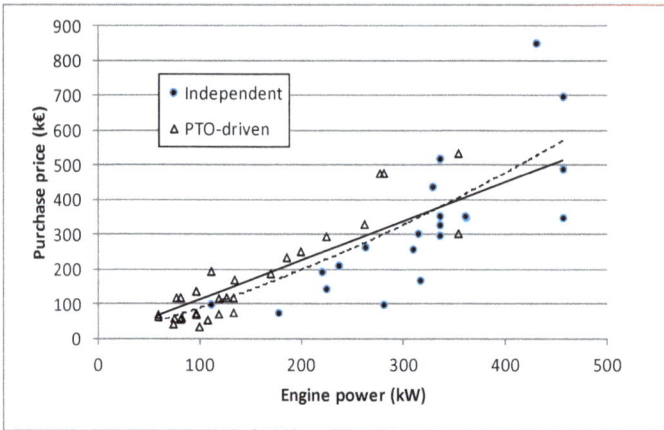

Figure 5. Purchase price (2017 values) for the complete chipping unit as a function of engine power.

Value retention was defined as the relationship between resale price and revaluated purchase price, expressed as a percent ratio. Value retention decreased with use, as one would expect. Among the many indicators for use, number of years in service and use intensity (h year^{-1}) proved the best predictors (Table 2). The total number of hours clocked in by a machine was half as good a predictor as the total number of years in service, a tendency that was already reported for harvesters and forwarders in previous studies [16,17]. Interestingly enough, neither engine power nor chipper type (i.e., independent-engine or PTO-driven) had any significant effect on value retention. The value losses were highest over the very first years and became increasingly smaller with extended service life (Figure 6), as typical of most machinery [18].

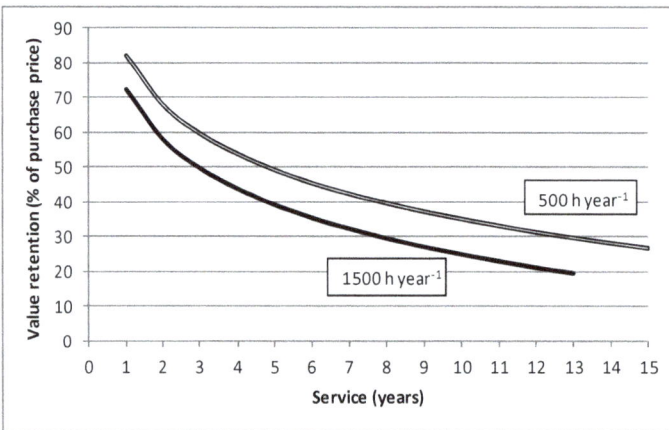

Figure 6. Value retention as a function of service and use intensity.

3.3. Service Life and Use Intensity

Determining reliable benchmark figures for service life was among the primary goals of this study. The average service life figures obtained from this study were not representative of total service life because they just described how old were the sample machines at the time of the survey, regardless of whether they were about to be replaced or not [19]. The maximum value in the range or the upper quintile offered a more representative estimate, which could be taken as a proxy for the total life one may reasonably expect from a chipper. In that case, total service life could be estimated anywhere above 10 years or 6000 h, with a maximum as high as 24 years or 20,000 h (Table 3). Similarly, the production expected over the lifetime of a chipper was likely higher than 100,000 t, with a maximum close to half a million t.

Table 3. Reference figures for service, production and use intensity ($n = 50$).

		Mean	SD	Min	Max	UQ
Service	years	6.7	4.4	1.3	24.3	≥ 10
Service	h	3588	3878	210	19,500	>6000
Production	kt	52.7	60.6	10.5	487.5	>100
Annual use	h year^{-1}	585	493	30	2040	>1000
Annual production	t year^{-1}	8876	10,188	148	37,647	>18,000

Notes: SD = standard deviation; UQ = upper quintile (20% of observations); t = metric green ton.

Mean use intensity was near to 600 h year^{-1} or 9000 t year^{-1}, but a large share of the contractors easily doubled these figures, as represented in the upper quintile. As expected, use intensity increased significantly with chipping unit price, indicating that more expensive and productive units were bought by contractors with more work (Table 4, Figure 7).

Table 4. Regression equations for estimating use intensity and productivity.

Use intensity = a + b €				
R^2 adjusted = 0.386; n = 50; F = 31.857; $p < 0.0001$				
	Coeff	SE	T	p
a	178.675	20.295	1.979	0.0536
b	0.002	3.048×10^{-4}	5.644	<0.0001
Productivity = a + b $P^{1.1117}$ + c P				
R^2 adjusted = 0.814; n = 50; F = 108.360; $p < 0.0001$				
	Coeff	SE	T	p
a	7.458	3.510	2.125	0.0389
b	0.246	0.088	2.806	0.0073
c	−0.434	0.178	−2.439	0.019

Where: Use intensity = h year^{-1}; € = Purchase price in €; n = number of valid observations; SE = Standard error; Productivity = t h^{-1}; P = engine power (kW).

3.4. Productivity and Cost

Long-term productivity was calculated from total production and total machine use, which was modelled as a function of relevant independent variables. Among these, engine power was the most influential and explained over 80% of the variability in the dataset. Chipper type had no significant effect on long-term productivity, supporting the notion of PTO-driven and independent-engine chippers being equally productive as long as engine power was the same. Productivity was affected by loader type and was ca. 2 t h^{-1} higher for those units that used a separate loader, instead of an integral one. However, the loader variable was borderline significant ($p = 0.067$) and therefore this result was suggestive, not conclusive. For this reason, the effect of loader type was not included in the final model. The final equation represented a power curve and described a clear scale effect, where engine power increments resulted in proportionally larger productivity increments (Figure 7). Once more, readers are warned against extrapolating the results of this model beyond the range of variation spanned by the original data.

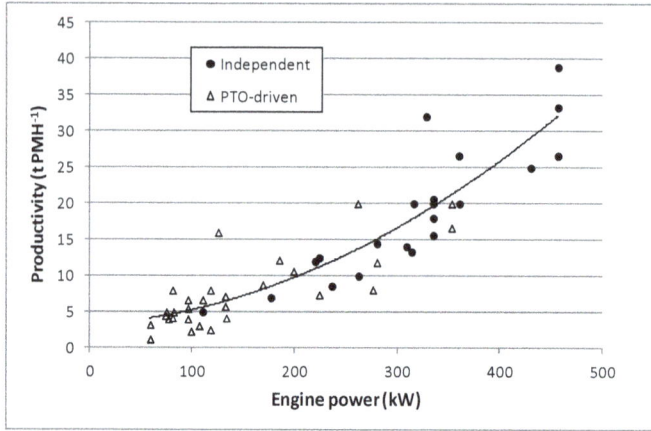

Figure 7. Chipping productivity as a function of engine power.

Raw chipping cost varied most commonly between 6 and 33 € t^{-1}, and it was represented by fixed cost for approximately 40%, without any significant differences between PTO-driven and independent-engine units ($p = 0.7020$). Obviously, the proportion of fixed cost decreased with use intensity, since the two were tied by a strictly deterministic relationship.

4. Discussion

4.1. Limitations of the Study

Before embarking in a proper discussion, it is just fair to state upfront the limitations of the study, namely: the national character of the sample, the reliance on company records and the largely subjective way of estimating resale value.

The sample certainly offers a good representation of Italian chipping business but may fall short of correctly representing the situation in other countries. However, it includes a wide range of machine types that are used throughout Europe, and may offer a good picture of biomass operations in general. In that regard, it is worth stressing that biomass chipping is a relatively simple process, which is less affected by terrain and forest characteristics than other steps in the forest value chain, such as felling, extraction or merchandising [20]. Certainly, the results of this study are valid for biomass operations only, and may not extend to pulp chip harvesting without much caution, because the technology used in America or Australia for producing pulp chips is different from that applied to European biomass operations [21].

The reliance on company records is another limitation, given the inherent variability of these records for what concerns accuracy and resolution. However, contractors were only asked to provide basic figures, generally recorded in a common way or easily calculated from more detailed records. Furthermore, contractors who did not have good records or felt uneasy about allowing access to their data could simply excuse themselves, which many actually did. Therefore, respondents had little reason to provide inaccurate or deceitful figures. The general consistency of the dataset and of derived figures, such as productivity, indicates that contractor responses are accurate. Gross inaccuracy would have been denounced by recurring outliers and by a low explanatory capacity of regression equations, neither of which did materialize.

Probably, the weakest element in this study was the estimate for resale value, which was often based on the individual appraisal of machine owners, since few of the machines in the sample had been actually sold back. Owners' appraisals are based on their subjective views of the market and the machines, and may reflect owner's expectations rather than an actual market value. However, owner figures were checked against market values for similar equipment reported on the most popular second-hand machinery on-line shops, which offer numerous quotes especially for loaders and carriers. Figures generally matched, corroborating the estimates provided by machine owners.

In any case, the knowledge gathered in this study is unique in its capacity to reflect chipping entrepreneurship. To our knowledge, there are no other studies that have gathered long-term data from so many chipping contractors, and the few available studies tap into the records of State companies and cooperatives, which often benefit from the existence of a formal administration office and a centralized infrastructure [22].

4.2. Long Live the Chipper

Chipping is hard work, and for this reason one may assume chippers to be short-lived, as they succumb to fatigue and rapid wear. This study contradicts the gloomiest expectations and points at a surprisingly long service life for a machine that receives so much punishment. The figures reported in this study are only marginally smaller than reported for harvesters and forwarders [16,17], and can easily exceed the 10,000 h threshold. However, the reported figures represent a technical service life and not an expected economical lifetime, which may be shorter (or longer yet).

If chipping is indeed brutal work, it is also true that modern chippers are designed for it by skilled manufacturers, who know how to build an efficient and durable piece of equipment. Furthermore, chippers are relatively simple machines, which supports longevity and facilitates maintenance. In fact, the study did not quantify maintenance cost, and therefore it could not establish if the exceptionally long service life of some machines in the sample pool was associated with an excessive increase of maintenance cost. If so, one would simply be observing a case of belated replacement, past optimum service life [23]. Evidence of increasing maintenance costs may be observed as a faster drop in value retention, compared with other machine types, but that is not the case either; in fact, used chipping units seem to have better value retention than dedicated cut-to-length equipment, such as harvesters and forwarders [16,17]. As far as one can tell, chippers are solid, durable machines.

On the other hand, the extremely simple design of most chippers may justify extending service life through the iterative renewal of worn parts, much like in Theseus' paradox [24]. In that case, the machine would be finally replaced only due to obsolescence, non-conformity with new safety regulations or changes in the owner's production target that would require purchasing a new machine. In the absence of direct evidence for the maximum rational service life of wood chippers, one has to fall back on indirect stochastic indicators, such as the quintile group. The number of respondents above the 4th quintile is large enough to mitigate the effect of the odd irrational owner, if any such owners exists in a group of experienced professionals. Further studies should aim at refining the current estimates for service life, addressing the issue of its interaction with maintenance cost.

Finally, the annual use figures reported in this study are significantly lower than those previously reported in a similar study of six Italian contractors [6]. In particular the present study estimates an annual use of 500 h for PTO-driven chippers and 700 h for independent-engine chippers, whereas previously published data are 900 h and 1200 h, respectively. This can be explained by differences in the respective samples: this study contains a wide sample that comprises both part-time chipping contractors and specialized full-time chipping contractors, while the previous study only includes the latter group.

4.3. Engine Power as the Main Predictor

One of the most important findings of this study is the strong association of engine power with the main figures of interest: purchase price, productivity and annual use. These relationships are logical and are already reported in previous chipper studies, at least for what concerns productivity [11]. What is surprising is rather the capacity of the variable "engine power" to displace all other predictors, despite the undeniable effect of other factors on the dependent variables of interest. That is largely due to the lack of information about these other variables, such as job characteristics (for productivity) or optional features (for purchase price). However, one may expect that the lack of such details would result in a much lower explanatory capacity for the estimated models: that it is not the case is a great comfort, and opens the possibility of issuing "quick and (not so) dirty" estimates based on simple fundamental information that should be easily available to any user. Engine power is a unifying predictor, capable of summarizing the most important effects recorded: the identification of a single powerful predictor lends additional merit to this study, and points at its capacity to dig deep into the fundamentals. For this same reason, the models in the study may return principal estimates of long-term performance but may not reflect as accurately specific work conditions, which makes them more suitable for strategic planning than for tactical planning.

4.4. Robust Productivity Model

The study yielded a simple and robust productivity model, which can be used as a benchmark for chipper productivity even in the absence of additional detail about piece size and job type. This model has a very high predicting capacity and offers a good match with previously published models, with the additional benefit of being simpler to use.

In that regard, it is important to stress that the time used as a denominator in the productivity equation is engine run time (hour meter), which deviates in some measure from both productive machine time (excluding all delays) and scheduled time (including all delays) [25]. The machine clock runs whenever the machine engine is on, and it will include the wide range of shorter delay events during which the operator does not deem practical to turn off the engine. In contrast, the operator is much likely to turn off the engine when the interruption is expected to last for a longer time, and certainly when performing machine maintenance. For this reason, the productivity estimates obtained from the model should best compare to the PSH_{15} productivity figures reported in many German studies. PSH_{15} stands for productive system hours, consisting of productive work time and all delay events lasting no more than 15 min. That seems the closest match to the description provided above, and should offer a viable proxy. Previous studies of chipper delays indicate that delay events with a maximum duration of 15 min account for only one-third of total delay time, and therefore the productivity estimates returned by the model are substantially higher than the actual productivity calculated on the basis of total scheduled time [9]. In that regard, one must notice that different machine meters record time in different ways: some simply measure the time when the engine is on—even if idle—while others only record the time when the engine runs above a certain rotational regime, thus excluding idle time. With so many different machine models from so many different manufacture years, the study sample is likely to contain a mix of both hour meter types, which certainly contributes to random variability. Nevertheless, this variability is not as large as to prevent building strong models and disclosing highly significant differences, as the study results demonstrate.

Finally, it is worth highlighting the form of the productivity curve estimated in this study. Since the exponent is larger than 1, the curve bends upwards and describes the typical phenomenon of increasing returns, which in this specific case has been interpreted as the scale effect of engine power. Conversely, curves relating productivity to piece size use exponents smaller than 1 and bend downwards, indicating diminishing returns as piece size increases [26]. That is important to notice, because it demonstrates that productivity is directly proportional to both engine power and piece size, and yet investments in engine power are more productive, as returns are higher and may easily offset limitations in piece size.

Another interesting result of this study is the lack of any principle differences between PTO-driven and independent-engine chippers. In fact, the only major difference is power, because PTO-driven chippers are limited by the power of existing farm tractors. However, for the same engine power, PTO-driven machines are as productive, durable and expensive as independent-engine machines. Of course, price equality is only true for the whole chipping unit, including loader and carrier: the chipper component alone is less expensive when no own engine is provided.

If the two unit types are technically and financially equivalent, one may wonder about what drives the choice towards one or the other option. Power availability is again a pivotal factor: if one needs a larger machine than a tractor can support, then an independent-engine chipper is the only solution. If power is not an issue, the data point at equipment flexibility as a main driver. Approximately half of the PTO-driven units make use of the carrier and the loader for additional tasks than just chipping. In those cases, a PTO-driven chipper allows disconnecting the tractor when it is needed for other jobs, and contributes to a better utilization of the tractor itself [27]. Even so, the question remains unanswered for the other half of the cases, where a tractor is permanently coupled with the chipper. Further research may address the decision factors leading to specific technology choices, which could improve our understanding of this business and help prospective users with their plans [28].

5. Conclusions

This study is currently the only one offering updated empirical data on the purchase price, value retention, economic life and annual use of chipping units used in biomass operations. Such data have been obtained from a large group of chipping contractors, representing the small and medium-scale enterprises that support the European chipping contracting sector. The information obtained from the study is essential to formulating reliable machine rate estimates. Results highlight a longer economic life and a better value retention than previously assumed for this equipment type, which may result from skilled machine design and professional use—both deriving from the experience gained over the years in this crucial sector. The study also reveals that engine power can be used as the main predictor for most of the parameters investigated in the study. Furthermore, it points at the basic equivalence between PTO-driven and independent-engine chippers, once differences in engine power are accounted for. Finally, the study offers a robust and simple productivity benchmark, which can be used for extracting fast and reliable estimates of expected long-term performance.

Acknowledgments: This project has been partly supported by the Bio Based Industries Joint Undertaking under the European Union's Horizon 2020 research and innovation program under grant agreement No. 720757 Tech4Effect and by the Swedish Energy Authority under grant agreement 41962-1. The Authors thank the 50 entrepreneurs who kindly provided the data necessary for this study, as well as Dott. M. De Stefano (Conaibo—Italian Association of Logging Contractors) for her support with data collection and verification.

Author Contributions: R.S., L.E. and N.M. conceived and designed the experiment; R.S. and N.M. performed the interviews and collected the data; R.S., L.E. and N.M. analyzed the data and wrote the paper.

Conflicts of Interest: The authors declare no conflict of interest.

References

1. Ackerman, H.; Belbo, L.; Eliasson, A.; de Jong, A.; Lazdins, A.; Lyons, J. The COST model for calculation of forest operations costs. *Int. J. For. Eng.* **2014**, *25*, 75–81.
2. Spinelli, R.; Magagnotti, N. A tool for productivity and cost forecasting of decentralised wood chipping. *For. Policy Econ.* **2010**, *12*, 194–198. [CrossRef]
3. Asikainen, A.; Pulkkinen, P. Comminution of Logging Residues with Evolution 910R chipper, MOHA chipper truck, and Morbark 1200 tub grinder. *J. For. Eng.* **1998**, *9*, 47–53.
4. Mitchell, D.; Gallagher, T. Chipping whole trees for fuel chips: A production study. *South. J. Appl. For.* **2007**, *31*, 176–180.

5.   Röser, D.; Mola-Yudego, B.; Prinz, R.; Emer, B.; Sikanen, L. Chipping operations and efficiency in different operational environments. *Silva Fenn.* **2012**, *46*, 275–286. [CrossRef]

6.   Spinelli, R.; Magagnotti, N. Determining long-term chipper usage, productivity and fuel consumption. *Biomass Bioenergy* **2014**, *66*, 442–449. [CrossRef]

7.   Nati, C.; Eliasson, L.; Spinelli, R. Effect of chipper type, biomass type and blade wear on productivity, fuel consumption and product quality. *Croat. J. For. Eng.* **2014**, *35*, 1–7.

8.   Eliasson, L.; von Hofsten, H.; Johanneson, T.; Spinelli, R.; Thierfelder, T. Effects of Sieve Size on Chipper Productivity, Fuel Consumption and Chip Size Distribution for Open Drum Chippers. *Croat. J. For. Eng.* **2015**, *36*, 11–17.

9.   Spinelli, R.; Visser, R. Analyzing and estimating delays in wood chipping operations. *Biomass Bioenergy* **2009**, *33*, 429–433. [CrossRef]

10. Howard, A. Improved accounting of interest charges in equipment costing. *J. For. Eng.* **1991**, *2*, 41–45. [CrossRef]

11. Spinelli, R.; Hartsough, B.R. A survey of Italian chipping operations. *Biomass Bioenergy* **2001**, *21*, 433–444. [CrossRef]

12. SAS Institute Inc. *StatView Reference*; SAS Publishing: Cary, NC, USA, 1999; pp. 84–93. ISBN 1-58025-162-5.

13. Spinelli, R.; De Francesco, F.; Eliasson, L.; Jessup, E.; Magagnotti, N. An agile chipper truck for space-constrained operations. *Biomass Bioenergy* **2015**, *81*, 137–143. [CrossRef]

14. ISTAT 2017—Rivaluta: Rivalutazioni e Documentazione su Prezzi, Costi e Retribuzioni Contrattuali. Available online: http://rivaluta.istat.it/Rivaluta/# (accessed on 4 November 2017).

15. Spinelli, R.; Visser, R.; Thees, O.; Sauter, U.H.; Krajnc, N.; Riond, C.; Magagnotti, N. Cable logging contract rates in the Alps: The effect of regional variability and technical constraints. *Croat. J. For. Eng.* **2015**, *36*, 195–203.

16. Spinelli, R.; Magagnotti, N.; Picchi, G. Annual use, economic life and residual value of cut-to-length harvesting machines. *J. For. Econ.* **2011**, *17*, 378–387. [CrossRef]

17. Malinen, J.; Laitila, J.; Väätäinen, K.; Viitamäki, K. Variation in age, annual ussage and resale price of cut-to-lenght machinary in different regions of Europe. *Int. J. For. Eng.* **2016**, *27*, 97–102.

18. Pudaruth, S. Predicting the price of used cars using machine learning techniques. *Int. J. Inf. Comput. Technol.* **2014**, *4*, 753–764.

19. González, E.; Epstein, L. Minimum cost in a mix of new and old reusable items: An application to sizing a fleet of delivery trucks. *Ann. Oper. Res.* **2015**, *232*, 135–149. [CrossRef]

20. Stampfer, K.; Kanzian, C. Current state and development possibilities of wood chip supply chains in Austria. *Croat. J. For. Eng.* **2006**, *27*, 135–145.

21. Ghaffariyan, M.; Brown, M.; Spinelli, R. Evaluating efficiency, chip quality and harvesting residues of a chipping operation with flail and chipper in Western Australia. *Croat. J. For. Eng.* **2013**, *34*, 189–199.

22. Irdla, M.; Padari, A.; Kurvits, V.; Muiste, P. The chipping cost of wood raw material for fuel in Estonian conditions. *For. Stud.* **2017**, *66*, 65–74. [CrossRef]

23. Butler, D.; Dykstra, D. Logging equipment replacement: A quantitative approach. *For. Sci.* **1981**, *27*, 2–12.

24. Plutarch. *Parallel Lives*; Dryden, J., Clough, A., Translators; Modern Library/Random House: New York, NY, USA, 1961; p. 1333.

25. Björheden, R.; Apel, K.; Shiba, M.; Thompson, M.A. *IUFRO Forest Work Study Nomenclature*; Swedish University of Agricultural Science, Department of Operational Efficiency: Garpenberg, Sweden, 1995; p. 16.

26. Jirousek, R.; Klvac, R.; Skoupy, A. Productivity and costs of the mechanized cut-to-length wood harvesting system in clearfelling operations. *J. For. Sci.* **2007**, *53*, 476–482.

27. Akay, A. Using farm tractors in small-scale forest harvesting operations. *J. Appl. Sci. Res.* **2005**, *1*, 196–199.

28. Spinelli, R.; Magagnotti, N.; Jessup, E.; Soucy, M. Perspectives and challenges of logging enterprises in the Italian Alps. *For. Policy Econ.* **2017**, *80*, 44–51. [CrossRef]

Article

Monitoring Cable Tensile Forces of Winch-Assist Harvester and Forwarder Operations in Steep Terrain

Franz Holzleitner [1,*], Maximilian Kastner [1], Karl Stampfer [1], Norbert Höller [2]
and Christian Kanzian [1]

[1]   Institute of Forest Engineering, Department of Forest and Soil Sciences, University of Natural Resources
     and Life Sciences, Vienna, Peter-Jordan-Str. 82/3, 1190 Vienna, Austria; maxkastner@gmx.net (M.K.);
     karl.stampfer@boku.ac.at (K.S.); christian.kanzian@boku.ac.at (C.K.)
[2]   Corporate Information Services, University of Natural Resources and Life Sciences, Vienna,
     Peter-Jordan-Str. 82/3, 1190 Vienna, Austria; norbert.höller@boku.ac.at
*   Correspondence: franz.holzleitner@boku.ac.at; Tel.: +43-1-47654-9122

Received: 21 December 2017; Accepted: 22 January 2018; Published: 24 January 2018

Abstract: The objective of this case study was to develop and test a specific survey protocol for monitoring tensile forces for winch-assisted harvesters and forwarders with a mounted or integrated constant-pull capstan winch technology. Based on the designed survey protocol, the interactions between work phases, machine inclination, and tensile forces in typical work conditions were analysed. The established workflow, including equipment and the developed analysis routines, worked appropriately and smoothly. The working load on the cable during the study did not exceed 50% of the maximum breaking strength. A maximum tensile force peak at 56 kN was observed during delays for the forwarder, and a peak of 75.5 kN was observed for the harvester, both of which are still within the safe working load when considering a safety factor of two.

Keywords: winch-assist; harvester; forwarder; tensile force; steep terrain harvesting; cut-to-length

1. Introduction

In steep terrain cable yarding combined with motor manual felling is still the most appropriate harvesting system to use [1]. However, winch-assisted harvesting machinery offers new opportunities in terms of cost efficiency and increased safety on steep terrain [2,3]. Reducing soil disturbance by decreased slip on the skid trail can also be achieved [4,5]. Holzleitner et al. [6] stated that machine utilization, as a well-known and major cost driver of expensive harvesting equipment, is not easy to reach. Mounted or integrated cable winches for harvesters or forwarders could offer entrepreneurs new possibilities to increase their portfolio and machine utilization due to providing a wider range of available harvesting operations.

The soil-tire interaction has been described in detail in terms of the interaction of forces and slip while driving on slopes according to different soil types and water content [7–10]. The maximum trafficable grades without winch assist for different soil conditions have been reported. Furthermore, Wijekoon [11] investigated the soil-tire interaction in terms of number of passes, the ground pressure, and rut depths of forest machines. Wismer and Luth [12] examined the climbing ability of forest machines and their maximum slope. The needed improvements for controlling a cable-towed vehicle to minimize slip control in timber harvesting operations were investigated by Salsbery and Hartsough [13]. Slip control was also one of the main goals in a study on tethered feller bunchers on tracks combined with the maximum gradeability, which ranged from 64% to 85% [14].

The Herzog Forsttechnik AG company was one of the first companies in Central Europe to experiment with winch-assisted forwarders, when in 1999 the company experimented with the prototype of a winch-assisted forwarder called Forcar FC150. Bombosch et al. [15] tested a mounted traction winch on a snow-grooming vehicle to support the driving of a forwarder on steep terrain with a grade of up to 85%. After their successful tests, they stated that a future solution for cost-effective harvesting operations on steep terrain could be a sensor-controlled winch, integrated or mounted on a harvester or forwarder, in order to simultaneously minimize soil compaction. In 2004, Herzog Forsttechnik AG successfully launched the first winch-assisted system for the market [16].

Since then, this technology was developed by entrepreneurs competing for contracts to improve machine utilization and decrease harvesting costs on steep terrain [2]. Regardless, the basic reason for using winch-assisted machinery was not to climb steep terrain; it was mainly invented and designed for supporting driving on soft soils to avoid rutting and to treat sensitive sites with care [16,17]. These results correspond to the assumptions made by Lindroos et al. [18] that existing machines can be adapted to local conditions and that innovation is triggered by the environment of contract competition.

Winch-assisted harvesters and forwarders enable fully mechanized operations on slopes steeper than 30%, which is essentially the limit for operating with ground-based machines depending on their extra equipment, such as chains or bogie tracks [19]. Two different winch-assist concepts are available: the integrated or mounted concept, where the winch is attached to the harvesting machine, or the self-propelled standalone anchoring winch that pulls the harvesting machine. In Central Europe, winch-assisted systems for wheeled or tracked harvesting machinery are mainly based on an integrated capstan or conventional drum winches with cable diameters between 14 and 22 mm. Compared to machinery operating overseas with cable diameters up to 28 mm [20], European equipment is lighter and therefore does not require such large cable dimensions.

Visser et al. [21] monitored tensile forces on winch-assisted machines in New Zealand. Compared to Central European machinery with a wheel-based chassis, the studied machinery in New Zealand had tracks. The authors outlined the importance of on-board monitoring systems for the tensile forces of cables. The use of tensile force monitoring systems for cable yarders goes back to 1988 and was intensified during the 1990s mainly due to safety issues, thereby increasing the lifetime of ropes and improving productivity due to the ability to maximize loads per cycle [22]. Dupire et al. [23] described the importance of knowing the tensile forces in cable yarding operations in order to reduce equipment wear and to decrease operating costs by improving safety.

A lack of detailed knowledge exists about how tensile forces behave in real world conditions. Therefore, it is unclear if current operations are adhering to safety standards. For machine operators and entrepreneurs, awareness about real values is crucial for safer timber harvesting operations, since there is no high resolution on-board information system installed in the cabin to accurately check the actual tensile forces on the cable.

The objective of this study was to develop a scientific approach for monitoring tensile forces for winch-assisted harvesting machinery with mounted or integrated systems. This approach should enable in-depth analysis of tensile forces and their behaviour during real operations with these machines. A specific survey protocol was developed and tested in the field during the case studies. Interactions between work phases, inclination, and tensile forces were analysed in order to determine typical work conditions.

2. Materials and Methods

2.1. Case Studies

The three study sites were located near the villages Gaflenz and Kirchberg/Pielach in Lower Austria and in Kapfenberg in Styria, Austria. Harvesting operations ranged from one study of commercial thinning and two others that were final fellings after storm damage. The dominating tree species was Norway spruce (*Picea abies*). All sites had even terrain with an inclination approximately up to 100%. The harvester operated uphill while the forwarder drove empty uphill and then downhill once loaded. All studied operations occurred during daylight conditions on five different days, and both drivers had more than 8000 h operating experience with the machinery in general, which included operation without a winch. The harvester was equipped with track bands on the front bogie axle and chains on the rear axle, whereas the forwarder used track bands on both bogie axles.

The two observed machines were a John Deere 1170 E harvester, where the winch is mounted in the front of the machine, and a John Deere 1110 E forwarder, with the winch integrated in the mid-section of the frame close to the crane (Figure 1). Both machines were equipped with a winch from the manufacturer Haas Maschinenbau GmbH Germany. Both winches were constant-pull traction winches, using capstan winch technology equipped with a storage drum and a working drum. The pulling force of the winch could be adjusted by the operator within certain limits and could be continuously set up to 9 tonnes. As a special feature, the winch was automatically synchronized with the driving unit independently of the chosen driving direction. With a cable 14 mm in diameter, the drum capacity was up to 500 m. Considering the maximum pulling force of the winch and the maximum breaking force of the cable of 181 kN, a safety factor of two was used. This corresponds to the actual draft version of ISO 19472-2, which is still under revision and not yet published. Both machines were equipped with an additional auxiliary winch holding a synthetic rope to facilitate the rigging process of the steel-made cable.

(a)   (b)

Figure 1. Observed John Deere 1170 E harvester with a mounted winch in the front (a) and a John Deere 1110 E forwarder with an integrated winch (b).

2.2. Measuring of Tensile Forces and Machine Inclination

In principle, tensile forces of cables can be directly measured in-line by installing load cells, dynamometers, or sheave pins between the cable and anchor, or indirectly via clamp-on devices using the linear relationship between the deflection in a cable and the resulting force on the central member [24]. During this study, we measured the tensile forces of the cable during operation with the CableBull cable tensile force measurement device from the German manufacturer Honigmann, based on the clamped-on tension monitoring system that measures the resulting forces on the central member for a given deflection (Figure 2).

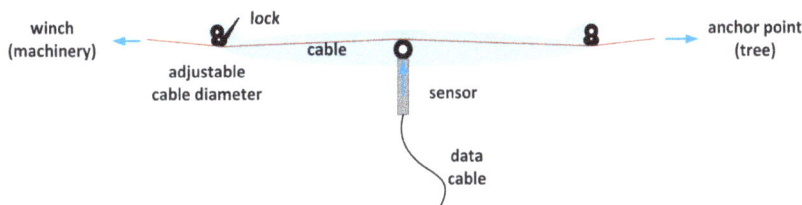

Figure 2. Simplified measuring principle for tensile forces, applied within the used CableBull device.

This device allows online monitoring and recording of tensile forces by clipping it onto the cable on the site while the machinery operates, instead of installing the device between the cable and the machine (Figure 3).

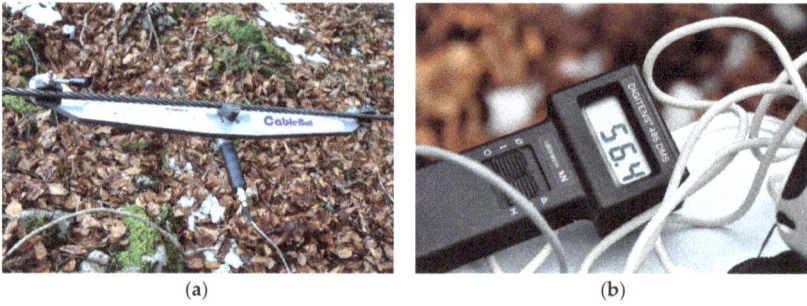

(a) (b)

Figure 3. The CableBull cable tensile force measurement device from the German manufacturer Honigmann based on the clamped-on tension monitoring system (**a**) and the handheld device for checking tensile forces (**b**).

For this study, a toolkit consisting of a Panasonic Toughbook with an extra battery for long-term measurements and data recording was used. In the field, simple data pre-analysis and quick visualization for plausibility checks were completed using the HCCEasy Software from Honigmann GmbH in Wuppertal, Germany. The sample rate for recording can be chosen in the range of 100 to 2000 Hz and was set to 100 Hz for this study. The selected sample rate was chosen according to previous studies [24], and was increased by a factor of 10 to ensure no peaks in tensile forces were missed.

Machine inclination was recorded with the Hobo Pendant G Acceleration Data Logger. The logger uses an internal three-axis accelerometer for measuring the dynamic and static acceleration of gravity in order to record and analyse machine inclination. The sensor was fixed on the rear of the machine's frame to measure machine inclination during operation (Figure 4). The data were stored on the data logger and read out with the corresponding HOBOware software from ONSET® in Bourne, MA, USA. Exported as a text file, the inclination over time can be used for detailed analysis. The data were transformed from degrees into a percentage after being read out. In this study, the sample rate for recording machine inclination was set to 0.03 Hz due to the data storage space required for long-term recording.

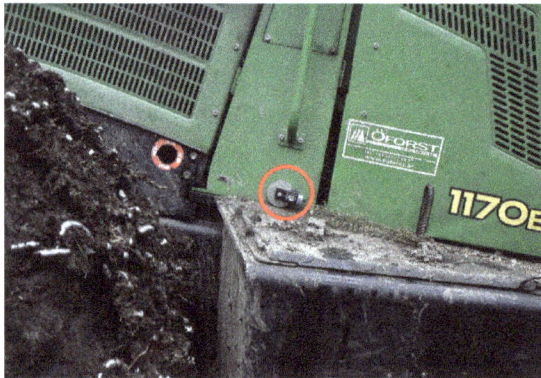

Figure 4. The location of the mounted acceleration data logger on the rear frame of the machinery.

Performing a detailed on-site time and motion study for the work phases was considered organizationally unfeasible and too risky. Therefore, the entire operation was recorded using two on-board video cameras to ensure the data was recorded in case one of the cameras failed, with the video cameras installed above the front windscreen at the protective grating. All videos were captured with 30 frames per second and stored on the memory card. The video capturing data was only limited by the battery life, which was around two hours (depending on the temperature). After the operation was finished, the video data was used to complete time and motion studies post-hoc in the office (Figure 5).

Figure 5. Installed video capturing technology at the protective grating above the front windscreen for recording harvesting activities. The second camera ensured data recording in case of failures.

The video data was analysed with a spreadsheet add-on written in Visual Basic for Applications (VBA) using a desktop computer. This add-on for Microsoft Soft Excel was developed by Lauren from LUKE (METLA). Because the add-on offers both options simultaneously, the working time was recorded using both the continuous timing method and the snap back timing method. After the time study was completed, the dataset was exported into an Oracle database table using an R script for further processing.

The tensile force, the inclination measurement equipment, and the video capturing under natural conditions were tested using a first in-field test. Based on this experiment and observations, the working phases were defined. For the harvester, four work phases were determined: (1) felling and processing starts with the fell cut and ends with the full opening of the head with the releasing of the top; (2) other crane work includes all crane work activities that are not related to handling a tree; (3) driving is any movement of the machine no matter if backward or forward; and (4) delays are defined as time not related to effective work.

Following the scheme of the harvester, forwarding activities were divided into seven work phases: (1) loading is when the forwarder starts moving the crane for loading logs onto the bunks without any movement of the machine; (2) unloading was defined by crane work for unloading logs from the bunks without any movement of the machine at the landing; (3) driving empty starts after unloading is finished, including driving to the next loading place, and ends with starting crane work for loading again; (4) driving-loading includes all activities of parallel loading and driving activities; (5) driving loaded starts with stopping crane work for loading and ends by arriving at the landing, ready for unloading; (6) other crane work includes all activities with the crane work not related to loading or unloading; and (7) delays are time not related to effective work.

2.4. Workflow Merging and Analysing Recorded Data

Based on precise time stamps from the captured video, the recorded working phases and machine inclination were assigned to the tensile force data by joins using structured query language (SQL) routines. Due to the chronological synchronizing of work phases and inclination with tensile force data, an in-depth analysis of tensile forces based on the current working situation was possible. For this in-depth analysis, a lower sample rate of the tensile force data with 25 Hz was used to reduce computation time. Compression of the tensile force data from 100 to 25 Hz was performed by calculating the mean and maximum values via an R script (Figure 6).

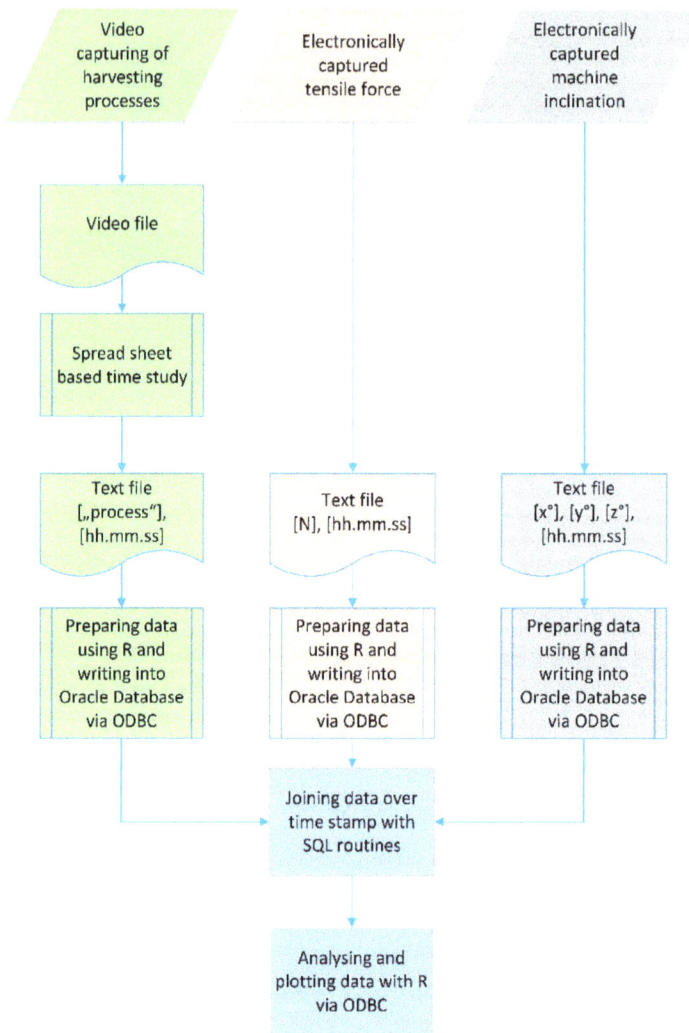

Figure 6. The developed workflow for preparing data from the time study, the tensile force measurements, and from the sensor recording machine inclination for writing via an Open Database Connectivity (ODBC) into the database and joined using the time stamps for further analysis. SQL: structured query language.

3. Results

3.1. Harvester

Altogether, four hours of harvesting activities including tensile forces were captured, which resulted in 357,475 measurements at the sample rate of 25 Hz. As expected, felling and processing of trees accounted for 41% of the total recorded time, followed by the activities for other crane work and driving. The maximum tensile force once reached a peak of 75.5 kN during felling and processing, which is still below the maximum safe working load of 50% considering the 14-mm diameter of the mounted cable. All activities showed an average in tensile force on the cable above 40 kN, and even close to 50 kN for the whole data set, indicating that the machine was always on steep terrain during operations without any flat sections (Table 1).

Table 1. Descriptive statistics for the recorded tensile forces in Newton (N) at the harvester divided into defined work phases.

Work Phase	Minimum	First Quartile	Median	Mean	Third Quartile	Maximum	Records
Felling and processing	12	43,140	46,553	47,797	53,517	75,647	146,250
Other crane work	359	43,943	47,930	49,501	57,373	68,779	100,627
Other work	21,603	39,308	45,838	44,763	51,792	60,103	2654
Driving	28,525	41,248	44,440	44,169	47,014	66,210	60,345
Delays ≤15 min	9083	38,810	45,194	45,817	55,990	68,001	37,236
Delays >15 min	29,363	44,224	44,793	44,524	45,281	55,139	10,363

The box plots clearly indicate that the pre-set tensile force that is chosen from the driver ranges from 40 to 60 kN and is set and changed constantly by the driver according to the need during operation (Figure 7). Overall, most of the activities used a working cable load of between 20% and 30% (Figure 8).

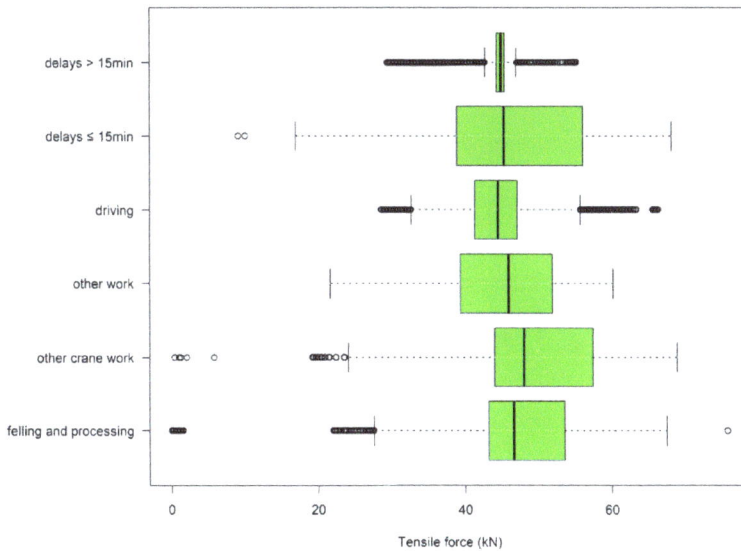

Figure 7. Tensile forces divided into defined work phases of the observed harvester for all captured data. Each box plot contains whiskers including data from 2.5% to 97.5%, boxes from 25% to 75%, and the black line denotes the median. Outliers are marked as points.

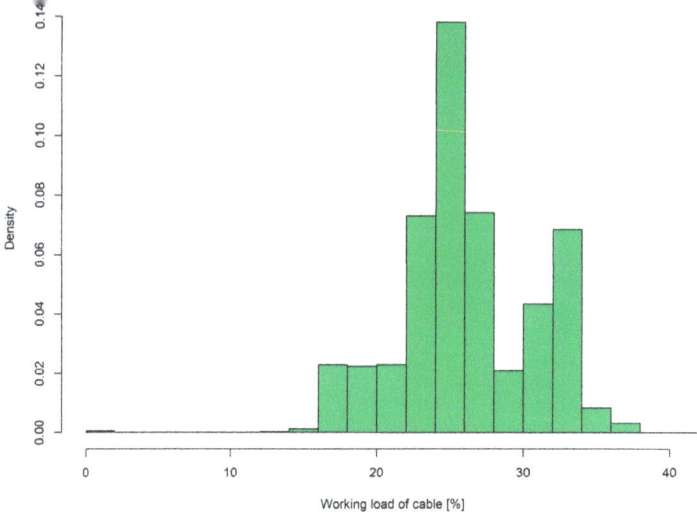

Figure 8. Working load of the cable used during harvesting activities for all captured data.

Compared to the forwarder, harvester activities on the slope did not indicate any cyclic force behaviour. Moreover, a harvester has less cable outhaul and inhaul than a forwarder. The tensile force graph consistently shows higher peaks during every working phase, mostly where the crane is active. This effect can be explained by harvesters handling whole trees after felling and processing (Figure 9).

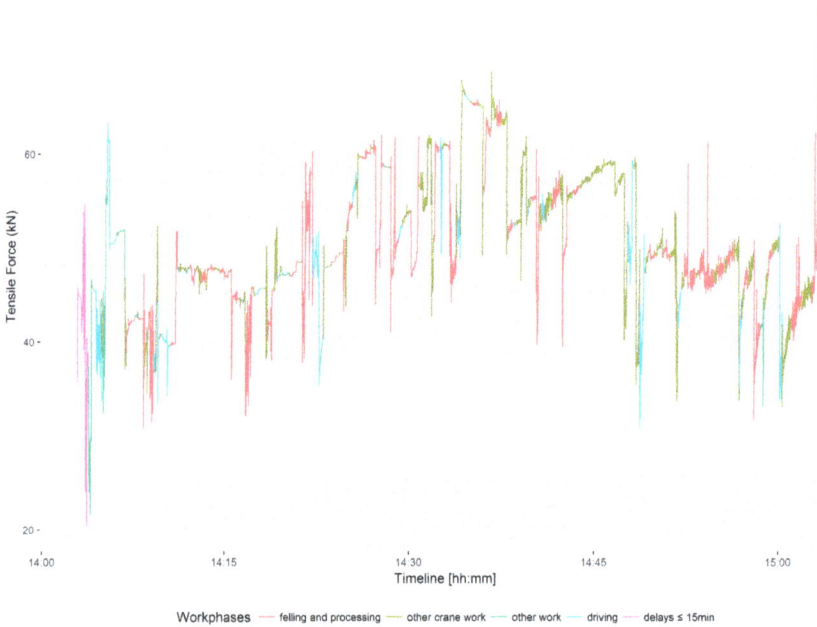

Figure 9. Tensile force graph showing the defined work phases of the observed harvester for a selected time period.

302

Unfortunately, the recorded dataset does not include machine inclination for the harvester.

3.2. Forwarder

The entire dataset consisted of 647,277 rows for the forwarder, which was already compressed using a factor of four representing the 25 Hz sample rate. In total, 7.2 h of work were recorded, representing 16 loads. Loading and unloading accounted for 58.5% of the time, and all driving activities consumed 28.4% of the total recorded time.

Mean tensile forces at loading were 97% higher than during the unloading phase at the landing. For all captured cycles, the average working load when driving empty compared to driving loaded had a difference of 12%. Maximum tensile forces were observed during delays with an observed peak of 56 kN, followed by driving empty during loading activities. The observed and captured peak corresponds to a working load of 30.9% of the maximum breaking strength of the cable used (Table 2).

Table 2. Descriptive statistics for the recorded tensile forces of the forwarder in Newton (N) divided into defined work phases.

Work Phase	Minimum	First Quartile	Median	Mean	Third Quartile	Maximum	Records
Loading	7137	31,159	39,260	34,642	41,931	54,708	240,781
Unloading	7538	14,957	15,807	17,563	16,969	49,517	137,668
Driving empty	5479	15,023	27,327	29,552	45,655	55,630	96,429
Driving and loading	7880	15,562	26,962	27,369	39,597	52,637	15,628
Driving loaded	4239	22,018	35,847	33,118	43,871	53,445	87,246
Delays ≤15 min	8742	12,927	14,508	19,148	14,562	56,259	39,559
Delays >15 min	8640	13,831	17,448	22,854	34,608	45,428	11,431
Other work	15,490	17,544	43,350	36,524	48,463	55,032	18,535

The working load on the cable did not exceed one-third of the maximum breaking strength during the entire study, and was most often between 5% and 11% and 22% and 28%. Analysing the density in terms of the workload based on the defined work phases showed that the majority of low working load was due to delays up to 15 min and the unloading phase. Higher working loads on the cable occurred during loading and driving activities and phases of other work (Figures 10 and 11).

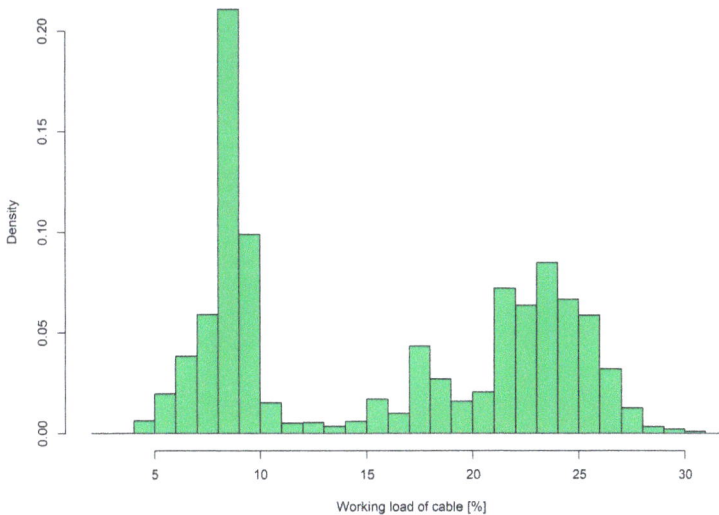

Figure 10. Working load of cable used during forwarding activities for all captured data.

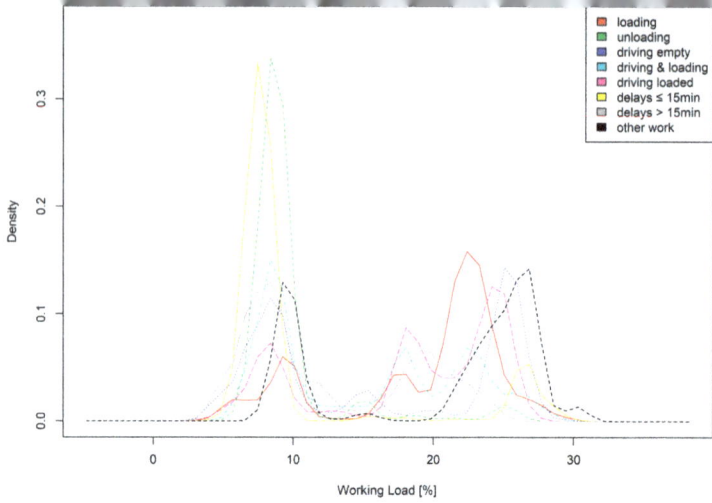

Figure 11. Working load for the cable used during forwarding activities divided into the defined work phases.

The graph of tensile force over time in Figure 12 shows one forwarder cycle divided into the defined work phases starting at the landing and finishing with unloading and driving empty to the next loading point. The pre-set pulling force from the winch, between 40 and 50 kN, that was chosen by the driver is visible and is three times higher than the tensile force during unloading. During unloading at the landing, the tensile force of the cable was lowered to approximately 20 kN, as chosen by the driver.

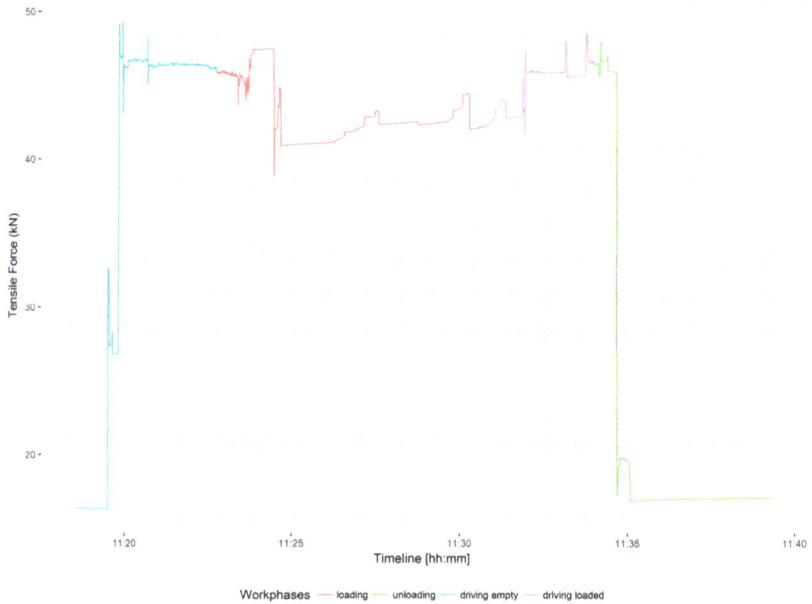

Figure 12. Tensile force graph showing one cycle of the observed forwarder over time with the assigned working phases.

304

A maximum machine inclination of 105% was reached during loading, unloading, and all driving activities. On average, machine inclination ranged from 40% to 58% for all activities in the stand (Figure 13).

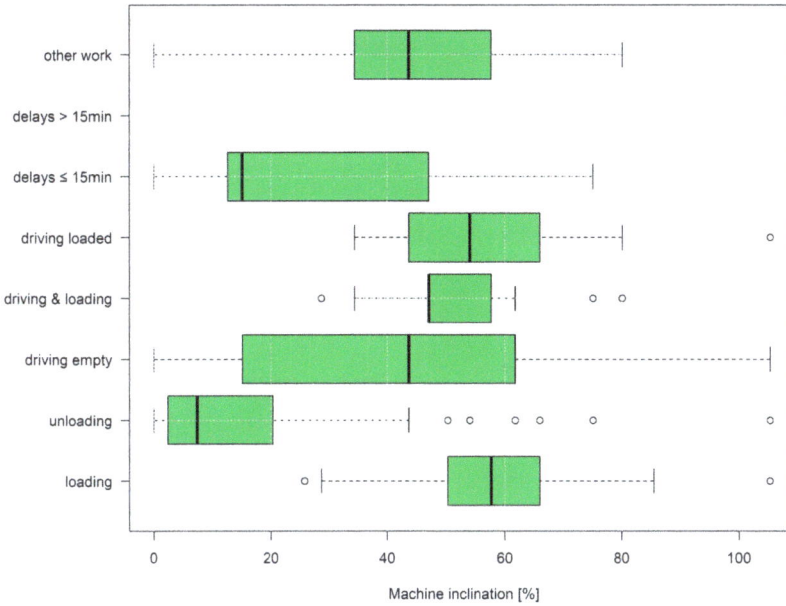

Figure 13. The recorded machine inclination for the forwarder during the study for a selected time period. Each box plot contains whiskers including data from 2.5% to 97.5%, boxes from 25% to 75%, and the black line denotes the median. Outliers are marked as points.

4. Discussion

We present the first detailed tensile force analysis study of winch-assisted fully mechanized harvesting operations in Europe. Based on the developed specific survey protocol interactions between recorded work phases, machine inclination and tensile forces were presented.

Notably, to handle the measuring device for tensile forces with its clamp-on system, if the cable tension were to abruptly release, the measuring device would fall off. This may result in the device being damaged. The purchase cost for the measurement equipment was close to €2900 for the sensor, €900 for the amplifier, and an additional €1500 for the software to check and transfer the data. However, the setup time was negligible due to the easy-to-use clamp-on system, which can be performed during an ongoing operation. Downloading data from the measurement device at the site was not needed, as the tensile force data was stored on a Panasonic Toughbook. This system would not be suitable for moving cables where the winch is anchored at the top of the hill. For such machinery, tensile force sensors, which are built-in shackles, would be an appropriate alternative.

For capturing machine inclination, an inertial measurement unit measuring roll, yaw, and pitch, that is connected to a single-board computer, would facilitate handling and offer additional benefits in terms of long-term data recording with a high sample rate for storage space and power supply. The video capturing of all harvesting activities was highly successful in this study. The data was stored on memory cards that were only changed when the operation or the skid trail was finished. Nevertheless, improvements could be made in terms of directly storing the video stream onto an external storage device and using a permanent power supply to avoid changing batteries every two hours.

The workflow used video to capture the machines' processes and record the tensile force on the cables, and machine inclination is divided into three subsets of data that was afterward joined using the video time stamps. Future research could combine all sensor data out in the field during data recording. This would reduce the effort required to determine the starting point for analysis. Furthermore, the use of sensors with a high sample rate for detecting activities could lead to semi-automatic recognition of work phases in harvesting operations or support long-term studies [25].

In this study, the winch from the harvester was mounted in the front section. The supplier of this product also offers a rear-mounted solution. This automatically affects the operation mode from uphill to downhill. However, the mounting cannot easily be changed in the forest according to changing conditions and requirements, especially considering the operator's workspace when it comes to steep terrain, where the levelling of the cabin is not possible and the driver is working downhill. Tensile forces could change due to smoother driving activities compared to when moving uphill with the harvester.

5. Conclusions

The objective of this study was to develop a scientific approach with a robust workflow for the in-depth monitoring and analysing of tensile forces for winch-assisted harvesters and forwarders, based on work phases and machine inclination. The main question was to determine if the machines in use are exceeding the safe working load of the cable being used under real working conditions.

The established workflow, including equipment and the developed analysis routines, worked well. During the study, the working load on the cable did not exceed 50% of the maximum breaking strength. A maximum tensile force peak was observed during delays for the forwarder at 56 kN, and a peak of 75.5 kN for the harvester, which are still within safe working loads, considering a safety factor of two.

Nevertheless, the study only covered a short period of four hours for one harvester and 7.2 h for one forwarder and did not cover a long-term period for different machines, drivers, or various conditions. Therefore, conclusions and findings regarding maximum peaks of tensile forces during harvesting on steep slopes should not be extrapolated to other machines, operators, or harvesting sites. Therefore, further detailed long-term studies are needed to cover different machinery, settings, and modes of operation in combination with productivity. Future studies could focus on the rope wear of this machinery.

Acknowledgments: This publication is a part of the national funded project "Fully mechanized timber harvesting with winch-assisted harvester and forwarder". The research leading to these results has received funding from the cooperation platform of Forst Holz Papier (FHP). The authors want to thank the entrepreneur Huber and Tazreiter GmbH, including his team, as well as the John Deere Austria partner (ÖFORST) for supporting the study. In addition, we also would like to thank the landowners for enabling the fieldwork. The authors also want to thank the editors and reviewers for their valuable input during the preparation of this original paper.

Author Contributions: Franz Holzleitner and Maximilian Kastner conceived and designed the study layout; Maximilian Kastner performed the field measurements; Franz Holzleitner, Christian Kanzian, and Norbert Höller analysed the data; Franz Holzleitner, Christian Kanzian, and Karl Stampfer wrote the paper.

Conflicts of Interest: The authors declare no conflict of interest.

References

1. Heinimann, H.R.; Stampfer, K.; Loschek, J.; Caminada, L. Perspectives on Central European Cable Yarding Systems. *Austrian J. For. Sci.* **2006**, *123*, 121–139.
2. Visser, R.J.M.; Stampfer, K. Expanding ground-based harvesting onto steep terrain: A review. *Croat. J. For. Eng.* **2015**, *36*, 321–331.
3. Axelsson, S.Å. The mechanization of logging operations in Sweden and its effect on occupational safety and health. *Int. J. For. Eng.* **1998**, *9*, 25–31.
4. Hartsough, B.R.; Miles, J.A.; Gaio, C.; Frank, A.A. Cable-towed vehicles for harvesting on mountainous terrain. In Proceedings of the International Mountain Logging and Pacific Northwest Skyline Symposium, Portland, OR, USA, 12–16 December 1988; pp. 54–77.

5.   Wratschko, B. Einsatzmöglichkeiten von Seilforwardern (Application of Cable Forwarders). Master's Thesis, Institute of Forest Engineering, University of Natural Resources and Life Sciences, Vienna, Vienna, Austria, 2006.
6.   Holzleitner, F.; Stampfer, K.; Visser, R.J.M. Utilization rates and cost factors in timber harvesting based on long-term machine data. *Croat. J. For. Eng.* **2011**, *32*, 501–508.
7.   Jacke, H.; Drewes, D. Kräfte, Schlupf und Neigungen—Ein Beitrag zur Terramechanik forstlicher Arbeitsmaschinen. Forces, Slip and Slopes—A Contribution for Terramechanics of Forestal Work Machines. *Forst Holz* **2004**, *50*, 259–262.
8.   Hittenbeck, J. Estimation of trafficable grades from traction performance of a forwarder. *Croat. J. For. Eng.* **2013**, *34*, 71–81.
9.   Weise, G. Seilgestützter Forstmaschineneinsatz am Hang. Einige Betrachtungen zum Abfangen des Hangabtriebs durch ein Stützseil (Employing winch-assisted forest machines on steep terrain. Some observations to resist downhill-slope force using a cable). *Forsttech. Inf.* **2004**, *9–10*, 113–116.
10. Weise, G. Traktionshilfswinden—Besser am Hang. *Forsttech. Inf.* **2016**, *5*, 7–14.
11. Wijekoon, M. Forest Machine Tire-Soil Interaction. Master's Thesis, KTH Industrial Engineering and Management, Machine Design, Stockholm, Sweden, 2012.
12. Wismer, R.D.; Luth, H.J. Off-road traction prediction for wheeled vehicles. *J. Terramech.* **1973**, *10*, 49–61. [CrossRef]
13. Salsbery, B.; Hartsough, B. Control of a cable-towed vehicle to minimize slip. *J. Terramech.* **1993**, *30*, 325–335. [CrossRef]
14. Sessions, J.; Leshchinsky, B.; Chung, W.; Boston, K.; Wimer, J. Theoretical stability and traction of steep slope tethered feller-bunchers. *For. Sci.* **2017**, *63*, 192–200. [CrossRef]
15. Bombosch, F.; Sohns, D.; Nollau, R.; Kanzler, H. Are Forest Operations on steep terrain (average of 70% slope inclination) with wheel-mounted forwarders without slipage possible? In Proceedings of the Austro2003-Symposium: High Tech Forest Operations for Mountainous Terrain, Schlägl, Austria, 5–9 October 2003.
16. Oberer, F. Symbiose aus Rad und Seil (Symbiosis of wheel and cable). *Wald Holz* **2012**, *1*, 28–30.
17. Wegmann, U. Kletterkünstler Hangforwarder (Climbing artist cable forwarder). *Wald Holz* **2019**, *1*, 30–31.
18. Lindroos, O.; La Hera, P.; Häggström, C. Drivers of advances in mechanized timber harvesting—A selective review of technological innovation. *Croat. J. For. Eng.* **2017**, *38*, 243–258.
19. Heinimann, H.R. Ground-based harvesting systems for steep slopes. In Proceedings of the International Mountain Logging and 10th Pacific Northwest Skyline Symposium, Corvallis, OR, USA, 28 March–1 April 1999; pp. 1–19.
20. Amishev, D. *Winch-Assist Technologies Available to Western Canada*; Technical Report No. 37; FPInnovations: Pointe-Claire, QC, Canada, 2016; p. 51.
21. Visser, R.J.M. *Tension Monitoring of a Cable Assisted Machine*; Harvesting Technical Note HTN05-11; Future Forests Research Limited: Rotorua, New Zealand, 2013.
22. Evanson, T. *Use of Tension Monitors to Estimate Payload*; Harvesting Technical Note HTN01-08; Future Forests Research Limited: Rotorua, New Zealand, 2009.
23. Dupire, S.; Bourrier, F.; Berger, F. Predicting load path and tensile forces during cable yarding operations on steep terrain. *J. For. Res.* **2016**, *21*, 1–14. [CrossRef]
24. Visser, R.J.M. Tensions Monitoring of Forestry Cable Systems. Ph.D. Thesis, Institute of Forest Engineering, University of Natural Resources and Life Sciences, Vienna, Vienna, Austria, May 1998.
25. Pierzchała, M.; Kvaal, K.; Stampfer, K.; Talbot, B. Automatic recognition of work phases in cable yarding supported by sensor fusion. *Int. J. For. Eng.* **2017**. [CrossRef]

Article
Postural Risk Assessment of Small-Scale Debarkers for Wooden Post Production

Raffaele Spinelli [1,2,*]**, Giovanni Aminti** [1]**, Natascia Magagnotti** [1,2] **and Fabio De Francesco** [1]

1   CNR Ivalsa, Via Madonna del Piano 10, 50019 Sesto Fiorentino FI, Italy; aminti@ivalsa.cnr.it (G.A.); magagnotti@ivalsa.cnr.it (N.M.); defrancesco@ivalsa.cnr.it (F.D.F.)
2   Australian Forest Operations Research Alliance (AFORA), University of the Sunshine Coast, Locked Bag 4, Maroochydore DC, Queensland 4558, Australia
*   Correspondence: spinelli@ivalsa.cnr.it; Tel.: +39-335-5429798

Received: 20 January 2018; Accepted: 28 February 2018; Published: 2 March 2018

Abstract: The study sampled six representative work sites in Northern and Central Italy, in order to assess the risk for developing musculo-skeletal disease due to poor work posture (postural risk) among the operators engaged in semi-mechanized post debarking operations. Assessment was conducted with the Ovako Working posture Analysis System (OWAS) on 1200 still frames randomly extracted from videotaped work samples. The postural risk associated with post debarking was relatively low, and varied with individual operations based on their specific set up. Postural risk was higher for the loading station compared with the unloading station, which makes a strong argument for job rotation. The study suggested that the infeed chute of small-scale debarkers might be too basic and should be further developed, in order to reduce postural risk. Obviously, better machine design should be part of an articulate strategy aimed at decreasing the postural risk and based on proper worksite organization and specific worker training.

Keywords: WMSD; OWAS; ergonomics; safety; logging

1. Introduction

For many decades, forest economy has been squeezed between decreasing product value and increasing labor cost, and this trend does not seem to be stopping any time soon—if at all. The classic solution is found in improving the efficiency of forest management, obtained by mechanizing operations through a considerable capital commitment. Against this background, economy of scale is the key to success, which conflicts with the decreasing size of many private forests as they get fragmented through heritage lines, often as part of the restitution process [1].

Many low-investment solutions have been proposed over time to increase labor productivity and yet small-scale technology cannot completely offset the efficiency gap with large-scale industrial operations [2]. In any case, technology improvements are designed to cut costs and they only tackle one side of the problem, while doing very little to address the other one, namely value recovery [3]. Most previous attempts to increase value recovery have focused on the manufacturing of high-value niche products that cannot be sold in large numbers on the commodity market. However, not all high-value products are niche products: in that regard, wooden posts represent a fortunate exception. They offer three important advantages: first, they carry a much higher price than any comparable small-size assortment; second, they can be manufactured from low-value small wood; third, they can be effectively produced with low-cost technology, especially suited to small-scale operations.

A large and expanding market for wooden posts is offered by activities such as mining, gardening, and agriculture. In particular, fruit growers offer attractive prices for quality posts, which are sold by the piece, and may attain the equivalent of 190 € m^{-3}, which is three times the price that one could obtain by selling the same material as firewood [4,5].

Furthermore, upgrading small wood to post standards is a straightforward operation that only requires sorting and debarking [6]. These operations are easily mechanized using low-cost equipment, generally a simple debarking machine powered by an industrial electric motor or by a farm tractor. Most small-scale debarkers are of the knife type, and employ a fast-spinning disk fitted with five or six radial knives that works very much like a planer. During work, the post is pushed with its sides against the turning disk, while a spiral-patterned feed roller makes it turn around its longitudinal axis so that all the external surface of the post will touch the spinning disk. The result is high-quality debarking, where most of the bark is shaved off the post surface, together with some of the wood on the eventual bumps and bends. As a result, the post is somewhat "straightened", adding to the aesthetic quality of the product, which is generally appreciated by customers.

This process is largely manual, which raises the question about the impact on operator safety and health. In particular, cycles repeat with some frequency, estimated at over 50 full cycles an hour [5]. Such a labor-intensive and repetitive job raises the obvious question about the potential strain on the musculo-skeletal system of workers, and makes a correct postural assessment especially important. Repeated over and over again, poor body postures may lead to musculo-skeletal disease, and that possibility defines postural risk. There is every reason to believe that the population potentially affected is relatively large. While no data are available on the actual numbers, one may produce a plausible estimate starting from the production volumes reported in the Eurostat database [7]. This offers a conservative estimate of around 500,000 m^3 per year, for EU 28—although much of the production is concentrated in France, Austria, Italy, Portugal, and Spain (in descending order of production). Dividing this figure by a mean debarking productivity equal to 1.5 m^3 per hour [5], one obtains over 333,000 h, or 450 full jobs, without considering all self-employed operations that go unrecorded. What is more, the demand for wooden posts is expanding and therefore the number of operators potentially affected is going to increase over the years.

Despite the large potential impact on the rural workforce, the ergonomic evaluation of post debarking has attracted little scientific interest so far. To our knowledge, the subject has never been addressed by any study—recent or old. This is a knowledge gap that needs filling, especially if one considers the large number of workers involved in post debarking and the wide variety of machines available for the task.

While a single study cannot fill this gap, it may still represent a good start and attract further attention by other research teams, eventually leading to an exhaustive ergonomic evaluation of this task. Post debarking consists of a fast and repetitive sequence of loading, pushing, and unloading, during which time operators will assume the same postures over and over again. That makes it worthwhile to investigate the risk for work-related musculo-skeletal disorders (WMSD), as a first step in the ergonomic evaluation of debarking duties. The fact that most operations are small-scale and are managed by small-medium enterprises (SMEs) makes the need for analysis more urgent, because working conditions in SMEs are known to be generally poorer than in larger and structured enterprises [8].

The ergonomic assessment of post debarking should cover a wide range of solutions, capable of representing the majority of operational set ups, characterized by various levels of refinement and ergonomic performance. Differences are likely to be important between machines that are equipped with a feeding deck and machines that are not.

Therefore, the goal of this study was to assess the postural risk associated with post debarking work, performed with a range of post debarking machines and set-ups capable of representing the main technical solutions currently available to small-scale operators. In particular, the study aimed to: (1) produce a benchmark for postural risk in small-scale post debarking work; (2) compare different machine designs and operation layouts in terms of postural risk; and (3) determine what specific work tasks incur the highest postural risk. The results of the study will inform recommendations for work safety supervision and machine design in the post debarking sector.

2. Materials and Methods

The study was conducted on six small-scale post debarking operations, taken as representative of the larger population of small-scale debarking operators. These operations were located in Northern and Central Italy, so as to cover a reasonably wide range of working conditions (Table 1). All operations were owned by small enterprises and all 12 workers involved in the trials were mesomorphic adult males, with an age between 30 and 55 years and a work experience of at least five years. Although different workers were employed for different operations, the skills and the anthropometric characteristics of the workers were considered generally representative of the workers in the region [9].

Table 1. Characteristics of the sample operations.

Operation	ID	A	B	C	D	E	F
Species		Chestnut	Chestnut	Chestnut	Chestnut	Chestnut	Chestnut
Machine	Make	Rabaud	Rabaud	Rabaud	Rabaud	Pribo	Neuhauser
Machine	Model	Robopel 250	Robopel 250	Robopel 250	Robopel 250	MSP 25	GS
Motor	Type	electric	electric	electric	tractor	electric	electric
Motor	kW	15	15	15	59	11	9
Load deck	Yes/No	Yes	No	No	Yes	Yes	Yes
Unload	H/V	H	H	V	H	V	V
Crew	n°	2	2	2	2	2	2
Pieces	n°	548	512	473	301	530	524
n° lenghts	n°	2	4	6	1	1	1
Length	cm	238	224	290	250	300	270
Diameter	cm	11	7.2	6.7	8	7.9	7.2
Post weight	kg	22	10	12	12	16	9
Total volume	m³ ub	12.7	5.2	5.4	4.1	8.4	6.4
Productive time	h	4.3	2	4.9	2.1	3.1	3.1
Delay time	h	1.8	0.7	1.6	1.3	1.7	1.4
Total worksite time	h	6.2	2.6	6.5	3.4	4.8	4.5
Delay	%	30	25	24	38	36	31
Productivity	m³ ub h⁻¹	3.0	2.6	1.1	2.0	2.7	2.1

Notes: H/V = Horizontal on a stack/Vertical against a wall; Diameter = post diameter at mid-length; ub = under bark; Productivity is calculated based on productive time only, and excluding delays.

The six operations represented three of the most popular debarking machines, as follows: Rabaud Robopel 250 (four units), Pribo MSP 25 (one unit), and Neuhauser GS (one unit). These machines were all disk type and processed one post at a time. All machines in the study were simple and inexpensive ($\leq$30,000 €), which made them especially suited for small-scale operations, compared with the large automated ring-, drum-, and rosser head-debarkers used at sawmills and pulpmills. Technical differences between the study machines were minor, and were found in some mechanical details. The most meaningful differences were in the way that the operations were set up, which implied a different work organization and different postural risk (Figure 1).

All units were stationed at wood yards and processed small logs sourced from chestnut (*Castanea sativa* L.) and locust (*Robinia pseudacacia* L.) coppice stands. Five units were powered by electric motors, whereas one was powered by a tractor, using the power take-off (PTO). Four units (A, D, E, F) were provided with a loading deck, in order to speed up loading and reduce worker effort. Decks consisted of a simple metal structure, designed to bring the posts to the same level as the debarker infeed chute and prevent the worker from bending down to lift the posts off the floor. When a loading deck was provided, posts were easily placed on the deck with a forklift. However, posts were moved from the deck to the debarker infeed chute manually, eventually using a sappie or a similar tool. Units B and C had no loading deck, which forced workers to manually lift and carry the posts from the piles and place them on the infeed chute.

Figure 1. Typical postures assumed during post debarking at the two main work stations: loading-push (**a**), loading-hold (**b**), unloading-pull (**c**), and unloading-stack (**d**).

All debarkers were manned by two workers: one for loading the undebarked posts and adjusting the machine during work, and the other for unloading the debarked posts and moving them to the appropriate stacks. Unloading was always manual, since no unloading deck was available. Sorting generally occurred during unloading, and consisted of placing the posts into different piles according to their characteristics. Debarked posts could be placed horizontally in stacks built on the yard floor (operations A, B, and D), or vertically against a wall (operations C, E, and F). In the latter case, the operator would not need to bend, as otherwise necessary when building a stack of horizontally placed posts.

Data collection took place in the period between the Winter of 2016 and the Spring of 2017. During that period, researchers visited each operation to record post debarking operations with two 12 megapixel digital cameras mounted on tripods, in order to simultaneously collect video data for both operators. Cameras were positioned to capture images of the entire body of each worker during the debarking cycle. The data pool consisted of 12 video files, as many as the operators surveyed in the study. Each video recording was at least 15 min in duration, and it captured at least 25 debarking cycles. Each loading and unloading cycle was subdivided into specific tasks, in order to determine if some tasks incurred a higher postural risk than the others, which could be used by machine manufacturers when designing new machine models (Table 2). Machine delays and rest breaks were excluded from the analysis.

Once in the laboratory, still frames were extracted from the video footage in the number of 100 stills for each worker and operation. Frame extraction was conducted at random intervals, whose duration averaged 30 s [10]. Random number tables were generated in order to assist with the sampling and avoid the risk of accidental synchronization between observation intervals and cyclic tasks [11]. The total number of videotape frames analyzed in this study was 1200, that is 100 stills × 12 workers.

Table 2. Description of main tasks.

Table 2. Description of main tasks.

Loading Station	
Carry	Carrying the post from the pile to the machine (it includes picking, if the pile is low on the floor)
Push	Pushing the post into the machine infeed
Hold	Holding the post while engaged by the machine
Wait	Waiting idle
Walk	Walking back to the pile to get another post
Unloading station	
Pull	Pulling the post off the machine out feed
Carry	Carrying the post from the machine to the stack
Stack	Placing the post on the stack
Wait	Waiting idle
Walk	Walking back to the machine

Operator postures on the still frames were attributed a postural risk index using the Ovako Working posture Analysis System (OWAS). Postural risk assessment can also be conducted with other methods, such as the Rapid Upper Limb Assessment (RULA) and the Rapid Entire Body Assessment (REBA). RULA offers additional information on wrist and elbow postures, which are not included within OWAS. However, RULA was developed for sedentary work and neglects lower body postures, whereas REBA was developed for the health care service and is less suitable for industrial work compared with OWAS [12]. In contrast, the OWAS method covers both the upper and lower body, and is relatively simple to apply [13]. Furthermore, OWAS is among the most widespread methods used to assess postural risk, which facilitates the comparison of study results with the data reported in the available literature on the subject [14]. The OWAS method is highly reliable [15] and it has been shown to match or surpass the performance of other newer and more detailed postural assessment methods [16].

Frame analysis was conducted using the ErgoFellow 2.0 software, developed by FBF SISTEMAS (Belo Horizonte, MG, Brazil). Using the interactive program interface, each frame was analyzed and classified according to the OWAS method, so that the software could return an Action Category (AC) score, indicating whether corrective action was necessary and how urgent it was. In particular, the OWAS method adopts four Action Categories, as follows: AC1 = Normal posture, no intervention required; AC2 = Slightly harmful posture, corrective action should be taken during the next regular review of work methods; AC3 = Distinctly harmful posture, corrective action should be taken as soon as possible; AC4 = Extremely harmful posture, corrective action should be taken immediately (i.e., right now!). The OWAS system attributes AC scores for each of the 252 combinations derived from three arm postures, four trunk postures, seven lower body postures, and three load weight classes. AC scores are based on a specific grid, developed by sector experts

Each frame was classified as one of the previously described work tasks in post debarking, which allowed us to determine whether specific task types involved a higher postural risk compared to the others.

Each work station, operation type, and task was also attributed an overall postural risk index (PRI), calculated as the frequency-weighted average of the AC scores recorded for that specific case, according to the equation below [17]:

$$PRI = (a \times AC1) + (b \times AC2) + (c \times AC3) + (d \times AC4) \tag{1}$$

where a, b, c, and d are the frequency of scores AC 1, 2, 3, and 4, respectively, represented as the percentage of total observations attributed a given score.

Data were analyzed with the Minitab 16 advanced statistics software (State College, PA, USA). Given the ordinal character of all variables, the statistical significance of the eventual differences between the AC scores for different operations, work stations, and work steps was checked with Pearson's chi-square (χ^2) test for sampling distributions. The elected significance level was $\alpha < 0.05$.

3. Results

The mean postural risk index for post debarking was 136, or 146 and 129 for the loading and unloading station, respectively. Loading incurred a higher postural risk than unloading, with twice as many observations in the AC2 class, compared with the general distribution (Figure 2). This distribution "anomaly" offered the largest contribution (70%) to the overall χ^2 score for the comparison of distributions, which was highly significant ($p < 0.001$).

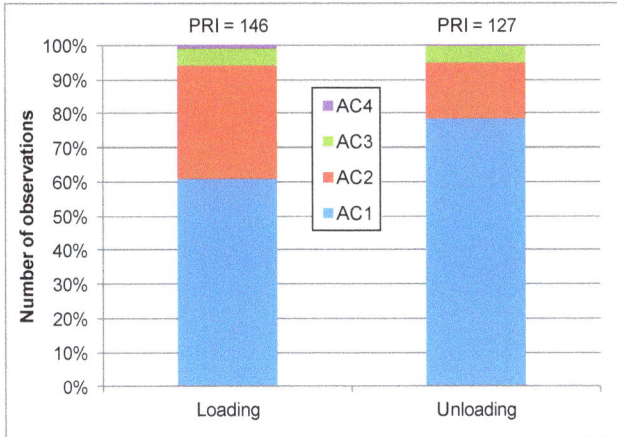

Figure 2. Percentage breakdown of observations among postural risk classes and overall postural risk index. Notes: PRI = postural risk index; AC = Action class, as estimated with the OWAS method; $n = 1200$; AC1 = Normal posture, AC2 = Slightly harmful posture, AC3 = Distinctly harmful posture, AC4 = Extremely harmful posture.

Overall, operations F and D performed the best, recording the lowest postural risk index for the whole group: 114 and 129, respectively (Table 3). In contrast, operation C raised the highest concern, with a postural risk index of 150. Differences were caused by a consistent drift of postural scores from AC1 to AC2, which turned out to be highly significant ($p < 0.001$) and contributed 80% to the overall χ^2 score.

Table 3. Percent distribution of observed frames among postural risk classes, by operation ($n = 1200$).

Operation	AC1	AC2	AC3	AC4	PRI	m^3 ub h^{-1}
A	65	24	8	3	149	3.0
B	78	17	5	1	129	2.6
C	55	39	6	0	150	1.1
D	70	25	5	0	136	2.0
E	65	30	6	0	142	2.7
F	87	12	1	0	114	2.1

$\chi^2 = 61.4$, $p < 0.001$, contribution bold = 79%, $R^2 = 0.0049$

Notes: AC = Action category; PRI = postural risk index; m^3 ub h^{-1} = productivity in m^3 under bark per hour, excluding delays; numbers in bold represent the largest contributors to the overall χ^2; bold = % contribution to the total χ^2 score given by the cells with bold characters; R^2 = regression coefficient for the linear relationship between PRI and productivity.

Postural analysis of different work tasks showed that the highest risk was incurred when carrying the undebarked posts to the machine infeed chute, when pushing them on the chute and towards the debarker disk, and when holding the post during debarking (Table 4). Carrying the debarked posts to the stacks also incurred a high postural risk, confirming that "carry" type tasks are inherently hazardous due to the handling of relatively heavy objects (10–20 kg apiece) that may force workers to adopt awkward postures. In contrast, the lowest postural risk was encountered with obviously neutral tasks, as waiting at rest and walking with no load, when operators were most likely to assume physiological postures. Distribution "anomalies" for these tasks explained most (85%) of the overall χ^2 score for the comparison of distributions, which was highly significant ($p < 0.001$).

Table 4. Percent distribution of observed frames among postural risk classes, by task type.

Station	Task	AC1	AC2	AC3	AC4	PRI
Loading	Carry	60	25	13	2	158
	Push	37	49	13	1	179
	Hold	58	39	2	1	147
$n = 600$	Wait	100	0	0	0	100
	Walk	100	0	0	0	100
Unloading	Pull	74	24	2	0	129
	Carry	68	15	16	1	151
	Stack	65	24	11	0	146
$n = 600$	Wait	100	0	0	0	100
	Walk	100	0	0	0	100

$$\chi^2 = 171.9, p < 0.001, \text{contribution bold} = 85\%$$

Notes: AC = Action category; PRI = postural risk index; numbers in bold represent the largest contributors to the overall χ^2; contribution bold = % contribution to the total χ^2 score given by the cells with bold characters.

A better appreciation of postural risk and of the mechanisms leading to given PRI scores was obtained by intersecting the postural risk index characterizing specific tasks with the incidence of these tasks over total cycle time. Taken together, the three tasks with the highest PRI ("carry", "push" and "hold") accounted for 83% of the loading cycle (Figure 3). Apparently, the loading cycle was divided between relatively risky and markedly neutral tasks ("walk" and "wait"), showing a somewhat polarized structure (i.e., high risk vs. no risk at all).

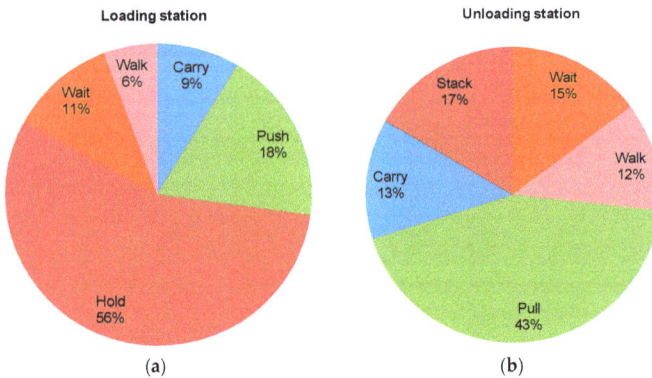

Figure 3. Percentage breakdown of observations among different tasks, by work station, (a) loading station, (b) unloading station.

The situation was better for the unloading cycle. First, neutral tasks were more frequent in the unloading cycle than in the loading cycle (27% vs. 13%), which was bound to reduce the overall postural risk of the unloading station. Second, over 40% of the unloading cycle was occupied by pulling the post from the outfeed chute (i.e., "pull"), which only incurred a moderate postural risk.

The effect of deck type was tested by comparing the postural risk index and the task frequency distribution of operations with and without a deck. In order to have a balanced dataset, the two operations without a deck were matched with two operations extracted from the larger pool of those that were fitted with a deck. Eventually, operations A, B, C, and D were selected for the comparison because they all used the same Rabaud Robopel 250 machine model.

Deck-fitted operations incurred a lower postural risk, and were characterized by a mean PRI of 144, versus 158 for the two operations that were not fitted with a deck. However, individual variations were high and the general difference between the two set ups (deck vs. no deck) was not significant ($p = 0.228$). In contrast, significant differences were found for both task distribution and postural risk between individual operations, and the worst performer was one of the two operations that were not fitted with a deck (Operation C). Yet, the next-worst was a deck-fitted operation (Operation D), which blurred the general picture. Both operations C and D were characterized by a very low incidence of neutral tasks (i.e., "wait" and "walk"), which may hint at a direct relationship between postural risk and how busy operators are (Table 5).

Similarly, the postural risk incurred at the unloading station was affected by how the debarked posts are arranged, i.e., horizontally on the floor or vertically against a wall. The analysis showed that the postural risk incurred by the two options was about the same (Table 6). However, the incidence of a neutral task was twice as high for the horizontal technique, compared with the vertical one, hinting at a somewhat more laborious routine for vertical stacking.

Table 5. Percent distribution of observed frames among task types (above) and postural risk classes (below) at the loading station, for operations with and without a loading deck ($n = 600$).

Loading		Deck	Deck	No Deck	No Deck
Operation	⇨	A	D	C	B
	Carry	11	0	9	0
	Push	11	22	15	50
Task	Hold	47	74	72	31
	Wait	25	1	2	18
	Walk	6	3	2	1
$\chi^2 = 122.9, p < 0.001$, contribution bold = 78%					
	AC1	80	52	36	67
	AC2	7	47	56	25
Risk	AC3	9	0	8	7
	AC4	4	1	1	1
	PRI	137	150	176	142
$\chi^2 = 73.9, p < 0.001$, contribution bold = 86%					

Notes: AC = Action category; PRI = postural risk index; numbers in bold represent the largest contributors to the overall χ^2; bold = % contribution to the total χ^2 score given by the cells with bold characters.

Table 6. Percent distribution of observed frames among task types (above) and postural risk classes (below) at the unloading station, for two stacking techniques: horizontal and vertical ($n = 600$).

Task	Horizontal	Vertical
Pull	35	52
Carry	12	14
Stack	17	16
Wait	22	8
Walk	14	10
$\chi^2 = 32.4$, $p < 0.001$, contribution bold = 93%		
Risk	**Horizontal**	**Vertical**
AC1	81	76
AC2	11	22
AC3	7	3
AC4	1	0
PRI	128	129
$\chi^2 = 14.5$, $p = 0.002$, contribution bold = 94%		

Notes: AC = Action category; PRI = postural risk index; numbers in bold represent the largest contributors to the overall χ^2; bold = % contribution to the total χ^2 score given by the cells with bold characters.

4. Discussion

The postural risk incurred during post debarking is relatively low. The worst case is represented by operation C and obtains a PRI of 150, which is much below that reported for material handling at factories (PRI = 236) [18] and for forest work (PRI = 250 ÷ 300) [17,19]. The best performers (operations B, D, and F) incur a lower musculo-skeletal risk than light manual packing work (PRI = 136) [20], and are basically safe. As far as small-scale forest operations are concerned, post debarking is characterized by a lower postural risk compared with traditional firewood processing work, which is the other activity typically endeavored by small-scale operators to increase the value of their products [21].

In fact, there are large differences between operations, which are difficult to explain. Neither the availability of a deck, nor the stacking technique, seem to have a univocal effect on postural risk. While it is true that the worst performer is not equipped with a loading deck, it is also true that the only other operation without a loading deck performs quite well and is actually second best (operation B). As for the stacking technique, there are indeed differences in task frequency between the horizontal and vertical stacking techniques, but these differences do not seem to affect postural risk.

Therefore, one must look for additional factors that may explain the differences between individual operations. These might be found in the specific characteristics of stacks and decks, such as their height, their exact position in the yard, and their distance from the machine. As for height, that may explain the good performance of operation B, which was not fitted with a loading deck but featured stacks that were replenished regularly, so as to remain near to the infeed chute and to stay always more or less at waist height. This way, the worker at the loading station did not need to bend over for picking posts off the floor, nor did he have to carry these posts to the infeed chute over any extended distance. On that note, it is important to stress that stack placement would likely impact the incidence of high-risk tasks such as "carry", and could be effectively manipulated in order to relieve postural risk.

Furthermore, one may speculate about the effect of the anthropometric characteristics of the workers. Simple machines like the debarkers observed in this study are relatively inflexible, and their structure only allows for minimal adjustments. Therefore, the same machine might be easier to negotiate for workers with specific anthropometric characteristics, and more difficult for other ones who are too tall or too short to work comfortably under a fixed infeed height setting, for instance. As a matter of fact, the workers at operation C—the worst performer—were especially tall, which may have forced them to assume unnatural body positions during some specific tasks.

Finally, it is possible that some of the differences between individual operations were related to training. While they were generally experienced, none of the workers in the study had received specific formal training, but they had all been taught on the job by older workers, as often occurs with small-scale part-time operations. Unfortunately, informal on-the-job training does not develop according to a set curriculum, but is based on the real-world experience of the older workers, which might be quite valid in some fields, but weak in others. Therefore, one may imagine that some workers had reached a better grasp of kinesiology than others, and therefore would use a more suitable work technique. Unfortunately, the study did not determine operator anthropometrics and training history, and therefore all these considerations cannot be substantiated by numbers, although they remain highly suggestive.

However, the study determined with some certainty the mean postural risk, as well as the specific postural risk, associated with individual tasks and specific work stations.

While mean postural risk has already been addressed in the opening of this section, here, one needs to make a specific remark on context. The mean postural risk figures reported in this study are valid for the operation of the debarker, but they do not include all ancillary operations that are part of post debarking duties. In particular, the study excluded any postures the operators assumed during delays. These delays easily represent 30% of the total worksite time [5], and generally consist of rest pauses or brief interruptions when moving away the processed product, or restocking the deck with new logs. During this time, workers are sedentary and they seldom perform any heavy physical activity, which is bound to reduce the overall postural risk.

In turn, that draws attention to the meaning of neutral tasks such as "wait" and "walk", and to their relation with productivity. In particular, "wait" implies inactivity, and the association of this task with a low postural risk may lead one to infer that postural risk increases with efficiency. In fact, classic delays that lead to inefficiency were excluded from the postural risk analysis, and therefore the tasks that do not describe a direct action on the machine or the work object are not necessarily implying an interruption of the production cycle. In fact, they simply indicate that the machine is producing without requiring a direct action from the worker, who might then be waiting until his intervention is again required. Similarly, walking to the machine or to the stacks occurs while the machine is engaging in its work, and does not imply any interruptions of the work cycle. Therefore, a higher frequency of "waiting" may simply describe an operation that is better organized and where the machine can work for a longer time without requiring direct worker action. That is clearly demonstrated by the absence of any meaningful correlation between productivity and PRI for the sample operations ($R^2 = 0.0049$).

At the same time, the association between specific tasks and postural risk offers some useful insights. The high risk associated with "carry" is expected, because moving around heavy objects is bound to involve some postural risk [22]. Less expected and more useful is the information about the high postural risk entailed by such tasks as "push" and "hold". Theoretically, these should be relatively easy tasks, as the post is supported by the debarker infeed chute. Yet, the study indicates that these tasks carry a significant postural risk, hinting at imperfect machine design. As a matter of fact, the infeed chute of small-scale debarkers is somewhat rudimentary and its architecture could be improved. Study results point to the need for an easily adjustable infeed chute, possibly equipped with a simple belt to assist with moving the post into the working mechanism, as available even on the cheapest firewood processors [23]. These features would certainly relieve worker effort and decrease the risk for WMSD.

Finally, the study offers compelling evidence for the different postural risk incurred at the two different work stations, stressing once more the importance of job rotation, which is made especially easy by the elementary skills required for manning both stations.

5. Conclusions

Post debarking involves a moderate postural risk, which is different for the two main work stations and varies with individual operations, based on their specific set up. However, debarker design could be improved, especially for what concerns the loading station, which is still too basic. Better machine design should be part of a more articulate strategy aimed at decreasing the postural risk of post debarking work sites. Future studies should also investigate the relationship between postural risk and worker anthropometrics, under the conditions of the current machine design with its rather inflexible settings, and possibly after developing improved designs that can be adjusted to fit operators with different anthropometric characteristics.

Acknowledgments: This project has been partly supported by the Bio Based Industries Joint Undertaking under the European Union's Horizon 2020 research and innovation program under grant agreement No. 720757 Tech4Effect. The Authors thank M. De Stefano (Conaibo—Italian Association of Logging Contractors) for recruiting suitable operators for the study.

Author Contributions: R.S. conceived and designed the experiment; R.S. and G.A. collected the data; R.S., F.D.F., and N.M. analyzed the data and wrote the paper.

Conflicts of Interest: The authors declare no conflicts of interest.

References

1. Mizaraite, D.; Mizaras, S. The formation of small-scale forestry in countries with economies in transition: Observations from Lithuania. *Small-Scale For.* **2005**, *4*, 437–450.
2. Spinelli, R.; Magagnotti, N.; Lombardini, C. Performance, capability and costs of small-scale cable yarding technology. *Small-Scale For.* **2010**, *9*, 123–135. [CrossRef]
3. Hanson, I.; Stuart, M. *Processing Trees on Farms: A Literature Review*; RIRDC Research Paper 97/20; Rural Industries Research and Development Corporation: Creswick, VIC, Australia, 1997; p. 58.
4. Lombardini, C.; Magagnotti, N.; Spinelli, R. Productivity and cost of industrial firewood processing operations. *For. Prod. J.* **2014**, *64*, 171–178. [CrossRef]
5. Spinelli, R.; Lombardini, C.; Aminti, G.; Magagnotti, N. Efficient debarking to increase value recovery in small-scale forestry operations. *Small-Scale For.* **2018**, in press. [CrossRef]
6. Lynch, D.; Mackes, K. *Opportunities for Making Wood Products from Small Diameter Trees in Colorado*; Research Paper RMRS-RP-37; USDA Forest Service, Rocky Mountain Research Station: Fort Collins, CO, USA, 2002; p. 11.
7. Eurostat. Database. Statistics on the Production of Manufactured Goods. Sold Volume Annual—2016. Available online: ec.europa.eu/eurostat (accessed on 4 January 2018).
8. Hermawati, S.; Lawson, G.; Sutarto, A. Mapping ergonomics application to improve SMEs working condition in industrially developing countries: A critical review. *Ergonomics* **2004**, *57*, 1771–1794. [CrossRef] [PubMed]
9. Spinelli, R.; Magagnotti, N.; Facchinetti, D. A survey of logging enterprises in the Italian Alps: Firm size and type, annual production, total workforce and machine fleet. *Int. J. For. Eng.* **2013**, *24*, 109–120.
10. Magagnotti, N.; Kanzian, C.; Schulmeyer, F.; Spinelli, R. A new guide for work studies in forestry. *Int. J. For. Eng.* **2013**, *24*, 249–253. [CrossRef]
11. Spinelli, R.; Laina-Relaño, R.; Magagnotti, N.; Tolosana, E. Determining observer method effects on the accuracy of elemental time studies in forest operations. *Balt. For.* **2013**, *19*, 301–306.
12. Genaidy, A.; Al-Shed, A.; Karwowski, K. Postural stress analysis in industry. *Appl. Ergon.* **1994**, *25*, 77–87. [CrossRef]
13. Kivi, P.; Mattila, M. Analysis and improvement of work postures in the building industry: Application of the computerized OWAS method. *Appl. Ergon.* **1991**, *22*, 43–48. [CrossRef]
14. Diego-Mas, J.; Poveda-Bautista, R.; Garzon-Leal, D. Influences of the use of observational methods by practitioners when identifying risk factors in physical work. *Ergonomics* **2015**, *58*, 1660–1670. [CrossRef] [PubMed]
15. De Bruijn, I.; Engels, A.; Van der Gulden, W. A simple method to evaluate the reliability of OWAS observations. *Appl. Ergon.* **1998**, *29*, 281–283. [CrossRef]
16. Kee, D.; Karwowski, K. A comparison of three observational techniques for assessing postural loads in industry. *Int. J. Occup. Saf. Ergon.* **2007**, *13*, 3–14. [CrossRef] [PubMed]
17. Zanuttini, R.; Cielo, P.; Poncino, D. Il metodo OWAS. Prime applicazioni nella valutazione del rischio di patologie muscolo-scheletriche nel settore forestale in Italia. *Forest@* **2005**, *2*, 242–255. (In Italian) [CrossRef]
18. Wahyudi, A.; Dania, W.; Silalahi, R. Work posture analysis of material handling using OWAS method. *Agric. Agric. Sci. Procedia* **2015**, *3*, 95–99. [CrossRef]

19. Calvo, A. Musculoskeletal disorders (MSD) risks in forestry: A case study to propose an analysis method. *Agric. Eng. Int.* **2009**, *11*, 1–9.
20. Lasota, A. Packer's workload assessment, using the OWAS method. *Logist. Transp.* **2013**, *18*, 25–32.
21. Spinelli, R.; Aminti, G.; De Francesco, F. Postural risk assessment of mechanized firewood processing. *Ergonomics* **2017**, *60*, 375–383. [CrossRef] [PubMed]
22. Olendorf, M.; Drury, G. Postural discomfort and perceived exertion in standardized box-holding postures. *Ergonomics* **2001**, *44*, 1341–1367. [CrossRef] [PubMed]
23. Manzone, M.; Spinelli, R. Efficiency and cost of firewood processing technology and techniques. *Fuel Process. Technol.* **2014**, *122*, 58–63. [CrossRef]

Article

Treatment of *Picea abies* and *Pinus sylvestris* Stumps with Urea and *Phlebiopsis gigantea* for Control of *Heterobasidion*

Kalle Kärhä [1],*, Ville Koivusalo [1,2], Teijo Palander [2] and Matti Ronkanen [1]

[1] Stora Enso Wood Supply Finland, P.O. Box 309, FI-00101 Helsinki, Finland;
    ville.koivusalo@storaenso.com (V.K.); matti.ronkanen@storaenso.com (M.R.)
[2] Faculty of Science and Forestry, University of Eastern Finland, P.O. Box 111, FI-80101 Joensuu, Finland;
    teijo.s.palander@uef.fi
* Correspondence: kalle.karha@storaenso.com; Tel.: +358-40-519-6535

Received: 31 January 2018; Accepted: 12 March 2018; Published: 15 March 2018

Abstract: *Heterobasidion* spp. root rot causes severe damage to forests throughout the northern temperate zone. In order to prevent *Heterobasidion* infection in summertime cuttings, stumps can be treated with urea or *Phlebiopsis gigantea*. In this study, the consumption of stump treatment materials and the quality of stump treatment work were investigated. A total of 46 harvesters were examined in May–November 2016 in Finland. The average stem size of softwood removal and softwood removal per hectare explained the consumption of stump treatment material. The quality of stump treatment work was good in the study. The best coverage was achieved with the stumps of 20–39 cm diameter at stump height (d_0). It can be recommended that the harvester operator self-monitors and actively controls his/her treatment result in cutting work and sets the stump treatment equipment in a harvester if needed. The results also suggested that when cutting mostly small- and medium-diameter ($d_0 \leq 39$ cm) conifers, the stump treatment guide bars with relatively few (<18) open holes are used, and at the harvesting sites of large-diameter trees, the guide bars with a relatively great (>27) number of open holes are applied.

Keywords: root rot; biotic factor; forest health; tree growth; stump protection; wood harvesting

---

1. Introduction

The root and butt rot fungus *Heterobasidion annosum* sensu lato (Fr.) Bref. is widely distributed in coniferous forests of the Northern Hemisphere, especially in Europe, North America, Russia, China and Japan [1]. There are three native *Heterobasidion annosum* species in Europe: (1) *Heterobasidion annosum* *sensu stricto* (*s.s.*) has a wide range of hosts and causes mortality to pines (*Pinus* spp.), especially Scots pine (*Pinus sylvestris* L.), and root and butt rot to Norway spruce (*Picea abies* (L.) Karst.) and Sitka spruce (*Picea sitchensis* (Bong.) Carr.). (2) *Heterobasidion parviporum* Niemelä and Korhonen causes root and butt rot to Norway spruce, and (3) *Heterobasidion abietinum* Niemelä and Korhonen causes disease to several Abies species in southern Europe [1,2].

Heterobasidion spp. root rot causes severe damage to forests throughout the northern temperate zone: In the European Union, annual losses attributed to growth reduction and degradation of wood are estimated at approximately €800 million [3,4]. In Finland, the damage caused by *Heterobasidion* spp. root rot for Norway spruce has been estimated to be approximately €40 million year^{-1} and some €5 million year^{-1} for Scots pine [5,6]. Climate change is thought to favor the living conditions and the spread of *Heterobasidion* spp. root rot [7,8]. In addition, shortening of winter lengthens the infection time of the spores of *Heterobasidion* spp. root rot and increases the proportion of summertime cuttings. Consequently, the prevention of *Heterobasidion* spp. root rot, as well as the obstruction of the spread of

Heterobasidion spp. root rot can be considered among the most significant challenges facing the modern forestry sector [9].

The pathogen of *Heterobasidion* spp. root rot infects fresh stumps after thinning and clear-cutting operations and spreads to neighboring trees via root-to-root contacts. In order to prevent *Heterobasidion* spp. root rot infection in summertime cuttings, stumps can be treated with urea that increases the pH of the stump surface, making it unsuitable for spore germination and preventing *Heterobasidion* spp. root rot from getting deeper into coniferous wood [10–16]. Alternatively, the stump surface can be covered with large amounts of the antagonistic fungus *Phlebiopsis gigantea* (Fr.) Julich, to prevent any pathogen spores that subsequently land on the stump surface to germinate [17–24].

According to the Plant Protection Product Register [25], four urea products are used in Finland: Moto-urea (license number: 3069), PS-kantosuoja-2 (1949), Teknokem Kantosuoja (3124) and Urea-kantokate (2928). Currently, the trademarks of biological control agents are Rotstop® (1648) and Rotstop® SC (2939) on the market in Finland [25]. The stump treatment areas have been annually 45,000–117,000 hectares in the 2010s in Finland [26,27].

The stump treatment with both urea and Rotstop reduces the basidiospore infection of *Heterobasidion* spp. root rot by an average of over 90% (cf. [28–33]). Achieving good pesticide efficacy requires careful treatment in order to wet the surface of the whole stump by spreading [31,34–37]. The effectiveness of prevention is reduced in relation to the uncovered area on the surface of the stump. Thus, the good coverage of stumps is an absolute prerequisite for high-quality stump treatment work [9].

According to the Government decree on the prevention of damage by *Heterobasidion* spp. root rot [38], *Heterobasidion* spp. root rot has to be prevented in mineral soils when the share of Norway spruce and Scots pine (i.e., conifers) of the total initial stand volume is more than 50% before wood harvesting operation and in peatland forests if the share of Norway spruce of the total initial stand volume is more than 50% before logging operation in Finland. In accordance with the Forest damage prevention act [39], the prevention of *Heterobasidion* spp. root rot must be carried out in thinnings and regeneration fellings in the risk zone of *Heterobasidion* spp. root rot between the beginning of May and the end of November in southern and central Finland (see Figure 1). Furthermore, the stump treatment has to be done for all conifer tree stumps of more than 10 cm in stump diameter (d_0) and the stump treatment material must cover at least 85% of the surface of each stump being treated [38]. Stump treatment is not required if any of the following conditions are met: (1) thermal growth season (i.e., the snow has melted in the opening places and the average daily air temperature has permanently raised more than +5 °C) has not started, (2) the air temperature of the wood harvesting day is below 0 °C, (3) there is a uniform snow cover on the ground, or (4) the lowest air temperature in the municipality of the harvesting site has been below 10 °C during the three-week period preceding the wood harvesting operation [38].

The stump treatment material is applied on the stump surface of coniferous trees using the harvester equipped with stump treatment facilities. Nowadays, the volumes of storage tanks in harvesters for the stump treatment material are typically around 100–150 dm³. The stump treatment material is pumped from the storage tank of a harvester to the harvester head whence it is discharged onto the stump surface of the conifer tree via holes spaced along the underside of the guide bar. There are pre-drilled (but not totally open) holes at a distance of 12–13 mm in the new stump treatment guide bar. Before bringing a new guide bar into use, the desired number of holes in the guide bar is opened by drilling with a 1.5 mm drill bit or hitting with a small spike. The number of pre-drilled holes in a guide bar depends on the length of the guide bar. For instance, the stump treatment guide bar of 75 cm in length has around 40 pre-drilled holes. When the length of the guide bar is 60 cm, the number of pre-drilled holes is typically less than 30 holes, and when the length of the guide bar is 90 cm, there are more than 50 holes in the guide bar.

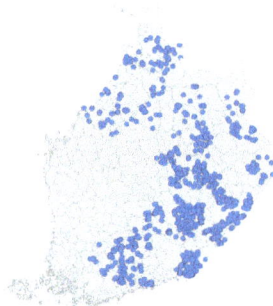

Figure 1. The distribution of harvesting sites (n = 1831) in the study. The gray color in the map displays the risk zone of the spread of *Heterobasidion* spp. root rot in Finland [39].

By means of the number and location of open holes in a guide bar and control systems for the treatment equipment of a harvester, the harvester operator can control the spraying of treatment material. Due to the variation in the stem size of removal in the forest stand, with smaller trees, some of the treatment materials often pass through the stump surface because the number of open holes in the guide bar usually has to be dimensioned according to the larger-diameter trees at a harvesting site [40].

There is only one report published in which the hectare-based consumption of stump treatment materials has been presented in Finland [41]. Mäkelä [41] estimated that the consumption of stump treatment material is around 40–60 dm^3 ha^{-1} in thinnings and approximately 50–90 dm^3 ha^{-1} in final cuttings. Mäkelä [41] forecasted his consumption figures of treatment product based on the number of stems cut and the total area of stump ends treated. The sales package labels of urea treatment products on the market promise that the consumption is 1.5–2.0 dm^3 m^{-2} of stump surface treated [42–45]. On the other hand, the sales package labels of Rotstop® and Rotstop® SC products give the following adequacy estimates: 0.33–0.68 dm^3 m^{-3} of softwood harvested or 25–150 dm^3 ha^{-1} [46,47].

Unfortunately, the current consumption figures presented in literature are not precise for using chemical and biological controls against *Heterobasidion* spp. root rot. Therefore, Stora Enso Wood Supply Finland (WSF) and the University of Eastern Finland carried out the study on stump treatment against *Heterobasidion* spp. root rot in Finland. The aims of the study were to produce more accurate information about stump treatment and to clarify the following:

- the consumption of stump treatment materials and
- the quality of stump treatment work (i.e., the coverage of stumps treated).

2. Materials and Methods

2.1. Data on the Consumption of Stump Treatment Materials

The consumption of stump treatment materials in 46 harvesters was collected in May–November 2016 in Finland at the harvesting sites of Stora Enso WSF. There were 25 Ponsse (Beaver, Ergo,

Fox, Scorpion and Scorpion King), 14 John Deere (1070D, 1070E, 1170E, 1270D, 1270E and 1270G), 5 Komatsu/Valmet (901, 901TX, 901TX.1, 911.4 and 911.5), 1 Logset (8H GTE) and 1 ProSilva (810) harvesters in the study. Since the harvesters of the study did not have the technology to perform automatic measuring of the consumption of stump treatment material, the consumption of treatment materials was manually measured by the harvester operators with recording forms. The measurement methods used by the operator differed between the harvesters of the study: Some operators measured the consumption of treatment materials when filling up the storage tank of a harvester by measuring the amount of substance added by a flow meter or by the signs in the storage tank. Some operators used a dipstick. All methods aimed at a minimum accuracy of five dm^3 measurement^{-1}.

There were 40 harvesters which used only urea as a stump treatment product in the study and only Rotstop® SC suspension was used in four harvesters. Furthermore, both urea and Rotstop® SC were used in two harvesters. In total, the stump treatment materials were measured to spread 309,427 dm^3 during the study period. Of this volume, three urea products (i.e., Moto-urea, PS-kantosuoja-2 and Teknokem Kantosuoja) accounted for 272,754 dm^3 (88.1%) and the share of Rotstop® SC was 36,673 dm^3 (11.9%).

The harvesting site-specific harvester production data (i.e., prd files [48]) provided the stand information, which was collected from the enterprise resource planning (ERP) system of Stora Enso WSF. The prd files were received for a total of 1831 harvesting sites. The prd files included the volume, number and average stem size of removal by tree species, as well as a cutting method. In addition, the hectare-based consumption figures for harvesting sites were calculated using the harvesting instruction maps of logging areas. If there was some indication of an abnormality in the implementation of the harvesting site cut in the prd file, the hectare-based consumption was not calculated for such harvesting sites. The geographical distribution of harvesting sites in the study is illustrated in Figure 1.

The total removal volume of softwood trees at the harvesting sites of the study was 587,120 m^3 solid over the bark (later only: m^3). The share of Norway spruce removal was 320,257 m^3 (54.5%) and the share of Scots pine was 266,863 m^3 (45.5%), and a total of 2,413,256 softwood trees were cut. Most of the softwood volume was cut from clear cuttings (59.3%) and later thinnings (27.9%). From first thinnings, softwood was felled 5.8% of the total softwood volume, 4.5% from seeding fellings and 2.3% from other fellings (i.e., cuttings of hold-over stands, shelterwood fellings and special cuttings).

GB, Iggesund, John Deere, Komatsu, Oregon and Ponsse guide bars were used in the harvesters of the study. The most commonly used guide bar trademark was the Iggesund by which in total 51.9% of the total softwood volume harvested was cut. The share of GB guide bars was 23.8% and with Oregon bars it was 15.3% of the total softwood removal cut in the study. The length of guide bars varied between 50 and 95 cm. From the total softwood removal, the majority (71.8%) was cut by the guide bars of 75 cm in length. The average number of open holes in stump treatment guide bars was 22.5 holes with the variation range of 3–41 holes. The study also detected the effect of the number of open holes in a stump treatment guide bar on the consumption of stump treatment material. In total, the harvester operators recorded the number of open holes in the guide bar for 1808 harvesting sites on the data collection forms. The volumes of softwood cut with the different number of open holes are described in Figure 2.

Moreover, the influence of the adjustment habits by harvester operator on the consumption of stump treatment material was investigated. The options for adjusting the stump treatment equipment (i.e., timing and duration in spraying and spreading pressures) in the interviews of harvester operators were as follows:

- By harvesting site,
- By cutting method,
- After detecting weak stump coverage in spraying or
- Never.

All harvester operators of the study ($n = 68$) were interviewed at the beginning of the study period (May 2016) and at the end of the study (October–November 2016). The adjustment habits of the operators, as well as the other study experiences and observations (i.e., Was it easy to measure the consumption of stump treatment material? Did the operator achieve the target accuracy set in his consumption measurements? In what kind of harvesting sites were there lots of problems with the coverage of stump surfaces in the treatment work?) were asked in the operator interviews. If the adjustment habits of the operators at the same harvester differed from each other, the harvester was classified into a group based on the harvester operator's response to most adjustments. The number of harvesters and harvesting sites in different adjustment classes are given in Table 1.

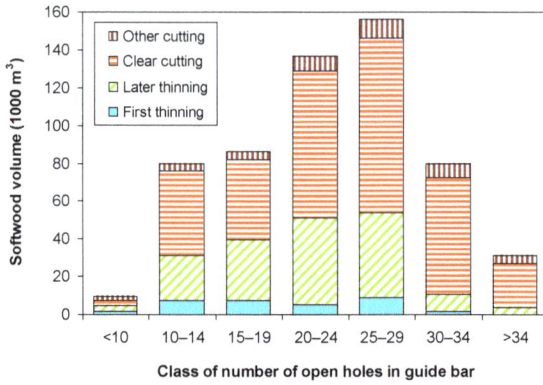

Figure 2. The distribution of the number of open holes in harvesters' stump treatment guide bars by cutting method in the study.

Table 1. The number of harvesters and harvesting sites in the different adjustment habit classes by harvester operator in the study.

Adjustment Habit Class	Number of Harvesters	Number of Harvesting Sites
By harvesting site	0	0
By cutting method	12	490
After detecting weak stump coverage in spraying	19	726
Never	15	615
Total	46	1831

2.2. Coverage Data

The quality of stump treatment work was evaluated with all harvesters of the study by inventorying the coverage of stump treatment on the stump surfaces of conifer trees cut after the stump treatment work. The goal was to make three coverage inventories for each harvester during the study period. Besides, the aim was to conduct one coverage inventory for each main cutting method (i.e., first thinning, later thinning and clear cutting) with each study harvester. The inventory of different cutting methods was done to ensure that the coverage of stump treatment would be valid on the stumps of different diameter within all harvesters involved in the consumption study.

The coverage of stump treatment material on the stump surface can be detected by the dye of the treatment material. The uncovered area of the entire stump surface by stump treatment material was estimated by using a transparent plastic measuring plate (Figure 3).

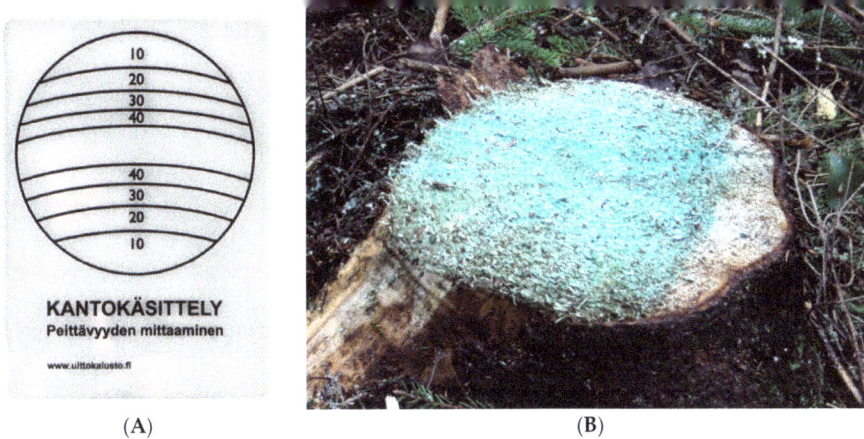

Figure 3. (**A**) A transparent measuring plate used in the study. By changing the distance of the transparent measuring plate above the stump, the focal length is selected by combining the edges of the stump and the ring of the measuring plate. Based on the relative proportions of the plate, it is possible to determine the relative proportion of the uncovered area of the stump surface. Photo courtesy of Uittokalusto Ltd. (**B**) The stump with the uncoverage rate of around 11–13% (not the blue area). Photo courtesy of Kalle Kärhä.

In each coverage inventory, the target was to measure 50 stumps [49,50]. In accordance with the Guidelines for inventorying the coverage of stump treatment prepared for the study, the stumps were measured via cluster sampling on the longest line of each logging area. From the line, the five closest conifer tree stumps were measured at the distance of ten meters from ten places, with a total sample size of 50 stumps. The stump diameter (d_0) and coverage percentage (i.e., coverage rate) of each stump selected for the inventory were recorded on the Inventorying form of the coverage of stump treatment (cf. [49,50]). The quality of stump treatment work was evaluated on the basis of the criteria of the Finnish Forest Centre [50], i.e., 85% or more of the stump surface of the approved stump should have been covered. Contrary to the consumption data, the quality inventories of stump treatment were carried out at a logging area-specific level (i.e., logging area may consist of one or several harvesting sites) instead of the harvesting site-specific measurements of consumption.

After inventorying the coverage of stumps, the percentages below 85% covered stumps were calculated on the form. When the sample was 50 stumps in the inventories, the deduction percentage was calculated by multiplying the number of uncovered stumps by two. The evaluation based on the deduction percentage was given to the quality of stump treatment work as follows:

- The deduction percentages of 0–9% marked a good level of coverage,
- 10–29% a satisfactory level and
- 30–100% marked an ineligible level of coverage [50].

The quality inventories of stump treatment were performed by a responsible wood harvesting officer at Stora Enso WSF for each study harvester. The quality inventories made by the harvester operators themselves were not used in the study. When all harvesters did not cut in the stands of all three main cutting methods (i.e., first thinning, later thinning and clear cutting), several inventories for the same cutting method were conducted with some harvesters. In total, 144 quality inventories (27 in first-thinning stands, 65 in later thinnings and 52 in clear cuttings) were carried out in the study. The final coverage data was 7042 stumps (Figure 4).

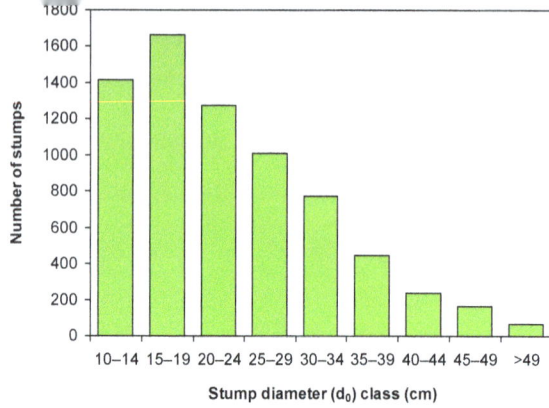

Figure 4. The frequency distribution of stumps ($n = 7042$) inventoried for the final coverage data of the study.

2.3. Analysis of Study Materials

The harvesting site-specific data on the consumption of stump treatment products, as well as the coverage data of the stumps inventoried were initially tested for normal distribution assumption by a Kolmogorv–Smirnov test. Based on the results of the test, the consumption and coverage data did not comply with normal distribution. Since the material was not distributed normally, the non-parametric methods were applied in the statistical analysis of the study. For a comparison of multiple samples in the study, a Kruskal–Wallis one-way ANOVA (χ^2) test was used and for comparison of two samples a Mann–Whitney (U) test was used.

The consumption (dm^3 m^{-3} of softwood, and dm^3 ha^{-1}) models of stump treatment material were formulated using regression analysis with the average stem size of softwood removal, softwood removal ha^{-1}, the density of softwood removal, treatment product dummy (1, if urea, 0, when Rotstop® SC), the number of open holes in a guide bar, and the dummy variables of operators' adjustment habits of treatment equipment (Adj_Dum_1: 1, if by cutting method, otherwise 0; Adj_Dum_2: 1, if after detecting weak stump coverage, otherwise 0; Adj_Dum_3: 1, if never, otherwise 0) as the independent variables. The different transformations and curve types were tested in order to achieve symmetrical residuals for the regression models and in order to ensure the statistical significance of the coefficients. All statistical analyses were conducted with IBM SPSS Statistics 21 software.

3. Results

3.1. Consumption of Stump Treatment Materials

The study results indicated that the consumption of stump treatment material depends significantly on the average stem size of softwood removal at the harvesting site (Figure 5). The consumption of stump treatment material was, on average, 1.09 dm^3 m^{-3} of softwood cut in first-thinning stands (the average stem size of softwood removal in the stand 83 dm^3), 0.72 dm^3 m^{-3} of softwood in later thinnings (154 dm^3), 0.39 dm^3 m^{-3} of softwood in clear-cutting stands (423 dm^3) and 0.43 dm^3 m^{-3} of softwood in other cuttings (i.e., seeding fellings, cuttings of hold-over stands, shelterwood fellings and special cuttings) (355 dm^3).

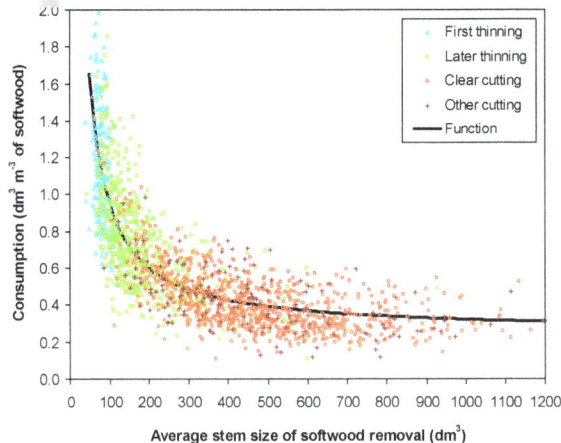

Figure 5. The consumption observations of stump treatment material as a function of the average stem size of softwood removal by cutting method, as well as the predicted consumption curve (cf. Table 2).

Table 2. Regression model for the consumption (dm^3 m^{-3} of softwood) of stump treatment material.

$y_1 = a + b/x_1$			
Adjusted R^2 = 0.625; F Value = 3056 ***; Standard Error of the Estimate of the Model = 0.215			
Coefficient	Estimate of Coefficient	Standard Error of Estimate	t-Value
a	0.260	0.008	31.469 ***
b	72.019	1.303	55.279 ***

Note: y_1 = consumption (dm^3 m^{-3} of softwood); x_1 = average stem size of softwood removal (dm^3); a = constant; b = coefficient of the variable; * $p < 0.05$; ** $p < 0.01$; *** $p < 0.001$.

In later thinnings and clear cuttings, the treatment product (i.e., urea and Rotstop® SC) used, the number of open holes in the stump treatment guide bar and the operators' adjustment habits of treatment equipment had a statistically significant effect on the consumption of stump treatment material in the study. The highest consumption was measured with urea, and when there were only a few open holes (<18 holes) in a guide bar and the harvester operator adjusted greatly (i.e., by cutting method) the stump treatment equipment in a harvester (Table 3). However, the impact of treatment product, the number of open holes, and the adjustment habits of operators on the consumption of treatment material was significantly lower than the influence of the average stem size and even lower than that of the cutting method (Figure 5, Table 3).

Table 3. The average consumption of stump treatment material by cutting method in the study.

Variable	Cutting Method			Statistically Significant Differences between the Variables by Cutting Method (FT, LT and CC)
	First Thinning (FT)	Later Thinning (LT)	Clear Cutting (CC)	
	Consumption (dm^3 m^{-3} of Softwood)			
Treatment product				
Urea	1.09	0.72	0.40	LT: *;
Rotstop® SC	1.14	0.71	0.31	CC: ***
Number of open holes in guide bars				
<18 (a)	1.11	0.82	0.41	LT: a–b ***, a–c ***;
18–27 (b)	1.06	0.68	0.41	CC: a–b *, a–c **
>27 (c)	1.31	0.64	0.35	
Adjustment habits by operator				
By cutting method (a)	1.13	0.80	0.40	LT: a–b **, a–c ***, b–c *;
After detecting weak coverage (b)	1.07	0.71	0.39	CC: a–c **
Never (c)	1.08	0.65	0.38	

Note: * $p < 0.05$; ** $p < 0.01$; *** $p < 0.001$.

When modelling the consumption (dm^3 m^{-3} of softwood) of stump treatment material, the average stem size of softwood removal in the stand best explained the consumption (Table 2). The coefficient of determination (adjusted R^2) of the consumption model was 62.5%. Other independent variables were also tested in the model, but they did not significantly increase the coefficient of determination of the consumption model (Table 2). The residuals of the model centered on zero and were symmetrical throughout the range of the average stem size observations.

The average hectare-based consumption of stump treatment material was 51.0 dm^3 ha^{-1} in first thinnings (the average softwood removal at the harvesting site 46 m^3 ha^{-1} and the average density of softwood removal 558 trees ha^{-1}), 44.6 dm^3 ha^{-1} in later thinnings (63 m^3 ha^{-1} and 402 trees ha^{-1}), 80.8 dm^3 ha^{-1} in clear cuttings (210 m^3 ha^{-1} and 491 trees ha^{-1}) and 58.9 dm^3 ha^{-1} in other cuttings (140 m^3 ha^{-1} and 409 trees ha^{-1}) (Figure 6).

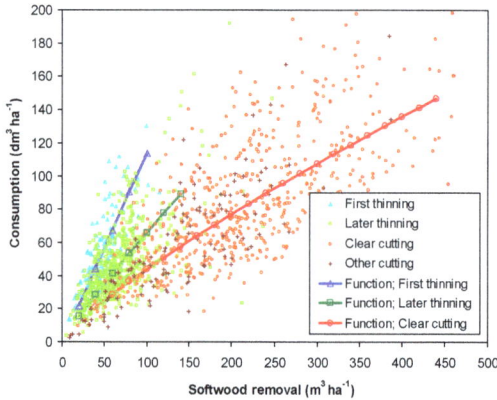

Figure 6. The hectare-based consumption observations of stump treatment material as a function of softwood removal per hectare and the predicted consumption functions by cutting method (cf. Table 4).

Table 4. Regression models for the hectare-based consumption of stump treatment material.

$y_2 = x_2{}^b$			
FT: Adjusted R^2 = 0.628; F Value = 226 ***; Standard Error of the Estimate of the Model = 0.312			
LT: Adjusted R^2 = 0.469; F Value = 446 ***; Standard Error of the Estimate of the Model = 0.397			
CC: Adjusted R^2 = 0.529; F Value = 625 ***; Standard Error of the Estimate of the Model = 0.342			
Cutting Method/Coefficient	Estimate of Coefficient	Standard Error of Estimate	t-Value
FT/b	1.027	0.007	145.803 ***
LT/b	0.909	0.004	213.822 ***
CC/b	0.819	0.003	302.811 ***

Note: y_2 = consumption (dm^3 ha^{-1}); x_2 = softwood removal (m^3 ha^{-1}); b = coefficient of the variable; FT = first thinning; LT = later thinning; CC = clear cutting; * $p < 0.05$; ** $p < 0.01$; *** $p < 0.001$.

The best hectare-based consumption models of stump treatment material by cutting method were achieved when the softwood removal hectare^{-1} was the independent variable in the models (Table 4). The residuals of the hectare-based consumption models also distributed symmetrically.

3.2. Quality of Stump Treatment Work

The coverage inventories showed that the quality of stump treatment work was good in the study: 72.2% of the coverage inventories indicated that the work quality was good. Correspondingly, 26.4% of stump treatment work was classed as satisfactory. Only 1.4% of the total stump treatment work inventories provided an ineligible result.

The proportion of less than 85% covered (i.e., not approved) stumps measured in the total coverage data was 6.6% and the proportion of 85% or better covered stumps was 93.4%. When analyzing the coverage by stump diameter class, it could be noted that the highest coverage was achieved with the stumps of 20–39 cm (Figure 7). The coverage of the smaller- (<20 cm) and larger-diameter (>39 cm) stumps inventoried was significantly lower (χ^2 = 35.5; $p < 0.001$) than the stumps of 20–39 cm.

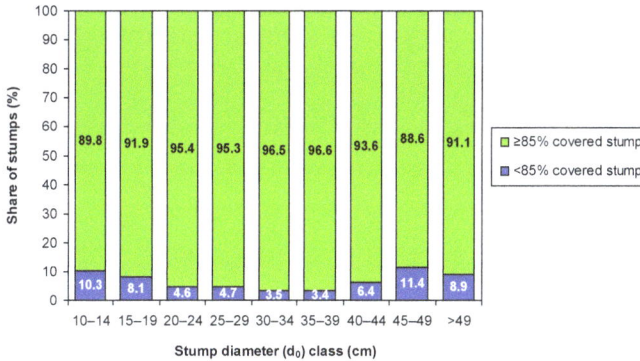

Figure 7. The shares of <85% and ≥85% covered stumps inventoried by stump diameter class.

In this study, the average coverage rate (i.e., the coverage percentage of all stumps inventoried) was 94.9% in first thinnings, 94.3% in later thinnings, and 95.1% in clear-cutting stands. The cutting methods differed significantly in the quality of stump treatment work for unequal stumps: In clear cuttings, the coverage rate with small-diameter (<20 cm) stumps was significantly lower (90.7%) than in first and later thinnings (94.4% and 93.8%, respectively) (Table 5). Correspondingly, in first-thinning stands, the coverage rate of stumps treated was good with both small (<20 cm) and medium-sized

(20–39 cm) stumps. With the larger-sized (>39 cm) stumps, the coverage rate was the highest (93.9%) in clear cuttings (Table 5).

Table 5. The average coverage rates by stump diameter class in the study.

| Variable | Stump Diameter (d_0) Class (cm) | | | Statistically Significant Differences between the Variables by Stump Diameter Class (S, M and L) |
| | 10–19 ("Small") | 20–39 ("Medium") | >39 ("Large") | |
	Coverage Rate (%)			
Cutting method				
First thinning (a)	94.4	96.8	-	S: a–c ***, b–c ***;
Later thinning (b)	93.8	95.1	91.2	M: a–b ***, b–c ***;
Clear cutting (c)	90.7	96.4	93.9	L: b–c **
Treatment product				
Urea	93.5	95.8	93.3	
Rotstop® SC	94.7	96.4	95.8	
Number of open holes in guide bars				
<18 (a)	93.9	96.5	92.5	M: a–b ***, a–c **;
18–27 (b)	93.3	95.3	91.5	L: a–c *, b–c **
>27 (c)	93.2	95.8	94.9	
Adjustment habits by operator				
By cutting method (a)	94.1	96.1	92.7	S: a–b ***, a–c ***, b–c ***;
After detecting weak coverage (b)	91.9	95.5	95.8	M: a–b **, b–c ***;
Never (c)	95.4	96.0	91.3	L: a–b ***, b–c ***

Note: S = "Small"; M = "Medium"; L = "Large"; * $p < 0.05$; ** $p < 0.01$; *** $p < 0.001$.

When clarifying the effect of the number of holes in a guide bar on the quality of treatment work, the best coverage rate was obtained with small- and medium-sized stumps when the guide bar was perforated with relatively few (<18) open holes, and with larger-sized (>39 cm) stumps when the guide bar was equipped with a relatively great (>27) number of open holes (Table 5). When investigating the influence of the operator's adjustment habits of treatment equipment, it could be noticed that the highest coverage rate was achieved as follows:

- with small (<20 cm) stumps when the harvester operator did not adjust the stump treatment equipment of the harvester at all (95.4%),
- with medium-sized (20–39 cm) stumps when the operator adjusted the treatment equipment in the harvester by cutting method (96.1%) and
- with large-diameter (>39 cm) stumps when the operator sets the treatment equipment after detecting weak stump coverage in spraying (95.8%) (Table 5).

4. Discussion and Conclusions

The data for the consumption of stump treatment material was almost 0.6 million m³ of softwood and more than 2.4 million softwood trees cut with 46 harvesters, and the stump treatment material was spread more than 300,000 dm³. The consumption data was hence relatively large. The study produced fresh data on the consumption of stump treatment materials. Among other things, novel consumption information is needed to define the equitable payments of stump treatment work for forest machine contractors. Besides, our consumption figures can be utilized when estimating and modelling the profitability of stump treatment against *Heterobasidion* spp. root rot [51–55].

In the study, measurement of the consumption of stump treatment material was challenging, as there was no technology for automatically measuring the consumption of treatment product in the study harvesters. The consumption of treatment products was measured using many measuring methods according to the alternative options used in the harvesters of the study, as well as the preferences of the operators. All methods aimed at a minimum accuracy of five dm³ per measurement. On the basis of operator interviews, each operator thought that he achieved a set target for the measurement accuracy. Nevertheless, in the near future, forest machine manufacturers should seriously consider equipping their harvesters with the automatic standard measurement system to verify the real-time and total consumption of stump treatment material at the harvesting site, as nowadays measuring the fuel consumption in modern harvesters is important.

Currently, the volumes of storage tanks in harvesters for the stump treatment material are typically 100–150 dm³. The storage tanks are sufficient for a single work-shift cutting in thinnings and clear cuttings (Table 6). However, efficient cutting in a double work-shift system calls for continuous cutting work, without visiting the roadside landing to fill up the stump treatment tank of a harvester between work shifts. In thinnings, the stump treatment tank of a harvester must be around 150 dm³ and in clear cuttings more than 150 dm³ for double work-shift cutting work (Table 6). Hence, forest machine manufacturers should construct larger storage tanks for the stump treatment material in harvesters in the future.

Table 6. Calculation of the sufficient volumes of storage tanks for the stump treatment material in harvesters working in one and double work shifts by cutting method.

	First Thinning	Later Thinning	Clear Cutting
Consumption of stump treatment material (dm³ m⁻³ of softwood) [1]	1.09	0.72	0.39
Cutting productivity (m³ of softwood SMH⁻¹)	7.5 [2]	12.5 [3]	28.5 [3]
Consumption of stump treatment material (dm³)			
In one work shift [4]	57.2	63.0	77.8
In two work shifts [5]	114.5	126.0	155.6

Note: [1] Average consumptions of stump treatment material in this study; [2] Cutting productivity in first thinnings = m³ per scheduled machine hour (SMH) by Kärhä et al. [56]; [3] Cutting productivity in later thinnings and clear cuttings by Eriksson and Lindroos [57]; [4,5] It was assumed that there are 7.0 SMHs in one work shift and 14.0 SMHs in two work shifts.

The study results showed that the average stem size of softwood removal in the stand has a significant effect on the consumption (dm³ m⁻³ of softwood) of stump treatment material. Furthermore, the softwood removal hectare⁻¹ by cutting method explained the hectare-based consumption of stump treatment material in the study. The average consumption of stump treatment material was 51 dm³ ha⁻¹ in first thinnings, 45 dm³ ha⁻¹ in later thinnings and 81 dm³ ha⁻¹ in clear cuttings. The results of the study were in line with the calculations by Mäkelä [41]: the consumption was 40–60 dm³ ha⁻¹ in thinnings and 50–90 dm³ ha⁻¹ in clear cuttings.

Many *Heterobasidion* researches [31,34–37] have pointed out that achieving good pesticide efficacy requires careful stump treatment in order to wet the surface of the whole stump by spreading, and the effectiveness of prevention work is reduced in relation to the uncovered area on the surface of the stump. Therefore, our target must invariably be a high-quality stump treatment. On the basis of the study results, it can be recommended that the harvester operator self-monitors and actively controls his/her treatment result in cutting work, especially operating in large-diameter forest stands, sets the stump treatment equipment in the harvester if needed, subsequently achieving a high-quality result in his/her stump treatment work.

It must be noted that, in this study, many harvester operators stated that they do not set stump treatment equipment in their harvesters at all. In fact, one-third of the harvesters were categorized in the group of "Never adjustments", i.e., no settings for the stump treatment equipment in a harvester (cf. Table 1). Thus, we need better education and communication concerning the significance of high-quality stump treatment work, active and continuous self-monitoring treatment result, and setting the stump treatment equipment of the harvester if needed. Oliva et al. [58] have underlined that it is essential to treat the large-sized stumps very carefully because the probability of stump-to-tree spread of *Heterobasidion* spp. root rot depends significantly on the diameter of the stump.

The coverage rate by cutting method was best in clear cuttings, but the difference between clear cuttings and thinnings was very small. Consequently, the stump treatment work can be considered successful and uniform with all cutting methods in the study. There was no significant difference between biological (Rotstop® SC) and chemical (urea) controls used in the coverage rates of stump treatment work. However, it must be noted that there were only six Rotstop® SC harvesters of the total 46 harvesters in the study, and from the total softwood volume cut in the study the proportion of Rotstop® SC was only 12%.

According to the statistics of the Finnish Forest Centre, the shares of good logging areas related to stump treatment work have been annually 73.9–76.7% in the coverage inventories in 2012–2016, and the shares of satisfactory and ineligible results in stump treatment work have been 17.1–24.3% and 2.9–7.1%, respectively [59]. In this study, the distribution of the treatment work results was as follows: good 72.2%, satisfactory 26.4% and ineligible 1.4%. Hence, in this study, the proportions of good and ineligible results were slightly lower and on the other hand the share of satisfactory logging areas was higher compared to the figures of the whole of Finland by Leivo [59] in coniferous forests in recent years.

Correspondingly, in this study, the share of less than 85% covered (i.e., not-approved) stumps measured was 6.6%. The Finnish Forest Centre has reported that the share of not-approved stumps of the total stumps inventoried was, on average, 10.2% in 2014 and 9.2% in 2015 in Finland [60,61]. Thus, the share of not-approved stumps in this study was smaller to the whole of Finland.

Based on the study results, the quality of stump treatment work can be found to be the best with the medium-sized (20–39 cm) stumps, and the coverage rate with the smaller (<20 cm) and larger (>39 cm) stumps was slightly lower than with the medium-sized stumps. The number of open holes in stump treatment guide bars had an impact on the quality of treatment work when cutting different sized coniferous trees. Accordingly, it can be concluded that in the stands of mostly small- and medium-diameter ($d_0 \leq 39$ cm) conifers, the treatment guide bars with relatively few (<18) open holes are used, and at the harvesting sites of large-diameter trees, the guide bars with a relatively great (>27) number of open holes are applied.

Several harvester operators interviewed underlined that the stump treatment is most difficult in the coniferous stands in which there is great variation in the stem size of removal. Especially in the case of larger-diameter clear cuttings, the stump treatment of small-sized stumps is very challenging. To sum up, since the adjustments of the controlling system of treatment equipment and the open holes in the treatment guide bar have to be decided in accordance with the dominant trees in the stand, nowadays there are difficulties to spray the divergent stumps perfectly. In the future, forest machine manufacturers could develop more advanced controlling systems of stump treatment for their harvesters, for instance self-adaptive spraying systems according to the stem size to be felled. This kind of self-adaptive spraying system requires, however, machine vision or mobile laser scanning systems on the harvesters to inform the controlling stump treatment system of the size of the next tree to be cut (cf. [62–65]).

Because the consumption data was measured as harvesting site-specific and the coverage data as logging area-specific, there were no possibilities to merge the materials and to compare more comprehensively the consumption and coverage data in the study. Consequently, a further study on the consumption and coverage could be performed to optimize the consumption of stump treatment material subjected to the high-quality coverage rate in the coniferous forests.

Acknowledgments: This study was supported by Stora Enso Wood Supply Finland. The research did not receive any other funding.

Author Contributions: K.K., V.K., T.P. and M.R. conceived and designed the experiments; V.K. performed the experiments; K.K. and V.K. analyzed the data; K.K., V.K. and T.P. wrote the paper.

Conflicts of Interest: The authors declare no conflict of interest.

References

1. Garbeletto, M.; Gonthier, P. Biology, Epidemiology, and Control of *Heterobasidion* Species Worldwide. *Annu. Rev. Phytopathol.* **2013**, *51*, 39–59. [CrossRef] [PubMed]
2. Korhonen, K.; Capretti, P.; Karjalainen, R.; Stenlid, J. Distribution of *Heterobasidion annosum* intersterility groups in Europe. In *Heterobasidion annosum: Biology, Ecology, Impact and Control*; Woodward, S., Stenlid, J., Karjalainen, R., Hüttermann, A., Eds.; CAB International: Wallingford, UK, 1998; pp. 93–104, ISBN 085199 275 7.
3. Woodward, S.; Stenlid, J.; Karjalainen, R.; Hüttermann, A. (Eds.) *Heterobasidion annosum: Biology, Ecology, Impact and Control*; CAB International: Wallingford, UK, 1998; 589p, ISBN 085199 275 7.
4. Asiegbu, F.O.; Adomas, A.; Stenlid, J. Conifer root and butt rot caused by *Heterobasidion annosum* (Fr.) Bref. s.l.. *Mol. Plant Pathol.* **2005**, *6*, 395–409. [CrossRef] [PubMed]
5. Müller, M.M.; Piri, T.; Hantula, J. Ilmaston lämpeneminen haastaa nykyistä tehokkaampaan juurikäävän torjuntaan (Global warming is challenging more effective control of *Heterobasidion* spp. root rot). *Metsätieteen Aikakauskirja* **2012**, *4*, 312–315. [CrossRef]
6. Finnish Forest Centre. Juurikäävän Torjunta (Prevention of *Heterobasidion* spp. root rot in Finland). Available online: https://www.metsakeskus.fi/juurikaavan-torjunta (accessed on 28 January 2018).
7. La Porta, N.; Capretti, P.; Thomsen, I.M.; Kasanen, R.; Hietala, A.M.; Von Weissenberg, K. Forest pathogens with higher damage potential due to climate change in Europe. *Can. J. Plant Pathol.* **2008**, *30*, 177–195. [CrossRef]
8. Müller, M.M.; Sievänen, R.; Beuker, E.; Meesenburg, H.; Kuuskeri, J.; Hamberg, L.; Korhonen, K. Predicting the activity of *Heterobasidion parviporum* on Norway spruce in warming climate from its respiration rate at different temperatures. *For. Pathol.* **2014**, *44*, 325–336. [CrossRef]
9. Kärhä, K.; Koivusalo, V.; Palander, T.; Ronkanen, M. Stump Treatment with Chemical and Biological Controls against *Heterobasidion* spp. Root Rot in Coniferous Forests of Finland. In Proceedings of the FORMEC 2017, 50th anniversary of the International Symposium on Forestry Mechanization, Brașov, Romania, 25–29 September 2017.
10. Vollbrecht, G.; Jørgensen, B.B. The effect of stump treatment on the spread rate of butt rot in *Picea abies* in Danish permanent forest yield research plots. *Scand. J. For. Res.* **1995**, *10*, 271–277. [CrossRef]
11. Brandtberg, P.-O.; Johansson, M.; Seeger, P. Effects of season and urea treatment on infection of stumps of *Picea abies* by *Heterobasidion annosum* in stands on former arable land. *Scand. J. For. Res.* **1996**, *11*, 261–268. [CrossRef]
12. Pratt, J.E.; Redfern, D.B. Infection of Sitka spruce stumps by spores of *Heterobasidion annosum*: Control by means of urea. *Forestry* **2001**, *74*, 73–78. [CrossRef]
13. Johansson, S.M.; Pratt, J.E.; Asiegbu, F.O. Treatment of Norway spruce and Scots pine stumps with urea against the root and butt rot fungus *Heterobasidion annosum*—possible modes of action. *For. Ecol. Manag.* **2002**, *157*, 87–100. [CrossRef]
14. Roy, G.; Laflamme, G.; Bussières, G.; Dessureault, M. Field tests on biological control of *Heterobasidion annosum* by *Phaeotheca dimorphospora* in comparison with *Phlebiopsis gigantea*. *For. Pathol.* **2003**, *33*, 127–140. [CrossRef]
15. Oliva, J.; Samils, N.; Johansson, U.; Bendz-Hellgren, M.; Stenlid, J. Urea treatment reduced *Heterobasidion annosum s.l.* root rot in *Picea abies* after 15 years. *For. Ecol. Manag.* **2008**, *255*, 2876–2882. [CrossRef]
16. Wang, L.Y.; Pålsson, H.; Ek, E.; Rönnberg, J. The effect of *Phlebiopsis gigantea* and urea stump treatment against spore infection of *Heterobasidion* spp. on hybrid larch (*Larix × eurolepis*) in southern Sweden. *For. Pathol.* **2012**, *42*, 420–428. [CrossRef]
17. Pratt, J.E.; Niemi, M.; Sierota, Z.H. Comparison of Three Products Based on *Phlebiopsis gigantea* for the Control of *Heterobasidion annosum* in Europe. *Biocontrol. Sci. Technol.* **2000**, *10*, 467–477. [CrossRef]
18. Pettersson, M.; Rönnberg, J.; Vollbrecht, G.; Gemmel, P. Effect of Thinning and *Phlebiopsis gigantea* Stump Treatment on the Growth of *Heterobasidion parviporum* Inoculated in *Picea abies*. *Scand. J. For. Res.* **2003**, *18*, 362–367. [CrossRef]
19. Vasiliauskas, R.; Lygis, V.; Thor, M.; Stenlid, J. Impact of biological (Rotstop) and chemical (urea) treatments on fungal community structure in freshly cut *Picea abies* stumps. *Biol. Control* **2004**, *31*, 405–413. [CrossRef]

20. Annesi, T.; Curcio, G.; D'Amico, L.; Motta, E. Biological control of *Heterobasidion annosum* on *Pinus pinea* by *Phlebiopsis gigantea*. *For. Pathol.* **2005**, *35*, 127–134. [CrossRef]

21. Nicolotti, G.; Gonthier, P. Stump treatment against *Heterobasidion* with *Phlebiopsis gigantea* and some chemicals in *Picea abies* stands in the western Alps. *For. Pathol.* **2005**, *35*, 365–374. [CrossRef]

22. Drenkhan, T.; Hanso, S.; Hanso, M. Effect of the Stump Treatment with *Phlebiopsis gigantea* against *Heterobasidion* Root Rot in Estonia. *Balt. For.* **2008**, *14*, 16–25.

23. Oliva, J.; Thor, M.; Stenlid, J. Long-term effects of mechanized stump treatment against *Heterobasidion annosum* root rot in *Picea abies*. *Can. J. For. Res.* **2010**, *40*, 1020–1033. [CrossRef]

24. Rönnberg, J.; Cleary, M.R. Presence of *Heterobasidion* infections in Norway spruce stumps 6 years after treatment with *Phlebiopsis gigantea*. *For. Pathol.* **2012**, *42*, 144–149. [CrossRef]

25. Finnish Safety and Chemicals Agency (Tukes). Kasvinsuojeluainerekisteri (Plant Protection Product Register). Available online: https://kasvinsuojeluaineet.tukes.fi/ (accessed on 28 January 2018).

26. Natural Resources Institute Finland. *Metsänhoito- ja Metsänparannustyöt Maakunnittain (Amounts of Silvicultural and Forest Improvement Work by Province)*; Statistics Database; Natural Resources Institute Finland: Helsinki, Finland, 2017.

27. Natural Resources Institute Finland. *Metsänhoito- ja Metsänparannustöiden Työmäärät (Amounts of Silvicultural and Forest Improvement Work)*; Statistics Database; Natural Resources Institute Finland: Helsinki, Finland, 2017.

28. Mäkelä, M.; Ari, T.; Korhonen, K.; Lipponen, K. *Stump Treatment in Mechanized Timber Harvesting*; Metsäteho Review 3; Metsäteho Oy: Helsinki, Finland, 2017.

29. Nicolotti, G.; Gonthier, P.; Varese, G.C. Effectiveness of some biocontrol and chemical treatments against *Heterobasidion annosum* on Norway spruce stumps. *Eur. J. For. Pathol.* **1999**, *29*, 339–346. [CrossRef]

30. Berglund, M.; Rönnberg, J.; Holmer, L.; Stenlid, J. Comparison of five strains of *Phlebiopsis gigantea* and two *Trichoderma* formulations for treatment against natural *Heterobasidion* spore infections on Norway spruce stumps. *Scand. J. For. Res.* **2005**, *20*, 12–17. [CrossRef]

31. Thor, M.; Stenlid, J. *Heterobasidion annosum* infection of *Picea abies* following manual or mechanized stump treatment. *Scand. J. For. Res.* **2005**, *20*, 154–164. [CrossRef]

32. Keča, N.; Keča, L. The Efficiency of Rotstop and Sodium Borate to Control Primary Infections of *Heterobasidion* to *Picea abies* Stumps: A Serbian Study. *Balt. For.* **2012**, *18*, 247–254.

33. Kenigsvalde, K.; Brauners, I.; Korhonen, K.; Zaļuma, A.; Mihailova, A.; Gaitnieks, T. Evaluation of the biological control agent Rotstop in controlling the infection of spruce and pine stumps by *Heterobasidion* in Latvia. *Scand. J. For. Res.* **2016**, *31*, 254–261. [CrossRef]

34. Berglund, M.; Rönnberg, J. Effectiveness of treatment of Norway spruce stumps with *Phlebiopsis gigantea* at different rates of coverage for the control of *Heterobasidion*. *For. Pathol.* **2004**, *34*, 233–243. [CrossRef]

35. Rönnberg, J.; Sidorov, E.; Petrylaitė, E. Efficacy of different concentrations of Rotstop® and Rotstop®S and imperfect coverage of Rotstop®S against *Heterobasidion* spp. spore infections on Norway spruce stumps. *For. Pathol.* **2006**, *36*, 422–433. [CrossRef]

36. Tubby, K.V.; Scott, D.; Webber, J.F. Relationship between stump treatment coverage using the biological control product PG Suspension, and control of *Heterobasidion annosum* on Corsican pine, *Pinus nigra* ssp. *laricio*. *For. Pathol.* **2008**, *38*, 37–46. [CrossRef]

37. Oliva, J.; Messal, M.; Wendt, L.; Elfstrand, M. Quantitative interactions between the biocontrol fungus *Phlebiopsis gigantea*, the forest pathogen *Heterobasidion annosum* and the fungal community inhabiting Norway spruce stumps. *For. Ecol. Manag.* **2017**, *402*, 253–264. [CrossRef]

38. Valtioneuvoston Asetus Juurikäävän Torjunnasta (Government Decree on Prevention of Damage by *Heterobasidion* spp. root rot) 264/2016. Available online: http://www.finlex.fi/fi/laki/alkup/2016/20160264 (accessed on 28 January 2018).

39. Laki Metsätuhojen Torjunnasta (Forest Damage Prevention Act in Finland) 1087/2013. Available online: http://www.finlex.fi/en/laki/kaannokset/2013/en20131087.pdf (accessed on 28 January 2018).

40. Mäkelä, M. Kantokäsittelyn Toteutus (Implementation of Stump Treatment); Metsäteho Opas. 2001. Available online: http://www.metsateho.fi/wp-content/uploads/2015/03/Kantokasittelyn_toteutus_opas.pdf (accessed on 28 January 2018).

41. Mäkelä, M. Kantokäsittelyn Tarkoitus (Purpose of Stump Treatment). 2011. Available online: http://docplayer.fi/5662252-Kantokasittelyn-tarkoitus.html (accessed on 28 January 2018).

42. Finnish Safety and Chemicals Agency (Tukes). Moto-urea, Kasvitautien Torjuntaan, Myyntipäällyksen Teksti (Moto-Urea, to Control Plant Diseases, Label on Sales Package). Available online: https://kasvinsuojeluaineet.tukes.fi/KareDocs/3069Myyntipaallyksenteksti.pdf (accessed on 28 January 2018).

43. Finnish Safety and Chemicals Agency (Tukes). PS-kantosuoja-2, Kasvitautien Torjuntaan, Myynti-Päällyksen Teksti (PS-kantosuoja-2, to Control Plant Diseases, Label on Sales Package). Available online: https://kasvinsuojeluaineet.tukes.fi/KareDocs/1949Myyntipaallyksenteksti.pdf (accessed on 28 January 2018).

44. Finnish Safety and Chemicals Agency (Tukes). Teknokem Kantosuoja, Kasvitautien Torjuntaan, Myynti-Päällyksen Teksti (Teknokem Kantosuoja, to Control Plant Diseases, Label on Sales Package). Available online: https://kasvinsuojeluaineet.tukes.fi/KareDocs/3124Myyntipaallyksenteksti.pdf (accessed on 28 January 2018).

45. Finnish Safety and Chemicals Agency (Tukes). Urea-kantokate, Kasvitautien torjuntaan, Myyntipäällyksen Teksti (Urea-Kantokate, to Control Plant Diseases, Label on Sales Package). Available online: https://kasvinsuojeluaineet.tukes.fi/KareDocs/2928Myyntipaallyksenteksti.pdf (accessed on 28 January 2018).

46. Finnish Safety and Chemicals Agency (Tukes). Rotstop, Kasvitautien Torjuntaan, Myyntipäällyksen Teksti (Rotstop, to Control Plant Diseases, Label on Sales Package). Available online: https://kasvinsuojeluaineet.tukes.fi/KareDocs/1648Myyntipaallyksenteksti.pdf (accessed on 28 January 2018).

47. Finnish Safety and Chemicals Agency (Tukes). Rotstop SC, Kasvitautien Torjuntaan, Myyntipäällyksen Teksti (Rotstop SC, to Control Plant Diseases, Label on Sales Package). Available online: https://kasvinsuojeluaineet.tukes.fi/KareDocs/2939Myyntipaallyksenteksti.pdf (accessed on 28 January 2018).

48. Skogforsk. StanForD. Standard for Forest Data and Communications. 2007. Available online: http://www.skogforsk.se/contentassets/b063db555a664ff8b515ce121f4a42d1/stanford_main-doc_070327.pdf (accessed on 28 January 2018).

49. Partanen, J.; Hostikka, A.; Kaikkonen, V.; Laukkanen, H.; Vuorenmaa, J. Suomen Metsäkeskuksen Maastotarkastusohje 2013 (Guidelines for Field Control by Finnish Forest Centre, 2013); Finnish Forest Centre. 2013. Available online: https://www.metsakeskus.fi/sites/default/files/smk-maastotarkastus-ohje-2013.pdf (accessed on 28 January 2018).

50. Leivo, J.; Partanen, J.; Nieminen, T.; Vuorenmaa, J.; Kuoppala, H.; Rahkola, S. Maastotarkastusohje (Guidelines for Field Control); Finnish Forest Centre. 2016. Available online: https://www.metsakes-kus.fi/sites/default/files/smk-maastotarkastusohje-2016.pdf (accessed on 28 January 2018).

51. Möykkynen, T.; Miina, J.; Pukkala, T.; von Weissenberg, K. Modelling the spread of butt rot in a *Picea abies* stand in Finland to evaluate the profitability of stump protection against *Heterobasidion annosum*. *For. Ecol. Manag.* **1998**, *106*, 247–257. [CrossRef]

52. Möykkynen, T.; Miina, J.; Pukkala, T. Optimizing the management of a *Picea abies* stand under risk of butt rot. *For. Pathol.* **2000**, *30*, 65–76. [CrossRef]

53. Möykkynen, T.; Miina, J. Optimizing the Management of a Butt-rotted *Picea abies* Stand Infected by *Heterobasidion annosum* from the Previous Rotation. *Scand. J. For. Res.* **2002**, *17*, 47–52. [CrossRef]

54. Pukkala, T.; Möykkynen, T.; Thor, M.; Rönnberg, J.; Stenlid, J. Modeling infection and spread of *Heterobasidion annosum* in even-aged Fennoscandian conifer stands. *Can. J. For. Res.* **2005**, *35*, 74–85. [CrossRef]

55. Thor, M.; Arlinger, J.D.; Stenlid, J. *Heterobasidion annosum* root rot in *Picea abies*: Modelling economic outcomes of stump treatment in Scandinavian coniferous forests. *Scand. J. For. Res.* **2006**, *21*, 414–423. [CrossRef]

56. Kärhä, K.; Rönkkö, E.; Gumse, S.-I. Productivity and Cutting Costs of Thinning Harvesters. *Int. J. For. Eng.* **2004**, *15*, 43–56.

57. Eriksson, M.; Lindroos, O. Productivity of harvesters and forwarders in CTL operations in northern Sweden based on large follow-up datasets. *Int. J. For. Eng.* **2014**, *25*, 179–200. [CrossRef]

58. Oliva, J.; Bendz-Hellgren, M.; Stenlid, J. Spread of *Heterobasidion annosum* s.s. and *Heterobasidion parviporum* in *Picea abies* 15 years after stump inoculation. *FEMS Microbiol. Ecol.* **2011**, *75*, 414–429. [CrossRef] [PubMed]

59. Leivo, J. *Juurikäävän Torjunnan Tulokset 2012–2016 Suomessa (Results of Stump Treatment Work in Finland, 2012–2016)*; Finnish Forest Centre, Statistics: Lahti, Finland, 2017.

60. Finnish Forest Centre. Kantokäsittely Juurikääpää Vastaan Tärkeää Kesäharvennuksissa (It Is Crucial to Treat Stumps against *Heterobasidion* spp. root rot in Summertime Thinnings). Available online: https://www.metsakeskus.fi/uutiset/kantokasittely-juurikaapaa-vastaan-tarkeaa-kesaharvennuksissa (accessed on 28 January 2018).

61. Finnish Forest Centre. Laadukas Kantokäsittely Tärkeää Kesäaikaisissa Harvennushakkuissa—Laho Romahduttaa Puukauppatilin (High-Quality Stump Treatment is Essential in Summertime Thinnings—A Root Rot Collapses the Account of Forest Owner in Timber-Sales Transactions). Available online: https://www.metsakeskus.fi/uutiset/laadukas-kantokasittely-tarkeaa-kesaaikaisissa-harvennushakkuissa-laho-romahduttaa (accessed on 28 January 2018).

62. Marshall, H.; Murphy, G. Economic evaluation of implementing improved stem scanning systems on mechanical harvesters/processors. *N. Z. J. For. Sci.* **2004**, *34*, 158–174.

63. Murphy, G.; Wilson, I.; Barr, B. Developing methods for pre-harvest inventories which use a harvester as the sampling tool. *Aust. For.* **2006**, *69*, 9–15. [CrossRef]

64. Murphy, G. Determining Stand Value and Log Product Yields Using Terrestrial Lidar and Optimal Bucking: A Case Study. *J. For.* **2008**, *106*, 317–324.

65. Kärhä, K.; Änäkkälä, J.; Hakonen, O.; Palander, T.; Sorsa, J.-A.; Räsänen, T.; Moilanen, T. Analyzing the Antecedents and Consequences of Manual Log Bucking in Mechanized Wood Harvesting. *Mech. Mater. Sci. Eng.* **2017**, *12*, 1–15. [CrossRef]

Article

Total Weight and Axle Loads of Truck Units in the Transport of Timber Depending on the Timber Cargo

Grzegorz Trzciński [1], Tadeusz Moskalik [1,*] and Rafał Wojtan [2]

[1]   Department of Forest Utilization, Faculty of Forestry, Warsaw University of Life Sciences—SGGW, Nowoursynowska 159, 02-776 Warsaw, Poland; grzegorz.trzcinski@wl.sggw.pl

[2]   Division of Dendrology and Forest Production, Faculty of Forestry, Warsaw University of Life Sciences SGGW, Nowoursynowska 159, 02-776 Warsaw, Poland; rafal_wojtan@sggw.pl

*   Correspondence: tadeusz.moskalik@wl.sggw.pl; Tel.: +48-22-593-8137

Received: 31 January 2018; Accepted: 22 March 2018; Published: 23 March 2018

Abstract: When transporting timber, the high variability of species, assortments and moisture content of the wood raw material does not allow the weight of the transported timber to be precisely determined. This often contributes to the excessive weight loading of the entire truck unit. The aim of the research is to present the variability of the total weight of truck units with wood cargoes (GVW—gross vehicle weight) depending on the weight of the empty unit and the transported timber load, as well as to analyze the changes in GVW, unit loads of wood and load on individual truck unit axles depending on the season. This study analyzes the total weight of truck units for 376 transports of Scots pine timber at different times of the year. The total weight of the truck units depends on the weight of an empty unit and the weight of the load. GVW was determined by using a weighbridge to weigh the vehicles and then the empty unit after unloading. The weight of the load was obtained as the difference between GVW and the tare. It was found that GVW differed significantly depending on the truck unit used, ranging from 43.60–58.80 Mg, often exceeding permissible limits for public roads. The individual axle loads for various truck units were also analyzed. The obtained results indicate that these loads are more equally distributed in the case of five-axle trucks compared to six-axle ones.

Keywords: timber transport; axle load; gross vehicle weight; timber load; empty vehicle weight

1. Introduction

Transport plays a vital role in the proper functioning of every economic system. The success of a company specializing in timber haulage depends not only on the transportation system, but also on the cooperation of all entities associated with its supply chain [1–3].

The share of costs relating to the transport of wood raw material in relation to the total costs of forestry activities is significant [4,5]. Its estimated amount is approximately 17%, and this is definitely a higher share compared to other sectors of the economy. The highest costs of transport operations are for hauling timber, accounting for 40–60% [6]. In order to increase their efficiency, companies operating in the timber haulage market try to reduce transport costs, as this is perceived as an important factor for increasing competitiveness. One of the methods of improving the efficiency of timber transport is to reduce the variability of loads and maximize the loads, in accordance with applicable public transport regulations, carried by truck units [5,7,8]. Research conducted in the USA indicates that by maintaining a uniform timber load weight, suppliers are able to achieve savings of 4–14% [9].

An important element in this respect are the legally-established limits of the permissible total weight of vehicles. In 1996, the Member States of the European Union adopted Directive 96/53/EC specifying permissible vehicle weights of 40 Mg and 44 Mg and weights on a single-axle of 100 kN and double axel of 160 kN [10]. In individual countries, institutions responsible for public transport and public roads have the power to limit the total weight of a truck unit (gross vehicle weight) with the simultaneous possibility of limiting permissible axle weights or increasing them in relation to those specified in EU regulations, as well as indicating the roads to be used by such vehicles [5,11–16]. In Poland, the permissible vehicle weight of a truck and trailer of over four axles has been set at 40 Mg or 44 Mg for a six-axle vehicle; a truck with a 40-ft intermodal container with a maximal single axle load of 80 kN and double axle load of 160 kN and for national roads at 100 and 115 kN [17]. In Finland, 7-, 8- and 9-axle truck units are allowed, with a maximal weight of 64, 68 and 76 tonnes, respectively, and 4.4 m in height [18]. The increase of the permissible weight of truck units in Finland contributed to the reduction of CO_2 and NO_x emissions and has resulted in economic benefits [19]. Analyses performed by Palander and Kärhä [20] on transporting timber at the increased permissible loads in Finland indicate the possibility of reducing the required number of vehicles and transport work conducted in tonne-kilometers. Estonian experiences show that increasing the weight limit from the current 44–60 Mg would reduce the share of transport costs in roundwood prices from 17.7–14.2% [8]. Vehicles traveling with a full load and with an increased weight can result in economic and environmental benefits in the long term [21,22].

In reality, however, when transporting timber, the high variability of the transported assortments made of different species and the variable moisture content of wood do not allow a precise determination to be made of the weight of the transported raw wood material [23–26]. These differences very often influence the excessive weight of the truck unit [27–30]. McDonnell et al. [31] specify that 58–80% of the vehicles exceeded the permissible truck unit weight, and Devlin [32] shows that the gross vehicle weight (GVW) was exceeded by 60% of the vehicles, of which 20% exceeded the permissible load capacity specified by the vehicle manufacturer. Research conducted in the USA also showed significant overloading of vehicles with wood; as much as 88% of the vehicles had a gross vehicle weight of more than 44 Mg [9].

The volume of the timber harvest in Polish forests has been systematically growing in recent years. In 2015, 40,247 million m^3 were harvested [33]. The units of the State Forests National Forest Holding plan to harvest over 42 million m^3 of timber in 2018. This wood raw material is delivered to several thousand recipients. This presents a major transport challenge, both in terms of logging operations and its transport. In most cases, approximately 90% of the transport is conducted by vehicles using high-tonnage five- or six-axle truck units [25].

The transportation of wood raw material is one of the key operations in forest management, representing a large share of total costs, and has significant potential for optimization [34–36]. A good diagnosis of the configuration of timber truck units, their own weight and their potential timber cargoes, given the large variation in cargo weight, can contribute to improving the efficiency of forest transport [5,9].

Determining the actual weights of transported timber loads and attempting to link them to GVW and individual axle loads will allow the overloading of vehicles and exceeded individual axle weights to be avoided by providing appropriate information to transport companies. The aim of the research is to present the variability of the total weight of truck units (GVW) with wood cargoes depending on the weight of the empty unit and the transported timber load, as well as to analyze the changes in GVW, unit loads of wood and load on individual truck unit axles depending on the season.

2. Materials and Methods

In order to conduct the research and perform the relevant analyses, deliveries were studied to one of the largest wood buyers in Poland, a sawmill supplied with large-sized timber, both long logs and shortwood. The transport was conducted by external companies commissioned by the sawmill. Data for 2016 were collected in four periods: January, at the end of March/beginning of April, July–August and October–November.

The gross weight of the truck unit (GVW) expressed in Mg is understood as the actual weight of the vehicle and trailer or truck unit and semi-trailer with all the equipment, the driver and timber load. GVW was determined based on weighing the entire truck unit on a weighbridge at the sawmill at the moment the wood raw material was delivered, and then after unloading, the empty unit was re-weighed. The actual weight of each load was obtained from the difference of GVW and the tare.

The volume of the load of transported timber, expressed as solid cubic volume (m^3 under bark), was determined on the basis of transport vouchers, on which the payments are based of the delivered timber from the supplier to the recipient. Large-sized logs were measured individually by each piece. It should also be noted that when selling timber in Poland and later when it is transported, the given volumes and conversion factors do not apply to the bark. The share of bark is significant and can range up to a dozen or so percent for pine logs [26], influencing the weight of the transported cargo. Therefore, the analysis refers to the weight of the entire load (wood, bark, water) or the weight of 1 m^3 of the load.

The load on the individual axles of high tonnage truck units was measured using Model DINI ARGEO WWSD portable truck scales with a 3590M309 weighing terminal with 0.01 Mg graduation. The scale system used is fully compliant with Polish regulations and allows vehicles in transit to be weighed. The loads on the individual wheel axles were measured successively for the whole unit: the vehicle and the trailer. During the measurement, the scales were placed on a flat maneuvering area of hardened concrete paving blocks, with no depressions. It was assumed that such a method of measuring the weight on a particular axle is close to the actual conditions of a forest road, whose longitudinal profile is never level, which in turn is the cause of increased wheel loads on the road surface. For this reason, the analyzed GVW was also determined based on weighing the vehicle on a weighbridge, and not on the sum of the load on individual axles. The weigh station was selected in such a way so as to maintain a level road scale, so that the measured axles were kept level.

The method of weighing vehicles used by the Polish Road Transport Inspectorate, which oversees compliance with permissible axle loads, assumes that measurements are taken with platform scales embedded in the surface, or by placing pads under unweighted axles, while maintaining a maximum allowed slope of 2%. The analysis was based on the results of measurements, taking into account a 5% allowable measurement error in accordance with the recommendations of the Polish Road Transport Inspectorate.

The obtained results were statistically analyzed with the STATISTICA 12 package. In all analyzed periods, the distribution of the variables for all parameters differed from the normal distribution. Therefore, the significance of the differences was mainly determined using the Kruskal–Wallis and Dunn tests.

3. Results

3.1. Characteristics of the Parameters of Truck Units and Transported Timber

The study analyzed 376 transports of pine timber (mainly shortwood, 3.7–5.0 m long) at different times of the year. The parameters characterizing the truck units are presented in Table 1. Timber was transported using truck units consisting of a truck and trailer (181 observations), truck and semi-trailer (166), truck and lightweight platform semi-trailer (17), as well as truck and dolly (12) (Figures 1 and 2).

It was found that the total weight of the truck units (GVW) differed significantly depending on type, ranging from 43.60 (platforms) to 58.80 Mg (trailers). The lowest average value for a given truck unit, amounting to 45.98 Mg, was noted for deliveries made by a tractor with a platform trailer, while the highest was for a truck with a dolly (51.44 Mg). The arithmetic mean for all truck units was 50.56 Mg. The largest variation for empty weight (tare), ranging from 13.8–23.7 Mg, was observed for the truck and semi-trailer, which are also the heaviest means of transport, with an average weight of 20.5 Mg. The average values of 29.07–30.29 m^3 for transported timber volume were similar regardless of the analyzed truck unit.

Table 1. Selected measurements by type of truck unit.

Truck Unit	Measure	Mean	SD	Min	Max	Q1	Median	Q3
Semi-trailer		50.436	2.473	45.000	56.750	48.850	50.225	51.750
Trailer	Gross vehicle	51.038	2.531	45.200	58.800	49.250	50.750	52.450
Platform	weight (Mg)	45.988	2.223	43.600	51.650	44.500	45.850	46.500
Dolly		51.438	1.810	49.400	54.400	50.100	50.600	53.250
Semi-trailer		20.500	1.112	13.800	23.700	19.800	20.500	20.950
Trailer	Tare (Mg)	20.224	0.879	17.900	22.000	19.700	20.300	20.900
Platform		14.941	0.311	14.220	15.250	14.800	15.000	15.200
Dolly		20.071	0.706	18.200	20.650	19.925	20.300	20.550
Semi-trailer	Volume of	29.074	1.769	23.960	35.040	28.100	28.710	30.140
Trailer	timber load	29.707	1.836	25.670	36.790	28.210	29.610	30.860
Platform	(m^3)	30.292	1.795	27.570	34.500	29.150	30.040	30.420
Dolly		29.982	0.710	28.940	30.940	29.340	30.000	30.620
Semi-trailer		1.030	0.055	0.869	1.159	0.995	1.028	1.071
Trailer	Weight of 1 m^3	1.038	0.053	0.828	1.179	0.998	1.042	1.077
Platform	of load (Mg)	1.025	0.034	0.957	1.075	1.005	1.035	1.045
Dolly		1.046	0.044	0.984	1.136	1.016	1.038	1.073

Notes: SD, standard deviation; Q1, first quartile; Q3, third quartile.

(a)   (b)   (c)   (d)

Figure 1. Stages of weighing the wheel axles of a truck unit hauling timber: (a) front axle, (b) second axle of the truck, (c) third axle of the truck, (d) rear axle.

(a)

(b)

(c)

(d)

Figure 2. Examples of truck units: (a) truck and trailer, (b) truck and semi-trailer, (c) truck and lightweight platform semi-trailer and (d) truck and dolly.

Due to the small number of measurements obtained for the units with a dolly (12) and platform trailer (17), all statistical analyses were only performed for the units with a trailer and semi-trailer.

The Mann–Whitney test was conducted to examine the significance of the selected features depending on the type of unit (semi-trailer or trailer). The results obtained are summarized in Table 2. The statistical analysis only showed a significant difference between the types of truck units for the weight of 1 m^3 of load (Table 2).

Table 2. Results of the analysis of significant differences between selected measures by type of truck unit.

Measure	Gross Vehicle Weight (Mg)	Tare (Mg)	Volume of Timber Load (m³)	Weight of 1 m³ of Load (Mg)	Load Weight (Mg)
p	0.0263	0.0183	0.0029	0.1586	0.0011

The Kruskal–Wallis test was performed for trailers and semi-trailers (due to the heterogeneity of variances) for the significance of GVW, tare, m³ load volume, weight of 1 m³ of load and load weight between individual measurement periods (a significance level of 0.05 was adopted for the analyses). The results are summarized in Table 3. In addition, the test of multiple comparisons of mean ranks was performed. The statistical analysis (Table 3) confirmed the differences between the values of the studied measures for specific types of truck units depending on when the measurement was taken.

Table 3. Results of the analysis of significant differences between selected measures by type of truck unit.

Type of Truck Unit	GVW (Mg)	Tare (Mg)	Volume of Timber Load (m³)	Weight of 1 m³ of Load (Mg)	Load Weight (Mg)
Semi-trailer	0.0000	0.0218	0.1316	0.0000	0.0002
Trailer	0.0000	0.0002	0.0312	0.0000	0.0000

The range of the results obtained for the studied measures depending on type of truck unit and time of measurement is presented in Figures 3–6. In the case of units with a semi-trailer, statistically-significant differences were found for GVW (Figure 3a), tare (June measurement results differ from those of November, and this is the only significant difference), load weights (Figure 4a) and weights of 1 m³ of load (Figure 5a).

Differences in the values of load volume between individual measurement periods turned out to be not statistically significant (Figure 6a). For units with a trailer, statistically-significant differences were found for all variables (Figures 3b, 4b, 5b and 6b), although in most cases, these differences related to November results. In the case of the m³ load volume, significant differences in comparison to other measurement dates were found only for data from September.

As shown in Figure 5, the differences are not very large, but they are statistically significant for the weight per cubic meter of transported round wood depending on the transport system used. However, in particular seasons of the year, not all differences have the same tendency. These differences could be explained by random factors, which may be related to various tree growth conditions influencing wood density or water content.

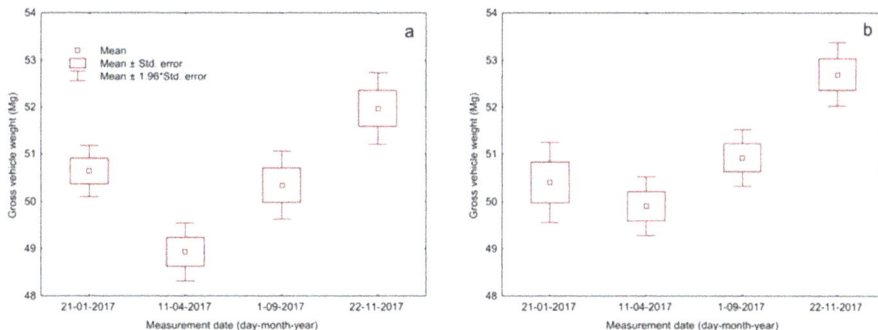

Figure 3. Comparison of average GVW values by measurement date for a unit with a semi-trailer (**a**) and a trailer (**b**).

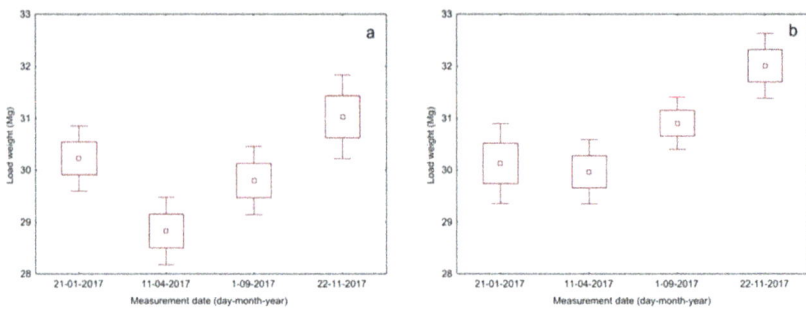

Figure 4. Comparison of average load weight values by measurement date for a unit with a semi-trailer (**a**) and a trailer (**b**).

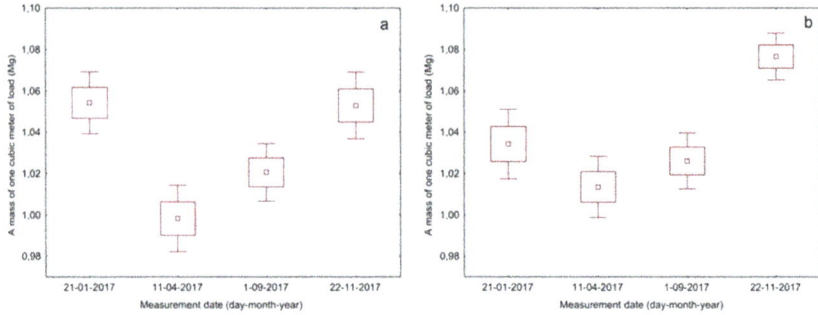

Figure 5. Comparison of average weight of 1 m³ of load values by measurement date for a unit with a semi-trailer (**a**) and a trailer (**b**).

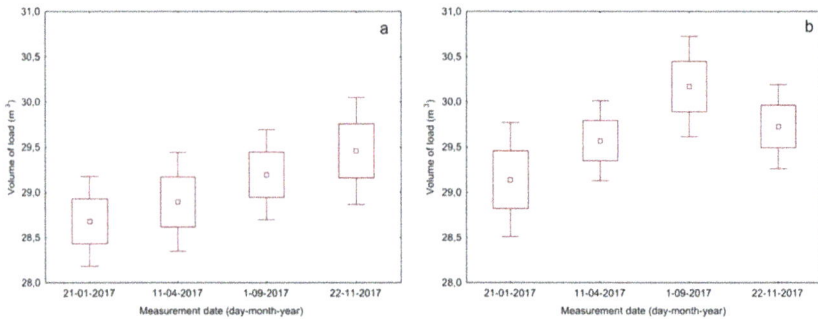

Figure 6. Comparison of average load volume values by measurement date for a unit with a semi-trailer (**a**) and a trailer (**b**).

3.2. Total Weight of Truck Units as a Function of the Tare and Volume of Transported Timber

When loading timber on a forest road, the driver is not sure how much wood to load so as to not overload the vehicle. Assuming that drivers know the mass of an empty unit (they are informed of this by the recipients after weighing) and knowing that the results of the measurement of the mass of 1 m³ of cargo are not statistically significant (Table 2), an attempt was made to develop a two-factor regression model for the GVW (Mg) values as a function of the tare of the truck unit Mg and load volume of the timber (m³). The presented linear model was developed using the least squares (OLS) method. The significance of the calculated coefficients was assessed using Student's t-test.

For the semi-trailers, it took the form of:

$$GVW = 1.04672 * Tare + 0.996916 * m^3 \tag{1}$$

The constant term was not taken into account because its value turned out to be not statistically significant. For the developed model, the standard error of the estimation was 1.59297, and the coefficient of determination R^2 was 0.9990. The assessment of the model parameters is presented in Table 4.

Table 4. Assessment of the model parameters for semi-trailers.

Parameter	Parameter Value	Standard Error	t-Statistic	p-Value
Tare	1.04672	0.0662784	15.7927	0.0000
Load volume m³	0.996916	0.0467613	21.3193	0.0000

In the case of a unit with a trailer, the regression model describing the GVW values depending on vehicle weight (tare) and load volume m³ took the form of:

$$GVW = 1.25897 * Tare + 0.860483 * m^3 \tag{2}$$

The offset also was not statistically significant in this case. The standard error of the estimation of the developed equation was 1.59736, and the coefficient of determination R^2 was 0.9990. The assessment of the model parameters is presented in the Table 5.

Table 5. Assessment of the model parameters for trailers.

Parameter	Parameter Value	Standard Error	t-Statistic	p-Value
Tare	1.25897	0.0821727	15.3211	0.0000
Load volume m³	0.860483	0.0559386	15.3826	0.0000

3.3. Distribution of the Axle Loads of the Truck Units

The analysis of axle loads was performed for 92 five-axle and 155 six-axle truck units consisting of a tuck and semi-trailer, as well as 46 five-axle and 135 six-axle units of a truck and trailer. The basic statistics characterizing the absolute load values of individual axles of the units are presented in Table 6. The smallest axle load for all units was on the first axle, spanning average values of 74–80 kN, with a range of results from 42–102 kN. The highest average loads were at the level of 106–113 kN, located on the second and third axles in the five-axle units, with a maximum single axle load of 170 kN occurring on the fifth axle.

The analysis performed with the Friedman test showed the existence of statistically-significant differences in the level of the vehicles' axle loads. This analysis was performed for groups of different types of units and different numbers of axles.

It should be noted that for both five- and six-axle truck units, the lowest load is on the first axle. In the case of five-axle units, the load on the remaining axles is fairly even. However, six-axle units have the highest load on the second axle (Figure 7), which then decreases in the direction of the last axle that is similarly loaded as the first axle. Despite the uneven axle load of the six-axle units, their maximum loads are less than in the case of the five-axle units.

Table 6. Assessment of the model parameters for trailers.

Unit	Axle	Axle Load Values (kN)			
		Mean	SD	Min	Max
five-axle semi-trailer	1	80.77	4.58	68.40	98.20
	2	111.02	11.64	86.60	143.40
	3	106.54	11.91	77.43	134.90
	4	107.60	19.89	75.90	157.70
	5	104.02	20.68	13.50	170.00
six-axle semi-trailer	1	80.55	6.01	42.56	102.03
	2	93.10	14.32	66.40	138.70
	3	89.28	11.84	63.65	137.75
	4	91.26	19.61	70.71	159.10
	5	84.95	13.21	53.30	126.70
	6	84.93	14.69	51.70	141.90
five-axle trailer	1	74.81	6.81	60.33	90.25
	2	113.07	10.18	90.25	135.19
	3	102.06	14.42	74.39	127.59
	4	97.93	15.04	73.34	141.08
	5	98.63	18.26	39.90	140.03
six-axle trailer	1	74.45	6.80	57.95	91.20
	2	106.94	9.99	83.60	129.20
	3	102.41	9.38	82.65	125.88
	4	80.85	8.74	62.70	104.69
	5	80.06	17.90	50.35	168.06
	6	72.18	9.51	36.77	101.65

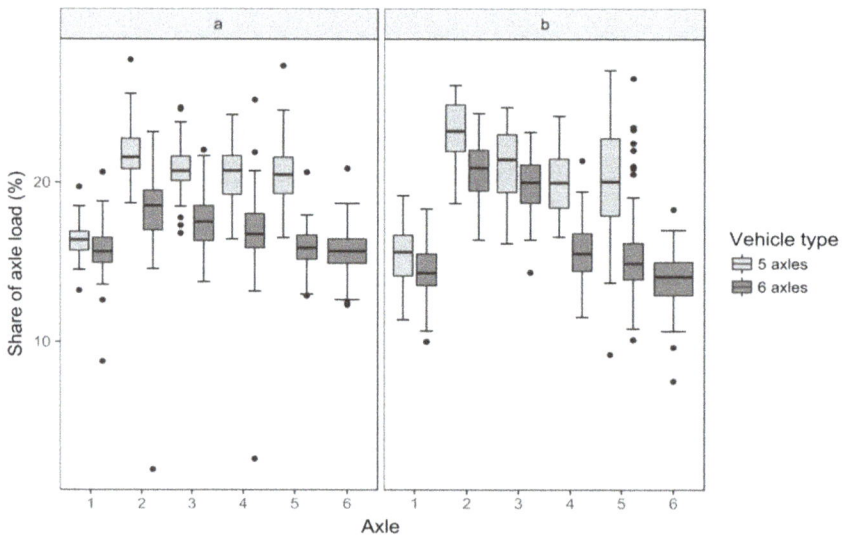

Figure 7. Distribution of the axle loads of the truck units with a semi-trailer (**a**) and a trailer (**b**).

4. Discussion

Transport plays an extremely important role in every area of the economy, including the forestry industry. The efficiency of transport depends on a number of factors, including, among others, the load capacity of the vehicle, driving time and fuel consumption [29,37]. When transporting timber, the high variability of species, timber assortment and moisture content of the wood influence the determination of the weight of the transported timber and GVW [15,27,29]. In the studies carried out for the same species of Scots pine, as well as the large-sized assortments, statistically-significant differences were found for the tested parameters in different periods of the year, as evidenced by the study results of Owusu-Ababio and Schmitt [30]. Research conducted by Tomczak et al. [26] indicates a large variation in the density of pine wood. On a longitudinal section of the trunk, the highest density was found for wood at breast height (0.816 Mg/m^3), whereas the lowest was from the section closest to the apex (0.707 Mg/m^3). This differentiation obviously affects the variability of the weight of the transported timber. Calculated per 1 m^3, the weight of the wood raw material was estimated at 0.979 Mg. The ratio of the actual density (0.979 Mg/m^3) to the density provided in the tables (0.740 Mg/m^3) amounted to 1.3:1 [26]. The actual weight of the wood raw material was therefore about 30% higher than the weight that can be estimated on the basis of the wood density tables developed for the needs of road transport. The results of the tests of the mass of 1 m^3 of cargo in the range of from 0.828–1.179 Mg/m^3 (Table 1) also indicate the need to verify the adopted table values. An erroneously-chosen indicator often contributes to overloading.

In the development of GVW regression equations, linear functions have been adopted, giving the possibility of their quick and easy application in the forest during timber loading. At the loading site, the driver obtains information about solid wood volume from the forester, who must be present during loading and issues a special receipt for each transport. In the case of large-sized wood, all loaded pieces are specified with their numbers and solid volumes in the receipt. For pulp wood, only one number and the general solid volume for every single stack are provided on the receipt. By applying the developed model, a driver can very quickly calculate GVW and determine in a simple way if the truck is overloaded or not. It seems to be a better solution than taking a tabulated value (0.740 Mg), which showed differences in real GVW. Of course, it is advisable to develop such dependencies for other species and assortments and verify the adopted tabulated conversion factors.

For the estimation of the gross vehicle weight model, the solid volume was applied. All obtained results refer to the large-sized logs measured individually by each piece. In the case of pulp wood, measured in stacked cubic volume (over bark), conversion factors must be applied to obtain the results in solid cubic volume (m^3 under bark). According to Polish regulations for pine wood, the conversion factors are 0.65 (for a length of wood of 1.0 or 1.2 m), 0.62 (for a length of 2.0; 2.4 m) and 0.6 (for a length of 5.0–7.0 m). In all cases, the load of 1 m^3 volume contains the wood itself, water and bark.

As was found in the research conducted, the total vehicle weights obtained from actual measurements differed significantly, ranging from 43.60–58.80 Mg, with an average value of 50.56 Mg (Table 1). The exceeded permissible truck unit weight (40 Mg in Poland) coincides with the results presented in the literature [9,28,30–32,37]. The actual GVW results obtained were compared to the calculated total vehicle masses, in which the mass of the load was determined based on the volume of transported timber (m^3) and the conversion value of 0.740 Mg/m^3 from the table in effect, which in this case provided GVW results of about 17% lower than the actual ones [38].

In recent years, a reduction in the weight of empty truck units allowing the loads of transported timber to be increased and the difference between a unit with a trailer (tare 17.9–22.0 Mg) and a truck with a semi-trailer (13.8–23.1 Mg) have been observed. The obtained results confirm the possibility of improving the efficiency of wood transport indicated by Sosa et al. [5] and Šušnjar et al. [39], in particular through the use of lightweight platform semi-trailers.

The increased total weights of truck units (GVW) are usually the reason that legal limits have been exceeded: 80 kN for single axles and 115 kN for tandem axles. In a larger percentage of truck units, we have seen that the permissible total weight is exceeded rather than the axle limits, which coincides with the results presented by Owus-Ababio and Schmitt [30] and Baumgras [40], or they were not exceeded at all [39]. In the timber truck units, we observed an increased load on the rear axles (Axles 2 and 3) of the vehicle or tractor, which is caused by the mounted loading device or the uneven distribution of the load and results from the demands on the driving axle of the vehicle. Transporting different lengths of wood, both shortwood and long logs, does not always allow the load to be spread evenly across all axles, which is confirmed by the results obtained for Axles 5 and 6 of the six-axle units.

5. Conclusions

The conducted research allowed statistically-significant differences to be confirmed between the total weights of truck units and load weights in particular seasons of the year. The average total weight for all units was 50.56 Mg, ranging from 43.60–58.8 Mg. Lower values were observed with the use of platform trailers; higher ones for trucks with trailers.

The high variability of load volume and mass of 1 m^3 of load in relationship to the accepted normative conversion value (0.740 Mg/m^3) results in greater actual GVW. In the majority of cases, these GVW values are over allowable limits.

In the analyzed period, the average weight of delivered timber loads, determined on the basis of weighing, was 29.76 Mg, oscillating in the range of 23.96–36.79 Mg. The relatively lightest loads, both in the case of trailers and semi-trailers, were transported in the early spring months, while the heaviest were in the autumn season. The decisive factor was the date when the wood was harvested and transported. During the dormant period of tree growth, the water content in wood tissue decreases, thus contributing to a reduction in the weight of the transported cargo.

In Poland, five- and six-axle units are most often used to transport wood. The load on individual axles of five-axle units is more or less even. However, for six-axle units, the highest load falls on the second axle, gradually decreasing on successive axles. From the point of view of achieving smaller maximum axle loads, timber should be transported with six axle units.

Vehicles are used for 90% of the transport of wood in Poland. As research has shown, in many cases, vehicles are excessively overloaded in relation to the road traffic regulations in force. In order to improve transport efficiency, it would be advisable in the future to increase the permissible load limits, using units with an increased number of axles, based on the example of the Scandinavian countries.

Author Contributions: G.T and T.M. conceived, designed, performed the experiments, and wrote the paper; R.W. analyzed the data.

Conflicts of Interest: The authors declare no conflict of interest.

References

1. Devlin, G.; Lyons, J.; Russel, F.; Joyce, M.; Forest Industry Transport Group. Managing Timber Transport Good Practice Guide. Solutions for a Growing Irish Forest Harvest. 2014. Available online: https://www.teagasc.ie/media/website/crops/forestry/advice/Managing-Timber-Transport---Good-Practice-Guide-Volume-1-2014-1.pdf (accessed on 25 November 2017).
2. Timber Transport Forum. Road Haulage of Round Timber Code of Practice Timber Transport Forum. Road Haulage of Round Timber Code of Practice. 2012. Available online: http://www.coford.ie/media/coford/content/links/transport/Road%20Haulage%20of%20Round%20Timber%20Code%20of%20Practice.pdf (accessed on 5 July 2017).
3. Sieniawski, W. Waloryzacja dostaw drewna do wybranych segmentów przemysłu drzewnego. Valorization of Wood Supply to Selected Segments of Timber Industry. Ph.D. Thesis, WULS-SGGW, Warsaw, Poland, June 2012.
4. Devlin, G.; McDonnell, K.; Ward, S. Timber haulage routing in Ireland: An analysis using GIS and GPS. *J. Transp. Geogr.* **2008**, *16*, 63–72. [CrossRef]
5. Sosa, A.; Klvac, R.; Coates, E.; Kent, T.; Devlin, G. Improving Log Loading Efficiency for Improved Sustainable Transport within the Irish Forest and Biomass Sectors. *Sustainability* **2015**, *7*, 3017–3030. [CrossRef]
6. Shaffer, R.M.; Stuart, W.B. A Checklist for Efficient Log Trucking. Virginia Cooperative Extension. 1998. USA. Available online: https://vtechworks.lib.vt.edu/bitstream/handle/10919/54904/420-094.pdf?sequence=1 (accessed on 10 December 2017).

7. Beardsell, M.G. Decreasing the Cost of Hauling Timber through Increased Payload. Ph.D. Thesis, Virginia Polytechnic Institute and State University, Blacksburg, VA, USA, 1986. Available online: https://vtechworks.lib.vt.edu/handle/10919/53617 (accessed on 9 September 2017).

8. Lukason, O.; Ukrainski, K.; Varblane, U. Economic benefit of maximum truck weight regulation change for Estonian forest sektor. Veokite täismassi regulatsiooni muutmise majanduslikud mõjud eesti metsatööstuse sektorile. *Est. Discuss. Econ. Policy* **2011**, *19*. [CrossRef]

9. Hamsley, A.K.; Greene, W.D.; Siry, J.P.; Mendell, B.C. Improving Timber Trucking Performance by Reducing Variability of Log Truck Weights. *South. J. Appl. For.* **2007**, *31*, 12–16.

10. Directive (EU) 2015/719 of the European Parliament and of the Council of 29 April 2015 Amending Council Directive 96/53/EC Laying Down for Certain Road Vehicles Circulating within the Community the Maximum Authorised Dimensions in National and International Traffic and the Maximum Authorised Weights in International Traffic (Text with EEA Relevance). Available online: https://publications.europa.eu/en/publication-detail/-/publication/22b313fc-f3bc-11e4-a3bf-01aa75ed71a1/language-en (accessed on 26 January 2017).

11. Chapter H-3.01 Reg 8. The Vehicle Weight and Dimension Regulations. 2010. Available online: http://www.qp.gov.sk.ca/documents/English/Regulations/Regulations/H3-01R8.pdf (accessed on 5 January 2017).

12. *FSVA Richtlinien für die Standardisierung des Oberbaues von Verkehrsflächen/RStO 01*; FGSV. 499. Forschungsges; für Strassen-und Verkehrswesen: Köln, Cologne, 2001; Volume 499, Available online: https://www.tib.eu/de/suchen/id/TIBKAT%3A334845300/Richtlinien-f%C3%BCr-die-Standardisierung-des-Oberbaues/ (accessed on 8 January 2018).

13. Pischeda, D. Technical Guide on Harvesting and Conservation of Storm Damaged Timber. 2004. Available online: http://videntjenesten.ku.dk/filer/Technical-Guide-on-harvesting-and-conservation-of-storm-damaged-timber.pdf (accessed on 6 January 2017).

14. Wilson, S. Permissible Maximum Weights of Lorries in Europe. Available online: https://www.itf-oecd.org/permissible-maximum-weights-lorries-europe (accessed on 26 January 2018).

15. Trzciński, G. *Analiza Parametrów Technicznych Dróg Leśnych w Aspekcie Wywozu Drewna Samochodami Wysokotonażowymi. Analysis of Technical Parameters of Forest Roads in Terms on Timber Haulage by High-Tonnage Vehicles*; Warsaw University of Life Sciences-SGGW: Warsaw, Poland, 2011; ISBN 978-83-7583-291-1.

16. Trzciński, G.; Tymendorf, Ł. Dostawy drewna po wprowadzeniu normatywnych przeliczników gęstości drewna do określenia masy ładunku. *Sylwan* **2017**, *161*, 451–459.

17. Regulation Regulation of the Minister of Infrastructure of 31 December 2002 on the Technical Conditions of Vehicles and Their Necessary Equipment.Rozporządzenie. Rozporządzenie Ministra Infrastruktury z Dnia 31 grudnia 2002 r. w Sprawie Warunków Technicznych Pojazdów Oraz Zakresu ich Niezbędnego Wyposażenia. Available online: http://prawo.sejm.gov.pl/isap.nsf/download.xsp/WDU20030320262/O/D20030262.pdf (accessed on 12 January 2018).

18. Oy, E.P. FINLEX®—Ajantasainen lainsäädäntö: Asetus ajoneuvojen käytöstä tiellä 1257/1992. Available online: https://www.finlex.fi/fi/laki/ajantasa/1992/19921257 (accessed on 26 January 2018).

19. Liimatainen, H.; Nykänen, L. *Impacts of Increasing Maximum Truck Weight—Case Finland*; Transport Research Centre Verne, Tampere University of Technology: Tampere, Finland, 2017; Available online: http://www.tut.fi/verne/aineisto/LiimatainenNyk%C3%A4nen.pdf (accessed on 10 January 2018).

20. Palander, T.; Kärhä, K. Potential Traffic Levels after Increasing the Maximum Vehicle Weight in Environmentally Efficient Transportation System: The Case of Finland. *J. Sustain. Dev. Energy Water Environ. Syst.* **2017**, *5*, 417–429. [CrossRef]

21. McKinnon, A.C. The economic and environmental benefits of increasing maximum truck weight: The British experience. *Transp. Res. Part Transp. Environ.* **2005**, *10*, 77–95. [CrossRef]

22. Knight, I.; Newton, W.; McKinnon, A.; Palmer, A.; Barlow, T.; McCrae, I.; Dodd, M.; Couper, G.; Davies, H.; Daly, A.; et al. Longer and/or Longer and Heavier Goods Vehicles (LHVs)—A Study of the Likely Effects if Permitted in the UK: Final Report. 2008. Available online: https://www.nomegatrucks.eu/deu/service/download/trl-study.pdf (accessed on 8 December 2017).

23. Letelier, O.B.; Cortada, W.B. Optimization of Load Distribution on Forest Trucks. Available online: http://www.fao.org/docrep/X0622E/x0622e0z.htm (accessed on 26 January 2018).

24. Shmulsky, R.; Jones, P.D. *Forest Products and Wood Science*, 6th ed.; Wiley-Blackwell: Chichester, UK; Ames, IA, USA, 2011; ISBN 978-0-8138-2074-3.

25. Trzciński, G.; Sieniawski, W.; Moskalik, T. Effects of Timber Loads on Gross Vehicle Weight. *Folia For. Pol. Ser. For.* **2014**, *55*, 159–167. [CrossRef]

26. Tomczak, A.; Jakubowski, M.; Jelonek, T.; Wąsik, R.; Grzywiński, W. Mass and density of pine pulpwood harvested in selected stands from the Forest Experimental Station in Murowana Goślina. *Acta Sci. Pol. Ser. Silvarum Colendarum Ratio Ind. Lignaria* **2016**, *15*, 105–112. [CrossRef]

27. Brown, M. The impact of tare weight on transportation efficiency in Australian forest operations. Harvesting and Operations Program, Research Bulletin 3. 2008. Available online: https://fgr.nz/documents/download/4740 (accessed on 8 December 2017).

28. Sieniawski, W.; Trzciński, G. Analysis of large-size and medium-size wood supply. In *Raport 12/2010*; Norwegian Forest and Landscape Institute: Honne, Norway, 2010; pp. 56–57.

29. Ghaffariyan, M.R.; Acuna, M.; Brown, M. Analysing the effect of five operational factors on forest residue supply chain costs: A case study in Western Australia. *Biomass Bioenergy* **2013**, *59*, 486–493. [CrossRef]

30. Owusu-Ababio, S.; Schmitt, R. Analysis of Data on Heavier Truck Weights. *Transp. Res. Rec. J. Transp. Res. Board* **2015**. [CrossRef]

31. McDonell, K.M.; Devlin, G.J.; Lyons, J.; Russell, F.; Mortimer, D. Assessment of GPS tracking devices and associated software suitable for real time monitoring of timber haulage trucks 2008. In *Annual Report*; COFORD: Ireland, UK; Available online: http://www.coford.ie/media/coford/content/publications/projectreports/annualreports/2008-AR-E.pdf (accessed on 18 February 2017).

32. Devlin, G.J. Applications and Development of Real-Time GPS Tracking Systems and On-Board Load Sensor Technology for Wood Transport in Ireland. COFORD, Ireland. 2008. Available online: http://www.coford.ie/media/coford/content/eventspresentations/events2008/GPSusagefor%20TimberTrucks.pdf (accessed on 7 April 2017).

33. GUS Forestry 2016. Available online: https://stat.gov.pl/en/topics/agriculture-forestry/forestry/forestry-2016,1,7.html (accessed on 26 January 2018).

34. Russell, F. Transportation networks in forest harvesting: Early development of the theory. In Proceedings of the Conference New Roles of Plantation Forestry Requiring An Appropriate Tending and Harvesting Operations, University of Washington, Seattle, WA, USA, 29 September–5 October 2002; Available online: http://faculty.washington.edu/greulich/Documents/IUFRO2002Paper.pdf (accessed on 7 April 2017).

35. Devlin, G.J.; McDonnel, K. Assessing Real Time GPS Asset Tracking for Timber Haulage. *Open Transp. J.* **2009**, *3*. [CrossRef]

36. McDonald, T.P.; Haridass, K.; Valenzuela, J. Mileage savings from optimization of coordinated trucking. In Proceedings of the 33rd Annual Meeting of the Council on Forest Engineering, Southern Research Station, Auburn, AL, USA, 6–9 June 2010; pp. 1–11. Available online: https://www.srs.fs.fed.us/pubs/ja/2010/ja_2010_mcdonald_002.pdf (accessed on 7 January 2017).

37. Šušnjar, M.; Horvat, D.; Zorić, M.; Pandur, Z. Axle Load Determination of Truck with Trailer and Truck with Semitrailer for Wood Transportation. *Croat. J. For. Eng.* **2011**, *32*, 379–388.

38. Trzciński, G.; Moskalik, T.; Wojtan, R.; Tymendorf, Ł. Variability of loads and gross vehicle weight in timber transportatio. *Sylwan* **2017**, *161*, 1026–1034.

39. Šušnjar, M.; Horvat, D.; Zorić, M.; Pandur, Z.; Vusić, D.; Tomašić, Ž. Comparison of real axle loads and wheel pressure of truck units for wood transportation with legal restrictions. In Proceedings of the FORMEC Conference, Pushing the Boundaries with Research and Innovation in Forest Engineering, Graz, Austria, 9–13 October 2011; Available online: https://www.formec.org/images/proceedings/2011/formec2011_paper_susnjar_etal.pdf (accessed on 8 January 2017).

40. Baumgras, J.E. Better Load-Weight Distribution is Needed for Tandem-Axle Logging Trucks. Research Paper, U.S. Forest Service. 1976. Available online: https://www.fs.fed.us/ne/newtown_square/publications/research_papers/pdfs/scanned/OCR/ne_rp342.pdf (accessed on 6 February 2017).

Article

A Spatially Explicit Method to Assess the Economic Suitability of a Forest Road Network for Timber Harvest in Steep Terrain

Leo Gallus Bont [1,*], Marielle Fraefel [2] and Christoph Fischer [3]

[1] Swiss Federal Institute for Forest, Snow and Landscape Research (WSL), Forest Production Systems Group, Zuercherstrasse 111, CH 8903 Birmensdorf, Switzerland
[2] Swiss Federal Institute for Forest, Snow and Landscape Research (WSL), GIS Group, Zuercherstrasse 111, CH 8903 Birmensdorf, Switzerland; marielle.fraefel@wsl.ch
[3] Swiss Federal Institute for Forest, Snow and Landscape Research (WSL), Scientific Service NFI, Zuercherstrasse 111, CH 8903 Birmensdorf, Switzerland; christoph.fischer@wsl.ch
* Correspondence: leo.bont@wsl.ch; Tel.: +41-44-739-26-67

Received: 31 January 2018; Accepted: 23 March 2018; Published: 27 March 2018

Abstract: Despite relatively high road density in the forests of Switzerland, a large percentage of that road network does not fulfill best practice requirements. Before upgrading or rebuilding the road network, harvesting planners must first determine which areas have insufficient access. Traditional assessment methods tend to only report specific values such as road density. However, those values do not identify the exact parcels or areas that are inaccessible. Here, we present a model that assesses the economic suitability of each timbered parcel for wood-harvesting operations, including tree-felling and processing, and off- and on-road transport (hauling), based on the existing road network. The entire wood supply chain from forest (standing trees) to a virtual pile at the border of the planning unit was captured. This method was particularly designed for steep terrain and was tested in the Canton of Grisons in Switzerland. Compared with classical approaches, such as the road density concept, which only deliver average values, this new method enables planners to assess the development of a road network in a spatially explicit manner and to easily identify the reason and the location of shortcomings in the road network. Moreover, while other related spatially explicit approaches focus only on harvesting operations, the assessment method proposed here also includes limitations (road standards) of the road network.

Keywords: harvesting; steep terrain; forest operations; forest road network; optimization heuristics

---

1. Introduction

For the efficient management of a forestry system, especially the harvesting and hauling of timber, a state-of-the-art forest road system is required. Those roads should accommodate large trucks, such as five-axle or 40-ton trucks in the case of Switzerland. Transport costs will be considerably higher, even with high forest road densities, if access is only possible with small trucks. As shown by Beck and Sessions [1], upgrading weak parts of the road network offers promising potential to increase the overall efficiency of harvesting and hauling operations. However, before the network can be upgraded or rebuilt, areas with insufficient access must be identified.

Classical approaches to assessing road networks in terms of timber production have been based on the "optimum road spacing/optimum road density" concept of Matthews [2]. Considering the various costs for on- and off-road transport as well as those for road construction, Matthews determined the road spacing for minimum overall costs. This approach was further expanded by Sundberg, Segebaden, Abegg, Thompson, and Heinimann [3–7] However, this concept of "optimum road density" assumes terrain conditions to be homogenous across the terrain or forest of interest, and the output is just an average value over the entire area. These problems make it difficult to identify sites with insufficient access.

Another area of research, the automatic planning of road networks, was triggered by the introduction of digital elevation models (DEMs) and their use in geographic information systems. Kirby [8] presented a linear programming approach that supports road network planning for different objectives, while Mandt [9] described a shortest-path application for building roads that connect two specific points. Later, Anderson and Nelson [10] developed a vector-based automatic road location model, in which a network was created by linking given landings with a shortest-path algorithm combined with heuristics algorithms. Compared to the previous approaches, this new method enabled the planners to implement a better representation of road links, which also made it possible to use the system in steep terrain. By mapping road-turning constraints and by introducing a generic road cost model based on a DEM and geotechnical layers, Stückelberger et al. [11–13] improved the approach of Anderson and Nelson. Another approach to the development of road layouts via DEM was later presented by Chung et al. [14]. While these approaches focus only on optimizing the road layout between certain points or landings, some methods have been developed that simultaneously optimize the harvest and the road network layout. The most common approach used in this context is PLANEX (Epstein et al. [15,16]), for which a Mixed Integer Linear Programming (MILP) model is used to minimize the costs of harvesting, machine installation, road construction and road transport. However, since real problems are very large, they must be addressed with a greedy heuristics algorithm. Diaz et al. [17] presented a tabu search metaheuristic, with significantly shorter computational times than PLANEX. CPLAN (Chung et al. [18]) also simultaneously optimized the assignment of cable-logging equipment and road link locations, based on a heuristics algorithm. Bont et al. [19] presented a modeling approach for a similar task, which could then be solved with a Mixed Integer Linear Programming formulation.

All those approaches were developed for planning new road networks and do not take into account existing infrastructure. There are few methods that consider redesigning of existing road networks. Henningson et al. [20] presented a model for redesigning forest road networks. They considered road upgrades to reduce the losses due to road closures caused by heavy rains or thawing. As this model is hard to solve, Flisberg et al. [21] presented an easier-to-solve implementation for this problem.

Before redesigning a road network, it is necessary to identify forest areas with insufficient access. This helps to set priorities, in particular if there are budget restrictions. Applying existing methods, assessment would only be possible within a limited area (catchment, less than 50 km^2), and not over large regions (whole regions or countries). Although the model Sylvaccess by Dupire et al. [22] automatically maps the accessibility of forests based on the three main logging techniques employed in France: skidder, forwarder, and cable yarder, it focuses only on harvesting operations and does not include the limitations of the forest road network or consider the hauling process.

Therefore, we have devised a new method that provides a spatially explicit overview of the need for road redevelopment. It can be used to assess the suitability of each timber parcel for economically efficient production (we defined 'timber parcel', or TP, as the smallest harvestable unit, here being 10 m × 10 m). This means that the entire forest area was partitioned into regular 10 m × 10 m parcels. The most inadequate parts of the road network are located in steep terrain [23,24], therefore the main focus here is on cable-based harvesting operations. The allocation is performed only using the existing road network. Our research goals were three-fold: (1) develop a method to assess each timber parcel for its economic suitability for the entire harvesting process, including transportation; (2) make this tool applicable on large scales, i.e., areas covering more than 5000 km^2; and (3) specifically take into account the characteristics of steep terrain. The main advantage of the proposed heuristic model over the operations research (OR)-based forest transportation models is that it is an easy-to-handle, computation-efficient model particularly tailored for assessing forest accessibility given the existing forest road network.

The first step in this endeavor was to review the current means for assessing road networks. After developing our representation model, we then applied it to the Canton of Grisons (Switzerland), and evaluated the results in cooperation with the cantonal forest service.

2. Materials and Methods

The term 'landing' is used to describe the transition point from off-road to on-road transport. A landing is a point on a road segment of a forest road on which, for example, a tower yarder can be installed. Generally, the wood has to be extracted to a landing before hauling. The term 'road segment' here refers to a segment of a forest road. It is assumed that landings can be installed on each road segment of the forest road network. Additionally, individual landings can also be set on the superordinate road network. Road segment length does not exceed 200 m. We further use the term 'collecting point', which refers to a virtual pile at the border of the observed site. It is the interface up to which we observe the timber production and is located on a railway station or on the superordinate road network. There could be more than one 'collecting point' in a project site.

Our model assesses the economic suitability of each TP for the wood-harvesting operation, including tree-felling and processing and off- and on-road transport (hauling), based on the existing road network. The entire wood supply chain from forest (standing trees) to the collecting points at the border of the planning unit was captured. The conceptual model, which is visualized in Figure 1, comprised three subsystems: (1) "Harvesting Options", which identified for each road segment the TPs that could be accessed with the available harvesting systems and then assessed the economic efficiency of the harvesting techniques; (2) "Hauling Route", which determined for each road segment the "best" path to the collecting points; and (3) "Heuristic Optimization Model", which evaluated the best combination of harvesting option and hauling route. Individual elements within this conceptual model are discussed below. An overview of the interactions between the model subsystems is given in Figure 2 (harvesting options), Figure 3 (hauling route) and Figure 4 (heuristic optimization).

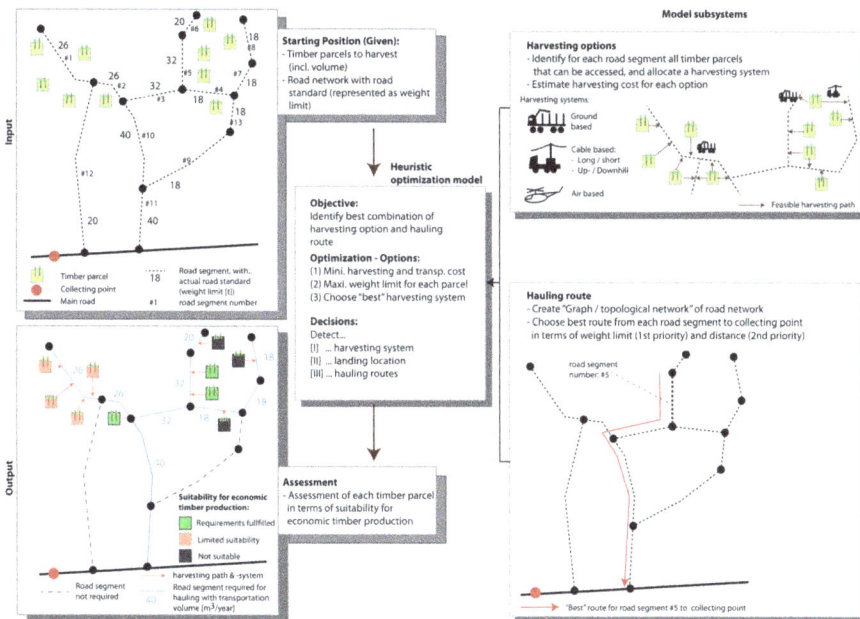

Figure 1. Conceptual model with input and output values (left) and subsystems (right).

351

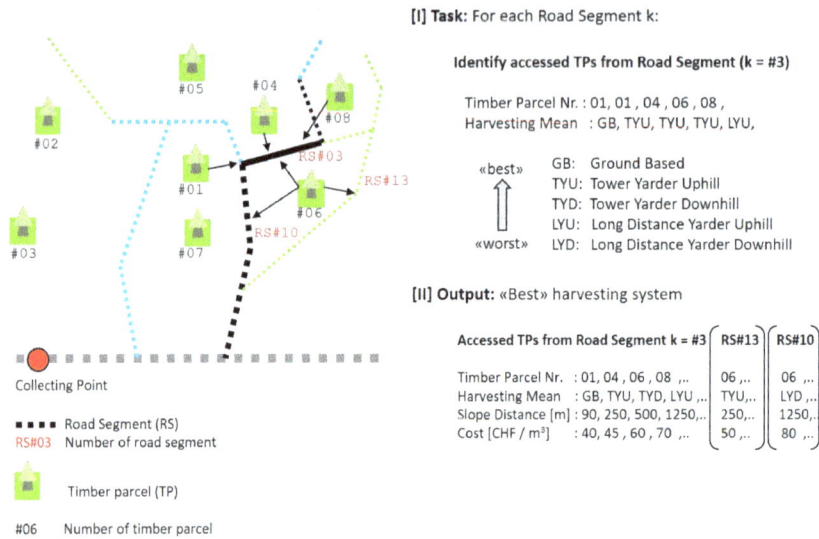

[I] Task: For each Road Segment k:

Identify accessed TPs from Road Segment (k = #3)

Timber Parcel Nr. : 01, 01 , 04 , 06 , 08 ,
Harvesting Mean : GB, TYU, TYU, TYU, LYU,

«best»   GB: Ground Based
        TYU: Tower Yarder Uphill
        TYD: Tower Yarder Downhill
        LYU: Long Distance Yarder Uphill
«worst» LYD: Long Distance Yarder Downhill

[II] Output: «Best» harvesting system

Accessed TPs from Road Segment k = #3	RS#13	RS#10
Timber Parcel Nr. : 01, 04 , 06 , 08 ,...	06 ,..	06 ,..
Harvesting Mean : GB, TYU, TYD, LYU ,..	TYU,..	LYD ,..
Slope Distance [m] : 90, 250, 500, 1250,..	250,..	1250,.
Cost [CHF / m³] : 40, 45 , 60 , 70 ,..	50 ,..	80 ,..

Collecting Point

▪▪▪▪ Road Segment (RS)
RS#03 Number of road segment

Timber parcel (TP)

#06   Number of timber parcel

Figure 2. Task and output of the model subsystem "harvesting options". TPs that could be accessed with the available harvesting systems are identified.

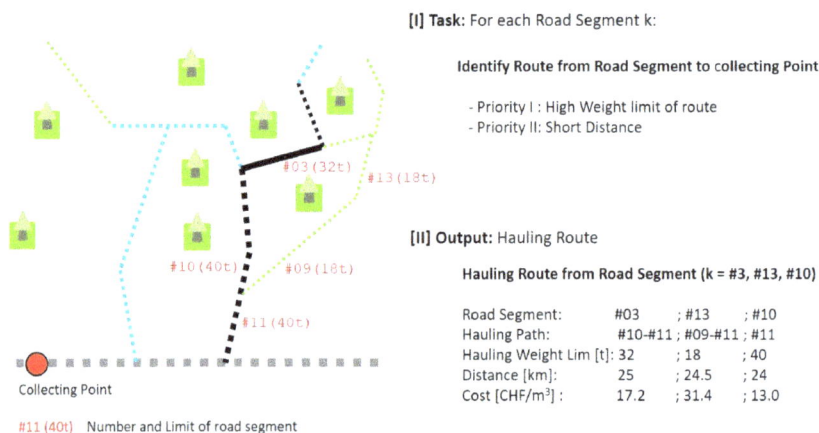

[I] Task: For each Road Segment k:

Identify Route from Road Segment to collecting Point

- Priority I : High Weight limit of route
- Priority II: Short Distance

[II] Output: Hauling Route

Hauling Route from Road Segment (k = #3, #13, #10)

Road Segment:	#03	; #13	; #10
Hauling Path:		#10-#11 ; #09-#11 ; #11	
Hauling Weight Lim [t]:	32	; 18	; 40
Distance [km]:	25	; 24.5	; 24
Cost [CHF/m³] :	17.2	; 31.4	; 13.0

Collecting Point

#11 (40t) Number and Limit of road segment

Figure 3. Task and output of the model subsystem "hauling route". The "best" path to the collecting point for each road segment is identified.

"Harvesting options" identifies for each road segment all TPs that could be accessed and allocates a harvesting system to each TP. In the example in Figure 2, TP #01, #04, #06, #08 can be accessed from road segment #03, with the harvesting techniques 'ground based' (GB), 'tower yarder uphill' (TYU), 'tower yarder downhill' (TYD) and 'long distance yarder uphill' (LYU). Further, the slope distance from the landing to the TP and a cost estimation are calculated. "Hauling routes" determines the "best" route to the collecting point for each road segment. "Best" route means that the weight limit of the route is as high as possible, whereas the hauling distance should be as small as possible. In the example in Figure 3, we listed the hauling routes from road segments #3, #13 and #10 in the export section. From road segment #3, the hauling route runs over road segment #10 and #11 with a weight limit of 32 t, a distance of 25 km and costs 17.2 CHF/m^3. The subsystem "heuristic optimization" takes the output from the other two subsystems as input and identifies for each TP up to three combinations of harvesting technique and hauling route, one of which guarantees the highest possible weight limit for hauling, another that yields the most efficient harvesting system and finally one with the lowest overall cost. It is possible that one detected combination fulfills several of these objectives. An example for TP #06 is given in Figure 4. Here, the combination that uses road segment #3 as landing fulfills objective III (lowest cost), while hauling via road segment #10 ensures the highest weight limit during hauling and via #13 the use of the most efficient harvesting technique.

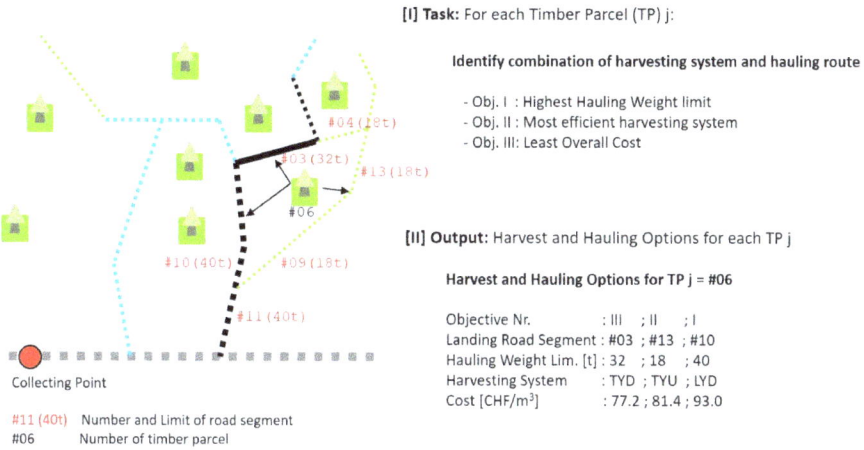

[I] Task: For each Timber Parcel (TP) j:

Identify combination of harvesting system and hauling route

- Obj. I : Highest Hauling Weight limit
- Obj. II : Most efficient harvesting system
- Obj. III: Least Overall Cost

[II] Output: Harvest and Hauling Options for each TP j

Harvest and Hauling Options for TP j = #06

Objective Nr.	: III	; II	; I
Landing Road Segment	: #03	; #13	; #10
Hauling Weight Lim. [t]	: 32	; 18	; 40
Harvesting System	: TYD	; TYU	; LYD
Cost [CHF/m³]	: 77.2	; 81.4	; 93.0

Figure 4. Task and output of the model subsystem "heuristic optimization", which evaluates favorable combinations of harvesting options and hauling routes for each timber parcel.

The result from this step forms the basis for the assessment (suitability for an economically efficient timber production) of each TP. The assessment is made on the basis of a combination of the hauling route weight limit and the applicable harvesting system. For example, in our case study, the best rating will be achieved if the hauling route weight limit is at least 28 t and concurrently a ground-based system or a tower yarder (downhill or uphill) can be used. In the example in Figure 4, the harvesting and hauling combination that uses road segment #3 as landing fulfills the requirements for the best rating ('Fulfills best practice requirements'), hence TP #06 also achieves the rating 'Fulfills best practice requirements'.

Given the assessment is an economic evaluation, it would be possible to consider only minimum cost as indicator. However, practical considerations mean that such an approach might not be the most accurate: First, hauling is often made by independent small enterprises. Their cost structure is a well-guarded secret, and therefore productivity models only roughly cover the actual cost structure. Furthermore, pricing depends on market conditions and is not constant; Second, in reality there are several potential destination points (several sawmills, heating and power plants, transshipment points to the railway, etc.), so the real length of the hauling route remains unknown; Third, harvesting productivity models generally consider particular conditions, but do not account for the whole range of properties usually found within a region. To take in account the full diversity, a range of productivity models would need to be used, and even then there would be no guarantee that all cases could be modeled. Fourth, even if such an 'overall' productivity model existed, it would be challenging to obtain reliable input parameters. For example, the extracted timber along a cable road is one of the most relevant input parameters for a cable yarder productivity model, however to make an estimation, knowledge of the silvicultural treatment scenario is required, which is not usually available over whole regions. Summing up, evaluating up to three different harvesting/hauling combinations allows a more robust suitability estimation to be obtained compared to the evaluation of only one alternative.

2.1. Input Data

The following 10 m resolution raster layers were input datasets: (1) digital elevation model (swissALTI³ᴰ [25]); (2) timber parcels to harvest (including volume); (3) obstacles to cable-yarding (e.g., high-voltage power lines, railways, superordinate roads, cable cars, and buildings); and (4) the soil property map (for soil trafficability). The timber parcel map indicates the forest area. To save memory and to increase the computing speed, the analysis is only performed for the defined TPs. The information about the timber volume can be used to estimate how much timber is transported along a certain road segment; however, this task is not covered in this paper. The digital elevation model provides terrain information when checking the feasibility of a cable road. Moreover, the digital elevation model, and the slope raster derived from it, are also used for the identification of the trafficable areas. The qualified landings (landings on the superordinate road network) and forest roads were imported as vector datasets, including information on the load-bearing capacity of the road (weight limit, i.e., the maximum gross vehicle mass of single trucks permitted on that road), representing the road standard. The road network was divided into segments, with a new segment starting either at junctions or wherever the road standard changed. Several timber collecting points (virtual piles) were designated on the superordinate road network.

2.2. Model Subsystem "Harvesting Options"

The primary objective of this study was to assess the quality of a given forest road network in steep terrain. Therefore, the main focus was on accurate analysis of cable-based harvesting systems. However, even mountainous regions have some areas that can be logged by ground-based systems. Here, the examination of such ground-based systems was simplified, but still adequate for predominantly steep terrain. Nevertheless, the method would have to be adapted if applied in areas mainly harvested with ground-based systems.

There were no constraints on the maximum volume to be extracted to each road segment. Further, it is possible to allocate more than one harvesting system to the same road segment from different TPs, and more than one potential harvesting system can be allocated to each TP.

2.2.1. Cable-Based Harvesting

The key objective of our model is to analyze the yarder accessibility of timber parcels, using the workflow described below and illustrated in Figure 5:

(a) Along each road segment, landings were placed approximately 30 m apart. At each landing, 32 (default value) radial lines were proposed (Figure 5a).

(b) For each line direction, we determined the maximum feasible distance when building a cable road with a given number of intermediate supports (five being the default) (Figure 5b). These maximum feasible distances represent potential endpoints that depend upon the properties of both terrain and yarding system (e.g., the maximum skyline length, breaking strength of the skyline, minimum clearance between skyline and ground, and any obstacles as mentioned above). We implemented the design approaches of Pestal and Zweifel [26,27], running the former by default because of its better calculation efficiency [28].

(c) Finally, we identified the TPs that were accessible from a section of cable road. If the center of a parcel was within a certain distance of the skyline (default: 30 m, or approximately one tree length), then the parcel was considered accessible. In the last step (Figure 5c), the accessible TPs associated with each single cable road were collected for each segment and stored with the following information that served as the basis for our cost estimation: yarding direction (uphill/downhill) and yarding distance (shortest-distance TP—road segment). If both up- and downhill yarding were possible for a particular TP, then the latter was preferred because it usually causes less damage to the remaining trees. If several different yarding technologies were analyzed concurrently, then steps (b) through (c) were repeated for each alternative.

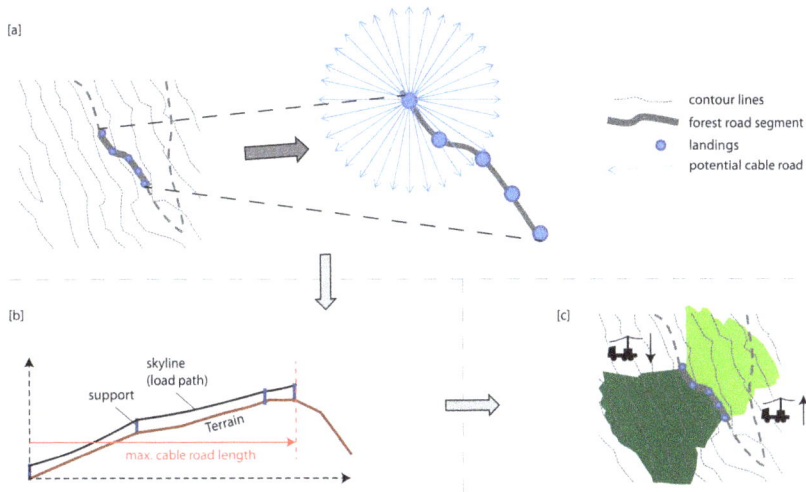

Figure 5. Workflow for analysis of yarder accessibility. (**a**) Potential landings are placed on road segments; at each landing, radial lines indicate potential cable roads; (**b**) Each potential cable road is checked for maximum feasible distance, based on terrain properties; (**c**) Accessible timber parcels along each cable road are collected and stored for each segment.

2.2.2. Ground-Based Harvesting

Ground-based harvesting requires trafficable terrain. The main factors that affect vehicle movement are slope, soil bearing capacity, and the frequency and size of obstacles [29]. Those factors can easily be measured on a single plot, but quantitatively describing their spatial and temporal variabilities is much more challenging [30]. For example, soil bearing capacity is not a constant value, but changes periodically because of meteorological disturbances such as rain or snowfall [31]. Terramechanical mobility models are usually utilized to calculate the maximum gradeability of vehicles. Some examples include work by Ashmore et al. [32] and Brixius [33] for wheeled vehicles, or the model of Ahlvin and Haley [29] for tracked machines. The use of winch-assisted vehicles was not considered here because winches should only be used to increase safety and not to extend the maximum gradeability. However, because spatial data about soil bearing capacity and obstacle frequency/size are usually not available on larger scales, the application of terramechanical mobility models is limited. As already mentioned, assessing the trafficability of TPs was a secondary task in our model, and since ground-based harvesting systems are of very little importance in our project area, terramechanical models were not used. Instead, a simpler method was chosen, for which we used a soil property map [34]. The soil property map shows geomorphological and pedological units, which were assessed according to their potential uses in agriculture and forestry (e.g., trafficability). The assessment of trafficability is based on the soil-mechanical and pedological properties of the mapping units, such as load-bearing capacity, plasticity, shear strength, skeletal content, permeability and watering of the soils. For example, a high skeleton content improves trafficability, whereas a large organic content in the soil makes trafficability more difficult with increasing water content. This map quantitatively predicted four classes of trafficability: (1) well trafficable; (2) trafficability limited if rain or percolation water appears; (3) only limited trafficability; and (4) trafficability heavily limited or impossible. We used maximum slope gradient limits of 35% (for soil class 1), 25% (class 2), 10% (class 3), and 0% (class 4) for estimating maximum gradeability.

We used the following approach: The open-source toolbox GDAL was used to generate a slope raster from the elevation raster. We then estimated for each TP the maximum gradeability for a vehicle on the basis of the soil property map, according to the classes of trafficability mentioned above. In order to determine whether a TP is trafficable, the slope was then compared with the trafficability class from the soil property map. If the slope on a TP was less than or equal to the maximum gradeability, then the TP was classified as trafficable. For example, if a TP has soil class 1 (well trafficable) and a slope of 30%, it is regarded as trafficable. However, if the TP has only soil class 2 and again a slope of 30%, it is considered not trafficable.

The derivation of a maximum gradeability based on the soil classification is expert based and therefore not precise and a potential source of error. However, these values matched those reported by Eichrodt [31] for sites with similar soil properties.

In addition, it was checked for all TPs that were classified as trafficable whether they were also connected to a forest road, in order to eliminate parcels that are classified as trafficable, but are surrounded by cells that are not trafficable. We checked for this through network analysis (shortest path algorithm), setting the maximum transport distance to a landing to 300 m.

2.2.3. Aerial-Based Harvesting

If neither cable-based nor ground-based harvesting systems can be established for an area, then one might consider the use of helicopters for particular TPs. However, this method is usually not economical and is therefore not included in the model in more detail.

2.2.4. Assignment of Harvesting Systems to Timber Parcels

The results of the "Harvesting Options" model informs planners which TPs along any given road segment can be accessed by certain harvesting systems, and can indicate the preferred logging direction (uphill/downhill). However, there might be several options possible for most TPs. Only the most efficient harvesting system for each TP was considered for further analysis. This was achieved by ranking harvesting systems according to their economic efficiency and then assigning each TP to the highest-ranked system that made the parcel accessible. In the case described here, we used two types of cable yarders—tower and long-distance yarder—as well as a ground-based system. These were ranked as follows, from most to least efficient: ground-based (1); tower yarder uphill (2); tower yarder downhill (3); long-distance yarder uphill (4); and long-distance yarder downhill (5).

The method of ranking different harvesting systems applied here was developed by Heinimann [35] for the investigation of cable crane deployment in the Swiss alps. This type of terrain evaluation is a synthesis of technical and functional site classification. Terrain parameters are recorded with regard to the selection of logging systems and logging systems are subsequently displayed directly on a map. Heinimann [36] and Lüthy [37] showed that this method can be incorporated into a Decision Support System as a component and deliver reliable results. This method delivers a terrain classification, and it does not propose a solution for operational management.

In application of this rule-based ranking system it is possible that, for example, a cable TP is surrounded by ground-based TPs, especially near the area of transition from trafficable areas to cable yarder areas as shown in Figure 6. This phenomenon has little impact on the final result (suitability map), nor does it pose a restriction for applying the harvesting map since firstly, the harvesting systems map is only an intermediate result. The processing steps and assumptions that follow will have a higher impact on the final result. Secondly, the transition between trafficable and yarder area will never be a hard border due to generalizations and uncertainties in the underlying data, such as the bearing capacity of the soil or the simplified slope map (10 m × 10 m resolution). In reality, there is rather a transition zone, and the isolated cable TPs reflect the range of this zone. However, to remove such isolated TPs, operational models such as described in Bont et al. [38] could be applied. This requires definition of a management scenario and is only applicable for small areas.

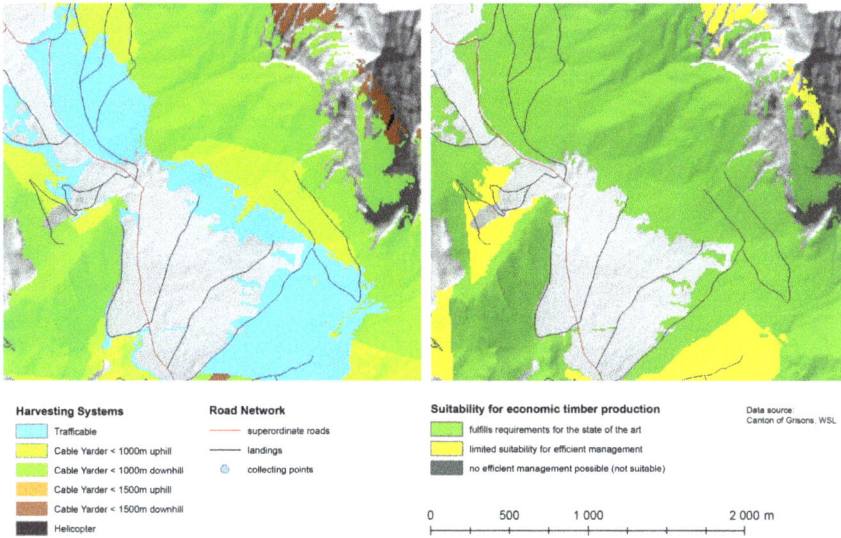

Harvesting Systems
- Trafficable
- Cable Yarder < 1000m uphill
- Cable Yarder < 1000m downhill
- Cable Yarder < 1500m uphill
- Cable Yarder < 1500m downhill
- Helicopter

Road Network
- superordinate roads
- landings
- collecting points

Suitability for economic timber production
- fulfills requirements for the state of the art
- limited suitability for efficient management
- no efficient management possible (not suitable)

Data source: Canton of Grisons, WSL

0   500   1 000   2 000 m

Figure 6. (**Left**) Map of harvesting systems located in the canton of Grisons, Switzerland. Especially near the transition from trafficable areas to cable yarder areas, there are some cable TPs that are isolated and surrounded by trafficable area; (**Right**) Suitability map.

At this point one can ask whether a verification of the 'ranking-based assignment of harvesting systems to the TPs' with an operational model makes sense. In the following this point will be discussed briefly. In contrast to the method presented above, an operational model such as described in Bont et al. [38] already proposes the spatial layout of the individual cable roads, an application of which is shown in Figure 7 [39]. However, our method classifies the same area almost exclusively as accessible by tower yarder (not displayed here). In principle, the result of the application of the Bont et al. [38] operational model is well in line with the terrain classification described above. The biggest difference lies in the distinction of the harvesting system 'Helicopter' (not accessible). Due to the spatial arrangement of cable roads, there are sometimes TPs between the individual lines which are classified as not accessible. This simply means that it is not economically worthwhile to set up cable roads for these few TPs, although these TPs could potentially be reached with a yarding system. Other areas at the edge of the forest were also classified as not accessible due to very small harvesting volumes making installation of additional cable roads non cost-effective. In addition, more harvesting systems, such as winches, were taken into account in the terrain classification approach presented here.

The comparison of an operational model with the ranking-based terrain classification shows that the operational model requires detailed input data (silvicultural objectives, definition of machines to be used) and clearly defined management objectives, but also delivers results with a high degree of precision. On the other hand, the results are not very robust. If the management objectives change, the solution may look completely different. The ranking-based terrain classification does not provide results accurate at pixel level, but the solutions are less dependent on predefined scenarios. In our case, the latter is more useful because a generally valid assessment that is as independent of specific management scenarios as possible is required. These requirements also make the verification of the ranking-based method by the operational model difficult.

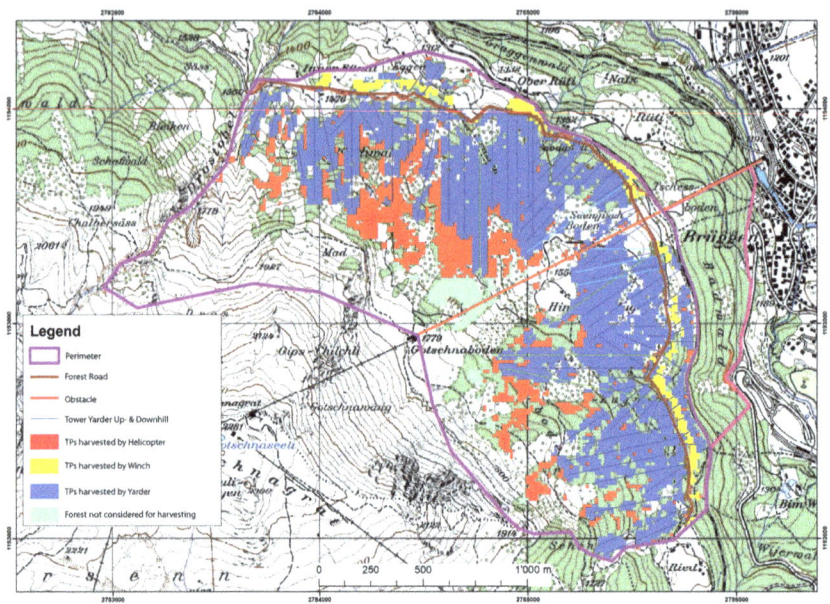

Figure 7. Application of an operational harvesting and cable road layout model (Breschan et al. [39]) in the Gotschna area in the canton of Grisons, Switzerland (topographic map © swisstopo (JA100118)).

2.3. Model Subsystem "Hauling Route"

The task of the model subsystem "Hauling route" was to identify the best route from the center of each road segment to one of the available collecting points (virtual pile at the interface of the planning unit). Both collecting point and route were selected in such a way that, as first priority, the weight limits during removal were as high as possible and, as second priority, the distance from the road segments to the "collecting point" was as short as possible.

2.3.1. Graph Representation

The road network was represented as a graph where the nodes were the intersections between road segments and the arcs were the segments. An example is displayed in Figure 8. In addition to the nodes of the road network, a virtual terminal node was inserted, which was linked with virtual arcs (with weight 0) to all collecting points. The idea behind introducing a virtual terminal node was to have only one terminal node in the representing network, instead of having several potential terminals (collecting points). This substantially simplifies the network analysis. The virtual node made it possible to identify the "best" path from any road segment to the collecting point, as well the "best" collecting point. This was accomplished by applying a shortest path algorithm [40] with the source node being any node in the road network and the terminal node being the virtual terminal node.

The weight of the arcs (road segments) are calculated ensuring that the roads chosen for hauling had the highest possible weight limits or load bearing capacities:

$$W = L * 100^{Ranking(T)}, \tag{1}$$

where W is the weight of the arcs, L is the length of a road segment (m), T is the weight limit of the road segment (t), and $Ranking(T)$ is a function that sorts the unique values of T and presents them in numerical order as the output. For example, if we had a set of $T = \{10, 40, 25, 18, 10, 28, 32, 18\}$, then the unique values were placed in descending order: {40, 32, 28, 25, 18, 10}. Thus, when applying *Ranking(40)*, the output would be 1; for *Ranking(25)*, it would be 4.

The results from this step in the analysis provided information about weight limit and hauling distance for each road segment, both of which were then used for calculating hauling costs. From this step, there is one best route for each TP.

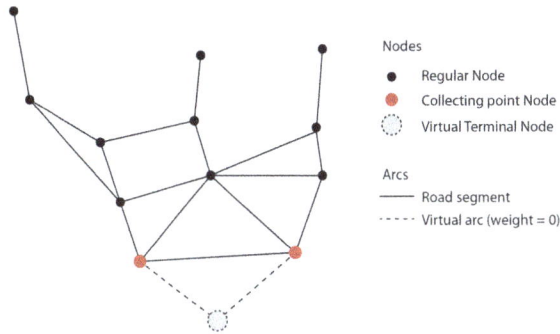

Figure 8. Conceptual representation of a road network, with regular nodes representing intersections between road segments. Collecting-point nodes are terminals for timber transportation. Best path and best collecting node are chosen by inserting a virtual terminal node that is linked with virtual arcs to collecting points.

2.3.2. Road Classification

Hauling costs depend upon the type of truck that can be driven on a forest road. Those trucks are usually classified according to the number of axles [41], which, in Switzerland, range from two to five (see truck properties in Table 1). For planning and design it is the axle load, and not the total weight of the truck, that is used when determining the dimensions of the road superstructure. Because axle loads are standardized, usually at approximately 8 to 10 t, the superstructure to accommodate a large truck does not have to be proportionally greater than that for small trucks, and normally is not restrictive. In contrast, artificial structures such as bridges often have a limiting effect because here the overall mass of the truck is relevant.

In many cases, the road geometry prohibits the movement of larger trucks either because of the curve radii or narrowness. To be suitable for a five-axle truck, a road should be at least 3.5 m wide (2.55 m for the truck itself plus a safety strip) and curves should have a minimum radius of 10 m with a road width of 5.5 m. Although analysis of the whole geometry of a complete road should be made, such data are not available because of the high cost of collection. Therefore, specifying the maximum permissible total weight (weight limit) often relies upon a synthesis of the limitations in artificial structures and road geometry if one is to describes the suitability for a corresponding truck type. Therefore, the attribute weight limit is used as an index of the road standard in our study.

Table 1. Main mass characteristics of different truck types used in hauling timber [41].

Number of Axles	Total Mass (t)	Mass of Truck (t)	Net Payload (t)
5	40	17	23
4	32	15	17
3	26	13	13
2	18	10	8

2.4. Heuristic Optimization Model: Selecting the Landing Site

So far the harvesting system that links a TP with the road segment has been determined in the subsystem "Harvesting Options" (Rule Based) and possible paths for transportation from the landing site (road segment) to the collecting node in the subsystem "Hauling route". Now the whole transportation chain from the TP across the landing site (road segment) to the collecting point is configured by applying a heuristic optimization algorithm.

A TP can be accessed from several road segments. The choice of the road segment (or landing site) from which the timber is to be moved determines the costs associated with different systems of harvesting and transport. The subsystem "Heuristic optimization model" determines the best way to harvest a particular TP and transport that timber to a collecting point. It involves selecting the harvesting system, identifying the road segment on which the landing is located, and choosing the hauling route. Three objective functions were implemented to determine the best solution: (option 1) maximize the weight limit for hauling (minimize hauling cost), (option 2) select the most efficient harvesting system (minimize harvesting cost), or (option 3) concurrently identify the most efficient harvesting and hauling methods (least cost). The heuristic algorithm was accomplished through the following steps:

- **Step 1:** Initialization

    (i) We assigned initial values for the attributes of each TP, including (1) highest weight limit for the hauling route (referred to as TP^W) (t); (2) Index number of road segment accessed from TP (TP^k) (no unit); (3) allocated harvesting system (TP^{HS}) (1 = ground-based, 2 = tower yarder uphill, 3 = tower yarder downhill, 4 = long-distance yarder uphill, or 5 = long-distance yarder downhill); and (4) total harvesting and hauling cost (TP^{TC}) (CHF). The initial values for all attributes and each TP were = −1, except for TP^{TC} and TP^{HS}, which were set = ∞.

- **Step 2:** Index the road segments with k from 1 to N (number of road segments)
- **Step 3:** For road segment $k = 1$:

    ○ (i) Evaluate T_k: the set of TPs j accessed by road segment k, as determined in the subsystem "harvesting options",

    w_k: the weight limit for hauling from road segment k to the collecting point (from subsystem "hauling option"), and

    H_{jk}: the harvesting system for accessing TP j from road segment k (from subsystem "harvesting options".

    ○ (ii) Check for each TP j in the set of T_k if attribute w_k exceeds TP^W. If so, then TP^W, TP^k, and TP^{HS} will be updated with the values from Step 3 (i): w_k, k, and H_{jk}.

- **Step 4:** Repeat step 3 for $k = k + 1$, until k equals N.

    If Option 2 was chosen to detect the best route, then Step 3 (ii) was modified to:

    - **Step 3 (ii Opt 2):** Check for each TP j in the set of T_k to determine whether attribute H_{jk}. is less than TP^{HS}. If so, then TP^W, TP^k, and TP^{HS} will be updated with the values from Step 3 (i): wk, k, and H_{jk}.

    If Option 3 was chosen, then Step 3 (ii) was modified to:

    - **Step 3 (ii a Opt 3):** Compute C_{jk}: total harvesting and hauling cost for each TP j in the set of T_k. This involved the hauling cost from road segment k to the collecting point as well as the harvesting cost of TP j with Harvesting system H_{jk} from road segment k.
    - **Step 3 (ii b Opt 3):** Check for each TP j of T_k to determine whether attribute C_{jk} is less than TP^{TC}. If so, then TP^W, TP^k, TP^{HS} and TP^{TC} will be updated with the values from Step 3 (i): w_k, k, H_{jk}; and C_{jk}.

From this we identified, for each TP, the road segment over which the timber would be transported (TP^k), the weight limit for hauling (TP^W), the harvesting system (TP^{HS}), and the costs for hauling and harvesting (TP^{TC}).

2.5. Assessing the Suitability for Efficient Timber Management

Before a forest road network can be evaluated, first "suitability for economical timber production" must be defined. The timber parcels were categorized into one of three classes, as defined by our project partner, the forest service of the Canton of Grisons: (1) Fulfills best practice requirements; (2) limited suitability for efficient management; or (3) no efficient management possible (not suitable). The suitability for economical timber production for any given site depended upon the harvesting system that was implemented and the weight limits during hauling (Table 2). At each proposed landing, conditions related to the harvest system and the weight limit during transport had to be fulfilled. For example, the only TPs placed within quality class 1 were those that could be reached with ground-based systems or a tower yarder and for which the timber could be transported with a 28-ton truck (or larger).

To conduct the analysis, we ran the "heuristic optimization model" for all three options and obtained the final attributes (TP^W, TP^{HS}) of each TP. We then assessed the suitability of the TPs for each option according to the classification in Table 2. For the final assessment, we considered the best assessment that had resulted from those three options for each TP.

Table 2. The suitability of an area for economical timber production depends upon the choice of harvesting system (4 options shown) as well as the weight limit during hauling (3 options shown), and can be described according to quality classes: (1) Fulfills best practice requirements; (2) limited suitability for efficient management; or (3) no efficient management possible (not suitable).

		Weight Limit During Hauling (t)		
		<18	$\geq$18, <28	$\geq$28
Harvesting system	Ground-based	(3)	(2)	(1)
	Tower yarder	(3)	(2)	(1)
	Long distance yarder	(3)	(2)	(2)
	Helicopter	(3)	(3)	(3)

2.6. Additional Output

Besides the final result (suitability map), the workflow produces additional results that might be of practical relevance:

(1) Map of the maximum weight limit for hauling, based on variable (TP^W), and optimization option 1;
(2) Map of the 'best' harvesting systems, based on variable (TP^{HS}), and optimization option 2; and
(3) Map of harvesting and hauling cost, based on variable (TP^{TC}), and optimization option 3.

Those maps were produced as an intermediate product during processing in the subsystem 'heuristic optimization'.

However, these maps should be used with caution, as some underlying data might be heavily simplified or based on many assumptions, such as the cost estimation. Further, the intermediate results are not suitable for operational planning. For example, a cable TP could potentially be surrounded by ground-based TPs (Figure 6), since each TP is assessed individually and no neighborhood relationships are taken into account. Neighborhood relationships are considered in operational models before making a decision [38]. In operational planning, treatment units are much smaller, and in those cases it should not be a problem to apply operational models.

The forest road network in the Swiss Canton of Grisons (total area of 7000 km², including 2023 km² in forests) does not, in many cases, meet the criteria for efficient and rational harvesting of wood. Moreover, not only the forest roads, but also the superordinate roads have restrictions that limit timber transport. The forest is located mainly in steep terrain, and more than 90% of the area would require cable-based harvesting systems, according to the National Forest Inventory [23]. Therefore, in our assessment, we considered ground-based harvesting as well as two yarder types: tower yarding (maximum skyline length of 1000 m) and long-distance yarding (maximum skyline length of 1500 m). To keep the cost calculations deliberately simple, we differentiated between up- and downhill yarding. The costs for harvesting by cable yarder are shown in Table 3; we assumed a flat cost of 40 CHF/m³ for the ground-based system. The costs in Table 3 are based on expert estimation and verified by HeProMo productivity models [42]. This procedure was chosen because in most cases productivity models are designed for particular conditions and a single productivity model is not well applicable to a heterogeneous region. Further, it would be challenging to obtain reliable input parameters for such models without knowing the silvicultural treatment scenario or the properties of the cut.

Table 3. Estimated costs of timber harvesting using two types of cable yarder over various slope distances (CHF/m³).

			Logging Distance (Slope Distance) (m)						
			0	250	500	750	1000	1250	1500
Cable system	Tower yarder	↑	40	45	50	55	60	-	-
		↓	50	55	60	65	70	-	-
	Long-distance yarder	↑	60	60	60	60	60	70	80
		↓	70	70	70	70	70	80	90

The hauling cost was computed based on the weight limit along the transportation route (Table 4). The canton was subdivided into five forest regions of 1000 to 2300 km² that were treated separately. Computations were done using Matlab with an i7-6700 CPU @ 3.40 GHz processor with 16 GB RAM.

Table 4. Hauling cost as a function of weight limit when hauling timber over a single distance.

Weight Limit (T) (t)	Hauling Cost (CHF/(m³·km))
$18 \leq T < 26$	1.28
$26 \leq T < 32$	0.91
$32 \leq T < 40$	0.69
$T \geq 40$	0.54

3. Results

Our research objectives were to develop a method that assesses timber parcels in terms of their suitability for economical production and is also applicable over large areas. Computation time was approximately six hours for the largest region, which covers approximately one-third of the canton. For the entire canton the calculations required about 20 h. The most time-consuming step was the modeling of the cable roads.

The main product of this analysis is the map of suitability of each TP for economical timber production (Figure 9, example of the Schiers/Schraubach region). In all, 30% of the total canton area falls within class 1 ("Fulfills best practice requirements"), but the proportions vary significantly among regions (Table 5). Whereas Region 2 is relatively well-developed, with 41% of its area qualifying for class 1, Region 2 is poorly developed, and only 16% of its area fulfills the best practice requirements for harvesting and transport (Table 5). Figure 9 and Table 5 also refer to the evaluation of limitations associated with both the forest and the superordinate roads. This evaluation demonstrated that the development of large parts of these forests is not up-to-date and cannot support competitive timber production (Table 5 and Figure 9). Furthermore, these results can now be used to indicate clearly which actions are most essential and to objectively identify target projects (Table 5, Figure 9).

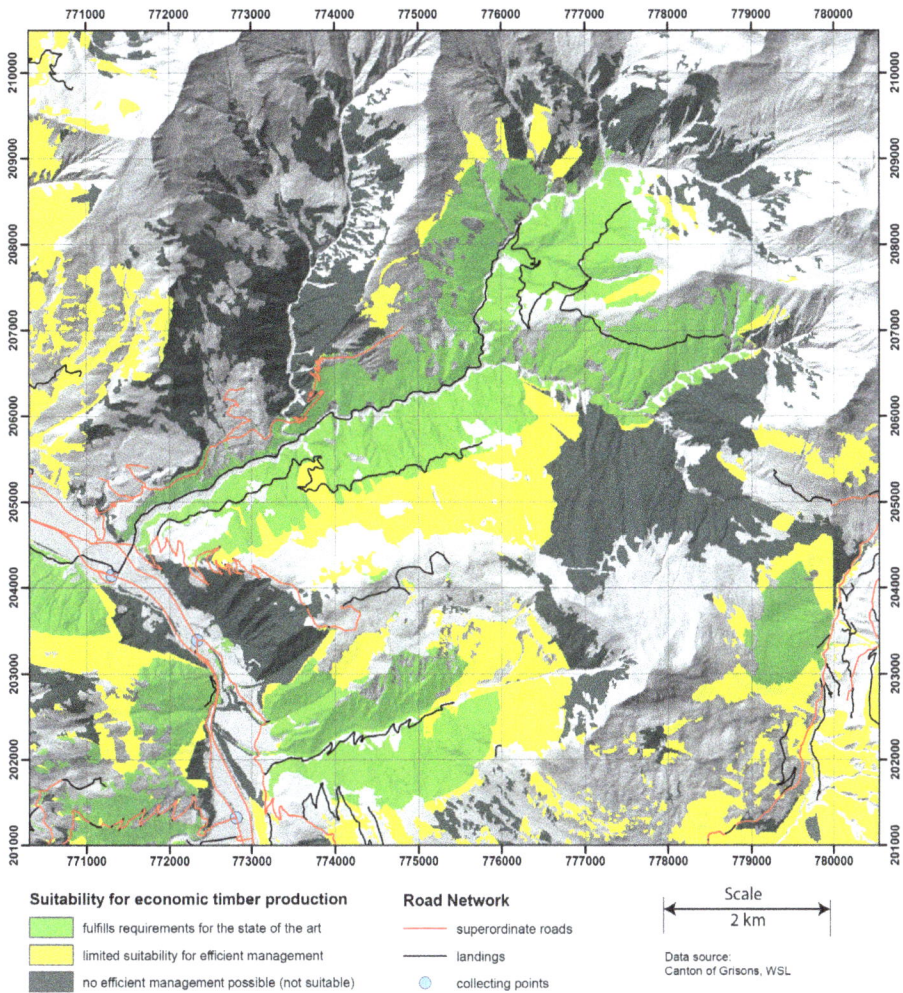

Figure 9. Extract from suitability map for region near Schiers in the canton of Grisons, Switzerland (hillshade dtm © swisstopo (JA100118)).

Table 5. Key figures used to determine the suitability of the timber parcels for economical timber production for the entire canton of Grisons and for Regions 2 and 3. (Class 1: Fulfills best practice requirements, Class 2: limited suitability for efficient management, and Class 3: no efficient management possible).

			Class 1	Class 2	Class 3	Total
Whole canton	area	(ha)	61,042	77,872	63,400	202,314
		(%)	30	38	32	100
	volume to harvest	(m³/year)	228,983	308,395	235,882	773,260
		(%)	30	40	30	100
Forest-region 3	area	(ha)	5375	17,642	9699	32,716
		(%)	16	54	30	100
	volume to harvest	(m³/year)	27,042	87,459	45,787	160,288
		(%)	17	55	28	100
Forest-region 2	area	(ha)	9104	6847	6172	22,123
		(%)	41	31	28	100
	volume to harvest	(m³/year)	39,755	28,918	277,46	96,419
		(%)	41	30	29	100

An additional outcome was the map of the harvesting method (Figure 10). The map differentiates between ground-based harvesting and 4 different types of cable yarding. The key results were the following: 5% of the forest area (over the whole canton) fell into the "ground-based" harvesting category, as opposed to 63% for which cable-yarding was the proposed harvesting method and 32% were classified as "not accessible". Further outcomes were the maps of the costs for harvesting and transport and the maximum hauling weight limit map (Figure 11), respectively.

These maps are useful to identify the reasons for the rating of an area. Areas in the left hand and the upper side of Figure 9 for example have a 'limited suitability' rating, which can be identified from the hauling weight limit map (Figure 11) as limited mainly due to the weight limit of the forest road.

Road densities were calculated for three road classes: roads with a weight limit between 18 and 27 t, 28 and 40 t and an overall class between 18 and 40 t (Table 6). This subdivision conforms with the subdivision of the suitability map. Therefore, we compare the road density for the '28–40 t class' with class 1 of Table 5 (Fulfills state-of-the-art requirements). In both approaches, perimeter 2 is identified as the best-accessed perimeter with a road density of 17 m/ha or 41% within class 1. However, although road density is at the upper end of the ideal range, the analysis shows that 28% of the surfaces are still uncovered (class 3) and 31% have limited suitability (class 2). Considering the '28–40 t class' for perimeter 3 and the whole canton, both have 9 m/ha road densities, therefore one might think that the share of well-accessible forest area should be about the same. However, the values for class 1 tell a different story with a share of 16% for perimeter 3 and 30% for the whole canton (Table 5). This shows that it is difficult to assess a forest road network using only the classical road density values. First, these values have a certain fuzziness and second they do not identify particular areas with an insufficient road network, so the location of the weak areas remains unknown. This outcome was expected, as the road density metric does not take into account the spatial variability in the terrain, the spatial distribution of the forest roads, obstacles that limit the use of a harvesting system (e.g., high tension power lines for cable yarders) or other limits of the forest road network, such as bridges with a limited bearing capacity.

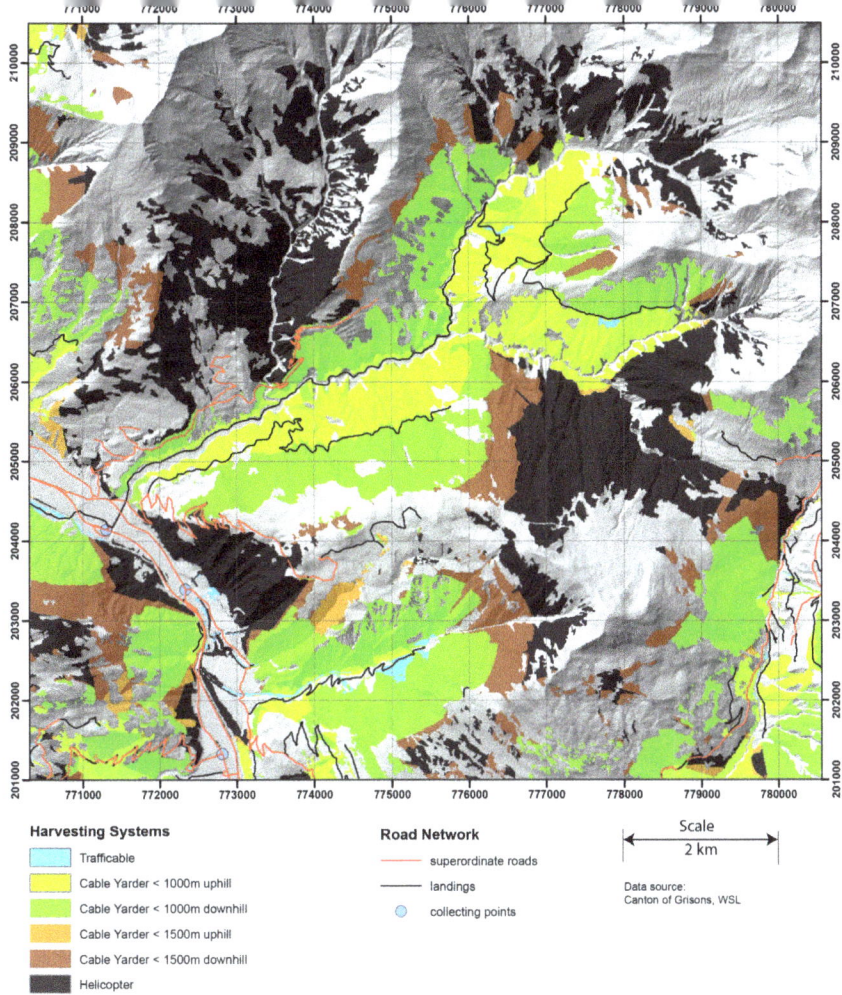

Harvesting Systems

- Trafficable
- Cable Yarder < 1000m uphill
- Cable Yarder < 1000m downhill
- Cable Yarder < 1500m uphill
- Cable Yarder < 1500m downhill
- Helicopter

Road Network

- superordinate roads
- landings
- collecting points

Scale
2 km

Data source:
Canton of Grisons, WSL

Figure 10. Extract of the harvesting system map (region near Schiers in canton of Grisons).

Table 6. Road Densities for the observed perimeters. The road densities were split into a class of roads with a weight limit between 18 and 27 t, 28 and 40 t and an overall class between 18 and 40 t.

Perimeter	Forest Area	Road Densities (m/ha)		
	(ha)	18–27 t	28–40 t	18–40 t
2	22,122	6	17	23
3	32,715	16	9	25
Whole Canton	202,312	8	9	17

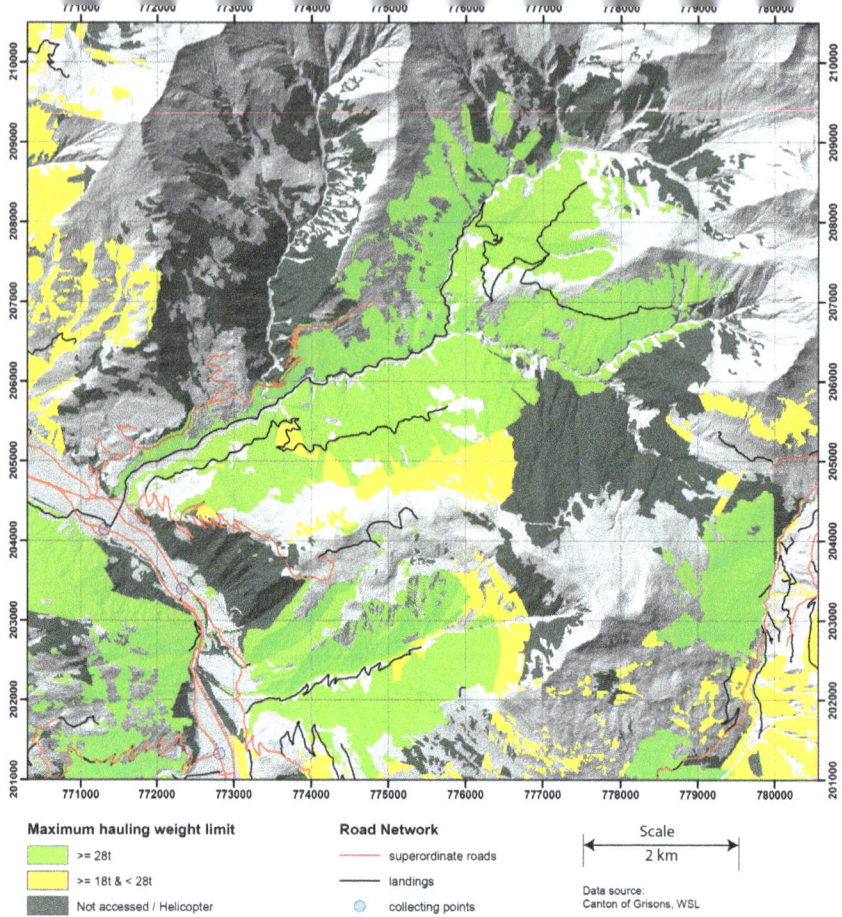

Figure 11. Extract of the maximum hauling weight limit map (region near Schiers in canton of Grisons).

4. Discussion

To verify the quality and usefulness of our new method, we requested that the forest service (local forest service and district officers) examine the results. This involved checking the suitability map as well as other outcomes such as the hauling weight limit map (based on TP^W) and the harvesting system map (based on TP^{HS}). The staff showed great interest in the maps and provided abundant feedback. Some classifications were reported as incorrect, mostly due to errors in the input data (mainly road standards), while other wrong classifications were caused by errors in the programming code. After those errors were rectified, the calculations were run again. This time no wrong classifications were reported by the forest service, and they deemed these maps very realistic. A second type of verification was performed by Schmid and Zürcher [43], who examined whether the intensity of the management in the protection forests (protection against natural hazards) was related to the classes depicted by the suitability map. They concluded that the suitability class correlated strongly with management in the protection forests. After these evaluations, the canton incorporated the results of this study into its forest development plan and declared that any future revisions of the road network must be in accordance with these findings.

The model is applicable on a large scale. In particular, these calculations of the entire test area, i.e., 7000 km^2, were completed within one day, and the calculation effort increased linearly with the size of the area analyzed. This allowed the computations to be run for a larger area within a reasonable amount of time.

Compared with classical approaches, e.g., the optimal road density concept [2], that deliver only average values, this new method enables planners to assess the quality of a road network in a spatially explicit manner. Moreover, while other related spatially explicit approaches (e.g., [22]) focus only on harvesting operations, the assessment method proposed here also includes the limitations associated with the road network, which are provided as an input. We are confident that a further development to cover a wider range of topographic conditions is feasible.

The approach is intended for steep terrain conditions, in which cable-based systems play a major role. If it were to be applied on flat terrain, trafficable areas would have to be modelled in more detail.

The boundaries between suitability classes are arbitrary. The 28-ton boundary between classes 1 and 2 was chosen at the explicit request of the project partner. This distinction does not correspond with truck classifications that are based on the number of axles. Here, a 32-ton boundary might make more sense. However, the classification boundaries as well as the number of classifications can easily be adapted for other surveys.

The hauling route was chosen in such a way that the weight limit was maximized. This worked well here because only a few different combinations were possible. However, it is conceivable that an alternative route with a slightly smaller limit would have a considerably shorter distance, making it more cost-effective. Under these circumstances, or if the transport destination (e.g., sawmill) were known, the "hauling route identification algorithm" might have to be modified.

The rule-based allocation of harvesting systems and the cost estimations were deliberately kept simple. More sophisticated models or methods would require many additional assumptions, e.g., the silvicultural treatment for each stand or the characteristics of each single cable road. To do so on this scale would require enormous effort in data collection and computation, with no guarantee that the final results would be more reliable. Furthermore, when conducting such an analysis, one must be aware that the quality of the output strongly depends upon manually evaluated input data (road segment standards evaluated by the local forest service). Finally, we are emphatic that the interpretation of results must be done carefully. For example, TPs identified as unsuitable do not necessarily require better development; the result simply indicates they are not ideal for timber production.

5. Conclusions

This new spatially explicit method was designed to assess the suitability of a forest road network for economical timber production. It accommodates steep terrain conditions, such as those found in the Central Alps, and is effective over large areas. The method was tested in the Swiss Canton of Grisons (7000 km^2), and all calculations were completed in less than one day. Its utility was verified by forest service personnel, who judged the outcome as very reliable and traceable. These results will form the basis for future revisions of road networks.

To our knowledge, this new approach is the first to factor in both production suitability and the appropriateness of harvesting techniques in steep terrain when developing a road network. It also assesses the entire harvesting operation, from tree-felling to processing and off- and on-road timber transport. We kept some elements, such as the rule-based allocation of harvesting systems or cost estimations, deliberately simple so that the method could be tested over a large geographical area. In contrast, operational harvest and transport optimization systems are not applied to large areas, but at a compartment or coupe level.

Our findings have several implications for practitioners: The method provides an easy-to-obtain, reliable and traceable basis that offers an overview of the development of a road network. This makes the method more efficient than current practices that still generally rely upon previous experience, anecdotal evidence, or personal observation.

Acknowledgments: We express our thanks to the Forest Service of the canton of Grisons (Switzerland), in particular Andreas Meier, Gian Cla Feuerstein, and Matthias Zubler, for providing the forest road network data, verifying the results and for funding the case study, and to Bronwyn Price for English editing.

Author Contributions: L.G.B. conceived and designed the method, performed the field study, L.G.B. and M.F. analyzed the data; L.G.B., M.F. and C.F. wrote the paper.

Conflicts of Interest: The authors declare no conflict of interest and the funding sponsors had no role in the design of the study; in the collection, analyses, or interpretation of data; in the writing of the manuscript, and in the decision to publish the results.

References

1. Beck, S.; Sessions, J. Forest road access decisions for woods chip trailers using Ant Colony Optimization and breakeven analysis. *Croat. J. For. Eng. J. Theory Appl. For. Eng.* **2013**, *34*, 201–215.
2. Matthews, D.M. Comments on. *J. For.* **1939**, *37*, 222–224.
3. Sundberg, U. Studier i skogsbrukets transporter. *J. Swed. For. Soc.* **1953**, *51*, 15–75.
4. Segebaden, G. *Von Studies of Cross-Country Transport Distances and Road Net Extension*; Skogshögskolan: Stockholm, Sweden, 1964.
5. Abegg, B. Schatzung der optimalen Dichte von Waldstrassen in traktorbefahrbarem Gelande. *Mitt. Eidgenoessische Anst. Fuer Forstl. Vers.* **1978**, *52*, 99–213.
6. Thompson, M.A. Considering overhead costs in road and landing spacing models. *J. For. Eng.* **1992**, *3*, 13–19. [CrossRef]
7. Heinimann, H.R. A Computer Model to Differentiate Skidder and Cable-Yarder Based Road Network Concepts on Steep Slopes. *J. For. Res. Jpn.* **1998**, *3*, 1–9. [CrossRef]
8. Kirby, M. An example of optimal planning of forest roads and projects. In *Planning and Decisionmaking as Applied to Forest Harvesting*; O'Leary, J.E., Ed.; Forest Research Laboratory, School of Forestry, Oregon State University: Corvallis, OR, USA, 1973; pp. 75–83.
9. Mandt, C.I. Network analyses in transportation planning. In *Planning and Decisionmaking as Applied to Forest Harvesting*; O'Leary, J.E., Ed.; Forest Research Laboratory, School of Forestry, Oregon State University: Corvallis, OR, USA, 1973; pp. 95–101.
10. Anderson, A.E.; Nelson, J. Projecting vector-based road networks with a shortest path algorithm. *Can. J. For. Res.* **2004**, *34*, 1444–1457. [CrossRef]
11. Stückelberger, J.A.; Heinimann, H.R.; Chung, W. Improved road network design models with the consideration of various link patterns and road design elements. *Can. J. For. Res.* **2007**, *37*, 2281–2298. [CrossRef]
12. Stückelberger, J.A.; Heinimann, H.R.; Chung, W.; Ulber, M. Automatic road-network planning for multiple objectives. In Proceedings of the 2006 Council on Forest Engineering (COFE) Conference, Coeur d'Alene, ID, USA, 22 July–2 August 2006; Chung, W., Han, H.S., Eds.; Council on Forest Engineering: Bangor, ME, USA, 2006; pp. 233–248.
13. Stückelberger, J.A.; Heinimann, H.R.; Burlet, E.C. Modeling spatial variability in the life-cycle costs of low-volume forest roads. *Eur. J. For. Res.* **2006**, *125*, 377–390. [CrossRef]
14. Chung, W.; Stückelberger, J.; Aruga, K.; Cundy, T.W. Forest road network design using a trade-off analysis between skidding and road construction costs. *Can. J. For. Res.* **2008**, *38*, 439–448. [CrossRef]
15. Epstein, R.; Morales, R.; Séron, J.; Weintraub, A. Use of OR Systems in the Chilean Forest Industries. *Interfaces* **1999**, *29*, 7–29. [CrossRef]
16. Epstein, R.; Weintraub, A.; Sapunar, P.; Nieto, E.; Sessions, J.B.; Sessions, J.; Bustamante, F.; Musante, H. A Combinatorial Heuristic Approach for Solving Real-Size Machinery Location and Road Design Problems in Forestry Planning. *Oper. Res.* **2006**, *54*, 1017–1027. [CrossRef]
17. Diaz, L.A.; Ferland, J.A.; Ribeiro, C.C.; Vera, J.R.; Weintraub, A. A tabu search approach for solving a difficult forest harvesting machine location problem. *Eur. J. Oper. Res.* **2007**, *179*, 788–805.
18. Chung, W.; Sessions, J.; Heinimann, H.R. An application of a heuristic network algorithm to cable logging layout design. *Int. J. For. Eng.* **2004**, *15*, 11–24.
19. Bont, L.G.; Heinimann, H.R.; Church, R.L. Concurrent optimization of harvesting and road network layouts under steep terrain. *Ann. Oper. Res.* **2015**, *232*, 41–64. [CrossRef]

20. Henningsson, M.; Karlsson, J.; Rönnqvist, M. Optimization Models for Forest Road Upgrade Planning. *J. Math. Model. Algorithms* **2007**, *6*, 3–23. [CrossRef]

21. Flisberg, P.; Frisk, M.; Rönnqvist, M. Integrated harvest and logistic planning including road upgrading. *Scand. J. For. Res.* **2014**, *29*, 195–209. [CrossRef]

22. Dupire, S.; Bourrier, F.; Monnet, J.-M.; Berger, F. Sylvaccess: Un modèle pour cartographier automatiquement l'accessibilité des forêts. *Rev. For. Fr.* **2015**, *70*, 111–126. [CrossRef]

23. Brändli, U.-B. (Ed.) *Schweizerisches Landesforstinventar. Ergebnisse der Dritten Erhebung 2004–2006*; Birmensdorf, Eidgenössische Forschungsanstalt für Wald; Schnee und Landschaft WSL; Bern, Bundesamt für Umwelt, BAFU: Birmensdorf, Switzerland, 2010; ISBN 978-3-905621-47-1.

24. Brändli, U.-B.; Fischer, C.; Camin, P. Stand der Walderschliessung mit Lastwagenstrassen in der Schweiz. *Schweiz. Z. Forstwes.* **2016**, *167*, 143–151. [CrossRef]

25. Federal Office of Topography swisstopo. *Das Hoch Aufgelöste Terrainmodell der Schweiz*; Federal Office of Topography swisstopo: Wabern, Switzerland, 2018.

26. Pestal, E. *Seilbahnen und Seilkräne für Holz-und Materialtransporte*; Georg Fromme & Co: Wien, Austria; München, Germany, 1961.

27. Zweifel, O. Seilbahnberechnung bei beidseitig verankerten Tragseilen. *Schweiz. Bauztg.* **1960**, *78*, 11.

28. Bont, L.; Heinimann, H. Optimum geometric layout of a single cable road. *Eur. J. For. Res.* **2012**, *131*, 1439–1448. [CrossRef]

29. Ahlvin, R.B.; Haley, P.W. *Nato Reference Mobility Model: Edition II. NRMM User's Guide*; US Army Engineer Waterways Experiment Station: Vicksburg, MS, USA, 1992.

30. Bekker, M.G. *Prediction of Design and Performance Parameters in Agro-Forestry Vehicles: Methods, Tests and Numerical Examples*; National Research Council Canada, Division of Energy: Ottawa, ON, Canada, 1983.

31. Eichrodt, A.W. *Development of a Spatial Trafficability Evaluation System*; vdf Hochschulverlag AG: Zürich, Switzerland, 2003; ISBN 3-7281-2905-4.

32. Ashmore, C.; Burt, E.C.; Turner, J.L. *Predicting Tractive Performance of Log-Skidder Tires*; American Society of Agricultural and Engineers: St. Joseph, MI, USA, 1985; pp. 85–159.

33. Brixius, W.W. *Traction Prediction Equations for Bias Ply Tires*; American Society of Agricultural and Engineers: St. Joseph, MI, USA, 1987.

34. Swiss Federal Office for Statistics (BFS). *Die Digitale Bodeneignungskarte der Schweiz [Digital Soil Property Map of Switzerland]*; BFS: Newcastle, Switzerland, 2004.

35. Heinimann, H.-R. *Seilkraneinsatz in den Schweizer Alpen: Eine untersuchung über die Geländeverhältnisse, die Erschliessung und den Einsatz Verschiedener Seilanlagen: Abhandlung zur Erlangung des Titels Eines Doktors der Technischen Wissenschaften der Eidgenoessischen Technischen Hochschule Zuerich*; Eidgenoessischen Technischen Hochschule: Zürich, Switzerland, 1986.

36. Heinimann, H.R. Conceptual Design of a Spatial Decision Support System for Harvesting Planning. In Proceedings of International NEFU/IUFRO/FAO/FEI Seminar on Forest Operations under Mountainous Conditions, Harbin, China, 24–27 July 1994; pp. 24–27.

37. Lüthy, D. *Entwicklung Eines Spatial Decision Support-Systems (SDSS) für die Holzernteplanung in Steilen Geländeverhältnissen*; vdf Hochschulverlag AG an der ETH Zürich: Zürich, Switzerland, 1998.

38. Bont, L.; Heinimann, H.R.; Church, R.L. Optimizing cable harvesting layout when using variable-length cable roads in central Europe. *Can. J. For. Res.* **2014**, *44*, 949–960. [CrossRef]

39. Breschan, J.; Maurer, S.; Bont, L.; Bolgè, R. *An Improved Workflow to Identify an Optimal Cable Road Layout for a Large Management Unit*; ETH Zurich: Zürich, Switzerland, 2017.

40. Dijkstra, E.W. A note on two problems in connexion with graphs. *Numer. Math.* **1959**, *1*, 269–271. [CrossRef]

41. Hirt, R. Wer hat Angst vor 40 Toennern. *Schweiz. Ing. Archit. SIA* **1997**, *49*, 4.

42. Eidg. Forschungsanstalt WSL. *Holzernte Produktivitätsmodelle HeProMo*; Eidg. Forschungsanstalt WSL: Birmensdorf, Switzerland, 2016.

43. Schmid, U.; Zürcher, S. *Analyse der Schutzwaldpflege im Kanton Graubünden Zwischen 2006 und 2015 im Hinblick auf die Erschliessungsgüte*; Fachstelle für Gebirgswaldpflege: Maienfeld, Switzerland, 2016; p. 17.

Article

Traffic-Induced Changes and Processes in Forest Road Aggregate Particle-Size Distributions

Hakjun Rhee [1,*], James Fridley [2] and Deborah Page-Dumroese [3]

[1]   Department of Environment and Forest Resources, Chungnam National University, 99 Daehak-ro, Yuseong-gu, Daejeon 34134, Korea

[2]   School of Environmental and Forest Sciences, University of Washington, Seattle, WA 98195-2100, USA; fridley@uw.edu

[3]   US Department of Agriculture Forest Service, Rocky Mountain Research Station, 1221 South Main Street, Moscow, ID 83843-4211, USA; ddumroese@fs.fed.us

*   Correspondence: hakjun.rhee@gmail.com; Tel.: +82-10-4899-8751

Received: 20 February 2018; Accepted: 30 March 2018; Published: 3 April 2018

Abstract: Traffic can alter forest road aggregate material in various ways, such as by crushing, mixing it with subgrade material, and sweeping large-size, loose particles (gravel) toward the outside of the road. Understanding the changes and physical processes of the aggregate is essential to mitigate sediment production from forest roads and reduce road maintenance efforts. We compared the particle-size distributions of forest road aggregate from the Clearwater National Forest in Idaho, USA in three vertical layers (upper, middle, and bottom of the road aggregate), three horizontal locations (tire track, shoulder, and half-way between them), and three traffic uses (none, light (no logging vehicles), and heavy (logging vehicles and equipment)) using Tukey's multiple comparison test. Light traffic appears to cause aggregate crushing where vehicle tires passed and caused sweeping on the road surface. Heavy traffic caused aggregate crushing at all vertical and horizontal locations, and subgrade mixing with the bottom layer at the shoulder location. Logging vehicles and heavy equipment with wide axles drove on the shoulder and exerted enough stress to cause subgrade mixing. These results can help identify the sediment source and define adequate mitigation measures to reduce sediment production from forest roads and reduce road maintenance efforts by providing information for best management practices.

Keywords: aggregate crushing; particle-size distribution; road aggregate; subgrade mixing; sweeping

---

1. Introduction

Aggregate (crushed rock) is one of the most common surfacing materials used on low volume forest roads [1,2]. It reduces both wheel load stress to the subgrade material and maintenance costs, and provides better driving comfort than native-surface roads [3]. It also helps reduce sediment production from forest roads [4–6]. However, road aggregate deteriorates due to traffic, weather, and material properties [7], and, as a consequence, increased sedimentation occurs. Aggregate material can also be deformed and lost from the road surface due to vehicle traffic which then requires additional road maintenance. Therefore, it is important to understand how forest road traffic changes aggregate properties and movement (i.e., traffic-induced processes that change forest road aggregate).

Traffic-induced processes have been speculated on, and occasionally discussed, as part of field-based observations. For example, Reid and Dunne [8] mentioned (1) "breakdown of the surfacing material" and (2) "forcing upward of fine-grained sediment from the roadbed as traffic pushes the surfacing gravels into the bed". Many other research studies have observed and discussed these processes, but often in varying terms. For example, the breakdown of road surfacing material (aggregate) was observed by Swift [4], described as "crushing" [9–11], and referred to as "powdering by traffic", "particle attrition", or "abrasion" [4,11,12]. It has also been described as "the mechanical degradation of surface aggregate under traffic loading" [13]. The second process, forcing upward of fine-grained sediment from the roadbed, was also observed by Swift [4], and has been further described as "piping" (enabling roadbed soil to pipe through the road prism; [6]), "pressing larger particles down through a matrix of fine sediment" [9], "churning" [10], "fine materials move through the pore spaces to the surface" [14], "pumping of fine particles onto the surface" [12], and "subgrade mixing" [13]. Our paper will refer to the breakdown of surfacing material as "crushing" and the forcing upward of fine-grained material as "subgrade mixing".

In addition to crushing and subgrade mixing, there is another traffic-induced process that alters forest road aggregate. Traffic moves loose aggregate material off the road by tire action [15], which is called "sweeping". Traffic-induced sweeping results in the movement of loose aggregate particles to the roadside and shoulder. This is particularly prevalent at sharp corners where road surface aggregate tends to accumulate at the roadside (Figure 1) [11,16,17]. Two extreme cases of sweeping can be considered: sweeping-out (losing large-size particles, such as gravel, near tire tracks) and sweeping-in (receiving large-size particles near the shoulder).

Figure 1. Sorted surfacing material on forest road. Left is the tire track with a coin with a diameter of 24 mm on it, and right is the shoulder.

It is difficult to investigate traffic-induced processes directly, because they can be related to many factors, such as the physical properties of road aggregate (e.g., particle size and strength) [13,18,19], tire pressure [20,21], road conditions (e.g., moisture content and compaction) [11,15,22], road gradient [6,12,23], road type (mainline or secondary road) [6], traffic use [6,8,11], and proximity to the root systems of neighboring trees [24]. However, the particle-size distribution (PSD) of road aggregate can be used as a surrogate to infer the traffic-induced processes of crushing, subgrade mixing, and sweeping.

Particle-size distribution is likely one of the most important characteristics of mineral soil and road aggregates and is routinely used to evaluate if the soil or aggregate is appropriate as an engineering material for road construction [2,25]. A PSD consists of a number of particle-size fractions (PSFs) for individual particle-size classes and is usually plotted as a cumulative frequency diagram in a semi-logarithmic scale (i.e., particle size using a logarithmic scale on the x-axis and percent passing on the y-axis). When traffic-induced processes such as crushing occur, smaller-size particles (fine sediment) are produced, changing the PSD. Therefore, the PSD can be used to infer the traffic-induced processes.

There have been few studies that have analyzed the soil and aggregate PSDs that are used for forest and gravel roads. In New Brunswick, Canada, surface and subgrade materials from two forest roads were analyzed using PSD [26], and two resurfacing methods were compared using PSDs from unpaved roads [16]. In addition, Foltz and Truebe [18,19] analyzed various aggregate material from the western USA and used PSD to evaluate aggregate quality. The PSD method has also been used to characterize surface and subgrade (base) materials in South Dakota, USA [27]. In Lithuania, PSDs were observed to become finer as gravel pavement wears away [28].

Understanding forest road PSDs can elucidate traffic-induced changes in aggregate size and distribution, and their processes. However, in previous studies, soil and aggregate materials were collected from limited cross-sectional locations. More importantly, previous studies did not investigate road aggregate PSD changes after traffic. To fully understand traffic effects on forest road aggregate, PSDs should be compared from various vertical and horizontal cross-sectional locations as well as after different traffic uses. We could not find any publications in which this is investigated. Understanding traffic effects on aggregate material can help reduce road maintenance efforts and costs, and help identify the sediment source (i.e., road aggregate or roadbed material) and define adequate mitigation measures to reduce sediment production by traffic from forest roads.

We hypothesize that all three traffic-induced processes (crushing, subgrade mixing, and sweeping) could alter the physical properties of forest road aggregate. Therefore, the goal of this study was to understand how traffic changes forest road aggregate PSD and, in particular, the magnitude of those changes for various traffic uses, thus contributing to a better understanding of forest roads, traffic, erosion, and their inter-relationships. The specific objectives were to (1) compare forest road aggregate PSDs with respect to vertical and horizontal location within a forest road cross-section and traffic use, and (2) infer how the traffic-induced processes changed the PSDs.

2. Materials and Methods

2.1. Study Area and Road Description

Study plots were located on three roads in the Clearwater National Forest area, about 39–43 km northeast of Moscow, Idaho (Figure 2; 47°03′12″ N, 116°40′47″ W; 47°04′07″, N 116°40′33″ W; and 47°02′57″ N, 116°43′58″ W, respectively). The study area has a normal precipitation of 658 mm year^{-1} [29] at a rainfall intensity of 18 mm hour^{-1} for 10-year return period [30,31]. It has dry summer months (average precipitation of 82 mm, July to September), on the basis of the closest weather station, located 13–17 km southwest of the study sites, in Potlatch, Idaho (46°57′36″ N, 116°51′18″ W; elevation 841 m). The normal annual mean temperature is 7.6 °C, ranging from −1.8 °C in December to 17.6 °C in August [29]. Each of the three roads used in this study had different traffic uses from the time they were constructed in 2002. The three roads were not specifically constructed for this study, but for actual forest management and log hauling. All three roads were primarily cut and fill roads, and were constructed using the same aggregate material (basalt) from the same quarry located 15–18 km south of the field sites. Each road was 5 m wide and had 0.1 m of aggregate applied on a fine subgrade of mineral soil in the same year, which was typical for forest road construction in the study area.

White Pine 3833 road was chosen to represent "no" traffic use (N). Since its construction this road was not used except for occasional light administrative traffic (about 20 passes per year). It had been closed to the public and no logging traffic passed prior to aggregate sample collection in 2005.

White Pine 3830-1 road was chosen for the "light" traffic use (L). After construction, this road was used by light trucks and automobiles, but with no log hauling. It also included approximately 20 passes of administrative vehicles per year. It is open to the public except during seasonal road closure when it is closed to large vehicles (larger than pickup trucks) from October to mid-June and closed to all vehicles from December to mid-May.

White Pine 377-M road was chosen for the "heavy" traffic use (H). It was used for log hauling during one dry season (June to September) before collecting aggregate samples. Log hauling was not allowed during wet road conditions because the load from heavy trucks and equipment could mix surface aggregate with subgrade and damage the road. Prior to use by logging traffic, the road had dust abatement applied by using calcium chloride ($CaCl_2$) flake at a rate of 0.8 kg m^{-2}. We estimated that approximately 375 round trips of a standard logging truck (one-way empty and one-way fully loaded) passed on the H road. Also, heavy logging equipment (e.g., tracked linkbelt yarder, high track caterpillar, excavator) passed on this road. Additionally, during logging, pickup-sized trucks made approximately 20 round trips per week for administrative and contractor use. The L and H roads were 12.8–22.8% more compacted than the N road when the aggregate samples were collected (Table 1).

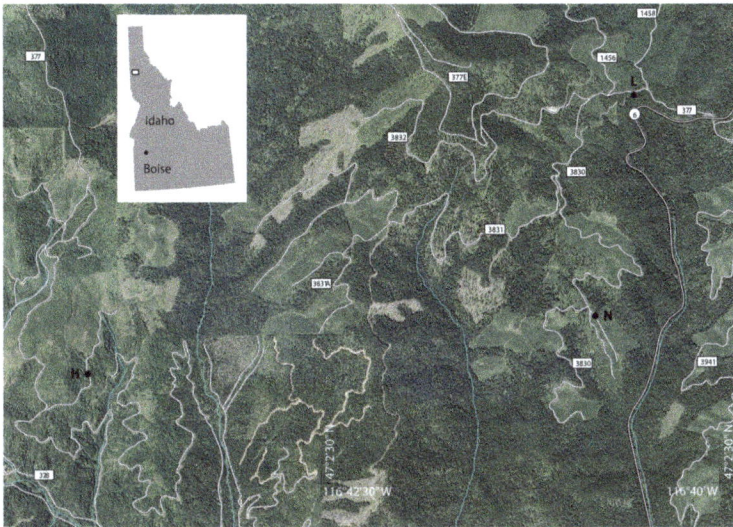

Figure 2. Location of the study plots on the White Pine roads in the Clearwater National Forest, Idaho. The map was modified from the 1:24,000 US topographic map (West Dennis, Idaho; 7.5-min quadrangle map) with orthoimage [32]. N, L, and H indicate the plot locations on no, light, and heavy traffic use roads. The N was used for log-hauls after collecting the samples, and the H, before collecting the samples in this image. White lines without a border indicate a forest road network; the white line with borders at the right side indicates the Idaho state highway 6; and blue lines indicate streams.

Table 1. Comparison of dry densities [1] (Mg m^{-3}) of forest road aggregate from different horizontal locations and traffic uses.

Traffic Use	Horizontal Location			
	Center	Tire Track	In-Between	Shoulder
No	A 1.606 b (0.095)	A 1.727 b (0.077)	A 1.671 c (0.055)	A 1.542 b (0.121)
Light	A 1.921 a (0.045)	A 2.053 a (0.176)	A 2.051 a (0.026)	N/A [2]
Heavy	A 1.893 a (0.009)	A 1.987 a, b (0.035)	A, B 1.885 b (0.073)	B 1.777 a (0.034)

[1] Dry densities were measured using the Troxler Model 3440 nuclear soil moisture density gauge [33] in the same year when the aggregate samples were collected. The values are mean dry density with standard deviation in parentheses. Mean values in a row preceded by the same superscript capital letter (A or B) are not significantly different between the different horizontal locations on the same traffic use ($p < 0.05$). Mean values in a column followed by the same lowercase letter (a, b, or c) are not significantly different between the different traffic uses at the same horizontal location ($p < 0.05$). [2] No dry density was measured at this site using the Troxler Model 3440 nuclear soil moisture density gauge, due to roadside vegetation.

2.2. Experimental Design

We used a total of 135 road aggregate samples collected during a single season. The 135 samples consisted of five replicates of 27 sample sets from the three vertical, three horizontal locations and three traffic uses. We analyzed the samples for PSD and compared them with respect to vertical, horizontal location, and traffic use. In addition, we compared PSDs on the N road and quarry samples to see if there were changes in PSD that should be attributed to road construction (i.e., when moving and applying aggregate). The PSD from the quarry was based on 12 aggregate samples collected at the time of road construction and was analyzed by the US Department of Agriculture, Forest Service, Palouse Ranger District office at the Clearwater National Forest.

2.3. Field Data Collection and Laboratory Analysis

For each road traffic treatment, we collected aggregate samples from three vertical layers (*upper* (U), *middle* (M), and *bottom* (B)) at three horizontal locations (*tire track*, *shoulder* and *in-between*) to be able to capture the trends in aggregate changes by traffic (Figure 3). *Tire track*, T, was located 1.2 m from the outside edge of the shoulder; *shoulder*, S, was located 0.3 m from the outside edge; and *in-between*, I, was located half-way between T and S. Vehicles and equipment, especially ones with trailers and wide axle widths, could pass over all three horizontal locations, including the shoulder. Samples were collected to an aggregate depth of 0.1 m, resulting in an aggregate sampling depth of 33.3 mm which was deep enough to include the largest aggregate particle (25.4 mm diameter). When a large particle was found near the 33.3 mm depth, it was included in the layer with a larger proportion of that particle size. Aggregate samples in the bottom layer were collected down to the subgrade layer. Replicate cross-sections on each of the three roads were approximately 0.5 m apart.

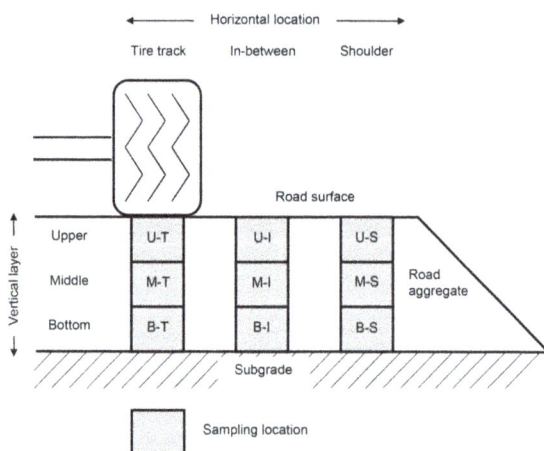

Figure 3. Sampling locations within a forest road cross section. The abbreviations indicate vertical layer and horizontal location of samples. U: upper; M: middle; B: bottom; T: tire track; S: shoulder; I: in-between.

The amount of aggregate material collected and analyzed was determined on the basis of the largest particle-size observed. The American Association of State Highway and Transportation Officials (AASHTO) T88-00 standard [34] and the American Society for Testing and Materials (ASTM) D422-63(2002) standard [35] suggested 2 kg of dry mass to analyze aggregate material with the largest particle of 25.4 mm diameter. A preliminary data analysis showed that 2 kg of dry mass required a volume of approximately 0.2 m length × 0.2 m width × 33.3 mm depth. Considering that a usual dry mass used for particle size analyses is about 0.5 kg (AASHTO T146-96 standard [36]), the collected 2 kg aggregate sample was divided into three or four sub-samples for ease in handling, analyzed following the AASHTO and ASTM procedures (AASHTO T88-00 standard [34] and ASTM D422-63(2002) standard [35]), and combined to make one PSD. We adopted a nested sampling method to reduce aggregate particle size analysis time [37]. For the nested method, we excavated and stored aggregate samples in two separate portions: *small portion* from 0.1 m × 0.1 m hole (0.5 kg), and *large portion* from 0.2 m × 0.2 m hole except the small portion (1.5 kg).

Particle size analyses followed the ASTM D2217-85 and D422-63(2002) standards [35,38] and Rhee et al. [37]. We conducted the *small portion* particle size analyses using 13 sieves: 25.4 mm (1 inch), 19.0 mm ($\frac{3}{4}$ inch), 12.7 mm ($\frac{1}{2}$ inch), 9.51 mm ($\frac{3}{8}$ inch), 6.35 mm ($\frac{1}{4}$ inch), 4.76 mm (US standard sieve No. 4), 3.36 mm (No. 6), 2.00 mm (No. 10), 1.00 mm (No. 18), 0.420 mm (No. 40), 0.250 mm (No. 60), 0.149 mm (No. 100), and 0.074 mm (No. 200). The *large portion* particle size analyses were conducted using only five sieves: 25.4 mm, 19.0 mm, 12.7 mm, 9.51 mm, and 6.35 mm. The *small* and *large portions* of the particle size analysis results were then combined to calculate a single particle size determination of each sampling location [37]. A particle size analysis finer than the 0.074 mm sieve was not conducted in the current study because a limited amount of crushing occurs below this particle-size in granular materials [39,40] and sieving is not used [35,40].

2.4. Comparison of Particle-Size Distributions

The PSD consists of particle-size fractions (PSFs) from each individual particle-size class. One way to compare PSDs is to compare the PSF results from each sieve size, which requires the same number and size of sieves. For example, 39 point-to-point comparisons are needed to compare three PSDs that consist of 13 sieve sizes. It is difficult to draw concise conclusions from many statistical comparisons for 13 individual sieve sizes. Instead, Hardin [39] suggested measuring particle breakage based on changes in entire PSD. He introduced breakage potential (B_p) as the area between the PSD curve and the 0.074 mm sieve in the cumulative frequency diagram (Figure 4). The B_p represents the total possible particle breakage if every particle were broken down smaller than the 0.074 mm-sieve size. We adopted the Hardin's B_p as a surrogate parameter to represent a PSD for general and statistical comparison purposes in this study. We also compared changes in the shape of PSDs because traffic-induced processes might alter the shape of PSDs while having the same B_p value.

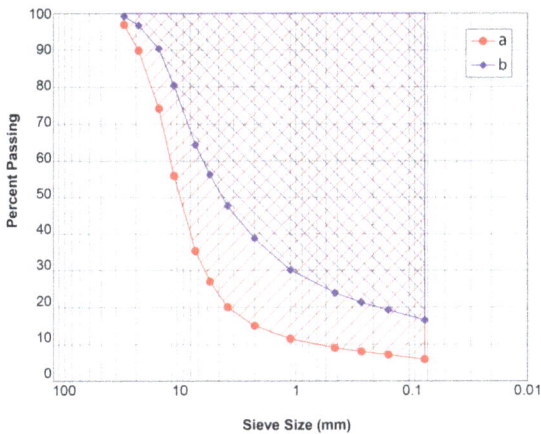

Figure 4. Hardin's Breakage Potential (B_p): (a) particle-size distribution with B_p of 186.77; and (b) particle-size distribution with B_p of 140.77. The unit of B_p is %·log (mm). Higher B_p value represents a particle-size distribution with coarser particles.

2.5. Statistical Analysis

We used Tukey's multiple comparison test (honestly significant difference test [41]) to compare PSDs among different treatments of vertical layer (U, M, and B), horizontal location (T, I, and S), and road traffic use (N, L, and H). In addition, ANOVA (Analysis of Variance [41]) was used to compare the PSDs from N with the PSD from the quarry. We used a significance level of $\alpha = 5\%$ for the statistical analyses.

3. Results

The PSDs from the collected aggregate were compared and analyzed with respect to different vertical layers and horizontal locations with the same level of traffic use and different traffic uses at the same cross-sectional location. In addition, we plotted the PSD data from the quarry and compared it with the other PSDs.

3.1. Particle-Size Distributions with No Traffic

The PSDs on the no traffic (N) road were uniform. There were no differences in PSDs from the different vertical layers or horizontal locations (Table 2 and Figure S1).

Table 2. No traffic road particle-size distributions from the vertical and horizontal locations using the breakage potential (B_p) method. The values are average B_p with standard deviation in parentheses. The unit of B_p is % log (mm).

Vertical Layer	Horizontal Location					Inferred Process
	Tire Track		In-Between		Shoulder	
Upper	177.84 (4.62)	=	174.42 (9.75)	=	186.61 (10.07)	N/A [1]
	‖		‖		‖	
Middle	174.32 (6.26)	=	171.74 (9.50)	=	176.20 (3.34)	N/A
	‖		‖		‖	
Bottom	170.35 (7.33)	=	164.74 (6.69)	=	175.91 (9.54)	N/A
Inferred process	N/A		N/A		N/A	

[1] No inferred process is available. "=" symbol between means indicates equality of values.

3.2. Particle-Size Distributions after Light Traffic Use

The PSDs on the light traffic (L) road changed depending on the vertical layers and horizontal locations (Table 3 and Figure S2). Comparing vertical layers showed that the PSD at U-T-L was finer than at M-T-L; and the PSD at B-T-L was finer than both (U-T-L < M-T-L < B-T-L). In addition, the PSD at U-I-L was finer than at M-I-L and B-I-L. At the shoulder location, the PSD at U-S-L was coarser than at M-S-L and B-S-L. Comparing horizontal locations showed that the PSD at U-S-L was coarser than at U-T-L and U-I-L. There were no differences in PSDs at the horizontal locations at M-L and B-L.

Table 3. Light traffic road particle-size distributions from the vertical and horizontal locations using the breakage potential (B_p) method. The values are average B_p with standard deviation in parentheses. The unit of B_p is %·log (mm).

Vertical Layer	Horizontal Location					Inferred Process
	Tire Track		In-Between		Shoulder	
Upper	153.33 (3.17)	=	155.28 (7.65)	<	186.77 (0.92)	Crushing/sweeping
	∧		∧		∨	
Middle	167.74 (7.26)	=	168.85 (7.73)	=	178.18 (4.66)	N/A [1]
	∧		‖		‖	
Bottom	176.65 (1.99)	=	174.27 (5.80)	=	173.32 (3.95)	N/A
Inferred process	Crushing		Crushing		Sweeping	

[1] No inferred process is available. "=/</>" symbols between means indicate respective significance.

3.3. Particle-Size Distributions after Heavy Traffic Use

On the heavy traffic (H) road, PSDs were uniform, except at B-S-H which had a finer PSD than the other vertical layers and horizontal locations (Table 4 and Figure S3). There were no differences in PSDs between different vertical layers at T-H and I-H. The PSD at B-S-H was finer than at U-S-H and M-S-H. There were no differences in PSDs at the horizontal locations at U-H and M-H. The PSD at B-S-H was finer than at B-T-H and B-I-H.

Table 4. Heavy traffic road particle-size distributions from the vertical and horizontal locations using the breakage potential (B_p) method. The values are average B_p with standard deviation in parentheses. The unit of B_p is %·log (mm).

Vertical Layer	Horizontal Location					Inferred Process
	Tire Track		In-Between		Shoulder	
Upper	154.84 (4.90)	=	153.31 (3.45)	=	156.44 (4.47)	N/A [1]
	‖		‖		‖	
Middle	163.47 (8.68)	=	156.37 (6.66)	=	157.89 (4.14)	N/A
	‖		‖		∨	
Bottom	154.97 (2.51)	=	151.31 (7.97)	>	140.77 (5.60)	Subgrade mixing
Inferred process	N/A		N/A		Subgrade mixing	

[1] No inferred process is available. "=/</>" symbols between means indicate respective significance.

3.4. Particle-Size Distributions for Different Traffic Uses at the Same Cross-Sectional Location

The PSDs on the N road were not the same as the PSD from the quarry and varied by cross-sectional locations (Table 5 and Figure S4) despite having no differences from different vertical layers and horizontal locations (Table 2 and Figure S1). The PSDs at U-T-N, U-S-N, and M-S-N were coarser than from the quarry. However, at B-I-N, the PSD was finer than from the quarry. As traffic increased from none to light to heavy, PSDs became finer, except at M-T (Table 5 and Figure S4).

Table 5. Particle-size distributions from different traffic uses at the same cross-sectional locations using the breakage potential (B_p) method. The values are average B_p with standard deviation in parentheses. The unit of B_p is %·log (mm).

Vertical Layer	Traffic Use							Inferred Process
	Quarry		No		Light		Heavy	
	Horizontal location: Tire track							
Upper	171.01 (4.88)	<	177.84 (4.62)	>	153.33 (3.17)	=	154.84 (4.90)	Crushing
	‖		‖		∧		‖	
Middle	171.01 (4.88)	=	174.32 (6.26)	=	167.74 (7.26)	=	163.47 (8.68)	N/A [1]
	‖		‖		∧		‖	
Bottom	171.01 (4.88)	=	170.35 (7.33)	=	176.65 (1.99)	>	154.97 (2.51)	Crushing
	Horizontal location: In-between							
Upper	171.01 (4.88)	=	174.42 (9.75)	>	155.28 (7.65)	=	153.31 (3.45)	Crushing
	‖		‖		∧		‖	
Middle	171.01 (4.88)	=	171.74 (9.50)	=	168.85 (7.73)	=	156.37 (6.66)	Crushing [2]
	‖		‖		‖		‖	
Bottom	171.01 (4.88)	>	164.74 (6.69)	=	174.27 (5.80)	>	151.31 (7.97)	Crushing
	Horizontal location: Shoulder							
Upper	171.01 (4.88)	<	186.61 (10.07)	=	186.77 (0.92)	>	156.44 (4.47)	Crushing
	‖		‖		∨		‖	
Middle	171.01 (4.88)	<	176.20 (3.34)	=	178.18 (4.66)	>	157.89 (4.14)	Crushing
	‖		‖		‖		∨	
Bottom	171.01 (4.88)	=	175. 91 (9.54)	=	173.32 (3.95)	>	140.77 (5.60)	Subgrade mixing

[1] No inferred process is available. [2] B_p of the no traffic use is statistically greater than the high traffic use. "=/</>" symbols between means indicate respective significance.

377

On the L and H roads, the PSDs were finer than on the N road at U-T and U-I. In addition, the PSDs on the H road were finer than on the L road at B-T, B-I, and shoulder locations. There were no differences between the PSDs at M-I-N and M-I-L, and at M-I-L and M-I-H; however, the PSD at M-I-H was finer than at M-I-N. The PSD at B-S-H was much finer than at B-S-N and B-S-L, and also finer than at other cross-sectional locations on the H road (Table 4).

4. Discussion

The study results help infer crushing, subgrade mixing, and sweeping-in processes by comparing PSDs from the quarry and forest road aggregate. We inferred that crushing occurred at U-T-L, U-I-L, and all cross-sectional locations on the H road, because the PSDs from these locations and traffic uses were finer than from the other locations. Crushing was only observed near the tire track and road surface (U-T and U-I) on the L road (Table 3) and was observed at all cross-sectional locations with increasing traffic use (H) (Table 4), indicating that the crushing occurred near the tire track on the road surface first, than in the deeper aggregate. Subgrade mixing might have occurred at B-S-H, because the PSD here was finer than the rest of the PSDs on H and the location of subgrade mixing (B-S-H) was the bottom layer (Table 4). Also the subgrade PSD was much finer than any of the road aggregate PSDs (Figure 5), confirming a strong likelihood of subgrade mixing. Considering that crushing occurred at the other cross-sectional locations on H, we inferred that both crushing and subgrade mixing occurred at B-S-H. We also inferred that sweeping occurred at U-S-L, because the PSD at U-S-L was coarser than at the other locations on the L road and the location of the sweeping-in (U-S-L) was the shoulder on the road surface (Table 3).

Our study results partially agree with the results from Toman and Skaugset [13] in which they observed that subgrade mixing did not occur on three forest roads in northwestern California and Oregon. However, our data indicate subgrade mixing might have occurred at one cross-sectional location on the H road. In addition, Toman and Skaugset [13] speculated that fine sediment came from either the surface aggregate itself or its breakdown by traffic, which agrees with our data that crushing is the dominant traffic-induced process influencing sediment production from forest roads.

No ruts were observed from the study roads and, therefore, 0.1 m deep aggregate was enough to prevent soil loss from the roadbed. Swift [4] reported that 0.05 m of aggregate depth was not enough, but 0.15 m prevented ruts and reduced soil loss from forest roads. When designing an aggregate-surfaced road, 0.1 m of minimum aggregate thickness is often chosen [1] and our study supports these previous findings.

Stress from vehicle loads causes the physical properties of forest road aggregate, such as PSD, to change. Compressive stress on a forest road is concentrated under the loading axis and is highest at or near the road surface [42,43]. We found that crushing occurred in the upper vertical layer at the tire track and in-between locations (U-T-L and U-I-L) where light traffic tires passed. Light traffic could pass on tire track and in-between locations, because light traffic vehicles have relatively narrow axle width (1.6–1.8 m) and the forest roads were relatively wide (5 m) leaving a large surface for driving. However, light traffic did not provide enough stress to crush the road aggregate below the upper layer (U). Likewise, with heavy traffic, crushing occurred at all cross-sectional locations, indicating that traffic passed on all horizontal locations including the shoulder. Heavy traffic vehicles have wide axle widths (1.8–2.2 m [44]) and many often have dual tires. Some heavy vehicles, such as Commercial Motor Vehicles (CMVs), are even wider (up to 2.6 m [45]). Overloaded vehicles used by the USDA Forest Service included a tracked loader with a width of 3.7 m and an axle load of 41 Mg [44]. In addition to heavy loads, tracked loaders do not use inflated tires but use steel tracks that likely crush the aggregate surface. Therefore, heavy logging vehicles and equipment can exert more structural damage (e.g., crushing) to a forest road than light traffic. For example, a passenger vehicle weighing 1.8 Mg with a tire pressure of 207 kPa (30 psi) would need to pass 528 times to cause the same structural damage to a forest road as a single pass by a standard 36 Mg log truck with a typical tire pressure of 550 kPa (80 psi [46]) [1]. Therefore, heavy traffic, including logging equipment, can provide enough stress to change the PSDs of all cross-sectional locations of forest road aggregate.

When there is enough stress, aggregate material may be crushed or moved down into the finer-textured subgrade material, but its movement is dependent on material strength, particle size of the aggregate, and road conditions such as water content and compaction. The material strength of cohesive soils and rocks can be approximated by using uniaxial compression strength [47]. Forest roads are usually well compacted with a uniaxial compression strength of 0.6–1 MPa (very stiff soil; very difficult to move with hand pick, pneumatic tool needed for excavation), ranging up to over 200 MPa (very strong rock; quartzite, dolerite, gabbro, and basalt). The normal and shear stress from traffic might not be enough to crush forest road aggregate particles, especially below the upper layer. However, the contact stress between tire and aggregate particle or between aggregate particles, might be enough to cause the crushing [48] because aggregate particles have polyhedron shapes with many flat faces and sharp edges [49], resulting in high contact stresses produced when the sharp edges are pressed. Farmani et al. [50] found that higher contact stress was distributed across the areas between large-size aggregate particles, indicating that this type of stress is more likely to break large-size particles. A material often loses its strength over time when experiencing stresses lower than the level that would cause instantaneous failure. This "fatigue" ultimately leads to fracturing if the stresses continue [51]. In nature, fatigue is a major factor in the physical weathering processes of rocks [52]. Erarslan and Williams [53] noted that static and cyclic loading due to vehicle-induced vibrations and traffic often caused rocks to fail at a lower stress. Vehicles used in forest operations, especially log trucks, have multiple axles and can exert repeated, cyclic loadings on the road aggregate. Therefore, traffic can cause the aggregate particles to break by fatigue failure. The roads we used had basalt as the aggregate material because this material is locally available and commonly used to surface forest roads throughout the Pacific Northwest [19]. There are a variety of aggregate tests developed by ASTM and AASHTO [18,19], including Los Angeles Abrasion (AASHTO T96 standard [54]) to measure aggregate's resistance to crushing (mechanical breakdown) from traffic; however, they are beyond the scope of this study. Even if the study roads had a stronger aggregate material, heavy traffic and logging equipment could change the PSDs of some or all of the cross-sectional locations to a 0.1 m depth.

The heavy traffic road (H) was subjected to application of calcium chloride for dust abatement. Calcium chloride is used to help hold fine particles on the road surface together, thus reducing dust, surface raveling (loose aggregate), washboarding (corrugations), and maintenance costs [55–57]. Therefore, the effects of calcium chloride on subsequent PSDs was likely limited to the road surface. Since our work indicates no significant differences in PSDs from the upper and middle layers on the H road, we could not detect the effect of calcium chloride on PSDs using an aggregate sampling depth of 33.3 mm. A different sampling method might be needed to investigate the effect of dust abatement on road aggregate PSDs.

Particle size is an important factor for aggregate crushing. Soil and rock (aggregate) strength increases with decreasing particle size [58–60]. Forest road aggregate usually consists of various-sized particles to meet the Forest Service's specifications for surface course aggregate [2]. Stress from traffic likely crushes larger-size particles into smaller ones within the road aggregate. Our data shows the changed PSDs did not get finer than a certain PSD limit, close to the particle-size distribution curves at U-I-H or B-I-H, except the particle-size distribution at B-S-H where subgrade mixing likely occurred in addition to crushing (Figure 5). This curve is the limit to which the stress from traffic can crush road aggregate. Traffic crushes and compacts road aggregate, and subsequently makes aggregate PSDs finer. Therefore, we can consider this PSD limit as the optimum compaction by crushing for a given road condition and aggregate material. A similar concept was noted by Fuller and Thompson [61] who described an idealized grading that represented the densest state of packing particles (Fuller packing) [62]. Once the aggregate reaches this limit, stress from traffic is delivered, without further changing the PSD, to the aggregate below or to the subgrade where more crushing or subgrade mixing may occur.

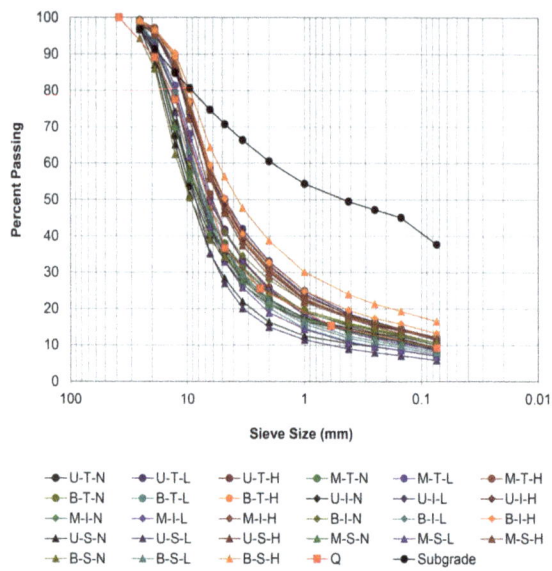

Figure 5. The particle-size distributions (PSDs) of the forest road aggregates and subgrade from the White Pine roads in the Clearwater National Forest and the quarry (Q) in Idaho. The abbreviations indicate different vertical layers (upper (U), middle (M), and bottom (B)), horizontal locations (tire track (T), in-between (I), and shoulder (S)), and traffic uses (no (N), light (L), and heavy (H)). The particle-size distributions changed by crushing did not get finer than a certain limit, close to the particle-size distribution curves at U-I-H or B-I-H, except the particle-size distribution at B-S-H where subgrade mixing likely occurred in addition to crushing. The road aggregate PSDs were based on average values of 5 replications [48]; the quarry, 12 replications; and the subgrade, 4 replications at B-S-H.

Aggregate is made by crushing rocks in a quarry and is delivered to a road construction site by vehicles such as dump trucks. It is then dumped out of the vehicle and spread on subgrade material. During this process, particle-size segregation might occur; as observed in some road aggregates [17,63]. While road aggregate is being delivered, vehicles vibrate and shake aggregate particles, which might potentially cause particle-size segregation. This particle-size segregation is often called the "Brazil-nut" effect which originated from the observation that shaking mixed nuts made the Brazil nuts (the biggest) move upward and end up on top of the mixed nuts [64]. This phenomenon occurs for granular materials having a wide size distribution, even at the molecular level, making large particles rise to the top of a mixture, and fine particles move downwards [64–66]. However, the shapes of the aggregate particles are polyhedrons [49], thus there can be interlocking of the aggregate particles which prohibits movement. Therefore, it is not known if particle-size segregation occurs during aggregate transport. Once aggregate arrives at the road construction site it is dumped out of the vehicle. When aggregate particles are dumped, large particles tend to roll down to the edge of the pile causing particle-size segregation. This is similar to a scree slope with rockfall sorting in which larger particles are at the bottom edge of the slope [67] and has been reported when granular materials with different particle properties (e.g., size, density, shape, resilience, angle of repose, and cohesiveness) are deposited [65,68]. Aggregate dumped and spread on subgrade likely has the same particle-size segregation occurring at road edges (i.e., shoulder). Interestingly, this study points out some inconsistencies. On the N road, PSDs were uniform (Table 2). However, at U-T-N, U-S-N, and M-S-N they were coarser than the PSD from the quarry (Table 5), indicating that particle-size segregation may occur from the quarry to the newly constructed road.

Root and soil interactions may also be another cause of changes in road aggregate, because vegetation root systems can loosen compacted road surface [69,70]. Also, roots from nearby trees can cause upheaval or displacement of the road surface [24]. This road surface change has particularly been noted on paved roads, but could also occur on unpaved roads, such as aggregate or native-soil-surfaced roads. However, we did not test for this on our road surfaces because of the relatively short timeframe that our study sites had been in use.

5. Conclusions

Knowledge of traffic-induced changes to forest road aggregate PSD is important for reducing road maintenance efforts and sediment production from forest roads. This study provides new data to help understand traffic-induced changes in PSD. Increased traffic changed the PSDs by all three processes we examined: crushing, subgrade mixing, and sweeping. We inferred these traffic-induced processes by comparing forest road aggregate PSDs from vertical and horizontal locations, and traffic uses. No differences in PSD were found on the no traffic road, indicating no processes occurred. Light traffic made the PSDs at U-T and U-I finer, indicating crushing occurred, and made the PSD at U-S coarser, indicating sweeping occurred. Heavy traffic and logging equipment with wide axles made all the PSDs finer, which indicates crushing occurred at all cross-sectional locations. In addition, the PSD at B-S was finer than the other locations and can be explained by subgrade mixing in the bottom layer at the shoulder.

Crushing appears to have been the dominant process on our study roads based on PSD comparisons. Crushing changed the PSDs up to a certain limit, but no finer. Having road aggregate close to the PSD limit would reduce aggregate crushing and fine sediment production. Crushing and subsequent compaction can cause the aggregate volume to change, resulting in permanent deformation of road surface with ruts, washboards, and potholes; requiring road maintenance. Therefore, having aggregate close to the PSD limit (i.e., the optimum compaction) will reduce road maintenance. Further, particle-size segregation of road aggregate may occur during transport, dumping, and surfacing the subgrade, but additional work is needed to investigate this segregation and to produce aggregate with more desirable PSDs.

Subgrade mixing can also be a dominant process in other geographical locations with different road conditions, aggregate, and subgrade properties (e.g., soft, weak subgrade, and wet road conditions). Understanding the physical processes on different road conditions will help mitigate sediment production from forest roads and reduce road maintenance efforts by providing information for best management practices. For example, strengthening the surface material (e.g., surface stabilization) is recommended if the dominant process is crushing; strengthening the subgrade (e.g., geotextile reinforcement on subgrade) if subgrade mixing is the dominant process; and collecting and recycling large aggregate particles on the shoulder and roadside for road resurfacing if sweeping occurs excessively. Future study is recommended to investigate traffic-induced processes in other locations where subgrade mixing or sweeping is the dominant process, and the effects of road treatments and management practices on the traffic-induced processes, for better road management.

Supplementary Materials: The following are available online at www.mdpi.com/link, Figure S1: The particle-size distributions from different vertical layers and horizontal locations on no traffic use (N): (a) the particle-size distributions from different vertical layers at tire track (T); (b) in-between (I); (c) at shoulder (S); (d) the particle-size distributions from different horizontal locations at upper (U); (e) at middle (M); and (f) at bottom (B). The particle-size distributions were based on average values of five replications [48], Figure S2: The particle-size distributions from different vertical layers and horizontal locations on light traffic use (L): (a) the particle-size distributions from different vertical layers at tire track (T); (b) in-between (I); (c) at shoulder (S); (d) the particle-size distributions from different horizontal locations at upper (U); (e) at middle (M); and (f) at bottom (B). The particle-size distributions were based on average values of five replications [48], Figure S3: The particle-size distributions from different vertical layers and horizontal locations on heavy traffic use (H): (a) the particle-size distributions from different vertical layers at tire track (T); (b) in-between (I); (c) at shoulder (S); (d) the particle-size distributions from different horizontal locations at upper (U); (e) at middle (M); and (f) at bottom (B). The particle-size distributions were based on average values of five replications [48], and Figure S4: The particle-size distributions from different traffic uses (no (N), light (L), and heavy (H)) at the same cross-sectional locations: (a) the particle-size distributions in upper layer at tire track location (U-T); (b) in upper layer at in-between location (U-I); (c) in upper layer at shoulder location (U-S); (d) in middle layer at tire track location (M-T); (e) in middle layer at in-between location (M-I); (f) in middle layer at shoulder location (M-S); (g) in bottom layer at tire track location (B-T); (h) in bottom layer at in-between location (B-I); and (i) in bottom layer at shoulder location (B-S). The particle-size distributions were based on average values of five replications [48].

Acknowledgments: The authors acknowledge many contributions from the following people: Randy Foltz (retired) and Benjamin Kopyscianski at the Rocky Mountain Research Station, USDA Forest Service for their guidance and help in the field and laboratory; David Ratliff, Dave Brady (retired), and Meg Foltz (retired) at the Palouse Ranger District, Clearwater National Forest, USDA Forest Service, and Dan Jones at the Potlatch Corporation for providing the information on White Pine roads and allowing us to access them; Peter Schiess (retired) at the School of Environmental and Forest Sciences, University of Washington, Hans-Erik Andersen at the Pacific Northwest Research Station, USDA Forest Service, and Finn Krogstad for their comments; and Natalie Copeland, Leah Kirkland, and Rebecca Shifler for analyzing the aggregate samples.

Author Contributions: Hakjun Rhee and James Fridley conceived and designed the experiments; Hakjun Rhee performed the experiments; Hakjun Rhee analyzed the data; and Hakjun Rhee and Deborah Page-Dumroese wrote the paper.

Conflicts of Interest: The authors declare no conflict of interest.

References

1. Bolander, P.; Marocco, D.; Kennedy, R. *Earth and Aggregate Surfacing Design Guide for Low Volume Roads*; Engineering Staff EM-7170-16; US Department of Agriculture, Forest Service: Washington, DC, USA, 1996; p. 24.
2. Turner, S.K.; Hutchinson, K. *Forest Service Specifications for Construction of Roads and Bridges*; Revised; Engineering Staff EM-7720-100; US Department of Agriculture, Forest Service: Washington, DC, USA, 1996; pp. 495–513.
3. Thompson, M.; Sessions, J. Optimal policies for aggregate recycling from decommissioned forest roads. *Environ. Manag.* **2008**, *42*, 297–309. [CrossRef] [PubMed]
4. Swift, L.W. Gravel and grass surfacing reduces soil loss from mountain roads. *For. Sci.* **1984**, *30*, 657–670.
5. Kochenderfer, J.N.; Helvey, J.D. Using gravel to reduce soil losses from minimum-standard forest roads. *J. Soil Water Conserv.* **1987**, *42*, 46–50.
6. Bilby, R.E.; Sullivan, K.; Duncan, S.H. The generation and fate of road-surface sediment in forested watershed in southwestern Washington. *For. Sci.* **1989**, *35*, 453–468.
7. Larcombe, G. *Forest Roading Manual*; Liro Forestry Solutions: Rotorua, New Zealand, 1999; pp. 168–173, 284.
8. Reid, L.M.; Dunne, T. Sediment production from forest road surfaces. *Water Resour. Res.* **1984**, *20*, 1753–1761. [CrossRef]
9. Luce, C.H.; Black, T.A. Effects of traffic and ditch maintenance on forest road sediment production. In Proceedings of the Seventh Federal Interagency Sedimentation Conference, Reno, NV, USA, 25–29 March 2001; Subcommittee on Sedimentation: Washington, DC, USA, 2001; pp. V67–V74.
10. Ziegler, A.D.; Sutherland, R.A.; Giambelluca, T.W. Interstorm surface preparation and sediment detachment by vehicle traffic on unpaved mountain roads. *Earth Surf. Proc. Landf.* **2001**, *26*, 235–250. [CrossRef]
11. Sheridan, G.J.; Noske, P.J.; Whipp, R.K.; Wijesinghe, N. The effect of truck traffic and road water content on sediment delivery from unpaved forest roads. *Hydrol. Process.* **2006**, *20*, 1683–1699. [CrossRef]
12. Ramos-Scharrón, C.E.; MacDonald, L.H. Measurement and prediction of sediment production from unpaved roads, St John, US Virgin Islands. *Earth Surf. Proc. Landf.* **2005**, *30*, 1283–1304. [CrossRef]
13. Toman, E.M.; Skaugset, A.E. Reducing sediment production from forest roads during wet-weather hauling. *Transp. Res. Rec.* **2011**, *2203*, 13–19. [CrossRef]
14. Foltz, R.B. Environmental impacts of forest roads: An overview of the state of the knowledge. In Proceedings of the Second International Forest Engineering Conference, Växjö, Sweden, 12–15 May 2003; Wide, M.I., Baryd, B., Eds.; Skogforsk: Uppsala, Sweden, 2003; pp. 121–128.
15. Foltz, R.B.; Evans, G.L.; Truebe, M. Relationship of forest road aggregate test properties to sediment production. In Proceedings of the Conference on Watershed Management and Operation Management 2000, Fort Collins, CO, USA, 20–24 June 2000; Flug, M., Donald, F., Watkins, D.W., Eds.; American Society of Civil Engineers (ASCE): Reston, VA, USA, 2000.
16. Gnanendran, C.T.; Beaulieu, C. On the behaviour of low-volume unpaved resource access roads: Effects of rehabilitation. *Can. J. Civ. Eng.* **1999**, *26*, 262–269. [CrossRef]

17. Johnson, G. Minnesota's experience with thin bituminous treatments for low-volume roads. *Transp. Res. Rec.* **2003**, *1819*, 333–337. [CrossRef]

18. Foltz, R.B.; Truebe, M. Effect of aggregate quality on sediment production from a forest road. In Proceedings of the Sixth International Conference on Low-Volume Roads, Minneapolis, MN, USA, 25–29 June 1995; National Academy Press: Washington, DC, USA, 1995; pp. 49–57.

19. Foltz, R.B.; Truebe, M. Locally available aggregate and sediment production. *Transp. Res. Rec.* **2003**, *1819*, 185–193. [CrossRef]

20. Foltz, R.B.; Burroughs, E.R. A test of normal tire pressure and reduced tire pressure on forest roads: Sedimentation effects. In Proceedings of the Forestry and Environment Engineering Solutions, New Orleans, LA, USA, 5–6 June 1991; Stokes, B.J., Rawlins, C.L., Eds.; American Society of Agricultural Engineers (ASAE) Publication: St. Joseph, MI, USA, 1991; pp. 103–112.

21. Foltz, R.B.; Elliot, W.J. Effect of lowered tire pressures on road erosion. *Transp. Res. Rec.* **1997**, *1589*, 19–25. [CrossRef]

22. Parsakhoo, A.; Lotfalian, M.; Kavian, A.; Hosseini, S.A. Assessment of soil erodibility and aggregate stability for different parts of a forest road. *J. For. Res.* **2014**, *25*, 193–200. [CrossRef]

23. Luce, C.H.; Black, T.A. Sediment production from forest roads in western Oregon. *Water Resour. Res.* **1999**, *35*, 2561–2570. [CrossRef]

24. Giuliani, F.; Autelitano, F.; Degiovanni, E.; Montepara, A. DEM modelling analysis of tree root growth in street pavements. *Int. J. Pavement Eng.* **2017**, *18*, 1–10. [CrossRef]

25. Ryżak, M.; Bieganowski, A. Methodological aspects of determining soil particle-size distribution using the laser diffraction method. *J. Plant Nutr. Soil Sci.* **2011**, *174*, 624–633. [CrossRef]

26. McFarlane, H.W.; Paterson, W.J.; Dohaney, W.J. *Use of the Benkelman Beam on Forest Roads*; Logging Research Reports LRR/49; Pulp and Paper Research Institute of Canada: Point-Claire, QC, Canada, 1973.

27. Selim, A.A.; Skorseth, O.K.; Muniandy, R. Long-lasting gravel roads: Case study from the United States. *Transp. Res. Rec.* **2003**, *1819*, 161–165. [CrossRef]

28. Zilioniene, D.; Cygas, D.; Juzenas, A.A.; Jurgaitis, A. Improvement of functional designation of low-volume roads by dust abatement in Lithuania. *Transp. Res. Rec.* **2007**, *1989*, 293–298. [CrossRef]

29. Climate Normals. Available online: http://www.ncdc.noaa.gov/data-access/land-based-station-data/land-based-datasets/climate-normals (accessed on 18 February 2018).

30. Miller, J.F.; Frederick, R.H.; Tracey, R.J. *NOAA Atlas 2: Precipitation-Frequency Atlas of the Western United States. Volume V—Idaho*; US Department of Commerce, National Oceanic and Atmospheric Administration (NOAA), National Weather Service: Silver Spring, MD, USA, 1973. Available online: http://www.nws.noaa.gov/oh/hdsc/PF_documents/Atlas2_Volume5.pdf (accessed on 18 February 2018).

31. Precipitation Frequency Data Server (PFDS). Available online: http://hdsc.nws.noaa.gov/hdsc/pfds (accessed on 18 February 2018).

32. US Geological Survey (USGS). The National Map. Available online: http://nationalmap.gov/index.html (accessed on 19 November 2013).

33. Model 3440 Surface Moisture-Density Gauge. Available online: http://www.troxlerlabs.com/products/3440.php (accessed on 4 May 2015).

34. American Association of State Highway and Transportation Officials. *AASHTO T88-00: Standard Method of Test for Particle Size Analysis of Soils*; AASHTO: Washington, DC, USA, 2004.

35. American Society for Testing and Materials. *ASTM D422-63: Standard Test Method for Particle-Size Analysis of Soils*; ASTM International: West Conshohocken, PA, USA, 2002.

36. American Association of State Highway and Transportation Officials. *AASHTO T146-96: Standard Method of Test for Wet Preparation of Disturbed Soil Samples for Test*; AASHTO: Washington, DC, USA, 2000.

37. Rhee, H.; Foltz, R.B.; Fridley, J.L.; Krogstad, F.; Page-Dumroese, D.S. An alternative method for determining particle-size distribution of forest road aggregate and soil with large-sized particles. *Can. J. For. Res.* **2014**, *44*, 101–105. [CrossRef]

38. American Society for Testing and Materials. *ASTM D2217-85: Standard Practice for Wet Preparation of Soil Samples for Particle-Size Analysis and Determination of Soil Constants*; ASTM International: West Conshohocken, PA, USA, 1998.

39. Hardin, B.O. Crushing of soil particles. *J. Geotech. Eng.* **1985**, *111*, 1177–1192. [CrossRef]

40. Lade, P.V.; Yamamuro, J.A.; Bopp, P.A. Significance of particle crushing in granular materials. *J. Geotech. Eng.* **1996**, *122*, 309–316. [CrossRef]

41. Zar, J.H. *Biostatistical Analysis*, 3rd ed.; Prentice Hall: Upper Saddle River, NJ, USA, 1996; pp. 179–217, ISBN 0130845426.

42. Söhne, W. Fundamentals of pressure distribution and soil compaction under tractor tires. *Agric. Eng.* **1958**, *39*, 276–281, 290.

43. Wong, J.Y. *Theory of Ground Vehicles*, 3rd ed.; John Wiley & Sons: New York, NY, USA, 2001; pp. 92–100, ISBN 0471354619.

44. Ritter, M.A. *Timber Bridges: Design, Construction, Inspection, and Maintenance*; Engineering Staff EM 7700-8; US Department of Agriculture, Forest Service: Washington, DC, USA, 1992; pp. 6-3–6-9.

45. Federal Highway Administration. *Federal Size Regulations for Commercial Motor Vehicles*; US Department of Transportation, Federal Highway Administration (FHWA), Office of Freight Management and Operations: Washington, DC, USA, 2004. Available online: http://ops.fhwa.dot.gov/freight/publications/size_regs_final_rpt/ (accessed on 18 February 2018).

46. Mills, K.; Pyles, M.; Thoreson, R. Aggregate surfacing design and management for low-volume roads in temperature, mountainous areas. *Transp. Res. Rec.* **2007**, *1989*, 154–160. [CrossRef]

47. Hoek, E.; Bray, J.W. *Rock Slope Engineering*, 3rd ed.; Elsevier Science: London, UK, 1981; p. 99, ISBN 0900488573.

48. Rhee, H. Inferring Traffic induced Sediment Production Processes from Forest Road Particle Size Distributions. Ph.D. Thesis, University of Washington, Seattle, WA, USA, 6 July 2006.

49. Anochie-Boateng, J.K.; Komba, J.J.; Mvelase, G.M. Three-dimensional laser scanning technique to quantify aggregate and ballast shape properties. *Constr. Build. Mater.* **2013**, *43*, 389–398. [CrossRef]

50. Farmani, M.B.; Memarian, H.; Hansson, J.; Dusseault, M.B. Contact forces in non-bonded pavement materials. *Road Mater. Pavement Des.* **2007**, *8*, 483–503. [CrossRef]

51. Mittemeijer, E.J. *Fundamentals of Materials Science: The Microstructure—Property Relationship Using Metals as Model. Systems*; Springer: Berlin, Germany, 2010; pp. 567–573, ISBN 9783642104992.

52. Selby, M.J. *Hillslope Materials and Processes*, 2nd ed.; Oxford University Press: Oxford, UK, 1993; pp. 144–145. ISBN 0198741839.

53. Erarslan, N.; Williams, D.J. Mechanism of rock fatigue damage in terms of fracturing modes. *Int. J. Fatigue* **2012**, *43*, 76–89. [CrossRef]

54. American Association of State Highway and Transportation Officials. *AASHTO T96: Standard Method of Test for Resistance to Degradation of Small-Size Coarse Aggregate by Abrasion and Impact in the Los Angeles Machine*; AASHTO: Washington, DC, USA, 2002.

55. Bolander, P.; Yamada, A. *Dust Palliative Selection and Application Guide*; US Department of Agriculture, Forest Service, Technology and Development Program: San Dimas, CA, USA, 1999.

56. Monlux, S. Stabilizing unpaved roads with calcium chloride. *Transp. Res. Rec.* **2003**, *1819*, 52–56. [CrossRef]

57. Monlux, S.; Mitchell, M. Chloride stabilization of unpaved road aggregate surfacing. *Transp. Res. Rec.* **2007**, *1989*, 50–58. [CrossRef]

58. Weibull, W. A statistical theory of the strength of materials. *Proc. R. Swed. Inst. Eng. Res.* **1939**, *151*, 1–45.

59. Lundborg, N. The strength-size relation of granite. *Int. J. Rock Mech. Min. Sci.* **1967**, *4*, 269–272. [CrossRef]

60. Tang, C.A.; Tham, L.G.; Lee, P.K.K.; Tsui, Y.; Liu, H. Numerical studies of the influence of microstructure on rock failure in uniaxial compression—Part II: Constraint, slenderness and size effect. *Int. J. Rock Mech. Min. Sci.* **2000**, *37*, 571–583. [CrossRef]

61. Fuller, W.B.; Thompson, S.E. The laws of proportioning concrete. *Trans. Am. Soc. Civ. Eng.* **1907**, *59*, 67–143.

62. Bardet, J.P. *Experimental Soil Mechanics*; Prentice Hall: Upper Saddle River, NJ, USA, 1997; pp. 12–13, ISBN 0133749355.

63. Chen, D.H.; Scullion, T.; Lee, T.C.; Bilyeu, J. Results from a forensic investigation of a failed cement treated base. *J. Perform. Constr. Fac.* **2008**, *22*, 143–153. [CrossRef]

64. Rosato, A.; Strandburg, K.J.; Prinz, F.; Swendsen, R.H. Why the Brazil nuts are on top: Size segregation of particulate matter by shaking. *Phys. Rev. Lett.* **1987**, *58*, 1038–1040. [CrossRef] [PubMed]

65. Thomson, F.M. Storage and flow of particulate solids. In *Handbook of Powder Science and Technology*, 2nd ed.; Muhammad, E.F., Otten, L., Eds.; Chapman & Hall: New York, NY, USA, 1997; pp. 389–486, ISBN 0412996219.

66. Breu, A.P.J.; Ensner, H.M.; Kruelle, C.A.; Rehberg, I. Reversing the Brazil-nut effect: Competition between percolation and condensation. *Phys. Rev. Lett.* **2003**, *90*, 014302. [CrossRef] [PubMed]

67. Gerber, E.; Scheidegger, A.E. On the dynamics of scree slopes. *Rock Mech.* **1974**, *6*, 25–38. [CrossRef]

68. Williams, J.C. The segregation of particulate materials: A review. *Powder Technol.* **1976**, *15*, 245–251. [CrossRef]

69. Kolka, R.K.; Smidt, M.F. Effects of forest road amelioration techniques on soil bulk density, surface runoff, sediment transport, soil moisture and seedling growth. *For. Ecol. Manag.* **2004**, *202*, 313–323. [CrossRef]

70. Foltz, R.B.; Rhee, H.; Yanosek, K.A. Infiltration, erosion, and vegetation recovery following road obliteration. *Trans. ASABE* **2007**, *50*, 1937–1943. [CrossRef]

Article

Spatial Distribution of Biomass and Woody Litter for Bio-Energy in Biscay (Spain)

Esperanza Mateos [1,*] **and Leyre Ormaetxea** [2]

1   Department of Chemical and Environmental Engineering, University of the Basque Country UPV/EHU, Rafael Moreno 'Pitxitxi', n 3, 48013 Bilbao, Spain
2   Department of Mathematics, Faculty of Science and Technology, University of the Basque Country UPV/EHU, Barrio Sarriena s/n, 48940 Leioa, Spain; leyre.ormaetxea@ehu.eus
*   Correspondence: esperanza.mateos@ehu.eus; Tel.: +34-946-104-343; Fax: +34-946-104-300

Received: 7 March 2018; Accepted: 1 May 2018; Published: 7 May 2018

Abstract: Forest management has been considered a subject of interest, because they act as carbon (C) sinks to mitigate CO_2 emissions and also as producers of woody litter (WL) for bio-energy. Overall, a sustainably managed system of forests and forest products contributes to carbon mitigation in a positive, stable way. With increasing demand for sustainable production, the need to effectively utilise site-based resources increases. The utilization of WL for bio-energy can help meet the need for renewable energy production. The objective of the present study was to investigate biomass production (including C sequestration) from the most representative forestry species (*Pinus radiata* D. Don and *Ecualyptus globulus* Labill) of Biscay (Spain). Data from the third and fourth Spanish Forest Inventories (NFI3-2005 and NFI4-2011) were used. We also estimated the potential WL produced in the forest activities. Our findings were as follows: Forests of Biscay stored 12.084 Tg of biomass (dry basis), with a mean of 147.34 Mg ha^{-1} in 2005 and 14.509 Tg of biomass (dry basis), with a mean of 179.82 Mg ha^{-1} in 2011. The total equivalent CO_2 in Biscay's forests increased by 1.629 Tg year^{-1} between 2005 and 2011. The study shows that the energy potential of carbon accumulated in the WL amounted to 1283.2 million MJ year^{-1}. These results suggest a considerable potential for energy production.

Keywords: carbon stock; woody litter; bioenergy potential; resources map; aboveground biomass, underground biomass

---

1. Introduction

In recent years there has been an increasing interest in the estimation of forest biomass, mainly in the context of the rules established in the Kyoto Protocol. According to this protocol, the CO_2 emissions limit for each nation must be estimated taking the carbon sinks and sources into account, including the carbon dioxide absorbed and stored by trees [1,2]. The carbon content in forests is the highest of all terrestrial ecosystems, being considerably higher than that in pastures and fields [3]. Some studies have shown that the simultaneous consideration of carbon sequestration in forest ecosystems and forest biomass production (e.g., wood and energy) in forest management could offer important means to reduce carbon emissions to the atmosphere in the future [4,5]. Temperate forests currently act as carbon sinks since they absorb more carbon from the atmosphere through photosynthesis than the carbon they produce through breathing. Notwithstanding, in a climate change environment, carbon dynamics may be altered. Hence, it is essential to engage in a sustainable forest management in which the annual average stock of carbon and its sequestration can be increased. On the other hand, the use of biomass for energy production has recently created a great deal of interest, which is partly due to environmental reasons. These reasons are mainly the problems caused by climate change and the need to search for a solution to the foreseeable exhaustion of fossil fuels.

The classical energy model based on the massive use of fossil fuels is unsustainable from both the environmental viewpoint and from the point of view of using up these resources. Thus, a new energy model must be established, based on the diversification of resources, consumption rationalization and efficiency, as well as environmental respect. With increasing energy costs, society is searching for alternative energy sources. The mean value of forest biomass used in Europe is 61%, while in Nordic countries or Austria this percentage is close to 90%. However, despite its potential forest biomass, Spain only exploits some 38% of the biomass that grows annually in its forests. In local terms, the goal for the Autonomous Community of the Basque Country (ACBC) in 2010 was to reach 795,000 MJ of biomass exploitation. As far as the contribution of forests and lands from the ACBC as carbon sink is concerned, the net drain effect has been estimated as 1.33×10^6 Mg CO_2 [6].

The use of forest biomass as an energy resource is closely related to employment, since for each job in the fossil fuel businesses, as many as fourteen jobs are generated with biomass. Especially in rural areas, this provides territorial equilibrium. Moreover, the use of this residue as an energy source helps to reach the compromises acquired by the European Union in the Kyoto protocol—by 2020, 20% of all energy consumption must come from renewable sources-[7]. In Spain, the 2011–2020 Renewable Energy Plan set the target of 20% of total primary energy needs to be met by renewable sources, and about 10% of these by bioenergy [8]. One of the most important advantages of biomass use is its low atmosphere-pollutant production when compared with conventional fuels: minimum production of SO_2 due to its low S content; the emission of NO_x is also significantly reduced since biomass combustion can take place at lower temperatures, almost without affecting its yield [9]. Apart from this, the use of forest biomass for energy purposes has a null CO_2 emissions balance since CO_2 emissions that occur as a result of its recovery as energy are offset by the amount absorbed by organisms for the production of biomass through photosynthesis.

Forest biomass fulfills a double aim from the environmental point of view: (a) capacity to produce renewable energy from it and (b) to keep an adequate degree of maintenance and cleanliness of our forests [10]. However, the energy valuation of forest biomass presents some problems due to its low energy density and the scattered production of the resource. As a result of this, the quantity of biomass becomes essential, since its supply to the energy plant must be guaranteed. Quantifying the amount of forest biomass is fundamental and essential in order to be able to calculate carbon storage, as well as to study climate change, health of forests, forestall productivity, and nutrients cycle [11].

Due to the quick change of forest masses (new hydraulic work, fires, etc.), it is difficult to establish the available biomass quantity at any given moment [12,13]. To overcome this problem, different techniques have been used in this project, the main ones being tele-detection and Geographic Information System (GIS). The use of these systems presents numerous advantages over traditional inventories since they allow very detailed spatial information to be gathered at a higher periodicity and a lower cost [14–18].

The methodologies used in recent years to determine the amount of forest biomass can be classified into two categories: (a) direct estimations in the field and (b) indirect methods [19,20]. In the direct methods, once the trees have been selected, they are subjected to destructive sampling: after being felled, the different parts of the tree considered (trunk, branches, leaves, etc.) are cut up into pieces. The weight of each of these parts is determined by means of different techniques in a number of sampling plots [21]. In the case of young trees, the complete parts are weighed once the tree has been felled. Nevertheless, for large-size trees this procedure is unviable and sampling techniques must be used [22]. After reviewing the different forest biomass estimation methods, the one chosen was the so-called indirect method, since this methodology provides similar results to those obtained through direct methods. Moreover, the indirect methodology is non-destructive, whereas the direct one is destructive and it involves a laborious procedure. Moreover, a higher quantity of quantitative information is used in indirect methods, and this methodology can be applied to data from future forest inventories [16,20].

In this study, the biomass accumulated in forests in Biscay is quantified using indirect methods based on forestry inventory data, basically using biomass equations or biomass expansion factors (BEFs) [23–27]. We used biomass equations in this research, because this method may provide more precise estimations than BEF and because it is used more frequently to estimate the biomass of forests [25,28,29].

In order to select the indirect method that best adapts to the model, an exhaustive analysis was made of those developed both at national and international levels. In recent decades, a compendium of equations gathered from different countries has been drawn up [30–34], but these are difficult to apply in our case due to a lack of data about their construction. Although the area analysed in this paper (Biscay) has great potential for the generation of forestry biomass, very few studies have been carried out on this area. Cantero et al. [35] did a preliminary study in order to establish growth models in plantations of *Pinus radiata* D. Don in Biscay, but new tools are still required to improve biomass generation predictions, taking the latest forestry inventories into account.

The objectives of this study were (a) to estimate biomass (aboveground and underground) and the carbon accumulated by the main forest species in Biscay, (b) to assess the annual woody litter (WL) for bio-energy obtained in the forestry treatments and its geographical distribution.

Among the numerous studies carried out in recent years about the estimation of forest biomass, some are located in areas with similar climatic characteristics and tree species to our study area, and therefore a number of equations used in these studies can be applied to the current project [36–38].

In the previous research undertaken in the study area, the estimation of WL was done considering the residues obtained after a ten-year average forestry management and silviculture periodicity. In each phase, 1/3 of the trees were cut for *P. radiata*. It was done according to data obtained in several parts of Spain [13]. In this previous research the influence of forest management and cutting rotations on carbon sequestration in forest biomass was not considered. In Biscay, the traditional method of forest management has usually been the production of timber (pulpwood and sawlogs). However, the current environmental concerns, in response to global warming caused by human beings, have forced a change in the traditional forest management of our forests to a more sustainable approach, attempting to balance economic performance with environmental aspects. Among these new approaches is the optimization of CO_2 sequestration in relation to forest biomass. In the current research, the quantity of forest biomass residues obtained in each rotation stage was estimated considering the periodicity, the age of the mass, and the number of trees cut according to a sustainable forest management, which optimizes the quantity of timber obtained and the carbon sequestration related to the process [4,39–41]. Pyörälä et al. [4] analyzed the effects of management on the economic profitability (NPV) of forest biomass production and the carbon neutrality of bioenergy use in Norway spruce stands. According to the study, maximizing the highest mean annual timber production and carbon neutrality of bioenergy use simultaneously was not possible. In general, higher carbon sequestration and carbon stock of the forest ecosystem provides higher carbon neutrality, but not higher NPV, and vice versa. In general, the net ecosystem CO_2 exchange is the highest at the younger stand age and starts to saturate after intermediate age, affecting the average carbon stock and biomass production over rotation. However, the mean annual carbon stock and carbon sequestration may be increased over a rotation by maintaining stocking higher than that currently recommended [39,41].

This study estimated the annual WL for bio-energy in Biscay, obtained in the forestry treatments and its geographical distribution. Data were taken from the Fourth National Forestry Inventory of the Province of Biscay (NFI4) [42].

2. Materials and Methods

2.1. Study Area

The study comprises the whole province of Biscay, one of the three provinces of the ACBC, located in the north of Spain at latitude 43''16' N and longitude 2''56' W (see Supplementary Material, Figure S1).

The mean annual rainfall is 1200 mm and the mean annual temperature is 12.5 °C. Frosts are infrequent and no physiological drought is apparent. Data from Fourth National Forest Inventory (NFI4 2011) depict that forest surface in this province is 131,748 ha, representing about 60% of its surface (221,232 ha). Thus, Biscay is considered the epicenter of Basque forest activity. The main forest species are *Pinus radiata* D. Don and *Eucalyptus globulus* Labill. Both are fast-growing forest species which have properly adapted to the climate conditions of the Atlantic coast of Biscay. *P. radiata* is the predominant species in this province, with 70,562 ha. This species is distributed in areas near the sea and at elevations lower than 360 m (Figure 1a). The production of timber on these plantations is primarily based on the rotational clear-cutting of even-aged stands [43]. The rapid growth and good productivity of *P. radiata* have made this species the dominant tree species in Biscay, providing 90% of timber production in this area. With 10,123 ha, *E. globulus* is an abundant species at low altitudes. Near the coast of Biscay (Figure 1b), forest managers of the province are are promoting the expansion of eucalyptus plantations to obtain biomass for the pulp and paper industry and for bioenergy.

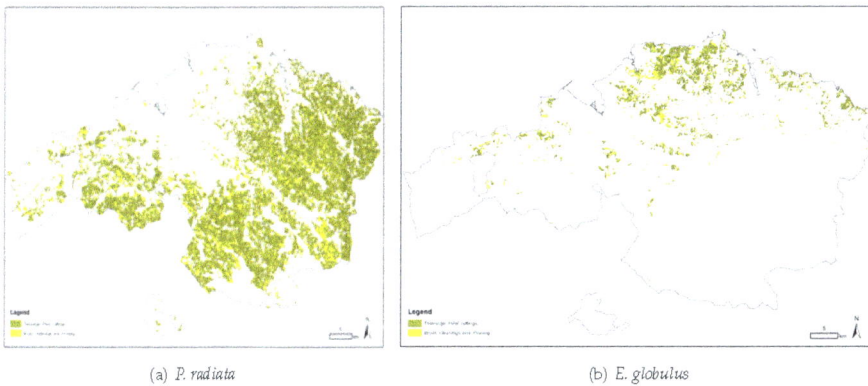

(a) *P. radiata*   (b) *E. globulus*

Figure 1. Plot map in Biscay in its different states.

2.2. Estimations of Biomass and Carbon Stock

Data from the Fourth National Forestry Inventory (NFI4) obtained in 2011 in Biscay were used in order to estimate the forest biomass fractions and the carbon sequestration (Figure 2). The NFI4 divides the province of Biscay into strata, which are defined as associations of areas with vegetation of similar characteristics. A stratum characterises the type or arboreal vegetation according to the species present in a particular zone, their states in terms of mass, and the fraction of tree cover per area [44]. Specific information concerning such strata is available at http://www.geo.euskadi.eus/mapa-forestal-del-pais-vasco-ano-2010/s69-geodir/es/. In this work we decided to select those species that fulfill two criteria: (a) high presence in the area of study, according to data obtained from the NFI-4, and (b) high potential for exploitable residue generation from the viewpoint of energy. Following both criteria, the species *P. radiata* and *E. globulus* were selected. Originally, we selected all the plots from the NFI4 whose areas were occupied by the species *P. radiata* and *E. globulus*, according to the forest map of Basque Country in 2010.

NFI4 (t2)   NFI3 (t1)

Stratum   Stratum

Number of trees ha⁻¹ (N)   Number of trees ha⁻¹ (N)

Forestal map | Biomass fractions Equations ([36], Table S1) | Biomass fractions Equations ([36], Table S1) | Forestal map

Associated area (S_i, ha) | Biomass per tree (W_i, kg dry mass) | Biomass per tree (W_i, kg dry mass) | Associated area (S_i, ha)

Biomass (B, Tg)   Biomass (B, Tg)

$$B_{t_2} = \sum_i \frac{N \cdot W_i}{10^6} S_i$$

$$B_{t_1} = \sum_i \frac{N \cdot W_i}{10^6} S_i$$

$$\Delta_{Annual\ Net\ B} = \frac{B_{t_2} - B_{t_1}}{t_2 - t_1}$$

$\Delta_{Annual\ Net}\ CO_2 =$

$\Delta_{Net} B \cdot CT \cdot MW \left(\frac{CO_2}{C}\right)$

Δ_{Annual} Withdrawals

$\Delta_{Annual\ Gross\ B} =$

$\Delta_{Net} B + \Delta_{annual\ withdrawals} B$

$\Delta_{Annual\ Gross}\ CO_2 =$

$\Delta_{Gross} B \cdot CT \cdot MW \left(\frac{CO_2}{C}\right)$

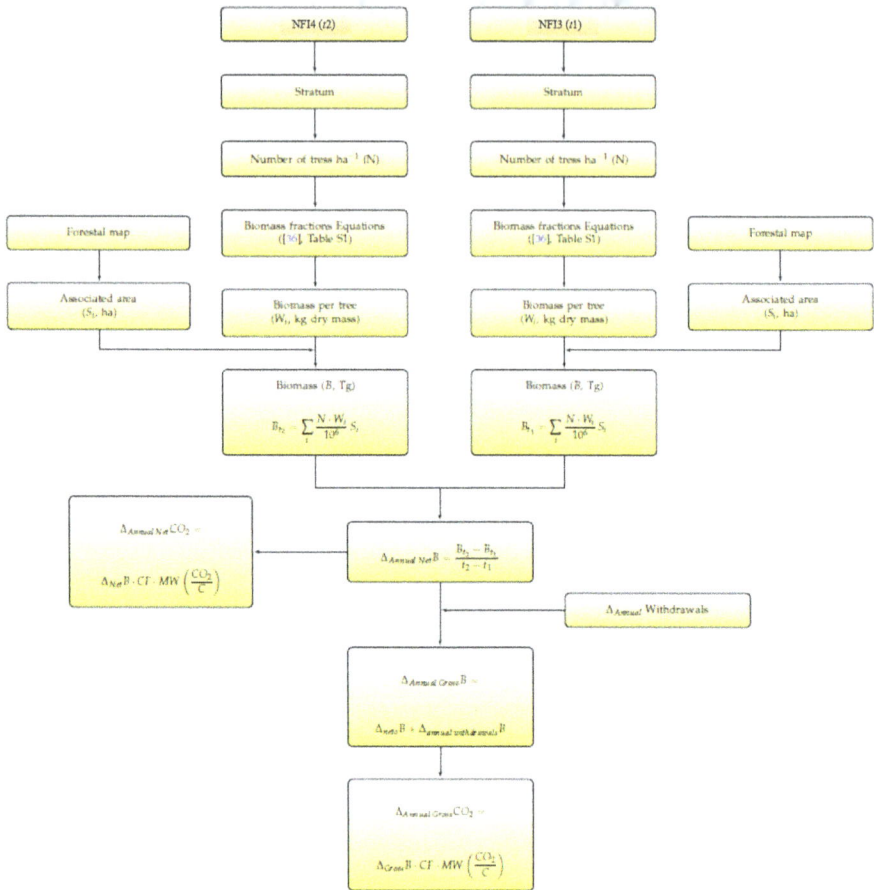

Figure 2. Schematic diagram of the method followed to determine biomass and carbon stock.

The usual methods proposed in the literature to determine the total biomass and its fractions use different independent variables such as tree diameter at breast height (DBH) measured 1.3 m above the ground; height basal area; circumference; or combinations of all of these. The general expression of these models is $W = a\,X^b$, where X is the variable which determines the dimension of the tree; W is the total biomass or the biomass of some elements (leaves, branches, etc.) expressed in kilograms (oven-dry weight); and a and b are the parameters to estimate [30,36,45,46].

After reviewing different estimation methods for forest biomass, we used allometric equations with DBH (cm) as the only explicative variable, because the DBH is the most used explanatory variable since it is very easy to measure and is related to the volume of the wood and the age of the tree [22,36,45,47–49].

In each selected plot, the biomass fractions of every *P. radiata* and *E. globulus*, W_i, tree were calculated using allometric equations by Montero et al. [36] (see Supplementary Material, Table S1). The biomass of each tree was analysed taking into account the expansion factor (EF) from the NFI4 which corresponds to that tree due to its diameter rating (see Supplementary Material, Table S2) as the plots in the NFI4 have variable radius. Consequently, a tree may or may not be measured depending on its diameter and its distance to the centre of the plot. Summing the biomass of all the analysed trees, we obtained the values of the biomass fractions, total aboveground and underground biomass in kg ha^{-1}. All the biomass data obtained are expressed on an oven-dry basis.

Total biomass was obtained using the forest map at http://www.geo.euskadi.eus/mapa-forestal-del-pais-vasco-ano-2010/s69-geodir/es/. Total biomass was converted into carbon by multiplying the oven-dry matter values by the carbon fraction (CF) of dry biomass. This value varies slightly depending on the forest species and the fractions (trunk, leaves, branches, roots, etc.) [50]. We used 51.3 for *P. radiata* [51] and 45.1 for *E. globulus* [52]. From the data about carbon stock [50], the accumulated equivalent CO_2 amounts were calculated considering the relationship of the molecular weights (MW(C/CO_2)). We compared the results of both 2005 (NFI3) and 2011 (NFI4) in order to calculate the annual increment of biomass and fixed CO_2. It was assumed that the loss of biomass was reduced due to extractions by cutting, without considering incidents such as pests, bacterial diseases, burnings, frosts, hailstorms, etc., which are assumed to be included in the expected rate of growth.

2.3. Estimation of Woody Litter (WL) for Bioenergy

At this stage of the study, we have not tried to estimate the total WL (non-timber) existing in the forests of Biscay, but that obtained after the forest exploitation and which shows nonexistent or very low commercial demand. Therefore, they are materials which can be considered as "final residues"—that is, residues which are no longer useful for any destiny other than their use as energy sources, and can be used for energetic applications owing to their excellent characteristics as fuel [8,12,53]. Potential residues (WL) consist of medium-sized branches (diameter range: 2–7 cm) and small branches (diameter: less than 2 cm) obtained from forest treatments. The NFI4 defines 12 strata in Biscay, among which those strata in which *P. radiata* and *E. globulus* were the main species were selected for this study. Thus, we selected those plots belonging to strata 1, 2, and 3, which have *P. radiata* as the main species; and stratum 9, with *E. globulus* as the main species (Table 1).

Table 1. Basic features of the strata from the Fourth National Forestry Inventory of the Province of Biscay (NFI4).

Stratum	Predominant Forest Species	Occupation * (%)	Mass Stage +	Forest Operation	Canopy Fraction ‡ (%)	Surface (ha)
1	*Pinus radiata*	≥70	Sawtimber, Poles	Thinning, Cutting	70–100	45,210
2	*Pinus radiata*	≥70	Sawtimber, Poles	Thinning, Cutting	5–69	5589
3	*Pinus radiata*	≥70	Saplings, Seedlings	Brush cleanings	40–100	12,382
9	*Eucalyptus* spp.	≥70; 30 < Esp. < 70	Sawtimber, Poles	Cutting	5–100	9183

* represents the percentage occupation of the predominant forest species. + represents the stage in the development of the referred species. ‡ represents the percentage of land covered by the horizontal projection of vegetation.

For estimation of the annual quantity of woody litter (Q_{WL}) in Biscay (Mg year^{-1}), two factors must be determined: (a) Forest residue per unit of surface and time derived from a forest mass (E_r, Mg ha^{-1} year^{-1}) in terms of estimation of the species and forest treatment each mass has been subjected to, and (b) the surface S_n (ha) occupied by the forest mass this residue will generate. The annual available quantities of dry biomass (expressed in Mg) are obtained as:

$$Q_{WL} = \sum_i S_i \, E_{r_i}.$$

(1)

The estimation of remains from WL was carried out in two different ways: stratum-by-stratum on the one hand, and all of them together on the other. In each situation, the estimation was accomplished by means of a confidence interval at a 95% level. Normality tests of the data were also carried out to determine if the values of the random variable E_r presented a normal distribution.

WL was obtained after a type of forestry management called rotation forest management (RFM) was used. This involves a sustainable forest management system in which repetitive cycles of silviculture consisting of several stages are applied, from planting or natural regeneration until felling [16]. This forestry management model is the one which is most widely used throughout the world, and is based on the intensive exploitation of rapid growth forestry species. It is done in 30-year rotations for *P. radiata* and 10-year rotations for *E. globulus* (Figure 3).

The evaluation method should consider the different phases across the complete rotation of a forest stand and the forest tasks performed in each phase [48]. In this way, and considering an adequate forest action, the WL production of a certain forest mass (Q_{WL}) can be predicted through its production cycle. Consequently, the different processes applied to stands in distinct cycles (e.g., brushings, first thinning, intermediate thinning, and regeneration fellings) generate different forest by-products.

With regard to the forestry applied in Biscay, the rotation cycle of *P. radiata* is an average of 30 years. After the final cutting, a reforestation with an initial average density of 1500 trees ha^{-1} is applied. In the tenth year, the first regeneration cutting takes place, removing 600 trees ha^{-1}. At this stage, all the usable biomass is aimed at woodchips, mainly for pulpwood or the cellulose pulp industry. The first commercial thinning takes place in the seventeenth year, removing 330 trees ha^{-1}. At this stage, 40% of the biomass is aimed at sawn wood and 60% at pulpwood or cellulose pulp. The second commercial thinning takes place in the twenty-fourth year, removing 220 trees ha^{-1}. At this stage, 20% of the usable biomass is used as woodchips for the pulpwood or cellulose pulp industry, 60% as sawn wood, and 20% as high-quality heavy timber for furniture or building. After 30 years, the last cutting takes place, removing 100% of the existing timber volume, whose products are about 15% for the pulp or cellulose industry, 20% for woodchips or sawn wood, and 65% for heavy timber destined for furniture or building. For *E. globulus*, a shift of 10 years was considered in this work. Clearing processes are not applied in the productive cycle of eucalyptus, and a one-time cut is made at rotation age (Table 2).

Table 2. Management regimens and main forestry by-products for *P. radiata* and *E. globulus*. DBH: diameter at breast height.

Forest Species	Stand Age	Forest Operation	By-Products	Equation	Source
P. radiata	10	Cleaning	Small trees DBH< 7.5 cm Branches	$e^{\frac{0.19327 \sigma^2}{2}} e^{-2.61093} D^{2.48}$ $e^{\frac{0.615400^2}{2}} e^{-4.12515} D^{2.1173}$	[36]
	17	First thinning	Branches		[37]
	24	Second thinning		$11.8224469\, D^{1.95}$	
	30	Final cuttings	Branches		
E. globulus	10	Final cuttings	Branches	$0.08459716\, D^{1.7564}$	[38]

The amounts of WL that might be obtained in each stratum considering such treatments were estimated. To sum up, the indirect methodology that was selected to obtain an annual forest residue estimator (E_r; Mg ha^{-1} year^{-1}) for each species from the tree stratum, considering the forest treatment applied, consists of determining the residual biomass of each forest species. Moreover, to know how and how often this residue will be produced, the turnover of the species and its corresponding forest treatment were determined. The forest biomass residue quantities that could be obtained in each stratum were estimated from those treatments (Figure 3).

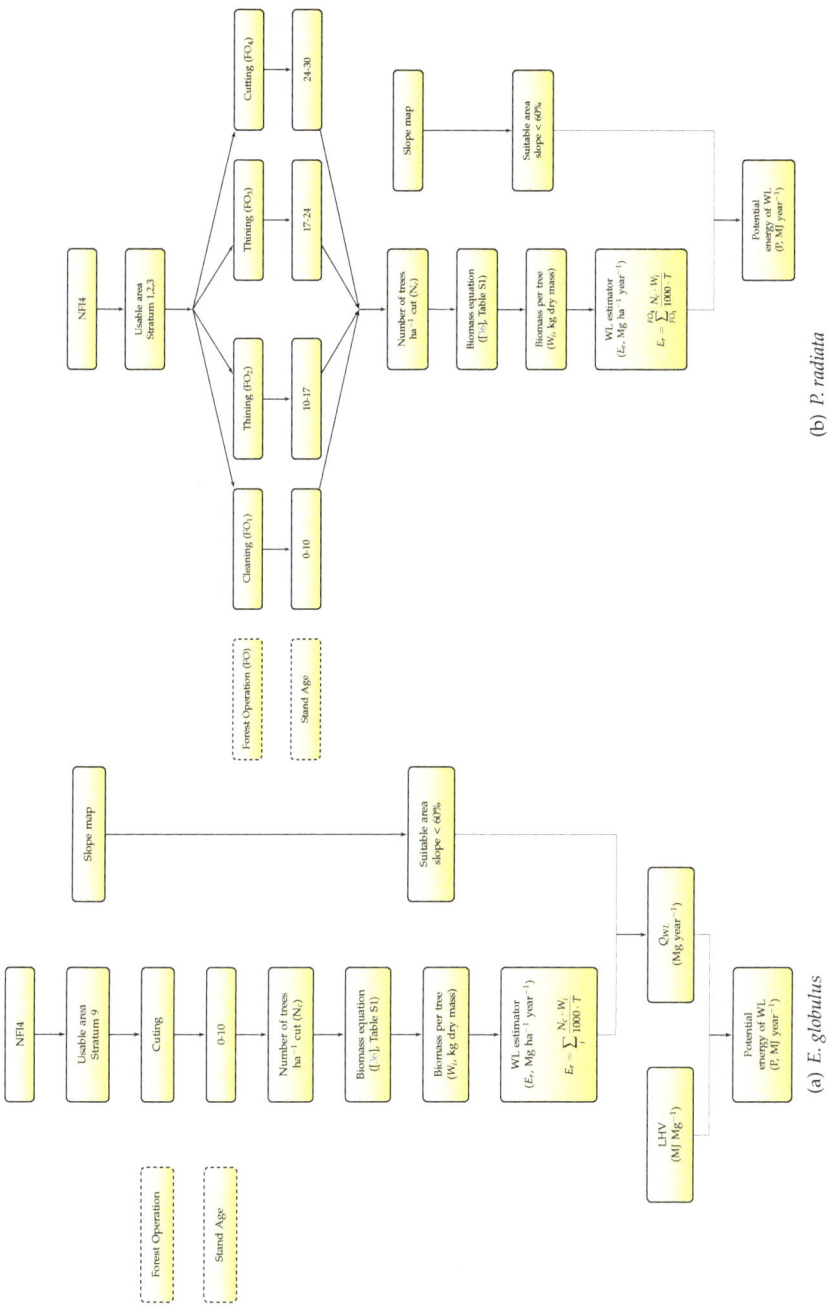

(a) *E. globulus*

(b) *P. radiata*

Figure 3. Methodological diagram for determining WL estimator and potential energy of WL.

The methodology selected to estimate forestry mass waste consisted of determining the residual mass of each tree sampled in the NFI4. This methodology is appropriate for larger-size trees ($D \geq 7.5$ cm). However, when acting on smaller trees, most of the tree is unusable for timber and as a result virtually the entire tree is made up of WL. Nevertheless, in this paper, a more conservative criterion has been considered, in which waste represents 80% of the total weight.

For estimation of the amount of annual biomass Q_{WL} (Mg year^{-1}) that might be generated by current forest masses in Biscay, the methodology applied uses a GIS. In order to carry it out, the forestall species distribution vectorial information is analyzed with a spatial resolution pixel of 2 m terms of its most characteristic species. In this way, it is possible to process an important amount of data and to manage the results obtained. The area occupied by each subregion and stratum was obtained from NFI4 and the corresponding subregion map, provided that this inventory includes the pertinent areas and stratum.

Once the potential quantities were calculated, the next step was the evaluation of constraints that can limit or reduce the harvestable amounts and energy utilization of such potential quantities For both environmental and economical reasons, the collection of WL should not be carried out in areas of steep slopes. The extraction of biomass was considered only for slopes less than 60%, since, in addition to not being economically viable, steeper terrain might involve erosion and soil loss problems. The slope map had to be previously digitised for its use by GIS (http://www.mapama.gob.es/es/cartografia-y-sig/visores/). The area of Biscay with slopes less than 60% was determined according to the following procedure: (a) the zone was reclassified starting from its slope layer in the GIS, assigning the value "1" to the areas involving slopes below slopes; (b) layers of slopes below 60%, which were obtained with the GIS from the basic topographical data, were merged [44].

The estimation of remains from WL was carried out in two different ways: stratum-by-stratum, and all strata together. In each situation, the estimation was accomplished by means of a confidence interval at a 95% level for the mean residue in metric Mg per year. Normality tests of the data were also carried out to determine if the values of the random variable E_r presented a normal distribution.

2.4. Biomass Potential Energy

The energy potential represents the total amount of forest residues from the selected species that is available for energy production purposes, and may be regarded as the upper limit for the value of energy that can be obtained from this kind of residues. Once the quantity of forest residue generated by the main forest species of Biscay was calculated in the strata in which those species are predominant, the potential energy that could be achieved with those residues, considering their sustainable exploitation, was estimated. The potential energy of the residues (P) is a function of the lower heating value (LHV) times the total residue for each species considered:

$$P = Q_{WL} \times LHV, \tag{2}$$

where P represents potential energy (MJ year^{-1}) and LHV represents the lower heating value in humid base (MJ Mg year^{-1}) of the forest residue obtained at the same humidity level at which productivity is considered. The humidity considered in this study was 30%, being the humidity of the WL after a few days of being on the soil. For this reason, it was necessary to obtain the LHV of each fuel at this humidity level. The LHV can be calculated from the higher heating value (HHV) [54–56].

2.5. Determination of Humidity, Chemical Property of Woody Litter (WL)

Fieldwork was focused on WL sample gathering from those forest treatments of the most representative species in Biscay. A random sampling per-stratum was carried out. The samples were collected during December 2011 through March 2012. In each of the sampling areas, the samples of forest biomass collected—roughly 2 kg per sample—came from forest treatments of branches (with a varying diameter ranging between 3 cm and 1 cm). The forest residue samples used in the experiments were taken from previously-chopped bulk samples which were introduced in polyethylene bags for transport to the laboratory, since they can be sealed and thus loss of humidity can be minimized. The time needed for taking the samples to the laboratory never exceeded 10 h. Hence, the level of humidity determined in the laboratory could be considered as the humidity of the sample itself. This fieldwork consisted of gathering samples of the most representative species in different locations of the province of Biscay.

With the aim of characterizing WL from the viewpoint of energy, elemental analyses (%C, H, and N), moisture, and higher heating value (HHV) of each sample were estimated. The samples were air-dried then oven-dried (65 °C, 24 h) to a constant weight and milled (0.25 mm). Moisture levels were determined by thermogravimetric analysis in a forced air convection oven (Digitronic-TFT, Selecta, Barcelona, Spain). The concentrations of total C, H, and N were determined by means of combustion in a LECO (Corporation St. Joseph, Michigan, MI, USA) automated analyzer. An adiabatic bomb calorimeter (IKA C 5012, IKA, Staufen, Germany) was used for determination of HHV. All determinations were carried out in triplicate.

2.6. Data Analysis

Once the fieldwork was carried out by obtaining the representative samples of the main forest species in the area, and after completing the necessary laboratory analyses, the project was fed into a personal geo-database (GDB), which in turn was implemented in a GIS from version ArcGIS TM 10.2, ESRI, Inc., Redlands, CA, USA. This enabled the combined use of the different data sources needed to quantify the biomass and its potential in the target area in a reliable and easy-to-update form.

Normality tests of the data were also carried out to determine if the values of the random variable E_r (estimator of residue in Mg ha^{-1} year^{-1}) presented a normal distribution. In order to do so, five different tests that belong to the R statistical software package Nortest were used (Anderson–Darling; Kolmogorov–Smirnov; Cramer–von Mises; Pearson; and Shapiro–Francia) [57,58]. The different statistical tests were carried out using statistical software packages in R.

3. Results and Discussion

3.1. Biomass and Carbon Stock Estimates

Table 3 shows the results of the estimations of biomass and CO_2 fixing in the forests of Biscay and the annual growth considering the difference between the forest inventories NFI4 (2011) and NFI3 (2005). The stock of total forest biomass (aboveground and underground) (TB) existing in the forests of the province of Biscay in the year 2011 amounted to 14.509 Tg of dry material, which implies a sequestration of 26.462 Tg of CO_2, of which 20.217 Tg of CO_2 correspond to P. radiata and 6.245 Tg to E. globulus. The estimated total aboveground biomass in 2011 was 10.352 Tg (dry material), of which 8.335 Tg was timber aboveground biomass susceptible to commercial exploitation. The annual net growth of total forest biomass obtained, comparing the values of both inventories, was 0.404 Tg year^{-1}. This value is not the real potential of biomass which could be used, but the increment which is not extracted and which is accumulated in the mass.

In order to estimate the annual gross growth of biomass and to identify the level of extractions (exploitation rate) which is being done, it was necessary to include the amount of biomass extracted by exploitation cuttings. To obtain this, the data related to the quantity of wood cut in Biscay between 2005 and 2011 was considered, as given by the county council as it is the administrative management entity of the forests in this province (Table 4). It was assumed that the loss of biomass was reduced to the extractions by cutting, without considering incidents such as pests, bacterial diseases, burnings, frosts, hailstorms, etc., which are assumed to be included in the expected rate of growth.

Data about forest cuttings showed that the amount of wood from E. globulus extracted during the period 2005–2011 was 920,092 m^3, 77% of which (710,884 m^3) corresponded to private forests and only 27% (209,208 m^3) to public ones. In relation to P. radiata, in the same period 2,720,509 m^3 of wood were extracted, of which 81.6% (2,220,197 m^3) corresponded to the exploitation of private forests and only 18.4% (500,312 m^3) to public ones. Consequently, the average extraction of *E. globulus* was 15.15 m^3 ha^{-1} year^{-1}, and of *P. radiata* was 6.426 m^3 ha^{-1} year^{-1}. Using the basic wood density as a conversion factor, it is possible to transform this data of the timber volume of this fraction into oven-dry weight [59,60]. Then, the percentages of weight were used in each biomass fraction of the forest species to obtain the amount of biomass of the different fractions extracted by cuttings [36]. The quantity of timber biomass taken annually by cuttings was 0.266 Tg year^{-1}, which added to the net growth result in a timber biomass annual gross growth of 0.479 Tg year^{-1} (Table 3). These values show that only 56% of the total annual growth of forest biomass was commercially exploited. The total annual gross growth of non-timber aboveground biomass was estimated as 0.124 Tg year^{-1}, susceptible to energetic valuation.

Table 3. Biomass fractions (Tg), annual increments (Tg year^{-1}), and accumulated equivalent CO_2 (Tg).

	Aboveground Biomass (AB)						Underground Biomass	Total Biomass (TB)
	Timber Biomass	Non-Timber Biomass				Total AB		
		b * > 7 cm	2 cm< b * <7 cm	b * < 2 cm	Needles			
Biomass 2005	7.059	0.313	0.674	0.484	0.184	8.714	3.370	12.084
Biomass 2011	8.335	0.360	0.751	0.579	0.327	10.352	4.157	14.509
Accumulated CO_2 (2011)	15.202	0.657	1.370	1.056	0.596	18.880	7.582	26.462
Δ_{net} annual biomass	0.213	0.008	0.013	0.016	0.024	0.273	0.131	0.404
Δ annual withdrawals	0.266	0.011	0.024	0.018	0.010	0.3306	0.159	0.489
Δ_{gross} annual biomass	0.479	0.019	0.037	0.034	0.034	0.603	0.290	0.893
Δ_{gross} annual CO_2	0.874	0.035	0.067	0.062	0.062	1.100	0.059	1.629

* b = branches.

Table 4. *P. radiata* and *E. globulus* cuttings carried out in Biscay between 2005 and 2011.

Year	E. globulus		P. radiata	
	Vpriv (m^3) *	Vpub (m^3) †	Vpriv (m^3)	Vpub (m^3)
2005	118,325	17,163	270,374	60,343
2006	96,484	26,857	39,864	120,446
2007	114,834	17,798	416,166	73,839
2008	126,992	25,143	287,474	65,079
2009	62,627	54,711	157,846	39,822
2010	83,577	31,971	345,713	71,961
2011	108,045	44,898	343,984	68,822
TOTAL	710,884	209,208	2,220,197	500,312

* Amount of timber extracted by cuttings in private forests. † Amount of timber extracted by cuttings in public forests.

3.2. Chemical and Energy Characterization

Table 5 shows the mean values obtained after the chemical and energy characterization of representative WL samples in Biscay. The results obtained show that WL had a similar composition of C in both forest species (*P. radiata* and *E. globulus*), close to 50%. With respect to N, it can be observed that *E. globulus* had a slightly higher percentage than *P. radiata*. Thus, the removal of forest residue of this species (mainly leaves) might increase the soil erosive processes, the avoidance of which would require fertilizers.

Table 5. Average values of moisture (wt. %) and heating values (MJ kg^{-1}) mass forestry of woody litter (WL). HHV: higher heating value.

Forest Species	C (% dm)	H (% dm)	N (% dm)	Moisture (bhcut)	HHV (MJ kg^{-1})	LHV (MJ kg^{-1})
P. radiata	51.768	6.078	0.974	44.0	21.2	19.8
E. globulus	51.046	6.422	1.197	52.5	21.1	19.7

3.3. Woody Litter (WL, Mg Year^{-1}) for Stratum 1

Studying normality by means of the above-mentioned five tests, it can be observed that the random variable E_r did not follow a normal distribution. The histogram of E_r suggests a certain positive asymmetry (see Supplementary Material, Figure S2), so the following transformation was accomplished $E_r' = \log(E_r + 1)$. In this case, the transformed variable E_r passed the five normality tests perfectly well (see Supplementary Material, Table S3). The annual mean WL obtained for stratum 1, with a 95% confidence interval, was (47,483, 51,335) Mg year^{-1}. The annual WL estimator (E_r) in this stratum took values of (1.050, 1.135) Mg ha^{-1} year^{-1} (dry mass) or (1.500, 1.622) (wet mass) (Table 6 and Figure 4a). As mentioned in previous sections, a humidity of 30% was considered when the WL for energy valuation was collected.

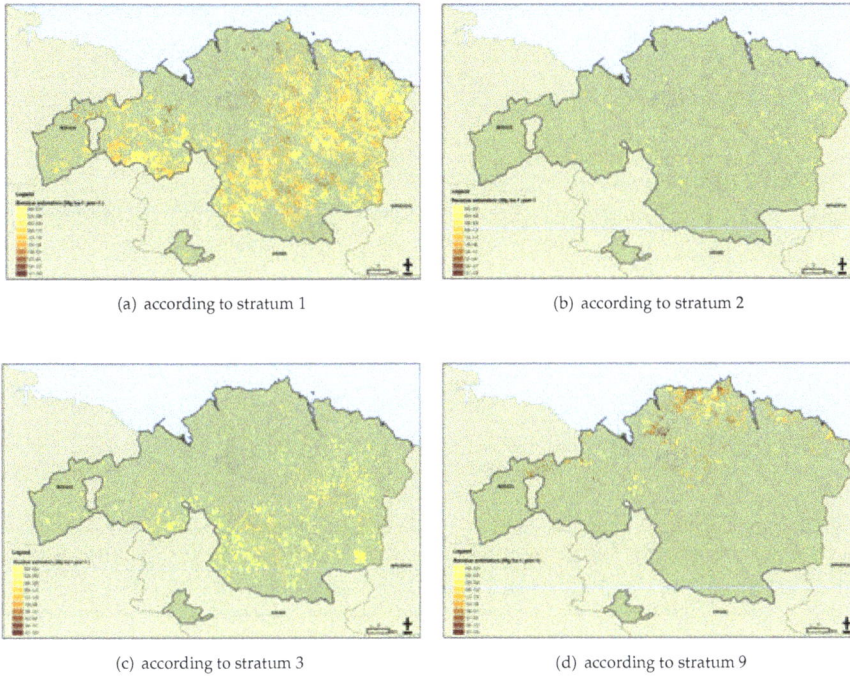

(a) according to stratum 1

(b) according to stratum 2

(c) according to stratum 3

(d) according to stratum 9

Figure 4. Annual WL estimators (E_r, Mg ha^{-1} year^{-1}).

Table 6. Average values of residue estimator on dry basis (E_r, Mg ha^{-1} year^{-1}). Annual woody litter (WL) usable for energy production on dry mass (WL, Mg year^{-1}) and theoretical and currently available energy potential in the province of Biscay (10^6 MJ year^{-1}).

Stratum	Forest Species	Area (ha)	E_r			Annual Woody Litter	Available Energy Potential
			Minimum	Medium	Maximum		
1	P. radiata	45,210	1.050	1.095	1.135	49,489.3	979.888
2	P. radiata	5589	0.272	0.312	0.353	1746.5	34.581
3	P. radiata	12,382	0.283	0.359	0.436	4446.5	88.041
9	E. globulus	9183	0.875	0.999	1.113	9172.1	180.690
Total		72,364				64,854.4	1283.200

3.4. Woody Litter (WL, Mg Year^{-1}) for Stratum 2

In this case, it was not necessary to carry out any transformation of the data since the E_r values adjusted to a normal distribution. Particularly, three out of the five normality tests used had a p value greater than 0.05 (see Supplementary Material, Table S3). The annual mean WL obtained, with a 95% confidence interval, was (1520, 1972) Mg year^{-1}. The values of the annual WL estimator (E_r) took values (0.272, 0.353) Mg ha^{-1} year^{-1} (dry mass) or (0.389, 0.504) Mg ha^{-1} year^{-1} (wet mass), with 30% humidity (Table 6 and Figure 4b).

3.5. Woody Litter (WL, Mg Year^{-1}) for Stratum 3

The values of E_r for stratum 3 did not follow a normal distribution, since they did not pass any of the five tests suggested. However, using the following transformation, root x, the data presented a normal aspect and passed all of the tests with high enough significance as indicated by p values (see Supplementary Material, Table S3). The annual mean WL obtained in stratum 3, with a confidence interval of 95%, was (3,496, 5396) Mg year^{-1}. The annual WL estimator (E_r) in this stratum took the values (0.283, 0.436) Mg ha^{-1} year^{-1} (dry mass) or (0.404, 0.623) Mg ha^{-1} year^{-1} (wet mass) (Table 6 and Figure 4c).

3.6. Woody Litter (WL, Mg Year^{-1}) for Stratum 9

The E_r values in stratum 9 followed a normal distribution, since they passed four out of the five tests with high enough significance ($p > 0.1$; see Supplementary Material, Table S3). The 95% confidence interval obtained for the annual mean WL was (8039, 10,305) Mg year^{-1}. Figure 4d shows the annual WL estimator (E_r) values obtained from mean values (0.875, 1.113) Mg ha^{-1} year^{-1} (dry mass), or 1.250–1.590 (wet mass) (Table 6 and Figure 4d).

3.7. Woody Litter (WL, Mg Year^{-1}) in Biscay Considering All Strata

When estimating the total residue considering all strata (1, 2, 3, 9) together, global data did not follow a normal distribution, but if data was transformed by means of $\log(x + 1)$, the p values obtained in the normality tests had a high enough significance (see Supplementary Material, Table S3). The estimation per 95% confidence interval for all the stratum together was (63,780, 69,542) Mg year^{-1}. Table 6 lists the amount of WL for each stratum usable for energy production and current energy potential in the province of Biscay. The results obtained after the statistical analyses of the data showed that the amount of mean forest biomass residue achieved with a 95% confidence interval was 64,854.4 Mg year^{-1}, from which 55,682.3 Mg ha^{-1} corresponded to *P. radiata* residue and 9172.1 Mg ha^{-1} to *E. globulus*. This means a potential energy supply of 1283.2 million MJ per year.

In this study, leaves and needles were excluded from the consideration of the energetic exploitation of WL. They are left on the ground, as they accumulate a high quantity of essential nutrients (N, P, K, Ca, and Mg) [37]. For this reason, and due to their lower calorific power in relation to the rest of the forest residue, leaves should not be extracted together with the rest of the forest biomass for its energetic use [20].

4. Conclusions

The main conclusions obtained in this work can be summarized as follows:

- *P. radiata* is still the main species in Biscay, although its extension has slightly fallen from 72,674 ha in 2005 to 70,562 ha in 2011 as a consequence of the drop in the demand of wood, the fall in prices, and cuttings caused by the economic crisis in this region, especially during 2008–2009 (Table 4). Thus, there has been an increase in the extension of plantings of overripe *P. radiata* of high height [61,62]. This led to an increase of 21% in this species' stock from 2005 to 2011, implying an annual growth of BT of 6.16 Mg ha^{-1} year^{-1}. In terms of the stock of C, it implies an annual growth of 3.01 Mg ha^{-1} year^{-1}.

E. globulus showed an increment of 12% in its stock from 1.134 Tg of BT (121.46 Mg ha^{-1}) to 1.269 Tg of BT (125.40 Mg ha^{-1}) in 2011, which represents an annual growth of BT of 0.63 Mg ha^{-1} year^{-1} in 2011. In relation to the stock of C, it represents an annual growth of just 0.328 Mg ha^{-1} year^{-1}. Overall, the total stock of carbon in these two species was 156.62 Mg ha^{-1}, but it could be increased by applying forest management practices such as increasing the age of rotation and/or decreasing the intensity of cuttings [22]. In relation to the Q_{WL} susceptible to energetic valuation, obtained after the forest treatments in the species *P. radiata* and *E. globulus* (Table 6) of 64,854.4 Mg year^{-1}, this quantity represents about 72% of the annual growth obtained from non-timber without leaves biomass and the 52% of that increment including the leaves (Table 3).

- *P. radiata* is mainly concentrated in stratum 1 in Biscay, where about 50,000 Mg year^{-1} residual biomass is annually obtained, which represents around 80% of the total forest residues, whereas in stratum 9, 9200 Mg year^{-1}1 are obtained, which represents almost the total amount of residual biomass from *E. globulus*. The estimated mean values of E_r generated every year from this forest species in this stratum were $(1.50, 1.62)$ Mg ha^{-1}year^{-1} (wet mass). These figures represent an amount of $(0.84, 0.91)$ Mg ha^{-1}year^{-1} (dry mass) biomass residue. The estimated values are similar to those obtained in previous studies carried out in other regions of Spain (e.g., Dominguez citepDominguez).
- *E. globulus* is mainly located in areas near the coast (see Supplementary Material, Figure S1b). These correspond to stratum 9 in Biscay. The estimated annual mean value of forest residue in the area was $(1.03, 1.32)$ Mg ha^{-1}year^{-1} (wet mass) or $(0.72, 0.92)$ Mg ha^{-1}year^{-1} (dry mass). Similar studies were carried out in other areas of Spain (e.g., Zabalo [20]).

The essential and traditional aim of forest management in the Basque Country has been to maximize the economic profits without risking the persistence of the mass. It is certain that carbon sequestration by forest masses has only recently been considered as another aim of management to be developed together with the economic factor. Former studies have shown that thinning intensification significantly increases the quantity of total biomass obtained (timber and residue biomass). However, this type of management can reduce carbon sequestration in forests [4]. For this reason, in this research an average forest management which optimizes economic (quantity of timber biomass) and environmental (CO_2 sequestration) factors has been assumed.

Supplementary Materials: Supplementary materials are available online at http://www.mdpi.com/s1. Figure S1: The target study area: province of Biscay (Spain). Table S1: Equations of forest biomass fractions expressed in kg (oven-dry-weight, $(102 \pm 2\,°C, 24\,h)$), coefficient of determination (R^2), and standard error of estimate (SEE). Table S2: Expansion factors (EFs) in plots from NFI4. Figure S2: Histograms of the values of E_r. Table S3: Normality test (p values) of the values of E_r.

Author Contributions: Esperanza Mateos developed conceptual ideas, designed the study, conducted data analysis in field experiments and wrote the paper. Leyre Ormaetxea developed conceptual ideas and designed the study.

Acknowledgments: This work was supported by the Basque Government and by the Office of Research of the University of the Basque Country grant by Project SAI10/147-SPE10UN90 and by Project NUPV14/11, respectively. The authors also thank the technical and human support provided by J.M. Eguskitza, Joseba M. González and José Miguel Edeso, Aitor Bastarrica and Leyre Torre for drawing some of the figures. Our sincere thanks to Hazi Fundazioa for providing essential data for this study.

Conflicts of Interest: The authors declare no conflict of interest.

References

1. Zhang, J.; Ge, Y.; Chang, J.; Jiang, B.; Jiang, H.; Peng, C.; Zhu, J.; Yuan, W.; Qi, L.; Yu, S. Carbon storage by ecological service forests in Zhejiang Province, subtropical China. *For. Ecol. Manag.* **2007**, *245*, 64–75.
2. Baul, T.K.; Alam, A.; Ikonen, A.; Strandman, H.; Asikainen, A.; Peltola, H.; Kilpeläinen, A. Climate Change Mitigation Potential in Boreal Forests: Impacts of Management, Harvest Intensity and Use of Forest Biomass to Substitute Fossil Resources. *Forests* **2017**, *8*, 455.

3. Noble, I.; Scholes, R. Sinks and the Kyoto Protocol. *Clim. Policy* **2001**, *1*, 5–25.

4. Pyörälä, P.; Peltola, H.; Strandman, H.; Antti, K.; Antti, A.; Jylhä, K.; Kellomäki, S. Effects of Management on Economic Profitability of Forest Biomass Production and Carbon Neutrality of Bioenergy Use in Norway Spruce Stands Under the Changing Climate. *Bioenergy Res.* **2014**, *7*, 279–294.

5. Canadell, J.G.; Raupach, M.R. Managing forests for climate change mitigation. *Science* **2008**, *320*, 1456–1457.

6. Basoa-Fundazioa. Guía de Buenas Prácticas en el Sector Forestal y de Transformación de la Madera. Available online: http://basoa.org/es/?option=com_k2&view=item&task=doc_download&id=133&Itemid=68 (accessed on 1 October 2007).

7. Lindner, M.; Karjalainen, T. Carbon inventory methods and carbon mitigation potentials of forests in Europe: A short review of recent progress. *Eur. J. For. Res.* **2007**, *126*, 149–156.

8. IDAE. Resumen del Plan de Energías Renovables 2011–2020. Available online: http://www.minetur.gob.es/energia/es-ES/Novedades/Documents/Resumen_PER_2011-2020.pdf (accesed on 1 November 2016).

9. Lapuerta, M.; Hernández, J. *Tecnologías de la Combustión*; Universidad de Castilla: Cuenca, Spain, 1998.

10. Jones, G.; Loeffler, D.; Calkin, D.; Chung, W. Forest treatment residues for thermal energy compared with disposal by onsite burning: Emissions and energy return. *Biomass Bioenergy* **2010**, *34*, 737–746.

11. Dong, L.; Zhang, L.; Li, F. Developing Two Additive Biomass Equations for Three Coniferous Plantation Species in Northeast China. *Forests* **2016**, *7*, 136.

12. Tolosana, E. *Manual Técnico para el Aprovechamiento y Elaboración de Biomasa Forestal*; Fucovasa y Mundi-Prensa: Madrid, Spain, 2009.

13. Mateos, E.; Garrido, F.; Ormaetxea, L. Assessment of Biomass Energy Potential and Forest Carbon Stocks in Biscay (Spain). *Forests* **2016**, *7*, 75.

14. Esteban, L.; Pérez, P.; Ciria, P.; Carrasco, J. *Evaluación de los Recursos de Biomasa Forestal en la Provincia de Soria. Análisis de Alternativas para su Aprovechamiento Energético*; Cuadernos de la SECF, CIEMAT: Madrid, Spain, 2004.

15. Panichelli, L.; Gnansounou, E. GIS-based approach for defining bioenergy facilities location: A case study in Northern Spain based on marginal delivery costs and resources competition between facilities. *Biomass Bioenergy* **2008**, *32*, 289–300.

16. Esteban, L.; García, R.; Ciria, P.; Carrasco, J. *Costs in Spain and Southern EU Countries. Clean Hydrogen-Rich Synthesis Gas Report, Chrisgras Fuels from Biomass*; Colección Documentos CIEMAT; CIEMAT: Madrid, Spain, 2009.

17. García, A.; Pérez, F.; de la Riva, J. Evaluación de los recursos de biomasa residual forestal mediante imágenes del satélite Landsat y SIG. *GeoFocus Rev. Int. Ciencia Technol. Inf. Geogr.* **2006**, *6*, 205–230.

18. García-Martín, A.; Galindo, D.G.; Pascual, J.; De la Riva, J.; Pérez-Cabello, F.; Montorio, R. Determinación de zonas adecuadas para la extracción de biomasa residual forestal en la provincia de Teruel mediante SIG y teledetección. *GeoFocus Rev. Int. Ciencia Technol. Inf. Geogr.* **2011**, *11*, 19–50.

19. Brañas, J.; González-Río, F.; Rodríguez Soalleiro, R.; Merino, A. Biomasa maderable y no maderable en plantaciones de eucalipto. Cuantificación y estimación. *CIS-Madera* **2000**, *4*, 72–75.

20. Zabalo, A. Modelo de Estimación del Potencial Energético de la Biomasa de Origen Forestal en la Provincia de Huelva. Ph.D. Thesis, Universidad de Huelva (Spain), Huelva, Spain, 2006.

21. Schreuder, H.T.; Gregoire, T.G.; Wood, G.B. *Sampling Methods for Multiresource Forest Inventory*; John Wiley & Sons: Hoboken, NJ, USA, 1993.

22. Balboa-Murias, M.A.; guez Soalleiro, R.R.; Merino, A.; Alvarez-González, J.G. Temporal variations and distribution of carbon stocks in aboveground biomass of radiata pine and maritime pine pure stands under different silvicultural alternatives. *For. Ecol. Manag.* **2006**, *237*, 29–38.

23. Brown, S. Measuring carbon in forests: current status and future challenges. *Environ. Pollut.* **2002**, *116*, 363–372.

24. Somogyi, Z.; Cienciala, E.; Mäkipää, R.; Muukkonen, P.; Lehtonen, A.; Weiss, P. Indirect methods of large-scale forest biomass estimation. *Eur. J. For. Res.* **2007**, *126*, 197–207.

25. Tobin, B.; Nieuwenhuis, M. Biomass expansion factors for Sitka spruce (Picea sitchensis (Bong.) Carr.) in Ireland. *Eur. J. For. Res.* **2007**, *126*, 189–196.

26. Tobin, B.; Black, K.; McGurdy, L.; Nieuwenhuis, M. Estimates of decay rates of components of coarse woody debris in thinned Sitka spruce forests. *Forestry* **2007**, *80*, 455–469.

27. Ruiz-Peinado, R.; Moreno, G.; Juarez, E.; Montero, G.; Roig, S. The contribution of two common shrub species to aboveground and belowground carbon stock in Iberian dehesas. *J. Arid Environ.* **2013**, *91*, 22–30.

28. Mäkelä, A.; del Río, M.; Hynynen, J.; Hawkins, M.J.; Reyer, C.; Soares, P.; van Oijen, M.; Tomé, M. Using stand-scale forest models for estimating indicators of sustainable forest management. *For. Ecol. Manag.* **2012**, *285*, 164–178.

29. Intergovernmental Panel on Climate Change (IPCC). *Good Practice Guidance for Land Use, Land-Use Change and Forestry*; IPCC National Greenhouse Gas Inventories; Cambridge University Press: Cambridge, UK, 2003.

30. Parresol, B. Assessing tree and stand biomass: A review with examples and critical comparisons. *For. Sci.* **1999**, *45*, 573–593.

31. Araujo, T.; Higuchi, N.; Andrade, J. Comparison of formulae for biomass content determination in a tropical rain forest site in the state of Pará, Brazil. *For. Ecol. Manag.* **1999**, *117*, 43–52.

32. Keller, M.; Palace, M.; Hurtt, G. Biomass estimation in the Tapajos National Forest, Brazil: examination of sampling and allometric uncertainties. *For. Ecol. Manag.* **2001**, *154*, 371–382.

33. Jenkins, J.C.; Chojnacky, D.C.; Heath, L.S.; Birdsey, R.A. National-Scale Biomass Estimators for United States Tree Species. *For. Sci.* **2003**, *49*, 12–35.

34. Berner, L.T.; Alexander, H.D.; Loranty, M.M.; Ganzlin, P.; Mack, M.C.; Davydov, S.P.; Goetz, S.J. Biomass allometry for alder, dwarf birch, and willow in boreal forest and tundra ecosystems of far northeastern Siberia and north-central Alaska. *For. Ecol. Manag.* **2015**, *337*, 110–118.

35. Cantero, A.; Espinel, S.; Sáenz, D. *Un Modelo de Gestión para las Masas de Pinus Radiata en el País Vasco.* 1995. Available online: http://secforestales.org/publicaciones/index.php/cuadernos_secf/search (accessed on 4 February 2012).

36. Montero, G.; Ruiz-Peinado, R.; Muñoz, M. *Producción de Biomasa y Fijación de CO$_2$ en los Montes Españoles*; Instituto Nacional de Investigación y Tecnología Agraria y Alimentaria (INIA): Madrid, Spain, 2005.

37. Merino, A.; Rey, C.; Brañas, J.; Rodriguez, R. Biomasa arbórea y acumulación de nutrientes en plantaciones de Pinus radiata. *Investigaciones Agrarias Sistemas Recursos Forestales* **2003**, *12*, 85–98.

38. Silva, R.; Tavares, M.; Pascoa, F. Residual Biomass of Forest Stands. *Pinus pinaster* Ait. and *Eucalyptus globulus* Labill. In Proceedings of the 10th World Forestry Congress, Paris, France, 17–26 September 1991.

39. Alam, A.; Kilpeläinen, A.; Kellomäki, S. Impacts of initial stand density and thinning regimes on energy wood production and management-related CO$_2$ emissions in boreal ecosystems. *Eur. J. For. Res.* **2012**, *131*, 655–667.

40. Profft, I.; Mund, M.; Weber, G.E.; Weller, E.; Schulze, E.D. Forest management and carbon sequestration in wood products. *Eur. J. For. Res.* **2009**, *128*, 399–413.

41. Garcia-Gonzalo, J.; Peltola, H.; Gerendiain, A.Z.; Kellomäki, S. Impacts of forest landscape structure and management on timber production and carbon stocks in the boreal forest ecosystem under changing climate. *For. Ecol. Manag.* **2007**, *241*, 243–257.

42. Fourth National Forestry Inventory. 2011. Available online: http://www.euskadi.eus/inventario-forestal-2011/web01-a3estbin/es/ (accessed on 27 November 2016).

43. Kim, T.J.; Bullock, B.P.; Wijaya, A. Spatial Interpolation of Above-Ground Biomass in Labanan Concession Forest in East Kalimantan, Indonesia. *Math. Comput. For. Nat. Resour. Sci.* **2016**, *8*, 26.

44. López-Rodríguez, F.; Pérez Atanet, C.; Blázquez, F.; Ruiz Celma, A. Spatial assessment of the bioenergy potential of forest residues in the western province of spain, Cáceres. *Biomass Bioenergy* **2009**, *33*, 358–366.

45. Álvarez González, J.G.; Balboa Murias, M.; Merino, A.; Rodríguez Soalleiro, R. Estimación de la biomasa arbórea de *"Eucalyptus globulus"* y *"Pinus pinaster"* en Galicia. *Recursos Rurais* **2012**, *1*, 21–30.

46. Gertrudix, R.R.P.; Montero, G.; del Rio, M. Biomass models to estimate carbon stocks for hardwood tree species. *For. Syst.* **2012**, *21*, 42–52.

47. Ter-Mikaelian, M.T.; Korzukhin, M.D. Biomass equations for sixty-five North American tree species. *For. Ecol. Manag.* **1997**, *97*, 1–24.

48. Gil, M.V.; Blanco, D.; Carballo, M.T.; Calvo, L.F. Carbon stock estimates for forests in the Castilla y León region, Spain. A GIS based method for evaluating spatial distribution of residual biomass for bio-energy. *Biomass Bioenergy* **2011**, *35*, 243–252.

49. Zianis, D.; Mencuccini, M. On simplifying allometric analyses of forest biomass. *For. Ecol. Manag.* **2004**, *187*, 311–332.

50. Thomas, S.C.; Martin, A.R. Carbon Content of Tree Tissues: A Synthesis. *Forests* **2012**, *3*, 332–352.

51. Crecente-Campo, F.; Dieguez-Aranda, U.; Rodriguez-Soalleiro, R. Resource communication. Individual-tree growth model for radiata pine plantations in northwestern Spain. *For. Syst.* **2012**, *21*, 538–542.

52. Diéguez-Aranda, U.; Alboreca, A.R.; Castedo-Dorado, F.; González, J.Á.; Barrio-Anta, M.; Crecente-Campo, F.; González, J.G.; Pérez-Cruzado, C.; Soalleiro, R.R.; López-Sánchez, C.; et al. Herramientas selvícolas para la gestión forestal sostenible en Galicia. *Forestry* **2009**, *82*, 1–16.

53. García, A. Estimación de Biomasa Residual Mediante Imágenes de Satélite y Trabajo de Campo. Modelización del Potencial Energético de los Bosques Turolenses. Ph.D. Thesis, Universidad de Zaragoza, Zaragoza, Spain, 2009.

54. Pérez, S.; Renedo, C.; Ortiz, A.; Mañana, M.; Silió, D. Energy evaluation of the Eucalyptus globulus and the Eucalyptus nitens in the north of Spain (Cantabria). *Thermochim. Acta* **2006**, *451*, 57–64.

55. Alvarez, J.; Soalleiro, R.R.; Rojo, A. A management tool for estimating bioenergy production and carbon sequestration in eucalyptus globulus and eucalyptus nitens grown as short rotation woody crops in north-west Spain. *Biomass Bioenergy* **2011**, *35*, 2839–2851.

56. Demirbaş, A. Calculation of higher heating values of biomass fuels. *Fuel* **1997**, *76*, 431–434.

57. Wolfram, S. *The Mathematica Book*; Wolfram Media: Champaign, IL, USA, 2003.

58. Spector, P. *Data Manipulation with R*; Springer Science & Business Media: Berlin, Germany, 2008.

59. Gutiérrez, A.; Fernández-Golfín, J. Cálculo de la densidad y de las variaciones dimensionales de la madera. Equivalencias numéricas entre valores. *Montes* **1997**, *49*, 28–33.

60. Lehtonen, A.; Mäkipää, R.; Heikkinen, J.; Sievänen, R.; Liski, J. Biomass expansion factors (BEFs) for Scots pine, Norway spruce and birch according to stand age for boreal forests. *For. Ecol. Manag.* **2004**, *188*, 211–224.

61. Albizu-Urionabarrenetxea, P.; Tolosana-Esteban, E.; Roman-Jordan, E. Safety and health in forest harvesting operations. Diagnosis and preventive actions. A review. *For. Syst.* **2013**, *22*, 392–400.

62. Domínguez, J. *Los Sistemas de Información Geográfica en la Planificación e Integración de Energías Renovables*; CIEMAT: Madrid, Spain, 2003.

MDPI

St. Alban-Anlage 66

4052 Basel

Switzerland

Tel. +41 61 683 77 34

Fax +41 61 302 89 18

www.mdpi.com

Forests Editorial Office

E-mail: forests@mdpi.com

www.mdpi.com/journal/forests

www.ingramcontent.com/pod-product-compliance
Lightning Source LLC
Chambersburg PA
CBHW051706210326
41597CB00032B/5387